DEFINITIONS OF TRIGONOMETRIC RATIOS

$P(x, y)$ is any point on terminal side of θ

$r = \sqrt{x^2 + y^2}$ and $r > 0$

$\sin \theta = \dfrac{y}{r}$	$\csc \theta = \dfrac{r}{y}, \; y \neq 0$
$\cos \theta = \dfrac{x}{r}$	$\sec \theta = \dfrac{r}{x}, \; x \neq 0$
$\tan \theta = \dfrac{y}{x}, \; x \neq 0$	$\cot \theta = \dfrac{x}{y}, \; y \neq 0$

SIGNS OF TRIGONOMETRIC RATIOS

Quadrant	$\sin \theta$ or $\csc \theta$	$\cos \theta$ or $\sec \theta$	$\tan \theta$ or $\cot \theta$
I	+	+	+
II	+	−	−
III	−	−	+
IV	−	+	−

REFERENCE ANGLE

$f(\theta) = f(\alpha)$ or $f(\theta) = -f(\alpha)$

I II III IV

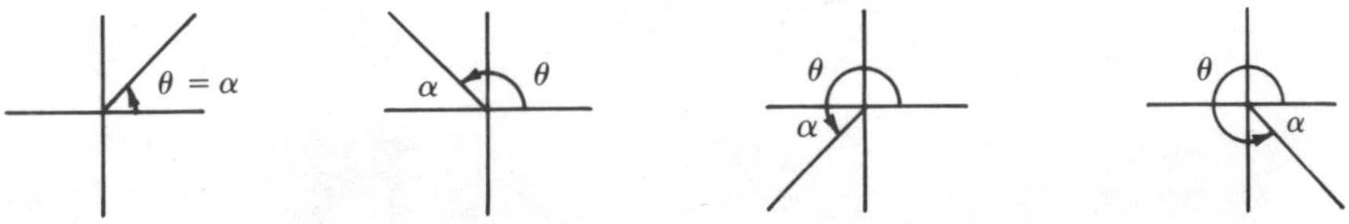

$\alpha = \theta$ $\alpha = \pi - \theta$ $\alpha = \theta - \pi$ $\alpha = 2\pi - \theta$

Understanding Algebra and Trigonometry

Understanding Algebra and Trigonometry

Gene R. Sellers
Sacramento City College

CHARLES E. MERRILL PUBLISHING COMPANY
A Bell & Howell Company
Columbus Toronto London Sydney

Published by
Charles E. Merrill Publishing Company
A Bell & Howell Company
Columbus, Ohio 43216

This book was set in Times Roman.
The production editor was Jo Ellen Gohr.
Cover by Dodrill Design.

Library of Congress Catalog Card Number: 78–70484
International Standard Book Number: 0–675–08306–0
Printed in the United States of America
2 3 4 5 6 7 8 9 10—85 84 83 82 81 80 79

Preface

What preparation must a student have to succeed in a course such as calculus? Furthermore, how much depth should be given to each topic selected for the course designed to provide this preparation? The variety of responses from mathematics educators to these two questions could probably be put into a one-to-one correspondence with the counting numbers.

Although this text includes what many instructors believe are essential topics for a precalculus course, I have not given all topics equal treatment. Topics such as trigonometry and analytic geometry are developed virtually from the foundation of the subject. Some areas, such as solving equations and inequalities in one variable, are approached with the assumption that students have had one or more years of algebra.

The goal that I have attempted to achieve throughout the text has been a *clear presentation for the student*. Consequently, I have given more space to worked examples than to proofs of theorems. The text is designed to be more practical than theoretical.

Consistent with this purpose, I have included questions in the margins of the text to remind students that they are responsible for reading the material comprehensively. That is, the questions asked are examples of the types of questions students should be asking themselves as they read and study the text. Furthermore, the questions provide me with justification for discussing *one more time* concepts that I believe require further explanation or amplification.

As an additional aid to the student, each chapter contains a summary of the definitions, axioms, and theorems presented in the chapter. I have also provided a self-checking exercise for each chapter, with each exercise keyed to the section of the chapter in which a particular skill or concept is developed. Answers to these exercises immediately follow the questions in the text. Each section of every chapter also includes an exercise set, for which selected answers are given at the end of the book.

Because I believe that the hand-held calculator is useful in the teaching of mathematics, I have included—whenever appropriate—examples and exercises in the text which require use of the calculator. Such examples are clearly indicated with a calculator symbol displayed in the margin. Instructions given for the calculator are typical of many models currently on the market and variations in the techniques of operation can be dealt with easily on an individual basis. Because variations between table values and calculator values can occur in the third and fourth decimal places, I have always stated answers in table values. However, when answers are rounded to two decimal places, table and calculator responses are generally the same.

The mathematical backgrounds and maturity of students are frequently very diverse. Students with a solid background can sometimes skip an entire chapter in the text. Other students require more explanation and exposure than the text provides. In an attempt to write a text for the average student, I have included many worked examples and exercises. To help those students who need more exposure than the text provides, a student study guide is available. The study guide contains (1) a brief review of each chapter, (2) additional worked exercises from the text, (3) additional practice exercises with answers, (4) a section on the techniques of factoring, (5) a section on the theory of sets, and (6) a section on mathematical induction.

A complete solutions manual and a test bank are also available for the instructor. The manual can be placed in mathematics laboratories designed to tutor students, duplicated in part by the instructor to share with the class, or used exclusively by the instructor. The test bank contains three chapter tests (forms A, B, and C) for each chapter. Form A can be used as a chapter review, form B as an assigned test, and form C for make-up tests. Each question in the test is keyed to the corresponding chapter and section in the text.

This text could not have been produced without the considerable assistance I received from several people. First, I wish to thank Andrea Posner, administrative editor for Charles E. Merrill Publishing Company, for signing the project initially and her work in helping to produce the best product possible. I am deeply indebted to Jo Ellen Gohr, production editor for this book, for the work she has devoted to the project. I wish to thank Joan Mahmud, Bergen Community College; Dr. Jerry Shyrock, Western Illinois University; Jacqueline Underwood, Knoxville College; and Jo Ward, University of Houston, Downtown College, for their helpful suggestions and criticisms in reviewing the manuscript.

I dedicate this book to my wife Linda, with all the love and devotion I feel for her. She has supported me physically, emotionally, and spiritually in more ways than words can express. This book could not have been written without her encouragement and help.

Contents

7 TRIGONOMETRIC FUNCTIONS 223

8 TRIGONOMETRIC IDENTITIES AND CONDITIONAL EQUATIONS 275

9 INVERSE TRIGONOMETRIC FUNCTIONS 311

10 APPLICATIONS OF TRIGONOMETRY 341

1 The Real Number System

A course in advanced mathematics must begin with a review of the fundamental topics that are, to some degree, familiar to the student. The quantity of review necessary to bring all students to some common level of proficiency in these topics is, of course, dependent on the individual. It is therefore the responsibility of each student to utilize the review material to achieve this basic level of proficiency. Specifically, in this chapter we review the subsets of the real numbers and the axioms assumed for the operations of addition and multiplication. We also review the relations of equality and order. We then develop the process of solving linear equations and inequalities and the simplification of rational and complex expressions.

THE REAL NUMBERS 1-1

Consider the set of numbers

$$A = \{-7, 18, \sqrt{5}, 0, 9/2, \sqrt{-4}, -0.03, 0.13113 \ldots, \sqrt[3]{8}, 0.3535 \ldots\}$$

Identify those numbers in set A that are **integers.** Now pick out the numbers in set A that are **irrational.** Finally, select the numbers that are not **real numbers.**

In the mathematical world all numbers are not considered to be the same. We have found it useful to classify numbers based on very specific definitions. In this section we will define six sets of numbers and discuss the relationships among the numbers in the sets. We will stress the identifying characteristic of the numbers in each set so that we will be able to place any number in its respective set. We use the terminology and symbolism of sets throughout the text. A review of the symbolism and terminology of sets is given in the student study guide for this text.

Natural or Counting Numbers

Ask someone to name ten numbers. Probably they will simply name the counting numbers one through ten. The natural or counting numbers are the numbers we learn first and use most frequently. The following definition lists the elements of this set:

Natural numbers

> **Definition:** The set of natural numbers is the set N, and
>
> $$N = \{1, 2, 3, \ldots\}$$

Notice that the set N is an **infinite set,** that is, there is no largest natural number. Furthermore, given that we have knowledge of the operations of addition and multiplication, we should note that set N is **closed** under these operations. We will discuss the property of closure more completely in a later section, but for now let's simply discuss what the property means for set N. If a and b are natural numbers ($a \in N$ and $b \in N$), then the sum of a and b and the product of a and b are also natural numbers ($(a + b) \in N$ and $ab \in N$). Thus, if we restrict ourselves to the operations of addition and multiplication, the natural numbers provide answers to any problem involving only elements of N.

Whole Numbers

The number 0 is a significant number in mathematics. We cannot divide by it; whenever we multiply by it we always get 0; any nonzero number with a zero exponent is 1; and adding 0 to any number b results in a sum b. Is it any wonder then that a new set of numbers is defined when 0 is included in the set of natural numbers?

Whole numbers

> **Definition:** The set of whole numbers is the set W, and
>
> $$W = \{0, 1, 2, 3, \ldots\}$$

Aren't all natural numbers also whole numbers?

Every element of N is also an element of W, but W contains 0, which is not included in N. Therefore, *N is a proper subset of W,* and we write $N \subset W$.

Integers

To see the necessity to expand the set of whole numbers, consider the operation of subtraction. If $a \in W$ and $b \in W$, then $(a - b) \in W$ if and only if a is greater than or equal to b. To provide an answer for the case when b is greater than a, a new number is needed, that is, a negative whole number. The set of positive and negative whole numbers and zero is called *integers*.

Integers

> **Definition:** The set of integers, J, is equal to the set of positive and negative whole numbers and zero:
>
> $$J = \{\ldots, -3, -2, -1, 0, 1, 2, 3, \ldots\}$$

Notice that J is both positively and negatively infinite. Furthermore, if $a \in J$ and $b \in J$, then $(a + b) \in J$, $ab \in J$, and $(a - b) \in J$. That is, the sum, product, and difference of integers are also integers.

Rational Numbers

The set of integers does not always provide an element for the division of any two integers. Specifically, $\left(\frac{a}{b}\right) \in J$ *if and only if a is a multiple of b*. The number a is a multiple of b if and only if $a = bn$ and n is a natural number. For example, 21 is a multiple of 3, since $21 = (3)(7)$ and 7 is a natural number.

Rational numbers

> **Definition:** A rational number is one that can be written as $\frac{a}{b}$, where a and b are integers and $b \neq 0$. The symbol for the set of rational numbers is R.

The elements of R cannot be listed, for although any two rational numbers can be *ordered,* the value of the next largest rational number greater than a given rational number cannot be determined.

Set R is closed under the operations of addition, subtraction, multiplication, and division by any nonzero number. That is, if $a \in R$ and $b \in R$, then $(a + b) \in R$, $(a - b) \in R$, $ab \in R$ and $\frac{a}{b} \in R$ provided $b \neq 0$. Since set R provides an answer for each of these four operations, we sometimes refer to addition, subtraction, multiplication, and division as **rational operations.**

Because the term *fraction* is not well defined in mathematics, we cannot call rational numbers fractions. However, a rational number is specifically defined as the **ratio of two integers.** Thus, $\frac{\sqrt{3}}{2}$ can be called a fraction, but it is definitely not a rational number since $\sqrt{3}$ is not an integer.

Couldn't we simply call rational numbers fractions?

If $\frac{a}{b}$ is any rational number, then the decimal equivalent of $\frac{a}{b}$ must either terminate or repeat a pattern of digits. The set of digits that repeat is called the **period** of the number, and for the number $\frac{a}{b}$ the period can be no more than $b - 1$ digits long.

EXAMPLE 1–1 Express the following rational numbers as decimal fractions.

(a) $\frac{43}{8}$ (b) $\frac{153}{37}$

Solutions:

(a)
```
      5.375
  8)43.000
    40
     30
     24
      60
      56
       40
       40
```

(b)
```
        4.135
  37)153.000
     148
       50
       37
       130
       111
        190
        185
          5
```

Thus, $\frac{43}{8} = 5.375$, and the decimal equivalent *terminates*.

Thus, $\frac{153}{37} = 4.135135\ldots$. We write $4.\overline{135}$, and the decimal equivalent is **periodic** with a period = 135.

If the decimal equivalent of a number terminates or is periodic, then the number can be expressed as the ratio of two integers.

Example 1–2 Express the following rational numbers as the ratio of two integers.

(a) 3.68 (b) $4.\overline{27}$

Solutions:

(a) $3.68 = 3.68\left(\frac{100}{100}\right) = \frac{368}{100} = \frac{92}{25}$

(b) Let $N = 4.\overline{27} = 4.272727\ldots$. Since the period contains two digits, we multiply both sides of the equation by 100, and $100N = 427.272727\ldots$. Therefore, if $100N = 427.272727\ldots$ and $N = 4.272727\ldots$, then the difference is

$$99N = 423.0000\ldots$$
$$N = \frac{423}{99} = \frac{47}{11}$$

Why did you multiply the first equation by 100?

The period in the number $4.\overline{27}$ contained two digits. Therefore, when the equation $N = 4.2727\ldots$ is multiplied by 100, the resulting equation has shifted the digits two places to the left, and the subtraction of the two equations yields a string of zeros. If the period had contained three digits, we would have multiplied both sides of the equation by 1000.

Irrational Numbers

Not all decimal fractions terminate or are periodic. For example,

$$0.13113111311113\ldots \quad \text{and} \quad 0.173256983105\ldots$$

are decimal fractions that do not terminate and are not periodic. The set of all such numbers belong to the set of irrational numbers.

Irrational numbers

> **Definition:** The set of irrational numbers, I, is the set of numbers whose decimal equivalent does not repeat a period and does not terminate.

This definition is weak in that it does not tell us what an irrational number is, but rather what it is not. However, it does enable us to recognize an irrational number. Thus, we classify 0.151151115 . . . as an irrational number, since the pattern of digits is continuously changing.

Other forms of irrational numbers are obtained by considering the roots of numbers. A thorough development of radicals is given in Chapter 2, and rational and irrational numbers are again considered. You may recall that π is an irrational number. Another important irrational number is e, the base for natural logarithms.

Real Numbers

Notice that the sets of rational and irrational numbers are *disjoint*. That is, no number can be an element of the rationals and irrationals simultaneously, for either its decimal expression is periodic (a rational number) or it is not periodic (an irrational number). The *union* of the sets R and I is the set of real numbers.

Real numbers

> **Definition:** The set of real numbers, $\mathcal{R}$, is the union of the sets of rational and irrational numbers. That is,
>
> $$R \cup I = \mathcal{R}$$

In Figure 1-1, a Venn diagram representation of the set of real numbers shows all the subsets of $\mathcal{R}$ discussed in this section.

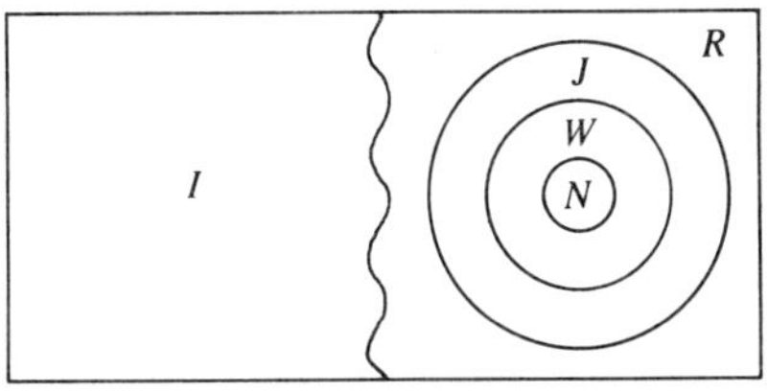

FIGURE 1-1 Venn diagram of $\mathcal{R}$, the set of real numbers

An important property of the elements of $\mathcal{R}$ is that *the square of any real number is nonnegative*. That is, if $x \in \mathcal{R}$, then x^2 is greater than or equal to 0 ($x^2 \geq 0$). For example, $3^2 \geq 0$, $(-5)^2 \geq 0$, and $0^2 \geq 0$.

As a matter of fact, there are numbers (numbers that are not real numbers) whose squares are negative. We will study these numbers — called *imaginary numbers* — in a later chapter.

Do you mean there are numbers whose squares are negative?

It is impossible to list the members of $\mathcal{R}$; therefore, we use a geometric line to display the real numbers. We assert that for every point on the line there exists a unique real number called the **coordinate** (or name) for the point. Furthermore, for every real number there exists a point on the line called the **graph** (or picture) for the number. Such an association between points and numbers — called a **one-to-one correspondence** — yields a **real number line** or, simply, a number line.

Figure 1-2 shows a number line with several integers indicated. The distance between the arbitrarily located points with coordinates 0 and 1 is called the **unit length.** By reproducing the unit length to the right of 1, the positive integers are located. The negative integers are located by reproducing the unit length to the left of 0.

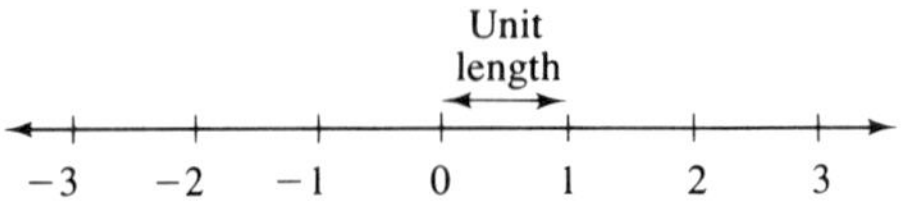

FIGURE 1-2 Graph of real number line

Exercise Set 1–1

1. Identify the sets to which the numbers in set A (Section 1-1) belong.
2. In the following table, place a check in the column if the number is a member of that set.

	N	W	J	R	I	$\mathcal{R}$
2.6						
$\sqrt[3]{2}$						
$\frac{33}{3}$						
$\frac{-19}{6}$						
3						
-1						
0						
$2.\overline{2}$						
$3\frac{5}{6}$						
$-\sqrt{5}$						
1.15115 . . .						
$-4.\overline{1}$						
$-2\frac{2}{3}$						
$\sqrt{25}$						
$\frac{1}{2}+\frac{1}{3}$						
$-0.\overline{25}$						

State the set that is the answer to the following operations.

3. $N \cap W =$
4. $\mathcal{R} \cup \varnothing =$
5. $J \cap N =$
6. $I \cap R =$
7. $I \cup R =$
8. $W \cup J =$
9. $J \cap \mathcal{R} =$
10. $I \cap \varnothing =$

11. Why must the decimal equivalent of any rational number repeat a pattern or terminate?
12. Describe the decimal equivalent of an irrational number.

Find the decimal equivalent of the following rational numbers to one full period if the number is periodic.

13. $\frac{1}{3}$
14. $\frac{1}{6}$
15. $\frac{6}{7}$

16. $\frac{17}{11}$ 17. $5\frac{2}{3}$ 18. $\frac{23}{13}$

19. $3\frac{3}{8}$ 20. $\frac{17}{101}$

Show that the following numbers are rational by writing them in the form $\frac{a}{b}$, where a and b are integers.

21. $0.\overline{13}$ 22. $3.\overline{2}$ 23. $4.\overline{261}$

24. 1.37 25. $\frac{0.2}{6}$ 26. $0.3\overline{1725}$

Sketch a real number line and find the approximate location of the following real numbers.

27. $\frac{22}{9}$ 28. $\frac{-1}{3}$ 29. $\frac{3}{5}$

30. 4.37146 . . . 31. π 32. $\frac{-19}{8}$

AXIOMS OF EQUALITY, ORDER, ADDITION, AND MULTIPLICATION 1–2

In this section we discuss the fundamental relationships that exist between any two real numbers a and b. We also discuss the meaning of the operations of addition and multiplication. We have all met the symbols $=$, $<$, $+$, $\times$, etc., in previous mathematics courses. However, these symbols frequently have been used without a basic knowledge or understanding of them. The following list of **axioms** (statements accepted as true) is given to validate the operations we will perform in subsequent chapters.

Axioms of Equality and Order

One of the principal parts of a sentence is the verb. In mathematics we have verbs that relate two numbers, two geometric figures, two sets, and so forth. The three verbs studied here are equals ($=$), is less than ($<$), and is greater than ($>$). The following axiom states that these three verbs alone are sufficient to relate any two real numbers a and b.

Axiom 1: Law of Trichotomy If a and b are any two real numbers, then one and only one of the following statements is true:

(1) $a < b$ (2) $a = b$ (3) $a > b$

The following axioms state the properties of the symbols $<$, $=$, and $>$.

Axiom 2: Reflexive Property of Equality For any real number a, $a = a$.

Axiom 3: Symmetric Property of Equality For any real numbers a and b, if $a = b$, then $b = a$.

Axiom 4: Transitive Property of Equality For any real numbers a, b, and c, if $a = b$ and $b = c$, then $a = c$.

Axiom 5: Substitution Property of Equality For any real numbers a and b, if $a = b$, then in any expression containing a, b may be substituted without changing the value of the expression.

Axiom 6: Transitive Property of Order For any real numbers a, b, and c, if $a < b$ and $b < c$, then $a < c$. Furthermore, if $a > b$ and $b > c$, then $a > c$.

We use the number line to provide a geometric model for the relations $<$, $=$, and $>$. In Figure 1-3 points A, B, C, and D are the graphs of the real numbers a, b, c, and d, respectively. Based on the relationships between pairs of points on the line, we make the following comparisons between pairs of numbers:

1. A is to the left of B; therefore, $a < b$.
2. C is to the right of B; therefore, $c > b$.
3. B is between points A and C; therefore, $a < b < c$.
4. D and B are the same point; therefore, $d = b$.

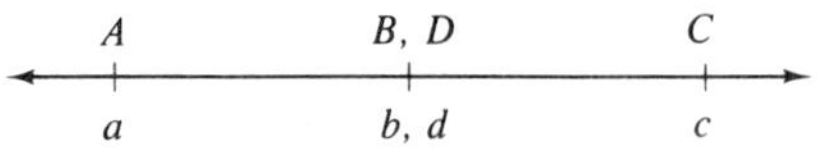

FIGURE 1-3 Graph of points A, B, C, and D

We'll try to explain *reflexive, symmetric,* and *transitive* further by using the following example.

EXAMPLE 1–3 Discuss the properties of N, where N is the relationship "lives next door to."

Solution: Let P, Q, and R be three people, such that Q lives between P and R, and the three people live in a row as shown in Figure 1-4.

FIGURE 1-4

(a) N is a **reflexive relationship** if and only if the statement P N P is meaningful, that is, P "lives next door to" P. Generally speaking, we do not find a person living in adjoining houses; thus, N is not reflexive.
(b) N is a **symmetric relationship** if and only if the statement P N Q is true whenever Q N P is true. That is, P "lives next door to" Q whenever Q "lives next door to" P. Since this is generally true, we accept N as a symmetric relationship.
(c) N is a **transitive relationship** if and only if the statement P N Q and Q N R implies that P N R is true. That is, if P "lives next door to" Q and Q "lives next door to" R, then we may conclude that P "lives next door to" R. Obviously the conclusion does not follow; therefore, N is not a transitive relationship.

EXAMPLE 1–4 Use the correct relationship between the following pairs of numbers:
(a) -3, -5 (b) 15/3, 5 (c) 7/3, 5/2

Solutions:
(a) Since the graph of (-5) is *to the left of* the graph of (-3), then $(-5) < (-3)$.
(b) Since the graph of 15/3 is *the same as* the graph of 5, then $15/3 = 5$.
(c) Since $7/3 = 14/6$ and $5/2 = 15/6$, then the graph of 7/3 is *to the left of* the graph of 5/2, and $7/3 < 5/2$.

Axioms of Addition and Multiplication

If a, b, and c are any real numbers, then we can state the following axioms.

Axiom 7: Closure $a + b$ and ab are real numbers.

Axiom 8: Commutative Addition and Multiplication

$$a + b = b + a \quad \text{and} \quad ab = ba$$

Axiom 9: Associative Addition and Multiplication

$$a + (b + c) = (a + b) + c \quad \text{and} \quad a(bc) = (ab)c$$

Axiom 10: Identity Elements for Addition and Multiplication There exist unique elements in $\mathcal{R}$, namely 0 and 1, such that for each a in $\mathcal{R}$,

$$a + 0 = 0 + a = a \quad \text{and} \quad a(1) = 1(a) = a$$

Axiom 11: Inverse Elements For every element a in $\mathcal{R}$, there exists a unique element $(-a)$ called the **additive inverse,** or **negative** of a, such that $a + (-a) = (-a) + a = 0$. For every element a in $\mathcal{R}$ $(a \neq 0)$, there exists a unique element called the **multiplicative inverse,** or **reciprocal** of a, such that $a(1/a) = (1/a)a = 1$.

Axiom 12: Distributive Property of Multiplication with Respect to Addition

$$a(b + c) = ab + ac \quad \text{and} \quad (a + b)c = ac + bc$$

The following examples demonstrate the use of some of these axioms.

EXAMPLE 1–5 Reduce the rational number $\frac{24}{36}$ to lowest term.

Solution: Since $24 = 2(12)$ and $36 = 3(12)$, then

$$\frac{24}{36} = \frac{2(12)}{3(12)} = \left(\frac{2}{3}\right)\left(\frac{12}{12}\right) = \frac{2}{3}(1) = \frac{2}{3}$$

by the identity axiom for multiplication.

EXAMPLE 1–6 Simplify $13k + 26k$.

Solution: By the distributive axiom of multiplication with respect to addition, we have

$$13k + 26k = (13 + 26)k = 39k$$

To complete this section we consider the operations of subtraction and division. Unlike addition and multiplication, these operations are *defined*.

Subtraction

Definition: For any real numbers a and b, $a - b = a + (-b)$.

The definition states that to subtract b from a, we add the negative of b to a.

EXAMPLE 1–7 Perform the following subtractions:

(a) $-15 - 8$ (b) $23 - (-10)$

Solutions:
(a) Since $b = 8$, $-b = -8$, and $-15 - 8 = -15 + (-8) = -23$.
(b) With $b = -10$, $-b = -(-10) = 10$, and $23 - (-10) = 23 + 10 = 33$.

Division

Definition: For any real numbers a and b, with $b \neq 0$,

$$\frac{a}{b} = a\left(\frac{1}{b}\right)$$

The definition states that to divide a by b, we multiply a by the reciprocal of b.

We can easily verify that subtraction and division are *not commutative operations,* they are *not associative operations,* and they *do not have an identity element*. When we define them in terms of addition and multiplication, they become "well-behaved operations."

Why are subtraction and division defined in terms of addition and multiplication?

Exercise Set 1–2

Name the axiom that justifies each of the following statements.

1. If $8 \neq 7$ and $8 \not< 7$, then $8 > 7$.
2. $7 + (-7) = 0$
3. If $a < 4 + 6$ and $4 + 6 < 13$, then $a < 13$.
4. $\frac{1}{7}(7) = 1$
5. $\left(\frac{11}{18}\right)\left(\frac{18}{11}\right) = \left(\frac{18}{11}\right)\left(\frac{11}{18}\right)$
6. $4w + 3w = (4 + 3)w$
7. $2a(b + 3) = (b + 3)2a$
8. $(3x) + 0 = 3x$
9. $7(2) \in \mathcal{R}$
10. $84(8 \cdot 1) = 84(8)$

Which of the following sets are closed under (a) addition and (b) multiplication?

11. $\{0, 1\}$
12. Positive odd integers
13. Rationals
14. Negative integers

Find the additive inverse or opposite of the following real numbers.

15. $\frac{3}{7}$
16. -4
17. 0
18. $-3\frac{5}{8}$

Find the multiplicative inverse or reciprocal of the following real numbers.

19. $\frac{-1}{8}$
20. $-2\frac{4}{7}$
21. $\frac{19}{26}$
22. 4

Name the axiom that justifies each step of the following equalities.

23. $(7a + 8) + 5a = 7a + (8 + 5a)$ (a) ____________
$\quad = 7a + (5a + 8)$ (b) ____________

$= (7a + 5a) + 8$ (c) ____________

$= (7 + 5)a + 8$ (d) ____________

$= 12a + 8$ (e) ____________

24. $(a + b)(c + d) = (a + b)c + (a + b)d$ (a) ____________

$= ac + bc + ad + bd$ (b) ____________

$= ca + cb + da + db$ (c) ____________

$= c(a + b) + d(a + b)$ (d) ____________

$= (c + d)(a + b)$ (e) ____________

25. $78 + (73 + (-78)) = 78 + ((-78) + 73)$ (a) ____________

$= (78 + (-78)) + 73$ (b) ____________

$= 0 + 73$ (c) ____________

$= 73$ (d) ____________

Use $<$, $>$, or $=$ to express the correct relationship between the following pairs of numbers.

26. $3, \frac{18}{6}$

27. $\frac{-7}{3}, \frac{-5}{3}$

28. $\frac{56}{27}, \frac{55}{26}$

29. $-0.01, -0.02$

30. $0.\overline{3}, 0.3$

31. $\frac{-2}{7}, \frac{-1}{7}$

32. If a and b are any real numbers, with $a > 0$ and $b < 0$, compare a and b.

33. If a is a real number and $a > 0$, compare $(-a)$ with 0.

34. If b is a real number and $b < 0$, compare $1/b$ with 0.

35. If a and b are any real numbers and $a > b$, compare $(-a)$ and $(-b)$.

36. If c is a real number and $c < 0$, compare $(-c)$ with 0.

Determine whether the following set relations are (a) reflexive, (b) symmetric, or (c) transitive.

37. A R B where A and B are sets and the relation R is disjoint.

38. A R B where A and B are sets and the relation R is equal.

39. A R B where A and B are people and the relation R is "is a brother of."

40. A R B where A and B are straight lines in a plane and the relation R is "is parallel to."

A set under a given operation may satisfy the closure, commutative, associative, identity, and inverse properties. Such a set is called a **commutative group** for the given operation.

41. Are the real numbers a commutative group under the operation of addition?

42. Are the real numbers a commutative group under the operation of multiplication?

43. Set $S = \{a, b, c, d\}$. The operation * is defined by the following table.

*	a	b	c	d
a	a	b	c	d
b	b	a	d	c
c	c	d	a	b
d	d	c	b	a

(a) Is S closed under the operation *?
(b) Is * a commutative operation?
(c) Is * an associative operation?
(d) Does S contain an identity element? If yes, identify it.
(e) Does each element in S have an inverse element? If yes, identify the inverse of each element in S.
(f) Is S a commutative group under the operation *?

1–3 LINEAR EQUATIONS AND INEQUALITIES IN ONE VARIABLE

A mathematical sentence contains mathematical symbols in certain combinations together with a mathematical verb. The statement is true or false depending on the manner in which the symbols are combined. For example, $6 + 5(3) = 21$ is a true statement, and $8 - 3(2) = 10$ is a false statement.

The sentence is an **open sentence** if any one of the symbols is a **variable,** that is, a symbol that names an unspecified number in a set containing more than one number. For example,

$$i = 2.54c \quad \text{and} \quad d = 4.9t^2$$

are two open sentences.

In the given equations the variables represent specific quantities. In the first equation i and c represent *inches* and *centimeters,* respectively, and the equation states the relationship between a measurement in inches and one in centimeters. In the second equation d and t represent *distance* and *time,* respectively. The equation states the relationship between the distance an object falls from rest under the influence of gravity and the time t during which it falls.

Do the variables i, c, d, and t represent anything in particular?

An open sentence cannot be classified as true or false until the variables are replaced with elements from a replacement set. The following definitions pertain to an open sentence and the process of solving the open sentence.

Solution set

> **Definition:** The solution set (or truth set) contains those elements of the replacement set that make the open sentence true.

Solving an open sentence

> **Definition:** Solving an open sentence is the process of finding the truth set.

Equivalent open sentences

> **Definition:** Two open sentences are equivalent if their truth sets are equal.

In this section we review the process of solving the most elementary type of open sentence, an equation or inequality of degree one in one variable. Such an open sentence is of the form

$$ax + b = 0, ax + b < 0, ax + b > 0, ax + b \leq 0, ax + b \geq 0$$

where a and b are real numbers and $a \neq 0$.

The following axioms are the basic tools that enable us to solve such an equation or inequality. In these axioms a, b, and c are real numbers.

> **Axiom 13:** **Addition of Equality** If $a = b$, then $a + c = b + c$.
> **Addition of Inequality** If $a < b$, then $a + c < b + c$.

> **Axiom 14:** **Subtraction of Equality** If $a = b$, then $a - c = b - c$.
> **Subtraction of Inequality** If $a < b$, then $a - c < b - c$.

> **Axiom 15:** **Multiplication of Equality** If $a = b$, then $ac = bc$.
> **Multiplication of Inequality**
> (1) If $a < b$ and $c > 0$, then $ac < bc$.
> (2) If $a < b$ and $c < 0$, then $ac > bc$.

> **Axiom 16:** **Division of Equality** If $a = b$ and $c \neq 0$, then $\frac{a}{c} = \frac{b}{c}$.
> **Division of Inequality**
> (1) If $a < b$ and $c > 0$, then $\frac{a}{c} < \frac{b}{c}$.
> (2) If $a < b$ and $c < 0$, then $\frac{a}{c} > \frac{b}{c}$.

In Axioms 13 through 16, the order relationship $<$ can everywhere be replaced by $>$, and the resulting axioms are also true. This is valid since $a < b$ and $b > a$ are equivalent statements.

What is the difference between Equations (1) and (2) in the multiplication and division of inequalities?

Recall that when we multiply or divide by a negative number the order relation is reversed. For example, $3 < 5$, but $3(-2) > 5(-2)$, or $-6 > -10$. Similarly, $3 > -5$, but $3(-2) < (-5)(-2)$, or $-6 < 10$.

The following examples review briefly the method of solving linear equations in one variable.

EXAMPLE 1–8 If $x \in \{\text{integers}\}$, solve the equation $3x - 5 = x + 3$.

Solution: If $3x - 5 = x + 3$, then

$$\begin{aligned} 2x - 5 &= 3 && \text{(subtract } x \text{ from both sides of the equation)} \\ 2x &= 8 && \text{(add 5 to both sides of the equation)} \\ x &= 4 && \text{(divide both sides of the equation by 2)} \end{aligned}$$

Since $4 \in \{\text{integers}\}$, the solution set is $\{4\}$. We can check this result by replacing the variable in the given equation with an element of the solution set to see whether a true statement is obtained.

$$\text{If } x = 4, \quad \text{then} \quad 3(4) - 5 = 4 + 3$$
$$7 = 7 \qquad \text{(True)}$$

Thus, $\{4\}$ is the truth set of the given equation.

EXAMPLE 1–9 If $x \in \{\text{rationals}\}$, solve the equation

$$\frac{3}{8}(2x + 1) + \frac{2}{3}x + 1 = \frac{1}{6}(3x - 5) + 3$$

Solution: We first eliminate the fractions in the equation by multiplying both sides of the equation by 24, the least common multiple (L.C.M.) of 8, 3, and 6.

$$\begin{aligned} 24\left(\frac{3}{8}(2x + 1) + \frac{2}{3}x + 1\right) &= 24\left(\frac{1}{6}(3x - 5) + 3\right) && \text{(mult. of eq.)} \\ 9(2x + 1) + 16x + 24 &= 4(3x - 5) + 72 && \text{(dist. axiom)} \\ 18x + 9 + 16x + 24 &= 12x - 20 + 72 && \text{(dist. axiom)} \\ 34x + 33 &= 12x + 52 && \text{(comb. like terms)} \\ 22x &= 19 && \text{(subt. of eq.)} \\ x &= 19/22 && \text{(div. of eq.)} \end{aligned}$$

The solution set is $\{19/22\}$ since 19/22 is a rational number. The check is left for the reader.

Parentheses and brackets are used to indicate intervals that contain the solution sets of inequalities. The following examples illustrate the association between the inequality and the corresponding graphs and set notation.

Inequality	Graph	Set Notation
$a \le x \le b$	Figure 1-5(a)	$x \in [a, b]$
$a < x < b$	Figure 1-5(b)	$x \in (a, b)$
$x < a$	Figure 1-5(c)	$x \in (-\infty, a)$
$x \ge b$	Figure 1-5(d)	$x \in [b, \infty)$

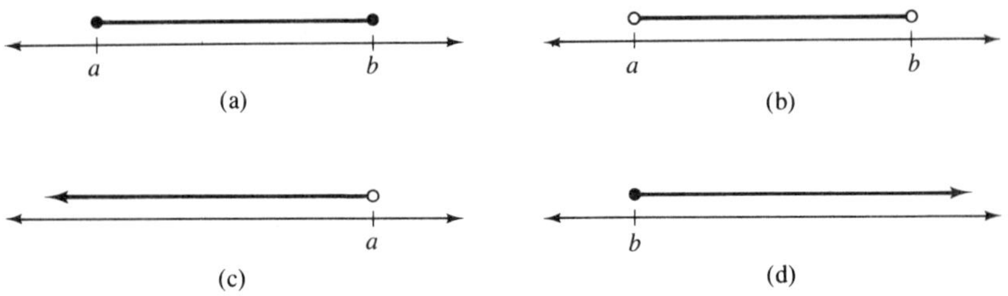

FIGURE 1-5

Notice that if the endpoint of an interval is included a solid dot is used on the graph and a bracket is used in the set notation. If the endpoint of an interval is excluded, a hollow dot is used on the graph and a parenthesis closes the set.

The symbol ∞ is used to represent **infinity,** and $-\infty$ represents **negative infinity.** Therefore, if $x \in (b, \infty)$, then the values of x can become *infinitely large through positive values*. Furthermore, if $x \in (-\infty, a)$, then the values of x can become *infinitely small through negative values.*

Solve the following inequalities and express the solution set in set notation and with a graph.

EXAMPLE 1–10 $5x - 4 < 3x + 8$

Solution: If $5x - 4 < 3x + 8$, then

$$2x < 12 \qquad \text{(addition and subtraction of inequality)}$$
$$x < 6 \qquad \text{(division of inequality)}$$

Thus, $x \in (-\infty, 6)$, and the graph is Figure 1-6.

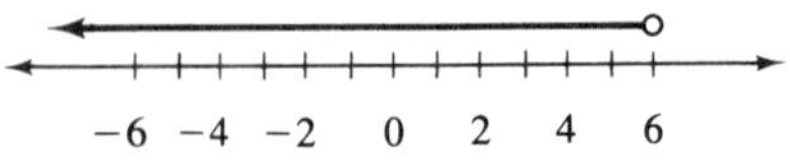

FIGURE 1-6 Graph of $(-\infty, 6)$

EXAMPLE 1–11 $\frac{3x}{2} + \frac{1}{3} \geq \frac{7x}{4} + \frac{7}{3}$

Solution: We multiply both sides of the inequality by 12, the L.C.M. of 2, 3, and 4.

$$12\left(\frac{3x}{2} + \frac{1}{3}\right) \geq 12\left(\frac{7x}{4} + \frac{7}{3}\right) \qquad \text{(multiplication of inequality)}$$
$$18x + 4 \geq 21x + 28 \qquad \text{(distributive axiom)}$$
$$-3x \geq 24 \qquad \text{(subtraction of inequality)}$$
$$x \leq -8 \qquad \text{(division of inequality by a negative number)}$$

Thus, $x \in (-\infty, -8]$, and Figure 1-7 is the graph.

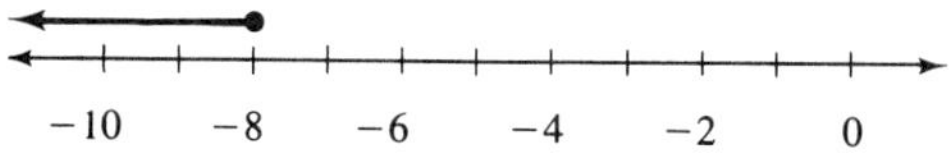

FIGURE 1-7 Graph of $(-\infty, -8]$

EXAMPLE 1–12 $-5 < 3x + 4 \leq 10$

Solution: The given inequality is a **compound statement** consisting of the inequalities $-5 < 3x + 4$ and $3x + 4 \leq 10$. The solution set is the **intersection** of the solution sets of both inequalities, since $3x + 4$ is simultaneously greater than -5 *and* less than or equal to 10. We can solve the given expression by operating on all three elements of the inequality at the same time as demonstrated below.

If $-5 < 3x + 4 \leq 10$, then

$$-9 < 3x \leq 6 \quad \text{(subtraction of inequality)}$$
$$-3 < x \leq 2 \quad \text{(division of inequality by a positive number)}$$

The solution set is $(-3, 2]$, and the graph is Figure 1-8.

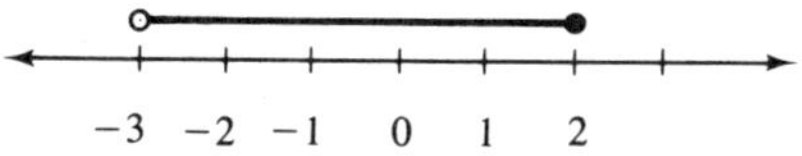

FIGURE 1-8 Graph of $(-3, 2]$

EXAMPLE 1–13 $5 - 3x \leq 11$ or $2x + 3 < -9$

Solution: The solution set of the given **compound inequality** contains the **union** of the solution sets of $5 - 3x \leq 11$ and $2x + 3 < -9$. We form the union of the two solution sets because a solution of either inequality — the first one or the second one — will make the compound inequality true.

If $5 - 3x \leq 11$ or $2x + 3 < -9$, then

$$-3x \leq 6 \quad \text{or} \quad 2x < -12$$
$$x \geq -2 \quad \text{or} \quad x < -6$$

Thus, $x \in (-\infty, -6) \cup [-2, \infty)$, and the graph of the solution set is given in Figure 1-9.

FIGURE 1-9 Graph of $(-\infty, -6) \cup [-2, \infty)$

Exercise Set 1–3

Find the solution set of the following equations where the replacement set is the set of integers.

1. $\frac{x}{3} = 8$
2. $2x = 11$
3. $\frac{x + 3}{3} = 7$

4. $3x + 2x = 6 - 86$

5. $17x + 2x^3 + 6 = 8x - (-2x^3 + 3)$

Find the solution set of the following equations where the replacement set is the set of rational numbers.

6. $3.6x + 0.48 = 8.2x - 0.9$

7. $x(x - 4) + 2 = x(x - 6) + 10$

8. $\frac{x}{5} + 2x = 57 - \frac{7x}{2}$

9. $\frac{-x}{4} + \frac{7}{3} = \frac{18}{5}$

10. $-0.36 + 0.1x - 0.02x = 0.2$

11. $-3(x - 6) + 2(2x - 3) = x + 10$

Solve the following inequalities where the replacement set is the set of real numbers. State the solution set in interval notation and graph on a number line.

12. $x - 4 > 1$

13. $\frac{x}{3} \geq -2$

14. $-3x > 6$

15. $\frac{x}{3} - \frac{1}{5} \leq x + 2$

16. $\frac{-x}{5} + 2 < \frac{x}{3}$

17. $2(x - 6) \leq 3(x - 4)$

18. $-3(x + 2) \geq 4(-x + 2) - 13$

19. $7x - (x - 3) \geq 4x + (3 + 2x)$

20. $3x \geq 4x$

21. $6 < 5x - 4 < 26$

22. $-3 < \frac{2x + 3}{5} < 1$

23. $3 < 3(x - 6) \leq 6$

24. $0 \leq \frac{2 - x}{3} < 2$

25. $2x + 3 \geq 9$ or $x - 5 < -6$

26. $\frac{x}{2} - 3 \leq 3$ or $\frac{x}{3} + 2 \geq 7$

27. $1 - 5x > 16$ or $3 - \frac{3x}{2} \leq 9$

28. $5(x - 6) + 2x < 3$ or $-2(x + 1) < 3(x - 4)$

1-4 OPERATIONS ON RATIONAL EXPRESSIONS

Many quantities in the physical world can be expressed mathematically with a **polynomial equation.** For example, the path followed by a projectile can, in most cases, be described by a polynomial equation. The analysis of certain trends begins by first obtaining a polynomial equation that most nearly approximates the observed data. We therefore consider the treatment of polynomials an important part of our mathematical education.

In previous courses in mathematics we have worked extensively with polynomials. We will consider in great detail the analysis of polynomial functions, but in this section we will use the polynomial expression only to define a **rational expression.** We will then review the rational operations of rational expressions.

Polynomial expression

Definition: If $a_0, a_1, \ldots, a_n$ are real numbers and $n, n-1, \ldots$ are positive integers, then a polynomial expression in x is one that can be written as

$$a_0x^n + a_1x^{n-1} + \cdots + a_n$$

Polynomial equation

Definition: $P(x)$ (read "P of x") is a polynomial equation in x if

$$P(x) = a_0x^n + a_1x^{n-1} + \cdots + a_n$$

and $a_0x^n + a_1x^{n-1} + \cdots + a_n$ is a polynomial expression.

Zero polynomial

Definition: $P(x)$ is the zero polynomial if $P(x) = 0$, for all x.

Rational expression in x

Definition: A rational expression in x is one that can be written as $\frac{P(x)}{R(x)}$, where $P(x)$ and $R(x)$ are polynomials and $R(x)$ is not the zero polynomial.

The following are examples of rational expressions in x:

(a) $\frac{3x+5}{x}$ (b) $\frac{x^2+1}{x+1}$ (c) $\frac{x^2+3x-5}{x^2-4}$

Each expression is a ratio of two polynomials. In example (a), $P(x) = 3x + 5$ because 3 and 5 are real numbers and the exponent on x is 1, a positive integer. Similarly, each of the other expressions is a polynomial in x. Notice that if $R(x) \neq 0$, then $x \neq 0, x \neq -1$, and $x \neq \pm 2$ in (a), (b), and (c), respectively. In all examples and exercises in the future, remember that any rational expression is restricted to values for which $R(x) \neq 0$.

Then $3x+5$, x^2+1, and so on must be polynomial expressions?

Negative of a rational expression

Definition: If $U(x)$ is a rational expression such that $U(x) = \frac{P(x)}{R(x)}$, then

$$-U(x) = \frac{-P(x)}{R(x)} \quad \text{or} \quad \frac{P(x)}{-R(x)}$$

A rational expression in more than one variable is an extension of the definition for a rational expression in one variable. Thus, if $P(x, y)$ and $R(x, y)$ are polynomials in x and y, and $R(x, y)$ is not the zero polynomial, then $\frac{P(x, y)}{R(x, y)}$ is a rational expression in x and y. The extension can be applied to polynomials of any finite number of variables. The following are examples of a rational expression in x and y:

$$\text{(a)}\ \frac{x-2y}{x+2y} \qquad \text{(b)}\ \frac{x^2+y^2}{xy}$$

Consistent with the agreement above, $x \neq -2y$, and $x \neq 0$ and $y \neq 0$, respectively, in these examples.

Simplified rational expression

> **Definition:** A rational expression $\frac{P(x)}{R(x)}$ $(R(x) \neq 0)$ is simplified if $P(x)$ and $R(x)$ have no common factors other than 1.

EXAMPLE 1–14 Simplify $\frac{3x^2 - x - 10}{x^2 + 3x - 10}$.

Solution:

$$P(x) = 3x^2 - x - 10 = (3x + 5)(x - 2)$$
$$R(x) = x^2 + 3x - 10 = (x + 5)(x - 2)$$

Thus, we have

$$\frac{P(x)}{R(x)} = \frac{(3x+5)(x-2)}{(x+5)(x-2)} = \frac{3x+5}{x+5}$$

Can't we also cancel the two 5's?

Since $3x + 5$ and $x + 5$ have no common factors other than one, $\frac{3x+5}{x+5}$ is the simplified form of the given rational expression. Because the fives are *terms*, not *factors*, they therefore may not be cancelled.

> **Theorem: Product of Rational Expressions**
>
> If $U(x) = \frac{P(x)}{R(x)}$ and $V(x) = \frac{S(x)}{T(x)}$, then
>
> $$U(x)\cdot V(x) = \frac{P(x)S(x)}{R(x)T(x)} \quad (R(x) \neq 0 \quad \text{and} \quad T(x) \neq 0)$$

EXAMPLE 1–15 Find $U(x)\cdot V(x)$ if $U(x) = \frac{3}{x+1}$ and $V(x) = \frac{x^2-1}{3x^3}$.

Solution:

$$\begin{aligned}
U(x)\cdot V(x) &= \left(\frac{3}{x+1}\right)\left(\frac{x^2-1}{3x^3}\right) && \text{(mult. of rational expressions)}\\
&= \frac{3(x+1)(x-1)}{(x+1)3x^3} && (x^2 - 1 = (x+1)(x-1))\\
&= 1(1)\left(\frac{x-1}{x^3}\right) && \left(\frac{3}{3} = 1 \text{ and } \frac{x+1}{x+1} = 1\right)\\
&= \frac{x-1}{x^3} && \text{(mult. property of 1)}
\end{aligned}$$

Theorem: Quotient of Rational Expressions

If $U(x) = \dfrac{P(x)}{R(x)}$ and $V(x) = \dfrac{S(x)}{T(x)}$, then

$$\frac{U(x)}{V(x)} = \frac{\dfrac{P(x)}{R(x)}}{\dfrac{S(x)}{T(x)}} = \frac{P(x)T(x)}{R(x)S(x)} \quad (R(x) \neq 0,\ S(x) \neq 0, \text{ and } T(x) \neq 0)$$

EXAMPLE 1–16 If $U(x) = \dfrac{x^2 - 4x}{6x^2 - x - 2}$ and $V(x) = \dfrac{x^2 - 3x - 4}{3x^2 + x - 2}$

find $\dfrac{U(x)}{V(x)}$.

Solution:

$$\frac{U(x)}{V(x)} = \left(\frac{x^2 - 4x}{6x^2 - x - 2}\right)\left(\frac{3x^2 + x - 2}{x^2 - 3x - 4}\right) \quad \text{(quotient of rational expressions)}$$

$$= \frac{x(x - 4)(3x - 2)(x + 1)}{(3x - 2)(2x + 1)(x - 4)(x + 1)} \quad \text{(factor each polynomial expression in numerator and denominator)}$$

$$= 1(1)(1)\left(\frac{x}{2x + 1}\right) \quad \left(\frac{x - 4}{x - 4} = \frac{3x - 2}{3x - 2} = \frac{x + 1}{x + 1} = 1\right)$$

$$= \frac{x}{2x + 1}$$

Theorem: Sum of Rational Expressions

If $U(x) = \dfrac{P(x)}{R(x)}$ and $V(x) = \dfrac{S(x)}{T(x)}$, and $R(x)$ and $T(x)$ have no common factors other than 1, then

$$U(x) + V(x) = \frac{P(x)T(x) + R(x)S(x)}{R(x)T(x)} \quad (R(x) \neq 0 \text{ and } T(x) \neq 0)$$

If $R(x)$ and $T(x)$ have a common factor other than one, then a least common multiple is determined in the manner used to find the L.C.M. of two rational numbers a and b.

EXAMPLE 1–17 If $U(x) = \dfrac{2x + 1}{x^2 - 1}$ and $V(x) = \dfrac{6x}{x + 1}$, find $U(x) + V(x)$.

Solution:

$$U(x) = \frac{2x + 1}{x^2 - 1} = \frac{2x + 1}{(x + 1)(x - 1)}$$

$$V(x) = \frac{6x}{x + 1} = \frac{6x(x - 1)}{(x + 1)(x - 1)} = \frac{6x^2 - 6x}{(x + 1)(x - 1)}$$

Since $U(x)$ and $V(x)$ now have a common denominator, we have

$$\begin{aligned} U(x) + V(x) &= \frac{2x + 1}{(x + 1)(x - 1)} + \frac{6x^2 - 6x}{(x + 1)(x - 1)} \\ &= \frac{2x + 1 + 6x^2 - 6x}{(x + 1)(x - 1)} \\ &= \frac{6x^2 - 4x + 1}{x^2 - 1} \end{aligned}$$

Because numerator and denominator have no common factor, the answer is in simplified form.

Subtraction of rational expressions

> **Definition:** If $U(x)$ and $V(x)$ are rational expressions, then
> $$U(x) - V(x) = U(x) + (-V(x))$$

EXAMPLE 1–18 If

$$U(x, y) = \frac{x + y}{x^2 - 5xy + 6y^2} \quad \text{and} \quad V(x, y) = \frac{2x - y}{2y^2 + xy - x^2}$$

find $U(x, y) - V(x, y)$.

Solution:

$$U(x, y) = \frac{(x + y)}{(x - 3y)(x - 2y)} \quad \text{and} \quad V(x, y) = \frac{(2x - y)}{(2y - x)(y + x)}$$

The denominators of $U(x, y)$ and $V(x, y)$ have a common factor that differs only in sign, since $(2y - x) = -(x - 2y)$. Thus,

$$-V(x, y) = \frac{(2x - y)}{(x - 2y)(x + y)}$$

Then we have

$$\begin{aligned} U(x, y) + [-V(x, y)] &= \frac{(x + y)}{(x - 3y)(x - 2y)} + \frac{(2x - y)}{(x + y)(x - 2y)} \\ &= \frac{(x + y)(x + y)}{(x - 3y)(x - 2y)(x + y)} + \frac{(2x - y)(x - 3y)}{(x - 3y)(x - 2y)(x + y)} \\ &= \frac{x^2 + 2xy + y^2 + 2x^2 - 7xy + 3y^2}{(x - 3y)(x - 2y)(x + y)} \\ &= \frac{3x^2 - 5xy + 4y^2}{(x - 3y)(x - 2y)(x + y)} \end{aligned}$$

EXAMPLE 1–19 Express as a single rational expression:

$$\left(\frac{3}{2(x - 1)} - \frac{1}{2(x + 1)}\right)\left(\frac{1}{2x} + \frac{1}{2(x + 2)}\right)$$

Solution: We first combine the pairs of rational expressions within the two pairs of parentheses:

$$\left(\frac{3}{2(x-1)}-\frac{1}{2(x+1)}\right)=\frac{3(x+1)-(x-1)}{2(x+1)(x-1)}$$
$$=\frac{3x+3-x+1}{2(x+1)(x-1)}$$
$$=\frac{(x+2)}{(x+1)(x-1)}$$

Similarly,

$$\left(\frac{1}{2x}+\frac{1}{2(x+2)}\right)=\frac{(x+2)+x}{2x(x+2)}$$
$$=\frac{x+1}{x(x+2)}$$

We now form the product of the two simplified expressions:

$$\left(\frac{(x+2)}{(x+1)(x-1)}\right)\left(\frac{x+1}{x(x+2)}\right)=\frac{(x+2)(x+1)}{(x+1)(x-1)(x)(x+2)}=\frac{1}{x(x-1)}$$

Exercise Set 1–4

Simplify the following rational expressions.

1. $\dfrac{5x-10}{15x-30}$

2. $\dfrac{x^2-4x}{x^3-4x^2}$

3. $\dfrac{4-3x}{3x-4}$

4. $\dfrac{x}{2x+x^2}$

5. $\dfrac{5x^2+7x-6}{15x^2-4x-3}$

6. $\dfrac{x^3-27}{(x-3)^2}$

7. $\dfrac{6x^2-21x+18}{12x^2-6x-36}$

8. $\dfrac{x^4-81}{3x^2y-27y}$

9. $\dfrac{6x^3+11x^2+4x}{2x^2-9x-5}$

10. $\dfrac{x^2+ax+bx+ab}{x+a}$

11. $\left(\dfrac{3x^2}{y^4}\right)\left(\dfrac{2y^2}{36x}\right)$

12. $\left(\dfrac{x^2-y^2}{4}\right)\left(\dfrac{6}{x+y}\right)$

13. $\left(\dfrac{7x^4y^3}{16xy^4}\right)\div\left(\dfrac{16xy^4}{7xy}\right)$

14. $\left(\dfrac{x^2-7x}{x^2-16}\right)\div\left(\dfrac{x^2-9}{x^2-7x+12}\right)$

15. $2x^2\left(\dfrac{x^3-1}{x^3-x}\right)\left(\dfrac{x+1}{x}\right)$

16. $\left(\dfrac{x^2-4}{x^2+4x+4}\right)\left(\dfrac{x^2+2x}{2-x}\right)$

17. $\left(\dfrac{3x+6y}{9x}\right)\left(\dfrac{12xy}{x^2-4y^2}\right)\div\left(\dfrac{18x^3y^3}{2x-4y}\right)$

18. $\left(\dfrac{30-11x+x^2}{9x-6x^2+x^3}\right)\div\left(\dfrac{x^2-9}{x^2+2x-15}\right)\left(\dfrac{x^2-3x}{25-x^2}\right)$

19. If $U(x)=\dfrac{x^2-4}{2x+4}$ and $V(x)=\dfrac{x^2+4x+4}{3x-6}$, find

(a) $U(x)\cdot V(x)$ (b) $\dfrac{U(x)}{V(x)}$ (c) $\dfrac{V(x)}{U(x)}$

20. If $U(x) = \dfrac{x^3 - 1}{x + 1}$, $V(x) = \dfrac{x^2 + 2x + 1}{x^2 + x + 1}$, and $W(x) = \dfrac{x^2 + 2x - 3}{x + 1}$, find

(a) $\dfrac{U(x) \cdot V(x)}{W(x)}$ (b) $\dfrac{U(x)}{V(x)} \cdot W(x)$ (c) $\dfrac{U(x)}{V(x) \cdot W(x)}$

21. $\dfrac{3}{x - 1} + \dfrac{2}{x - 1}$

22. $\dfrac{x}{x + y} - \dfrac{y}{x - y}$

23. $\dfrac{14}{x^2 - 9} - \dfrac{6}{x - 3}$

24. $x + \dfrac{x}{x - 2}$

25. $x - 1 + \dfrac{2}{x + 1}$

26. $\dfrac{x}{2} - \dfrac{x}{x - 1} + \dfrac{x + 1}{2x - 2}$

27. $\dfrac{x + 4}{x^2 - x + 2} + \dfrac{2x - 3}{x^2 - x + 2}$

28. $\dfrac{x + y}{x - y} - \dfrac{x - y}{x + y} - \dfrac{4xy}{x^2 - y^2}$

29. $\left(\dfrac{1}{x} - \dfrac{1}{x - 1}\right)\left(\dfrac{x^2 + x - 2}{x}\right)$

30. $\left(\dfrac{3}{x - 3} - \dfrac{1}{x + 1}\right)\left(\dfrac{3}{x + 3} - \dfrac{2}{x + 1}\right)$

1-5 COMPLEX FRACTIONS

Section 1-4 was designed to sharpen our skills in operating on rational expressions. The motivation was simply that we frequently encounter such expressions in applied problems in mathematics. The basis for this section is that we sometimes are faced with expressions that are categorically called *complex fractions*. We first define a complex fraction, then we consider two methods designed to reduce the expression to its simplest form.

Complex fraction in *x*

Definition: A complex fraction in x is the ratio of two expressions that contains at least one rational term in the numerator or denominator of the expression for which the denominator is not one.

The following are examples of complex fractions in the variable x:

(a) $\dfrac{x}{1 + \dfrac{1}{x}}$ (b) $\dfrac{x - \dfrac{1}{x}}{x^2 - 1}$ (c) $\dfrac{1 + \dfrac{2}{x}}{1 - \dfrac{2}{x}}$

We now extend the above definition to include complex fractions in more than one variable. The following are examples of complex fractions in the variables x and y; x, y, and z; and r, s, t, and u, respectively.

(a) $\dfrac{\dfrac{x}{y} + \dfrac{y}{x}}{x + y}$ (b) $\dfrac{\dfrac{1}{x} + \dfrac{1}{y} + \dfrac{1}{z}}{x + y + z}$ (c) $\dfrac{\dfrac{r}{s} + \dfrac{t}{u}}{\dfrac{rs}{tu}}$

Simplified complex fraction

> **Definition:** A complex fraction is simplified when it is expressed as a simplified rational expression.

Basically, there are two approaches that can be used to simplify a complex fraction. The first method is to convert both numerator and denominator of the fraction to a single rational expression and then simplify the ratio of the two rational expressions. The following example illustrates this method.

EXAMPLE 1–20 Simplify

$$\frac{\frac{a}{b} - \frac{b}{a}}{\frac{1}{b} - \frac{1}{a}}$$

First we combine $\frac{a}{b} - \frac{b}{a}$ into one expression and $\frac{1}{b} - \frac{1}{a}$ into one expression. We will then simplify the quotient of the two rational expressions.

Solution:

$$\frac{a}{b} - \frac{b}{a} = \frac{a^2}{ab} - \frac{b^2}{ab} = \frac{a^2 - b^2}{ab}$$

$$\frac{1}{b} - \frac{1}{a} = \frac{a}{ab} - \frac{b}{ab} = \frac{a - b}{ab}$$

Then we have

$$\frac{\frac{a}{b} - \frac{b}{a}}{\frac{1}{b} - \frac{1}{a}} = \frac{\frac{a^2 - b^2}{ab}}{\frac{a - b}{ab}} = \left(\frac{a^2 - b^2}{ab}\right)\left(\frac{ab}{a - b}\right) = \frac{ab(a + b)(a - b)}{ab(a - b)} = a + b$$

The alternate approach for simplifying a complex fraction requires the multiplication property of 1. First we determine the least common multiple of every denominator of a rational expression in the fraction. We then multiply each term of the complex fraction by the L.C.M. and simplify the result. The following examples illustrate the process.

EXAMPLE 1–21 Simplify

$$\frac{1 - \frac{1}{x + 1}}{\frac{2}{x - 1} + 1}$$

There are two rational expressions in the fraction, $\frac{1}{x + 1}$ and $\frac{2}{x - 1}$. The L.C.M. of the two denominators is the product $(x + 1)(x - 1)$. We therefore multiply the complex fraction by the ratio $\frac{(x + 1)(x - 1)}{(x + 1)(x - 1)}$ and simplify.

How many rational expressions does this fraction contain?

Solution:

$$\left(\frac{1-\dfrac{1}{x+1}}{\dfrac{2}{x-1}+1}\right)\left(\frac{(x+1)(x-1)}{(x+1)(x-1)}\right)=\frac{(x+1)(x-1)-(x-1)}{2(x+1)+(x+1)(x-1)}$$
$$=\frac{x^2-1-x+1}{2x+2+x^2-1}$$
$$=\frac{x(x-1)}{(x+1)^2}$$

This example illustrates how multiplication by the L.C.M. reduces the complex fraction to the ratio of two polynomials which may or may not simplify when factored.

EXAMPLE 1–22 Simplify

$$x+\frac{1}{x-\dfrac{1}{x}}$$

Solution: First we simplify the right term of the given fraction:

$$\left(\frac{1}{x-\dfrac{1}{x}}\right)\left(\frac{x}{x}\right)=\frac{x}{x^2-1}$$

Then we have

$$x+\frac{x}{x^2-1}=\frac{x(x^2-1)+x}{x^2-1}=\frac{x^3}{x^2-1}$$

Exercise Set 1–5

1. $\dfrac{2+\dfrac{1}{3}}{4}$

2. $\dfrac{x+\dfrac{3x}{y}}{y}$

3. $\dfrac{\dfrac{3y}{x}-x}{x+\dfrac{1}{x}}$

4. $\dfrac{\dfrac{x}{y}-\dfrac{y}{x}}{1-\dfrac{x}{y}}$

5. $\dfrac{x-\dfrac{x^2-1}{x}}{1-\dfrac{x+1}{x}}$

6. $\dfrac{x+1-\dfrac{2}{x}}{x+4+\dfrac{4}{x}}$

7. $\dfrac{1+\dfrac{4}{x}+\dfrac{3}{x^2}}{1+\dfrac{3}{x}}$

8. $\dfrac{x+\dfrac{4}{y}-\dfrac{2}{y^2}}{x-\dfrac{1}{xy}}$

9. $\dfrac{x+6+\dfrac{6}{x+1}}{x+2+\dfrac{3}{x+6}}$

10. $\dfrac{\dfrac{x-y}{x+y}}{\dfrac{x-y}{x+2y}}$

11. $\dfrac{\dfrac{1}{x^2-1}+1}{\dfrac{1}{x-1}+1}$

12. $\dfrac{\dfrac{1}{x-2}+\dfrac{1}{x+2}}{\dfrac{1}{x-2}-\dfrac{1}{x+2}}$

13. $\dfrac{\dfrac{1}{x}+2+\dfrac{3}{x+1}}{\dfrac{1}{x}-2-\dfrac{3}{x+1}}$

14. $\dfrac{\dfrac{x}{y}+\dfrac{y}{x}+2}{\dfrac{x}{y}-\dfrac{y}{x}}$

15. $\dfrac{\dfrac{x+y}{x-y}+\dfrac{x}{y}}{\dfrac{xy}{x^2-y^2}+\dfrac{x}{x+y}}$

16. $1+\dfrac{1}{1+\dfrac{1}{1+\dfrac{1}{x}}}$

17. $\dfrac{1-\dfrac{1}{x}}{1+\dfrac{1}{1-\dfrac{2}{x}}}$

18. $\dfrac{1+\dfrac{1}{1-\dfrac{2}{x}}}{1-\dfrac{3}{1-\dfrac{2}{x}}}$

19. $\dfrac{x+3-\dfrac{6}{x+2}}{x+\dfrac{2x-5}{x+2}}$

20. $\dfrac{x-\dfrac{1}{x-\dfrac{1}{x}}}{x^2-2}$

SUMMARY

Natural numbers

Definition: $N = \{1, 2, 3, \ldots\}$

Whole numbers

Definition: $W = \{0, 1, 2, 3, \ldots\}$

Integers

Definition: $J = \{\ldots, -2, -1, 0, 1, 2, \ldots\}$

Rational numbers

Definition: Rational numbers can be written as a/b, where a and b are integers and $b \neq 0$.

Irrational numbers

Definition: Irrational numbers have decimal equivalent values that do not terminate and do not repeat a period.

Real numbers

Definition: The set of real numbers is the union of rational and irrational numbers.

In Axioms 1–12, unless otherwise restricted, a, b, and c are any real numbers.

Axioms of equality and order

Axiom 1: Law of Trichotomy Only one of the following statements is true:

(1) $a < b$ (2) $a = b$ (3) $a > b$

Axiom 2: Reflexive Property of Equality $a = a$

Axiom 3: Symmetric Property of Equality If $a = b$, then $b = a$.

Axiom 4: Transitive Property of Equality If $a = b$ and $b = c$, then $a = c$.

Axiom 5: Substitution Property of Equality If $a = b$, then in any expression containing a, b may be substituted without changing the value of the expression.

Axiom 6: Transitive Property of Order If $a < b$ and $b < c$, then $a < c$. Furthermore, if $a > b$ and $b > c$, then $a > c$.

Axioms of addition and multiplication

Axiom 7: Closure $a + b$ and ab are real numbers.

Axiom 8: Commutative Addition and Multiplication

$$a + b = b + a \quad \text{and} \quad ab = ba$$

Axiom 9: Associative Addition and Multiplication

$$a + (b + c) = (a + b) + c \quad \text{and} \quad a(bc) = (ab)c$$

Axiom 10: Identity Elements for Addition and Multiplication There exist unique elements in $\mathcal{R}$, namely 0 and 1, such that for each a in $\mathcal{R}$,

$$a + 0 = 0 + a = a \quad \text{and} \quad a(1) = 1(a) = a$$

Axiom 11: Inverse Elements For every element a in $\mathcal{R}$ there exists a unique element $(-a)$, called the additive inverse of a, such that $a + (-a) = (-a) + a = 0$. For every element a in $\mathcal{R}$ ($a \neq 0$) there exists a unique element called the multiplicative inverse of a, such that $a(1/a) = (1/a)a = 1$.

Axiom 12: Distributive Property of Multiplication with Respect to Addition

$$a(b + c) = ab + ac \quad \text{and} \quad (a + b)c = ac + bc$$

Subtraction

Definition: For any real numbers a and b, $a - b = a + (-b)$.

Division

Definition: For any real numbers a and b, with $b \neq 0$, $a/b = a(1/b)$.

More axioms of equality and order

Axiom 13: Addition of Equality If $a = b$, then $a + c = b + c$.
Addition of Inequality If $a < b$, then $a + c < b + c$.

Axiom 14: Subtraction of Equality If $a = b$, then $a - c = b - c$.
Subtraction of Inequality If $a < b$, then $a - c < b - c$.

Axiom 15: Multiplication of Equality If $a = b$, then $ac = bc$.
Multiplication of Inequality
(1) If $a < b$ and $c > 0$, then $ac < bc$.
(2) If $a < b$ and $c < 0$, then $ac > bc$.

Axiom 16: Division of Equality If $a = b$ and $c \neq 0$, then $a/c = b/c$.
Division of Inequality
(1) If $a < b$ and $c > 0$, then $a/c < b/c$.
(2) If $a < b$ and $c < 0$, then $a/c > b/c$.

Polynomial equation

Definition: $P(x)$ is a polynomial equation in x if

$$P(x) = a_0x^n + a_1x^{n-1} + \cdots + a_n$$

and $a_0, a_1, \ldots, a_n$ are real numbers and $n, n - 1, \ldots$ are positive integers.

Rational expression

Definition: A rational expression in x can be written as $\dfrac{P(x)}{R(x)}$, where $P(x)$ and $R(x)$ are polynomials and $R(x)$ is not the zero polynomial.

Theorems on rational expression

If $U(x) = \dfrac{P(x)}{R(x)}$ and $V(x) = \dfrac{S(x)}{T(x)}$, with $R(x) \neq 0$ and $T(x) \neq 0$, then

(a) $U(x) \cdot V(x) = \dfrac{P(x)S(x)}{R(x)T(x)}$

(b) $\frac{U(x)}{V(x)} = \frac{P(x)T(x)}{R(x)S(x)}$

(c) $U(x) + V(x) = \frac{P(x)T(x) + R(x)S(x)}{R(x)T(x)}$, and $R(x)$ and $T(x)$ have no common factors other than 1.

(d) $U(x) - V(x) = U(x) + (-V(x))$

Simplified rational expression

Definition: A rational expression $\frac{P(x)}{R(x)}$ where $R(x) \neq 0$ is simplified if $P(x)$ and $R(x)$ have no common factors other than 1.

Complex fraction

Definition: A complex fraction in x is the ratio of two expressions that contains at least one rational term in the numerator or denominator of the expression for which the denominator is not 1. A complex fraction is simplified when it is expressed as a simplified rational expression.

SELF-CHECKING EXERCISE

NOTE: The dual numeral in **boldface** at the end of each set of instructions refers to the section of the chapter that will help you to solve the exercises.

Identify all the subsets of the real numbers of which the following numbers are members. **(1-1)**

1. 3
2. $-9/4$
3. $-\sqrt{2}$
4. 3.1416

Convert the given rational numbers to a decimal fraction. **(1-1)**

5. 21/16
6. 23/12

Convert the given decimal fractions to the ratio of two integers. **(1-1)**

7. 0.025
8. 0.615615 . . .

Use $<$, $>$, or $=$ to express the relationship between the following pairs of numbers. **(1-2)**

9. $-7/4, -1.\overline{76}$
10. $\sqrt{16}, 36/9$
11. $\sqrt{5}, 29/11$

Identify the property or properties illustrated. **(1-2)**

12. $x + 0 = x$
13. $(-1/2)(-2x) = x$
14. $(7x + 8y) + z = 7x + (8y + z)$
15. If $3a + b = 2$, then $2 = 3a + b$.

Solve. **(1-3)**

16. $3(5 - 2x) = -2(x - 4) + 3$
17. $\frac{5(2x - 7)}{2} + 2(x + 5) = \frac{7}{4} + 9x$

Solve and graph the solution set. **(1-3)**

18. $3x + 7 < x + 13$
19. $4 - x \geq 5(x - 4)$
20. $3(1 - 2x) - 1 \leq 2(2x + 7) + 3$
21. $-3 < \frac{4 - 2x}{7} < -1$

Simplify. **(1-4)**

22. $\dfrac{16a^2 - 8ab + b^2}{12a^2 + 5ab - 2b^2}$

23. $\dfrac{x^2y + 3x^2 - xy - 3x}{x^3 + 2x^2 - 3x}$

24. $\dfrac{2 + 2x - 4x^2}{x^2 - 9} \cdot \dfrac{5x^2 + 12x - 9}{10x^2 - x - 3}$

25. $\dfrac{x^2 + x}{6x^2 - 17x + 7} \div \dfrac{x^2 - 1}{9x^2 - 42x + 49}$

26. $\dfrac{a}{2a - 1} + \dfrac{3}{a}$

27. $\dfrac{3}{b + 1} - \dfrac{6}{2b - 1}$

28. $\dfrac{y + 2}{y^2 - y - 2} + \dfrac{1 - y}{3y + 3}$

29. $\dfrac{x + 5}{2x - 2} + \dfrac{3}{4} - \dfrac{2x + 7}{x^2 + x - 2}$

30. Simplify the following expression. **(1-5)**

$$\frac{1 - \dfrac{2}{x} + \dfrac{1}{x^2}}{\dfrac{2}{x^2} - \dfrac{1}{x} - 1}$$

Answers to Self-Checking Exercise

1. $N, W, J, R, \mathcal{R}$
2. $R, \mathcal{R}$
3. $I, \mathcal{R}$
4. $R, \mathcal{R}$
5. 1.3125
6. $1.91\overline{6}$
7. 1/40
8. 205/333
9. $>$
10. $=$
11. $<$
12. Identity element for addition
13. Associative, multiplicative inverse, and identity element for multiplication
14. Associative
15. Symmetric property of equality
16. $\{1\}$
17. $\{-37/8\}$
18. $x < 3$

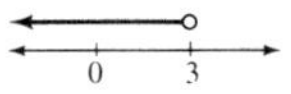

19. $x \leq 4$

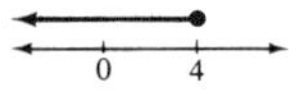

20. $x \geq \dfrac{-3}{2}$

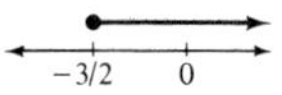

21. $\dfrac{11}{2} < x < \dfrac{25}{2}$

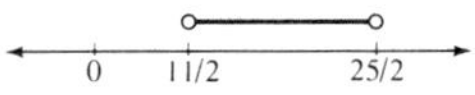

22. $\dfrac{4a - b}{3a + 2b}$
23. $\dfrac{y + 3}{x + 3}$
24. $\dfrac{2(1 - x)}{x - 3}$
25. $\dfrac{x(3x - 7)}{(2x - 1)(x - 1)}$
26. $\dfrac{a^2 + 6a - 3}{a(2a - 1)}$
27. $\dfrac{-9}{(b + 1)(2b - 1)}$
28. $\dfrac{4 + 6y - y^2}{3(y + 1)(y - 2)}$
29. $\dfrac{5x + 14}{4(x + 2)}$
30. $\dfrac{1 - x}{2 + x}$

Exponents and Radicals

2

A strong point of mathematics is at the same time a source of difficulty for those of us who try to master the subject. We are referring to the generous use of symbols to indicate quantities, operations, relationships, etc. We must certainly be in awe at times — if not overwhelmed — by the multitude of symbols that indicate what to do, or not to do, in a language that needs a constant interpreter.

In this chapter we review two of the more common symbols of mathematics: the exponent and the radical. We will note the similarity between the two symbols and practice changing from one symbolic form to the other. We view this chapter as an opportunity to improve our skills of manipulation with these important symbols of mathematics.

INTEGRAL EXPONENTS 2-1

In Chapter 1 we developed the several subsets of the real numbers by taking them in a natural sequence. We use the same method now in the study of exponents and begin by defining a *natural number exponent*. We then define a *zero exponent* and, finally, a *negative integral exponent*.

Integral exponent

> **Definition:** If b is any real number and n is a positive integer (that is, a natural number), then
>
> $$b^n = b \cdot b \cdot b \cdot \ldots \cdot b \qquad (n \text{ factors of } b)$$

In the expression b^n, b is called the **base** and n is the **exponent.**

Zero exponent

> **Definition:** For any nonzero real number b, $b^0 = 1$.

Does this mean that any number to the zero power is 1 ?

Any number *other than zero* with a zero exponent is one. Thus, we have $5^0 = 1$, $(-7)^0 = 1$, and $\left(\frac{12}{3}\right)^0 = 1$; but 0^0 is *undefined*.

Negative exponent

> **Definition:** If b is any nonzero real number and n is a positive integer, then
> $$b^{-n} = \frac{1}{b^n}$$

EXAMPLE 2–1 Write the following with only positive exponents.

(a) $\dfrac{3^0x^{-2}}{2^{-1}y}$ (b) $\dfrac{(-5a^2)^0}{c^{-3}d^2}$, $(a \neq 0)$

Solutions:
(a) Because $3^0 = 1$, $x^{-2} = 1/x^2$, and $2^{-1} = 1/2$, we have

$$\frac{3^0x^{-2}}{2^{-1}y} = \frac{1\left(\frac{1}{x^2}\right)}{\left(\frac{1}{2}\right)y} = \frac{\frac{1}{x^2}}{\frac{y}{2}} = \frac{2}{x^2y}$$

Notice that the negative exponents on the factors x and 2 change their position from numerator to denominator (or vice versa) in the given ratio.
(b) If $a \neq 0$, then $(-5a^2)^0 = 1$, and $c^{-3} = 1/c^3$. Therefore,

$$\frac{(-5a^2)^0}{c^{-3}d^2} = \frac{1}{\frac{d^2}{c^3}} = \frac{c^3}{d^2}$$

We state the following theorems for integral exponents to refresh our memories. The pattern of proof for the theorems is to prove them first for the positive integers by using the definition of an exponent. Then we use the definition of a negative exponent to extend the theorems to include all integers. We accept the theorems without proof.

In the following theorems, a and b are real numbers $(b \neq 0)$ and n and m are integers.

> **Theorem 1:** $(a^n)(a^m) = a^{n+m}$
>
> **Theorem 2:** $(a^n)^m = a^{nm}$
>
> **Theorem 3:** $\dfrac{b^n}{b^m} = b^{n-m}$
>
> **Theorem 4:** $(ab)^n = (a^n)(b^n)$
>
> **Theorem 5:** $\left(\dfrac{a}{b}\right)^n = \dfrac{a^n}{b^n}$

With negative exponents a rational expression $U(x)$ can be written as follows:

$$U(x) = \frac{P(x)}{R(x)} = P(x)(R(x))^{-1} \qquad (R(x) \neq 0)$$

Example 2–2 Simplify $3xy^{-1} + 6x^{-1}y$.

Solution: We know that $3xy^{-1} = 3x/y$ and $6x^{-1}y = 6y/x$. The least common denominator of the two rational expressions is xy; therefore, we have

$$\begin{aligned} \frac{3x}{y} + \frac{6y}{x} &= \frac{3x^2}{xy} + \frac{6y^2}{xy} \\ &= \frac{3x^2 + 6y^2}{xy} \end{aligned}$$

A practical use of integral exponents is found in writing numbers in scientific notation.

Scientific notation

> **Definition:** A rational number M is expressed in scientific notation if it is written as the product of a number N (where $1 \leq N < 10$) and a power of 10.

Example 2–3 Write in scientific notation:

(a) 5,900,000 (b) 0.000012

Solutions:

$$\text{(a) } 5{,}900{,}000\left(\frac{1{,}000{,}000}{1{,}000{,}000}\right) = 5.9 \times 1{,}000{,}000$$
$$= 5.9 \times 10^6$$

$$\text{(b) } 0.000012\left(\frac{100{,}000}{100{,}000}\right) = 1.2 \times \frac{1}{10^5}$$
$$= 1.2 \times 10^{-5}$$

In both parts (a) and (b) we simply moved the decimal point to obtain a number between 1 and 10. Then, the power of 10 indicated how many places the decimal point had been moved. Notice that a *positive exponent* indicated that the decimal point was moved *to the left,* and a *negative exponent* resulted when the decimal point was moved *to the right.*

Many hand-held calculators include a key labeled y^x. Depending on the capacity of the calculator, with this function key we can obtain the approximate value of the integral power x of any positive real number y.

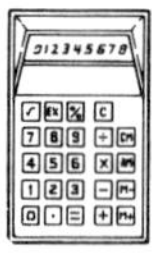

EXAMPLE 2–4 Use a hand-held calculator to find the approximate value of the following to five decimal places.*

(a) $\left(\frac{3}{7}\right)^5$ (b) $\left(\frac{1}{8}\right)^4$ (c) 13^{-3}

Solutions:

(a) Enter 3 and divide by 7. Display on calculator reads 0.42857. Enter 5 and press the y^x key. Display reads 0.01446.

(b) Enter 8 and depress the $1/x$ (or reciprocal) key. Display reads 0.12500. Enter 4, then press the y^x key. Display reads 0.00024.

(c) Enter 13. Now enter 3 and change the sign, then press the y^x key. Display reads 0.00046.

Exercise Set 2–1

Write the following expressions using only positive exponents.

1. 2^{-3}
2. $\frac{-1}{3^{-4}}$
3. $(-5)^{-2}$
4. $\frac{3^{-1}}{3^{-2}}$
5. $(6b)^{-2}$
6. $(-8b^{-4})^0$
7. $\left(\frac{x^{-1}y^{-2}}{z^{-3}}\right)^3$
8. $(-3x)^2(4x)^{-2}(5x)^{-3}$

Write the following as simplified rational expressions.

9. $x^{-1}y + xy^{-1}$
10. $\frac{3^{-2}5^2 + 3^2 5^{-2}}{15^{-2}}$
11. $x^{-1} + y^{-1}$
12. $\frac{x^{-2} - y^{-2}}{(xy)^{-1}}$
13. $(x^{-1} - y^{-1})(x^{-2} + y^{-2})$
14. $\frac{3^{-2} - 5^{-2}}{2^{-2}}$

Write the following in scientific notation.

15. 373
16. 37.3
17. 373,000
18. 0.00373
19. The distance from the earth to the moon is approximately 240,000 miles. Write this number in scientific notation.
20. Avogadro's number is given as 6.023×10^{23} in scientific notation. Write out this number.

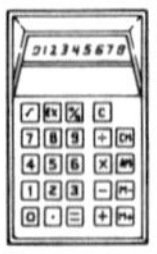

Use a hand-held calculator to find the values of the following expressions. When necessary, round off each answer to three nonzero digits.

*The instructions given are typical of many current models of hand-held calculators. If your calculator does not operate according to these instructions, consult the operation manual or ask your instructor for assistance.

21. 3^5 22. 2^8 23. 6^4

24. 5^{-3} 25. 7^{-2} 26. 15^{-4}

27. $\left(\frac{1}{3}\right)^4$ 28. $\left(\frac{1}{4}\right)^5$ 29. $\left(\frac{1}{9}\right)^6$

30. $\left(\frac{3}{8}\right)^3$ 31. $\left(\frac{5}{3}\right)^{-5}$ 32. $\left(\frac{13}{5}\right)^{-4}$

33. $\left(\frac{27}{15}\right)^4$ 34. $\left(\frac{23}{10}\right)^{-3}$ 35. $\left(\frac{7}{11}\right)^{-7}$

RATIONAL EXPONENTS 2–2

The exponent takes on a new interpretation as we extend the use of an exponent to include a rational number n/m. To introduce the rational exponent in a simplified form, we first consider the rational exponent with $n = 1$, and so $n/m = 1/m$.

Rational exponent $1/m$

> **Definition:** If a and b are positive real numbers and m is a nonzero integer, then $a^{1/m} = b$ if and only if $a = b^m$.

EXAMPLE 2–5 Evaluate b in the following equations:

(a) $b = 16^{1/2}$ (b) $b = 125^{1/3}$

Solutions:
(a) We know that $b = 16^{1/2}$ if and only if $b^2 = 16$. Since $4^2 = 16$, then $b^2 = 4^2$ and $b = 4$. Notice that $(-4)^2 = 16$, but the restriction that b is a positive real number yields the root 4 for the given equation.
(b) We know that $b = 125^{1/3}$ if and only if $b^3 = 125$. Since $5^3 = 125$, then $b^3 = 5^3$ and $b = 5$.

EXAMPLE 2–6 Evaluate b in the following equations:

(a) $b = 32^{-1/5}$ (b) $b = \left(\frac{1}{81}\right)^{-1/4}$

Solutions:
(a) $b = 32^{-1/5}$ if and only if $b^{-5} = 32$. Therefore,

$$\frac{1}{b^5} = 32 \quad \text{or} \quad b^5 = \frac{1}{32}$$

Since $\left(\frac{1}{2}\right)^5 = \frac{1}{32}$, then $b^5 = \left(\frac{1}{2}\right)^5$ and $b = \frac{1}{2}$.

(b) $b = \left(\frac{1}{81}\right)^{-1/4}$ if and only if $b^{-4} = \frac{1}{81}$ or $b^4 = 81$. Since $3^4 = 81$, then $b^4 = 3^4$ and $b = 3$.

Aren't the numbers used in Examples 2-5 and 2-6 rigged?

We used certain numbers so that the exact values of b could be determined by using our knowledge of the powers of natural numbers. When we introduce a new concept, we always use *rigged problems* so that we can check answers with our previous mathematical training. However, with a hand-held calculator we can find the *approximate value* of b when the numbers are not exact.

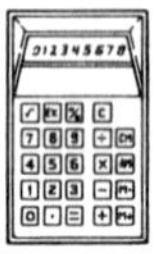

EXAMPLE 2–7 Use a hand-held calculator to find the value of b to five decimal places.

(a) $b = 10^{1/3}$ (b) $b = 20^{-1/2}$

Solutions:
(a) Enter 10. Enter 3 and press $1/x$ (or reciprocal) key, then press the y^x key. Display reads 2.15443. Thus, $b \approx 2.15443$ (read "b is approximately equal to 2.15443").
(b) Enter 20. Enter 2, press change sign key and $1/x$ key. Now press y^x key. Display reads 0.22361, and $b \approx 0.22361$.

We now extend the definition of a rational exponent to include the case for $n \neq 1$.

Rational exponent n/m

> **Definition:** For any positive real numbers a and b and integers n and m ($m > 1$), $a^{n/m} = b$ if and only if $a^n = b^m$.

Note that if $n/m < 0$, we assert that $n < 0$. Thus, we usually restrict m to the set of positive integers and $m > 1$.

Are a and b always restricted to only positive real numbers?

We do not always restrict a and b to positive real numbers, but if $a < 0$, then $a^{n/m} = b$ does not always yield a *real number* for b. (We assume, of course, that n and m have no common factors other than 1.) For now we simply assert that for $a^{n/m} = b$ ($a < 0$), *b is a real number for m, an odd integer,* and *b is not a real number for m, an even integer*. Therefore, for simplicity we restrict a and b to positive real numbers.

EXAMPLE 2–8 Evaluate b in the following equations:

(a) $b = 16^{3/4}$ (b) $b = \left(\frac{1}{4}\right)^{5/2}$ (c) $b = 9^{-3/2}$

Solutions:
(a) $b = 16^{3/4}$ if and only if $b^4 = 16^3$. Since $16^3 = (2^4)^3$ and $(2^4)^3 = (2^3)^4$, then $b^4 = (2^3)^4$ and $b = 2^3$, or 8.
(b) $b = \left(\frac{1}{4}\right)^{5/2}$ if and only if $b^2 = \left(\frac{1}{4}\right)^5$. But

$$\left(\frac{1}{4}\right)^5 = \left(\left(\frac{1}{2}\right)^2\right)^5 = \left(\left(\frac{1}{2}\right)^5\right)^2$$

Therefore, we have

$$b^2 = \left(\left(\frac{1}{2}\right)^5\right)^2 \quad \text{and} \quad b = \left(\frac{1}{2}\right)^5, \quad \text{or} \quad \frac{1}{32}$$

(c) $b = 9^{-3/2}$ if and only if $b^2 = 9^{-3}$. Since $9^{-3} = (3^2)^{-3} = (3^{-3})^2$, then $b^2 = (3^{-3})^2$ and $b = 3^{-3}$, or 1/27.

In Example 2–8 we again used *rigged numbers* so that the answers could be determined without a calculator or a set of tables. The following example shows how a hand-held calculator with the y^x key can be used to find the approximate values of any positive real number a raised to some rational number n/m.

EXAMPLE 2–9 With a hand-held calculator, find the value of the following numbers correct to five decimal places.

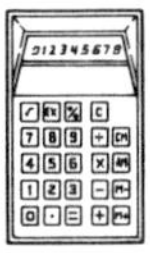

(a) $10^{2/5}$ (b) $30^{-3/4}$

Solutions:
(a) Enter 10. Enter 2 and divide by 5. Depress the y^x key. Display reads 2.51189.
(b) Enter 30. Enter 3, divide by 4, and press the change sign key. Now press the y^x key and the display reads 0.07801.

We restate the theorems for integral exponents to include rational exponents. The restrictions on a, b, n, m, p, and q from the above definitions are implied for each theorem.

Theorem 1: $$(a^{n/m})(a^{p/q}) = a^{(nq+mp)/mq}$$

Theorem 2: $$(a^{n/m})^{p/q} = a^{np/mq}$$

Theorem 3: $$\frac{a^{n/m}}{a^{p/q}} = a^{(nq-mp)/mq} \qquad (a \neq 0)$$

Theorem 4: $$(ab)^{n/m} = (a^{n/m})(b^{n/m})$$

Theorem 5: $$\left(\frac{a}{b}\right)^{n/m} = \frac{a^{n/m}}{b^{n/m}} \qquad (b \neq 0)$$

EXAMPLE 2–10 Evaluate the following:

(a) $(7^{3/2})(7^{-1/2})$ (b) $(64^{1/2})(64^{1/3})$

Solutions:
(a) $(7^{3/2})(7^{-1/2}) = 7^{3/2+(-1/2)} = 7^1 = 7$
(b) $(64^{1/2})(64^{1/3}) = 64^{1/2+1/3} = 64^{5/6}$
We know that $b^6 = 64^5$ and $64^5 = (2^6)^5 = (2^5)^6$. Thus, $b^6 = (2^5)^6$, and $b = 2^5$ or 32. Therefore, $64^{5/6} = 32$.

Exercise Set 2–2

Find the numerical value of each of the following. The answer is an integer or a ratio of integers.

1. $9^{1/2}$ 2. $64^{1/3}$ 3. $81^{1/4}$

4. $\left(\frac{1}{4}\right)^{1/2}$ 5. $\left(\frac{1}{27}\right)^{1/3}$ 6. $343^{1/3}$

7. $\left(\frac{49}{100}\right)^{1/2}$ 8. $\left(\frac{32}{243}\right)^{1/5}$ 9. $9^{-1/2}$

10. $\left(\frac{1}{27}\right)^{-1/3}$ 11. $16^{3/2}$ 12. $216^{2/3}$

13. $8^{2/3}$ 14. $\left(\frac{1}{32}\right)^{2/5}$ 15. $\left(\frac{4}{49}\right)^{3/2}$

16. $\left(\frac{-7}{17}\right)^{0}$ 17. $(7^{1/3})^3$ 18. $(4^{2/3})^{3/2}$

19. $(4^{1/3})(4^{2/3})$ 20. $\frac{16^{1/4}}{16^{3/4}}$ 21. $(64^{1/6})(64^{2/3})$

22. $27^{2/3}$ 23. $(2^{3/2})(8^{1/2})$ 24. $\left(\frac{2^{1.6}}{2^{-4.4}}\right)^{1/2}$

Using the theorems for rational exponents, write the following in as simple a form as possible. Assume $x > 0$ and $y > 0$.

25. $(x^{1/2})^{1/3}$ 26. $\left(\frac{-x^{3/2}}{x^{-2/3}}\right)^3$

27. $(x^{1/2})(x^{2/3})^6$ 28. $((x^{1/2})^{1/2})^{1/2}$

29. $(x^{1/2} + y^{1/2})(x^{1/2} - y^{1/2})$ 30. $(x^{1/2} + y^{1/2})^2$

31. $(x^{3y/2})(x^{2y/3})$ 32. $((2x^{1/2}y^{-3})^{1/2})^{-4}$

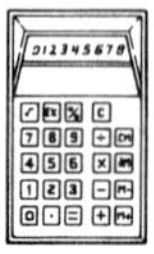

Use a hand-held calculator to find the values of the following expressions. When necessary, round the answer to three nonzero digits.

33. $8^{2/3}$ 34. $27^{4/3}$ 35. $625^{3/4}$

36. $10^{1/3}$ 37. $20^{1/4}$ 38. $15^{3/2}$

39. $\left(\frac{10}{3}\right)^{1/2}$ 40. $16^{-3/2}$ 41. $25^{-5/2}$

42. $\left(\frac{3}{2}\right)^{-1/2}$ 43. $\left(\frac{10}{9}\right)^{-2/5}$ 44. $\left(\frac{19}{10}\right)^{5/3}$

2-3 RADICALS

A mathematical symbol encountered almost as frequently as an exponent is the radical sign. In this section we first learn the meaning of the radical sign in terms of exponents; then we develop the arithmetic of radicals. The purpose of this section is to sharpen our skills so that radical expressions encountered in future mathematical work will be skillfully handled.

Relationship Between Radicals and Exponents

The symbol $\sqrt[m]{a}$ is called a *radical,* a is called the *radicand,* and m is the *root index*. Notice that m is never nonpositive. Furthermore, if $m = 2$, then $\sqrt[2]{a}$ is written as $\sqrt{a}$.

$\sqrt[m]{a^n}$

> **Definition:** If a is any real number ($a > 0$ for m even), where n and m are integers and $m > 1$, then
>
> $$\sqrt[m]{a^n} = (\sqrt[m]{a})^n = a^{n/m}$$

Why must a be positive when m is even?

Recall that as we developed the laws of exponents for rational numbers, we restricted a to a positive real number. There are two reasons for this restriction. First, as noted in Section 2-2, if $a < 0$ and m is even, then b is not a real number and $a^{1/m} = b$. Therefore, rather than introduce a new set of numbers to give meaning to expressions such as $(-8)^{1/2}$ and $(-81)^{1/4}$, we simply restrict a to nonnegative numbers. Secondly, the y^x key on most (if not all) hand-held calculators operates for only positive values of y.

Later we will introduce the set of numbers that provide answers to *even roots of negative numbers*. For now, however, let's agree to restrict the variables in radicands to positive values whenever the root index is even.

EXAMPLE 2–11 Write the following in radical form:

(a) $(32y)^{3/2}$ (b) $(-28z)^{5/3}$

Solutions:

(a) $(32y)^{3/2} = \sqrt{(32y)^3}$ or $(\sqrt{32y})^3$

(b) $(-28z)^{5/3} = \sqrt[3]{(-28z)^5}$ or $(\sqrt[3]{-28z})^5$

Note that since $m = 3$ and 3 is odd, the negative radicand in (b) represents a real number.

If $a > 0$ and m is even, then $\sqrt[m]{a} = a^{1/m}$ is a real number b such that $b^m = a$. For example, $\sqrt{16} = b$ such that $b^2 = 16$, and $\sqrt[4]{16} = b$ such that $b^4 = 16$. For both examples, b can be a positive or a negative number in that

$$4^2 = (-4)^2 = 16 \quad \text{and} \quad 2^4 = (-2)^4 = 16$$

We call the value of b that is *positive* the **principal nth root of a.** Thus, the principal roots of $\sqrt{16}$ and $\sqrt[4]{16}$ are 4 and 2, respectively. In general, if a is any real number and m is odd, then there is only one value of b such that $b^m = a$.

EXAMPLE 2–12 Write the following in exponential form:

(a) $(\sqrt[3]{10y})^2$ (b) $\sqrt[5]{-3z^2}$

Solutions:

(a) With $m = 3$ and $n = 2$, $(\sqrt[3]{10y})^2 = (10y)^{2/3}$.

(b) With $m = 5$ and $n = 1$, $\sqrt[5]{-3z^2} = (-3z^2)^{1/5}$.

Theorems for Radicals

The following theorems for radicals are a result of the theorems for rational exponents. For each theorem, n and m are positive integers ($m > 1$); and if m is even, then $a > 0$ and $b > 0$.

Theorem 1: $\sqrt[m]{ab} = (\sqrt[m]{a})(\sqrt[m]{b})$

For Theorem 1 we offer the following proof. By definition,

$$\begin{aligned}\sqrt[m]{ab} &= (ab)^{1/m}\\ &= (a^{1/m})(b^{1/m})\\ &= (\sqrt[m]{a})(\sqrt[m]{b})\end{aligned}$$

An extension of the theorem is the following equality:

$$(\sqrt[m]{ab})^n = (\sqrt[m]{a})^n(\sqrt[m]{b})^n$$

(The proofs of the remaining theorems are included in the exercises.)

Theorem 2: $\left(\sqrt[m]{\dfrac{a}{b}}\right)^n = \dfrac{(\sqrt[m]{a})^n}{(\sqrt[m]{b})^n}$ or $\dfrac{\sqrt[m]{a^n}}{\sqrt[m]{b^n}}$

Theorem 3: $\sqrt[n]{\sqrt[m]{a}} = \sqrt[nm]{a}$

Theorem 4: $(\sqrt[m]{a})^m = a$

Theorem 5: *Case 1.* $\sqrt[m]{a^m} = a$ for $a > 0$, and for $a < 0$ and m odd
Case 2. $\sqrt[m]{a^m} = |a|$ for $a < 0$ and m even

EXAMPLE 2–13 Use the radical theorems to rewrite the following expressions:

(a) $\sqrt[3]{2x^3}$ (b) $\sqrt[4]{5y^4}$ (c) $(\sqrt{x} + \sqrt{y})^2$

Solutions:

(a) $\sqrt[3]{2x^3} = (\sqrt[3]{2})(\sqrt[3]{x^3}) = \sqrt[3]{2}(x)$ or $x\sqrt[3]{2}$

(b) $\sqrt[4]{5y^4} = (\sqrt[4]{5})(\sqrt[4]{y^4}) = |y|\sqrt[4]{5}$

Since the exponent on y is even, $y^4 \geq 0$ for any y. Thus, $\sqrt[4]{y^4} = y$ if $y \geq 0$, and $\sqrt[4]{y^4} = -y$ if $y < 0$. We therefore write $\sqrt[4]{y^4} = |y|$ to represent both answers.

Why did you put the y in absolute values?

(c) $(\sqrt{x} + \sqrt{y})^2 = (\sqrt{x})^2 + 2(\sqrt{x})(\sqrt{y}) + (\sqrt{y})^2$
$= x + 2\sqrt{xy} + y$

[Note: $(\sqrt{x} + \sqrt{y})^2 \neq x + y$]

Simplification of Radicals

Mathematicians at times exhibit a zealous ambition for stating expressions in what is termed *simplified form*. For example, although 12/8 = 3/2, the latter is considered a preferred form, since the common factor 4 has been eliminated. The following definition states the preferred form of a radical expression.

Simplified radical form

Definition: The expression $\sqrt[m]{a^n}$ is in simplified form if all of the following are true:

(1) a contains no polynomial factors of degree greater than m.

(2) m and n contain no common factors other than 1.

(3) a is a rational expression with denominator 1.

(4) A radical does not occur in the denominator of any rational expression.

EXAMPLE 2–14 Express the following as simplified radicals:

(a) $\sqrt{320}$ (b) $\sqrt[4]{4x^2}$

(c) $\sqrt{\frac{7}{3}}$ (d) $\dfrac{x^2 - y^2}{\sqrt{x} + \sqrt{y}}$

Solutions:

(a) $\sqrt{320} = \sqrt{(64)(5)} = \sqrt{64}\sqrt{5} = 8\sqrt{5}$

(b) $\sqrt[4]{4x^2} = \sqrt[4]{(2x)^2} = \sqrt{2x}$

(c) $\sqrt{\frac{7}{3}} = \left(\dfrac{\sqrt{7}}{\sqrt{3}}\right)\left(\dfrac{\sqrt{3}}{\sqrt{3}}\right) = \dfrac{\sqrt{21}}{3}$

Note that the simplification shown in (c) is frequently referred to as *rationalizing the denominator*, that is, converting the irrational number $\sqrt{3}$ to the rational number 3.

(d) $\left(\dfrac{x^2 - y^2}{\sqrt{x} + \sqrt{y}}\right)\left(\dfrac{\sqrt{x} - \sqrt{y}}{\sqrt{x} - \sqrt{y}}\right) = \dfrac{(x^2 - y^2)(\sqrt{x} - \sqrt{y})}{x - y}$

$$= \frac{(x + y)(x - y)(\sqrt{x} - \sqrt{y})}{(x - y)}$$

$$= (x + y)(\sqrt{x} - \sqrt{y})$$

Notice that since the radical in the denominator of (d) is a binomial, we multiplied by the **conjugate** of $\sqrt{x} + \sqrt{y}$, namely $\sqrt{x} - \sqrt{y}$, to simplify the given radical.

Operations on Radicals

The distributive property that permits us to combine like terms also allows us to combine **like radicals.** That is, since $5x + 7x = 12x$, then $5\sqrt{3} + 7\sqrt{3} = 12\sqrt{3}$. By like radicals we mean radical terms that agree both in root index and radicand.

> If x and y are any real numbers, then
> $$x\sqrt[m]{a} + y\sqrt[m]{a} = (x + y)\sqrt[m]{a}$$
> and
> $$x\sqrt[m]{a} - y\sqrt[m]{a} = (x - y)\sqrt[m]{a}$$

EXAMPLE 2–15 Combine like radicals.

(a) $\sqrt{18x} + 2\sqrt{8x} - 5\sqrt{2x}$ (b) $\sqrt[3]{81} + \sqrt{147} - \sqrt[3]{24} - \sqrt{108}$

Solutions:

(a) $\sqrt{18x} + 2\sqrt{8x} - 5\sqrt{2x} = 3\sqrt{2x} + 4\sqrt{2x} - 5\sqrt{2x}$

$= (3 + 4 - 5)\sqrt{2x} = 2\sqrt{2x}$

(b) $\sqrt[3]{81} + \sqrt{147} - \sqrt[3]{24} - \sqrt{108} = 3\sqrt[3]{3} + 7\sqrt{3} - 2\sqrt[3]{3} - 6\sqrt{3}$

$= \sqrt[3]{3} + \sqrt{3}$

Notice that the two terms in (b) cannot be combined, since the root indices are not alike.

EXAMPLE 2–16 Multiply or divide the following and simplify:

(a) $(2\sqrt{5x})(3\sqrt{10x})(\sqrt{x})$ (b) $\dfrac{\sqrt{21} + \sqrt{30}}{\sqrt{3}}$

Solutions:

(a) $(2\sqrt{5x})(3\sqrt{10x})(\sqrt{x}) = 6\sqrt{50x^3}$

$= 6\sqrt{(25x^2)(2x)}$

$= 30|x|\sqrt{2x}$

(b) $\dfrac{\sqrt{21} + \sqrt{30}}{\sqrt{3}} = \dfrac{\sqrt{21}}{\sqrt{3}} + \dfrac{\sqrt{30}}{\sqrt{3}} = \sqrt{\dfrac{21}{3}} + \sqrt{\dfrac{30}{3}} = \sqrt{7} + \sqrt{10}$

Notice that both terms of the numerator in (b) contain 3 as a factor. Therefore, the expression is divided into two terms and the simplification follows. A longer, yet equally valid method is to rationalize the denominator and simplify.

$$\left(\frac{\sqrt{21}+\sqrt{30}}{\sqrt{3}}\right)\left(\frac{\sqrt{3}}{\sqrt{3}}\right)=\frac{\sqrt{63}+\sqrt{90}}{3}=\frac{3\sqrt{7}+3\sqrt{10}}{3}$$
$$=\frac{3(\sqrt{7}+\sqrt{10})}{3}$$
$$=\sqrt{7}+\sqrt{10}$$

You should be aware that more than one approach is possible to a given example, and although one method may be faster than another, they should all eventually lead to the same result. Be bold and experiment with different ways to solve a given exercise.

Exercise Set 2–3

Write the following in radical form.

1. $16^{1/3}$
2. $3^{1/2}$
3. $(x^2y^3)^{2/5}$
4. $(-7)^{4/3}$

Write the following in exponential form.

5. $\sqrt{2xy}$
6. $\sqrt[3]{7^2}$
7. $\sqrt[4]{\frac{x}{y^3}}$
8. $\sqrt[5]{30x^3y^2}$

Write each of the following in simplified radical form.

9. $\sqrt{8}$
10. $\sqrt{1056}$
11. $\sqrt[3]{54}$
12. $\sqrt[4]{32}$
13. $\sqrt{27x^3}$
14. $\sqrt{98x^2y^3}$
15. $\sqrt{17}\sqrt{34}$
16. $\sqrt[4]{2x}\sqrt[4]{16x^3}$
17. $\sqrt[4]{4x}\sqrt[4]{4x}$
18. $\frac{\sqrt{2}}{\sqrt{3}}$
19. $\sqrt[3]{\frac{4}{9}}$
20. $\frac{2\sqrt{3}}{3\sqrt{2}}$
21. $\sqrt[5]{\frac{1}{a^3}}$
22. $\frac{1}{\sqrt{x+1}}$
23. $\frac{1}{\sqrt{x}+1}$
24. $\frac{1}{\sqrt{x}-1}$
25. $\frac{1}{2\sqrt{x}-3}$
26. $\frac{1}{\sqrt[3]{x}+1}$
27. $\frac{2}{\sqrt[3]{x^2}+2\sqrt[3]{x}+4}$
28. $\sqrt{x^3y^{-3}}$
29. $\sqrt[5]{64x^5y^{-4}}$
30. Prove Theorem 2.
31. Prove Theorem 3.
32. Prove Theorem 4.
33. Prove Theorem 5.

Perform the indicated operations and leave all radicals in simplified form.

34. $\sqrt{20}+\sqrt{45}+\sqrt{125}$
35. $2\sqrt[3]{16}+4\sqrt[3]{192}$
36. $2\sqrt[3]{7}-3\sqrt[3]{56}$
37. $x\sqrt[4]{16x^5}-3\sqrt[4]{x^9}-\sqrt[4]{81x^9}$
38. $\sqrt{\frac{3}{2}}+\sqrt{24}$
39. $\sqrt[6]{\frac{27}{8}}-2\sqrt{\frac{3}{2}}$
40. $(\sqrt{3}-\sqrt{2})^2$
41. $(\sqrt{2})(3\sqrt{3})(4\sqrt{2})(\sqrt{12})$
42. $\sqrt{2}(\sqrt{6}-\sqrt{14})(\sqrt{6}+\sqrt{14})$
43. $\frac{\sqrt{39}-\sqrt{13}}{\sqrt{13}}$
44. $(\sqrt[3]{7}+\sqrt[3]{3})(\sqrt[3]{49}-\sqrt[3]{21}+\sqrt[3]{9})$
45. $\frac{2\sqrt{5}}{\sqrt{10}+2\sqrt{2}}$

46. $\dfrac{\dfrac{1}{\sqrt{x}} + x}{\dfrac{1}{\sqrt{3x}} - 1}$

47. $\sqrt{\dfrac{1}{9} + \dfrac{1}{64}}$

48. $\sqrt{\dfrac{1}{9}} + \sqrt{\dfrac{1}{64}}$

49. $\sqrt{36 + 144}$

50. $\sqrt{36} + \sqrt{144}$

SUMMARY

Integral exponent

Definition: If b is any real number and n is a positive integer, then $b^n = b \cdot b \cdot b \cdot \ldots \cdot b$ (n factors of b).

Zero exponent

Definition: For any nonzero real number b, $b^0 = 1$.

Negative exponent

Definition: If b is any nonzero real number and n is a positive integer, then $b^{-n} = \dfrac{1}{b^n}$.

Theorems on exponents

In Theorems 1–5, a and b are real numbers ($b \neq 0$) and n and m are integers.

Theorem 1: $(a^n)(a^m) = a^{n+m}$

Theorem 2: $(a^n)^m = a^{nm}$

Theorem 3: $\dfrac{b^n}{b^m} = b^{n-m}$

Theorem 4: $(ab)^n = (a^n)(b^n)$

Theorem 5: $\left(\dfrac{a}{b}\right)^n = \dfrac{a^n}{b^n}$

Scientific notation

Definition: A rational number M is expressed in scientific notation if it is written as the product of a number N(where $1 \leq N < 10$) and a power of 10.

Rational exponent 1/*m*

Definition: If a and b are positive real numbers and m is a nonzero integer, then $a^{1/m} = b$ if and only if $a = b^m$.

Rational exponent *n*/*m*

Definition: If a and b are positive real numbers and n and m are integers ($m > 1$), then $a^{n/m} = b$ if and only if $a^n = b^m$. (*Note:* The exponent theorems stated for integral exponents also apply to rational exponents n/m and p/q.)

$\sqrt[m]{a^n}$

Definition: If a is any real number ($a > 0$ for m even), where n and m are integers and $m > 1$, then $\sqrt[m]{a^n} = (\sqrt[m]{a})^n = a^{n/m}$.

Theorems on radicals

In Theorems 1–5, $a > 0$ and $b > 0$, and n and m are positive integers ($m > 1$).

Theorem 1: $\sqrt[m]{ab} = (\sqrt[m]{a})(\sqrt[m]{b})$

Theorem 2: $\left(\sqrt[m]{\dfrac{a}{b}}\right)^n = \dfrac{(\sqrt[m]{a})^n}{(\sqrt[m]{b})^n}$ or $\dfrac{\sqrt[m]{a^n}}{\sqrt[m]{b^n}}$

Theorem 3: $\sqrt[n]{\sqrt[m]{a}} = \sqrt[nm]{a}$

Theorem 4: $(\sqrt[m]{a})^m = a$

Theorem 5: *Case 1.* $\sqrt[m]{a^m} = a$ for $a > 0$ or for $a < 0$ and m odd
Case 2. $\sqrt[m]{a^m} = |a|$ for $a < 0$ and m even

Simplified radical form

Definition: $\sqrt[m]{a^n}$ is in simplified radical form if all of the following are true:

(1) a contains no polynomial factors of degree greater than m.

(2) m and n contain no common factors other than 1.

(3) a is a rational expression with denominator 1.

(4) A radical does not occur in the denominator of any rational expression.

SELF-CHECKING EXERCISE

Simplify the following expressions. **(2-1)**

1. $25^{-3/2}$
2. $(x^{-3/5} \cdot x^{2/5})^5$
3. $(x^{-2} - y^{-2})$
4. $(5x + 3)^0, x \neq \frac{-3}{5}$
5. $(9^{4/8} + 27^{2/6})^2$
6. $((64)^{1/2})^{1/3}$

Write the following numbers in scientific notation. **(2-1)**

7. 37,000,000
8. 0.000059

Find the numerical value of each of the following. **(2-2)**

9. $27^{2/3}$
10. $16^{5/4}$
11. $(-125)^{2/3}$
12. $64^{-2/3}$
13. $\left(\frac{27}{8}\right)^{-1/3}$
14. $\left(\frac{-1}{32}\right)^{-1/5}$
15. $(4^{3/2})(8^{5/3})$
16. $(10^{3/2})(10)^{-3/2}$
17. $\frac{10^{1.2}}{10^{-1.8}}$

Write the following in simplified form. Assume $x > 0$ and $y > 0$. **(2-2)**

18. $(x^{3/4})^{2/3}$
19. $(x^{1/2}y^{1/3})^6$
20. $\left(\frac{2x^{5/3}}{y^{3/2}}\right)^{-6}$
21. $(16x^8y^{-6})^{-1/2}$
22. $(2x^{-3/4})^4(2^{-1}x^{1/4})^8$
23. $(x^{-2} + y^{-2})^2$

Write the following in radical form. **(2-3)**

24. $(2x)^{1/2}, x > 0$
25. $(3x^2y)^{2/3}$

Write the following in exponential form. **(2-3)**

26. $5\sqrt{xy^3}, x > 0$ and $y > 0$
27. $\sqrt[3]{\frac{125x^2}{y}}, y \neq 0$

Simplify the following radical expressions. **(2-3)**

28. $\sqrt[3]{162x^4y^6}$

29. $\dfrac{1}{\sqrt{17} - 3}$

30. $\sqrt{45} + \sqrt{\dfrac{2}{27}} - \sqrt{20}$

31. $\sqrt{8}(\sqrt{32} - \sqrt{98})$

32. $\sqrt{\dfrac{1}{9} + \dfrac{1}{25}}$

33. $\left(\dfrac{1 - 3\sqrt{2}}{\sqrt{3}}\right) + \left(\dfrac{2\sqrt{3} - 1}{\sqrt{2}}\right)$

Answers to Self-Checking Exercise

1. 1/125
2. $1/x$
3. $\dfrac{y^2 - x^2}{x^2y^2}$
4. 1
5. 36
6. 2
7. 3.7×10^7
8. 5.9×10^{-5}
9. 9
10. 32
11. 25
12. 1/16
13. 2/3
14. -2
15. 256
16. 1
17. 10^3, or 1000
18. $x^{1/2}$
19. x^3y^2
20. $\dfrac{y^9}{64x^{10}}$
21. $\dfrac{y^3}{4x^4}$
22. $\dfrac{1}{16x}$
23. $\dfrac{1}{x^4} + \dfrac{2}{x^2y^2} + \dfrac{1}{y^4}$, or $\dfrac{(x^2 + y^2)^2}{x^4y^4}$
24. $\sqrt{2x}$
25. $\sqrt[3]{9x^4y^2}$
26. $5x^{1/2}y^{3/2}$
27. $5x^{2/3}y^{-1/3}$
28. $3xy^2\sqrt[3]{6x}$
29. $\dfrac{\sqrt{17} + 3}{8}$
30. $\sqrt{5} + \dfrac{\sqrt{6}}{9}$
31. -12
32. $\dfrac{2\sqrt{30}}{15}$
33. $\dfrac{2\sqrt{3} - 3\sqrt{2}}{6}$, or $\dfrac{\sqrt{3}}{3} - \dfrac{\sqrt{2}}{2}$

Quadratic Equations, Inequalities, and Polynomials of Degree *n*

3

Mathematics as an art is sufficient reason for its existence, and the beauty of its structure is reward enough for pursuing the study of the subject. The fact that mathematical solutions can be obtained for physical problems only emphasizes the value of comprehending such a body of knowledge.

In this chapter we explore some of the ways that we can find solutions to equations and inequalities of a particular classification — namely, polynomials of degree n. Some word problems are introduced to show that the content of this chapter can be applied to actual physical problems you may encounter in science, engineering, business, and similar courses.

QUADRATIC EQUATIONS 3–1

Parabolic shapes are evident in many places. The gentle arch of a thrown ball and the cross section of many optical lenses are two examples of shapes that are approximately that of a parabola. The mathematical equation that describes a parabola is a polynomial equation of the form

$$P(x) = ax^2 + bx + c$$

where a, b, and c are real numbers and $a \neq 0$. We will study the parabola in Chapter 4 when we study functions. In this section we will consider the quadratic equation for $P(x) = 0$, that is,

$$ax^2 + bx + c = 0$$

where a, b, and c are real numbers and $a \neq 0$.

If $ax^2 + bx + c$ can be readily factored, then we can solve the given equation with the *zero-product theorem*. Recall that if a and b are real numbers with $ab = 0$, then either $a = 0$ or $b = 0$.

EXAMPLE 3–1 Solve the equation $2x^2 - 7x + 6 = 0$.

Solution: If $2x^2 - 7x + 6 = 0$, then

$$(2x - 3)(x - 2) = 0 \quad \text{(factored form of } 2x^2 - 7x + 6)$$
$$2x - 3 = 0 \quad \text{or} \quad x - 2 = 0 \quad \text{(zero-product theorem)}$$
$$x = \frac{3}{2} \quad \text{or} \quad x = 2$$

The solution set is $\{3/2, 2\}$.

Another method for solving a quadratic equation is to "complete the square" and then determine the roots of the equation. We review the process of completing the square with Example 3–2. Simultaneously, we solve the general quadratic equation $ax^2 + bx + c = 0$ in Example 3–3. The result of Example 3–3 is the **quadratic formula** that can be used to solve any quadratic equation.

EXAMPLE 3–2 Solve $2x^2 - 5x - 1 = 0$.

EXAMPLE 3–3 Solve $ax^2 + bx + c = 0$.

Solution: (Example 3–2, left column)

Solution: (Example 3–3, right column)

Step 1. Subtract the constant term from both sides of the equation.

$$2x^2 - 5x = 1 \qquad\qquad ax^2 + bx = -c$$

Step 2. Divide by the coefficient of the x^2 term.

$$x^2 - \frac{5}{2}x = \frac{1}{2} \qquad\qquad x^2 + \frac{b}{a}x = \frac{-c}{a}$$

Step 3. Add the square of 1/2 the coefficient of the x term to both sides of the equation.

$$x^2 - \frac{5}{2}x + \frac{25}{16} = \frac{1}{2} + \frac{25}{16} \qquad\qquad x^2 + \frac{b}{a}x + \frac{b^2}{4a^2} = \frac{-c}{a} + \frac{b^2}{4a^2}$$

Step 4. Factor the left side of the equation as a perfect square.

$$\left(x - \frac{5}{4}\right)^2 = \frac{1}{2} + \frac{25}{16} \qquad\qquad \left(x + \frac{b}{2a}\right)^2 = \frac{-c}{a} + \frac{b^2}{4a^2}$$

Step 5. Simplify the right side of the equation.

$$\left(x - \frac{5}{4}\right)^2 = \frac{33}{16} \qquad\qquad \left(x + \frac{b}{2a}\right)^2 = \frac{b^2 - 4ac}{4a^2}$$

Step 6. Take the square root of both sides of the equation.

$$x - \frac{5}{4} = \frac{\pm\sqrt{33}}{4} \qquad\qquad x + \frac{b}{2a} = \frac{\pm\sqrt{b^2 - 4ac}}{2a}$$

Step 7. Subtract the constant term from both sides of the equation.

$$x = \frac{5}{4} \pm \frac{\sqrt{33}}{4} \qquad\qquad x = \frac{-b}{2a} \pm \frac{\sqrt{b^2 - 4ac}}{2a}$$

Step 8. Simplify the right side of the equation.

$$x = \frac{5 \pm \sqrt{33}}{4} \qquad\qquad x = \frac{-b \pm \sqrt{b^2 - 4ac}}{2a}$$

The result of Example 3–3 is stated in the following theorem.

> **Theorem: Quadratic Formula** The roots of any quadratic equation $ax^2 + bx + c = 0$ (where a, b, and c are real numbers and $a \neq 0$) are of the form
>
> $$\frac{-b \pm \sqrt{b^2 - 4ac}}{2a}$$

This theorem permits us simply to identify a, b, and c in any quadratic equation, substitute these values in the formula, and obtain the roots. The following example illustrates the process.

EXAMPLE **3–4** Solve $3x^2 - 2x - 4 = 0$.

Solution: For the given equation, $a = 3$, $b = -2$, and $c = -4$. Therefore,

$$x = \frac{-(-2) \pm \sqrt{(-2)^2 - 4(3)(-4)}}{2(3)}$$

$$= \frac{2 \pm \sqrt{52}}{6} = \frac{2 \pm 2\sqrt{13}}{6}$$

$$= \frac{1 \pm \sqrt{13}}{3}$$

To check these roots in the original equation would require more arithmetic than we like to deal with. The following theorem states an alternate procedure for checking the roots that is simpler.

> **Theorem:** If r and s are roots of the quadratic equation $ax^2 + bx + c = 0$, then
>
> $$r + s = \frac{-b}{a} \quad \text{and} \quad rs = \frac{c}{a}$$

Proof. If r and s are roots of $ax^2 + bx + c = 0$, then $x = r$ and $x = s$, or $x - r = 0$ and $x \quad s = 0$. Then we have

$$(x - r)(x - s) = 0 \quad \text{and} \quad x^2 - (r + s)x + rs = 0$$

But $ax^2 + bx + c = 0$, or

$$x^2 + \frac{b}{a}x + \frac{c}{a} = 0$$

If

$$x^2 + \frac{b}{a}x + \frac{c}{a} = x^2 - (r + s)x + rs$$

then

$$-(r+s) = \frac{b}{a} \quad \text{and} \quad rs = \frac{c}{a}$$

$$r+s = \frac{-b}{a} \quad \text{and} \quad rs = \frac{c}{a}$$

EXAMPLE 3–5 Check the roots of Example 3–4.

Solution: If $3x^2 - 2x - 4 = 0$, then $a = 3$, $b = -2$, and $c = -4$. Therefore, $-b/a = 2/3$ and $c/a = -4/3$. If

$$r = \frac{1+\sqrt{13}}{3} \quad \text{and} \quad s = \frac{1-\sqrt{13}}{3}$$

then

$$r + s = \frac{1+\sqrt{13}+1-\sqrt{13}}{3} = \frac{2}{3} = \frac{-b}{a}$$

and

$$rs = \left(\frac{1+\sqrt{13}}{3}\right)\left(\frac{1-\sqrt{13}}{3}\right) = \frac{1-13}{9} = \frac{-4}{3} = \frac{c}{a}$$

Since both requirements for r and s in the root theorem are met, the answers for Example 3–4 are correct.

Does the quadratic formula always give us two real roots?

We get different kinds of roots for quadratic equations. We can get one or two **real roots** that can be rational or irrational, or two roots that are **not real.** To determine the kinds of roots a given quadratic equation has, we need simply to evaluate $b^2 - 4ac$, the **discriminant** of the equation.

> **Theorem:** If r and s are the roots of $ax^2 + bx + c = 0$, then
>
> (1) r and s are *real* and *unequal* if $b^2 - 4ac > 0$.
>
> (2) r and s are *real* and *equal* if $b^2 - 4ac = 0$.
>
> (3) r and s are *not real* and *unequal* if $b^2 - 4ac < 0$.

EXAMPLE 3–6 Determine the kind of roots for the following equations:

(a) $5x^2 + 7x + 3 = 0$ (b) $9x^2 + 25 = 30x$

Solutions:
(a) With $a = 5$, $b = 7$, and $c = 3$, then

$$b^2 - 4ac = 49 - 4(5)(3) = -11$$

Since $b^2 - 4ac < 0$, the roots are not real and unequal.
(b) If $9x^2 + 25 = 30x$, then $9x^2 - 30x + 25 = 0$. With $a = 9$, $b = -30$, and $c = 25$, then

$$b^2 - 4ac = 900 - 4(9)(25) = 900 - 900 = 0$$

Since $b^2 - 4ac = 0$, the roots are real and equal.

Example 3–7 A rectangular grass plot is surrounded by a walk one yard wide. The length of the grass area is 3 yards more than its width. The ratio of the grass area to total area is 13/18. Find the dimensions of the grass area.

Solution: Let x and $x + 3$ equal the width and length, respectively, of the grass area. Then $x + 2$ and $x + 5$ are the width and length, respectively, of the total enclosed area. Thus, we have

$$\frac{x(x + 3)}{(x + 2)(x + 5)} = \frac{13}{18}$$

From this we get

$$18x(x + 3) = 13(x + 2)(x + 5)$$

$$18x^2 + 54x = 13x^2 + 91x + 130$$

$$5x^2 - 37x - 130 = 0$$

$$(x - 10)(5x + 13) = 0$$

$$x = 10 \quad \text{or} \quad x = \frac{-13}{15}$$

Since x represents a measurement, we reject the negative value. The grass area is 10 by 13 yards and the total enclosed area is 12 by 15 yards.

Exercise Set 3–1

Find the solution set of the following quadratic equations. Check all solutions by the roots theorem. If a given equation has no real roots, write "none" for the solution set.

1. $x^2 + x - 3 = 0$
2. $x^2 + 6x + 1 = 0$
3. $2x^2 - 3x + 2 = 0$
4. $2x^2 + 3x - 2 = 0$
5. $x^2 - 6x = 6$
6. $7x^2 = 3x - 9$
7. $3x^2 + 2x - 1 = 0$
8. $9x = x^2 - 3$
9. $-2x^2 + 5x = -12$
10. $x^2 - 11 = 2x$
11. $x^2 - 5x = 0$
12. $3x^2 - 3x + 7 = 0$
13. $2x^2 - 8x = 6$
14. $2x^2 - 18 = 0$
15. $\frac{4}{5}x - 19x^2 = 3$
16. $x^2 = \frac{-34x}{7} + \frac{5}{7}$
17. $(x - 5)(x - 1) = 21$
18. $x(x + 3) = -2$
19. $x^2 - \sqrt{6}\,x + 1 = 0$
20. $5x(x + 2) = 4(3x + 1)$

Without finding the roots, determine the kinds of roots for the following equations.

21. $2x^2 - 8x + 8 = 0$
22. $x^2 - 3x + 5 = 0$
23. $x^2 - 7x - 3 = 0$
24. $3x^2 + 18x + 27 = 0$
25. $x^2 - 14x + 63 = 14$
26. $2x^2 + 3x - 5 = 0$

Given the following roots, determine a quadratic equation that has these roots as solutions.

27. 3, 2
28. 5, 5
29. $2 \pm \sqrt{3}$
30. $3 \pm \sqrt{10}$
31. $\frac{17}{2}, \frac{4}{3}$
32. $\pm\sqrt{5}$

33. A student works in a store one summer and earns \$180. At a rate of \$2 less per day, it would have taken one more day to earn the same amount. Find the original number of days the student worked to earn \$180.
34. The area of a triangle is 133 square feet. If the altitude is 5 feet shorter than the base, find the base and altitude of the triangle.
35. A yard in the shape of a rectangle is to have a rectangular lawn surrounded by colored gravel strips of uniform width. If the yard is 20 by 50 feet and the lawn area is 400 square feet, how wide should the gravel strips be?
36. A number is 6 larger than seven times its reciprocal. Find the number.
37. A monorail car travels 100 miles in 50 minutes less time than it takes an automobile to travel the same distance. Find the speeds of the two vehicles if the monorail goes 20 miles per hour faster than the automobile.
38. Two carpenters work together to build a garage that is to be added onto a house after the foundation has been poured. If one could do the entire job in 4 days and the other in 7 days, how long does it take them to do the job together?
39. A boat pulling water skiers can travel 3 miles upstream and then travel 4 miles downstream in ¼ hour. If the rate of the current is 5 miles per hour, find the rate of the boat in still water to the nearest mile per hour.
40. The area of the picture in a square, framed picture is 3/4 of the total area. If the frame is 2 inches wide, find the dimensions of the frame to the nearest tenth of an inch.

3-2 EQUATIONS QUADRATIC IN FORM

Frequently we encounter equations that are not of the form $ax^2 + bx + c = 0$, but in the process of solving we obtain an equation that is quadratic. Or, sometimes we make a substitution to convert a given equation into one that is quadratic. As we shall see, the quadratic formula can be used to solve equations of the following forms:

$$\frac{x}{x+1} - \frac{3}{x-1} = 2$$

$$x^{2/3} - 9x^{1/3} + 8 = 0$$

$$\sqrt{6x-4} = \sqrt{x-1} + \sqrt{2x}$$

Example 3–8 Solve for x:

$$\frac{x}{x+1} - \frac{3}{x-1} = 2$$

Solution: Since $\frac{x}{x+1}$ and $\frac{3}{x-1}$ are expressions for which the denominator cannot equal zero, $x \neq \pm 1$. Therefore, if $x \neq \pm 1$, then

$$(x-1)(x+1)\left(\frac{x}{x+1} - \frac{3}{x-1}\right) = 2(x-1)(x+1)$$

and

$$\begin{aligned} x(x-1) - 3(x+1) &= 2(x^2 - 1) \\ x^2 - x - 3x - 3 &= 2x^2 - 2 \\ 0 &= x^2 + 4x + 1 \end{aligned}$$

With $a = 1$, $b = 4$, and $c = 1$, we have

$$\begin{aligned} x &= \frac{-4 \pm \sqrt{16 - 4(1)(1)}}{2(1)} \\ &= \frac{-4 \pm 2\sqrt{3}}{2} \\ &= -2 \pm \sqrt{3} \end{aligned}$$

Since neither value obtained is 1 or -1, both solutions are potential roots of the given equation. If $x = -2 + \sqrt{3}$, then

$$\begin{aligned} \frac{-2+\sqrt{3}}{(-2+\sqrt{3})+1} - \frac{3}{(-2+\sqrt{3})-1} &= 2 \\ \left(\frac{-2+\sqrt{3}}{-1+\sqrt{3}}\right)\left(\frac{-1-\sqrt{3}}{-1-\sqrt{3}}\right) - \left(\frac{3}{-3+\sqrt{3}}\right)\left(\frac{-3-\sqrt{3}}{-3-\sqrt{3}}\right) &= 2 \\ \frac{-1+\sqrt{3}}{-2} - \frac{-9-3\sqrt{3}}{6} &= 2 \\ \frac{3-3\sqrt{3}+9+3\sqrt{3}}{6} &= 2 \\ 2 &= 2 \end{aligned}$$

In a similar manner the other solution also checks, and the solution set is $\{-2 \pm \sqrt{3}\}$.

In the checking process in Example 3–8 we needed to rationalize two denominators to verify the root. Notice also that we used the multiplication property of one when the first ratio was multiplied by $-3/-3$ to get the denominator 6 in the third to the fourth step.

Example 3–9 Solve for x:

$$x^{2/3} - 9x^{1/3} + 8 = 0$$

Solution: Let $u = x^{1/3}$; then $u^2 = x^{2/3}$. We then write the given equation in terms of u:

$$u^2 - 9u + 8 = 0 \quad \text{and} \quad (u - 8)(u - 1) = 0$$

We then have $u = 8$ or $u = 1$. Therefore,

$$x^{1/3} = 8 \quad \text{or} \quad x^{1/3} = 1$$

$$x = 512 \quad \text{or} \quad x = 1$$

Check. If $x = 512$, then

$$512^{2/3} - 9(512)^{1/3} + 8 = 0$$

$$64 - 72 + 8 = 0 \qquad \text{(True)}$$

If $x = 1$, then

$$1^{2/3} - 9(1)^{1/3} + 8 = 0$$

$$1 - 9 + 8 = 0 \qquad \text{(True)}$$

Thus, the truth set is $\{1, 512\}$.

EXAMPLE 3–10 Solve and check $\sqrt{x - 3} = x - 5$.

Solution: To eliminate the square root term, we square both sides of the equation. Then we obtain a quadratic equation.

$$(\sqrt{x - 3})^2 = (x - 5)^2$$

$$x - 3 = x^2 - 10x + 25$$

$$0 = x^2 - 11x + 28$$

We can factor the quadratic equation as follows:

$$(x - 4)(x - 7) = 0$$

Thus, we have $x - 4 = 0$ or $x - 7 = 0$. Therefore, $x = 4$ or $x = 7$.

Check. If $x = 4$, then

$$\sqrt{4 - 3} = 4 - 5$$

$$1 = -1 \qquad \text{(False)}$$

If $x = 7$, then

$$\sqrt{7 - 3} = 7 - 5$$

$$2 = 2 \qquad \text{(True)}$$

The solution set is $\{7\}$. Notice that we reject the **potential root** 4, since a false statement is obtained when 4 is substituted into the original equation. Such a root is called **extraneous.**

An extraneous root can be obtained in solving a radical equation because of the step in which we square both sides of the equation. To illustrate, let's consider the equation $x = 2$ with solution set $\{2\}$. If we square both sides of the equation, then $x^2 = 4$, and the solution set of this equation is $\{-2, 2\}$. The root -2 of the new equation is extraneous in the given equation.

EXAMPLE 3–11 Solve and check $\sqrt{6x - 4} = \sqrt{x - 1} + \sqrt{2x}$.

Solution: Since the equation contains square root radicals, we square both sides of the given equation:

$$6x - 4 = (x - 1) + 2\sqrt{2x(x - 1)} + 2x$$
$$3x - 3 = 2\sqrt{2x^2 - 2x}$$

We again square both sides of the equation because the equation still contains a square root term:

$$9x^2 - 18x + 9 = 4(2x^2 - 2x)$$
$$x^2 - 10x + 9 = 0$$

Factoring the quadratic equation yields

$$(x - 1)(x - 9) = 0$$

Therefore, $x = 1$ or $x = 9$.

Check. If $x = 1$, then

$$\sqrt{6(1) - 4} = \sqrt{1 - 1} + \sqrt{2(1)}$$
$$\sqrt{2} = \sqrt{2} \qquad \text{(True)}$$

If $x = 9$, then

$$\sqrt{6(9) - 4} = \sqrt{9 - 1} + \sqrt{2(9)}$$
$$\sqrt{50} = \sqrt{8} + \sqrt{18}$$
$$5\sqrt{2} = 2\sqrt{2} + 3\sqrt{2} \qquad \text{(True)}$$

Thus, the solution set is $\{1, 9\}$. Notice that even though we squared the given equation twice, we did not obtain any extraneous roots.

Exercise Set 3–2

Solve and check the following equations.

1. $\dfrac{x^2}{3} - \dfrac{x}{12} - \dfrac{1}{24} = 0$
2. $\dfrac{1}{30} - \dfrac{x}{15} - \dfrac{x^2}{2} = 0$
3. $\dfrac{8}{x} + 6 = \dfrac{5}{x}$
4. $\dfrac{6}{x^2} = \dfrac{5}{x} - 1$
5. $\dfrac{1}{x + 3} = \dfrac{1}{3 - x} + 2$
6. $\dfrac{x + 4}{x - 1} = 1 + \dfrac{x + 2}{x + 1}$
7. $\dfrac{2y + 7}{2y + 1} - 3 = \dfrac{3 - 2y}{y - 2}$
8. $\dfrac{5y - 5}{y^2 + 1} = \dfrac{4}{y + 1}$
9. $\dfrac{1 - 3a}{4a + 5} = \dfrac{a + 2}{2a - 3} - \dfrac{10a^2 + 2a + 13}{8a^2 - 2a - 15}$

10. $\dfrac{2}{a^2 + 2a - 15} - \dfrac{5}{a + 5} = \dfrac{-3a}{a^2 + 3a - 10}$

11. $\sqrt{2x + 3} = 5$
12. $\sqrt{3x - 5} = 8$
13. $\sqrt{5x - 1} = -3$
14. $-5 = \sqrt{7 - 2x}$
15. $\sqrt[3]{11x - 2} = 4$
16. $\sqrt[3]{5x + 7} = -2$
17. $\sqrt[4]{x + 1} = 3$
18. $\sqrt[4]{3x - 8} = -1$
19. $\sqrt[5]{x - 1} = -2$
20. $4 - \sqrt{4x - 4} = x$
21. $\sqrt{2y - 5} = y - 4$
22. $\sqrt{6 - 2y} + 3 = y$
23. $\sqrt{y^2 + 9} = y - 1$
24. $\sqrt{4 - y^2} = y + 2$
25. $\sqrt{3b + 1} = \sqrt{2b - 1} + 1$
26. $\sqrt{b + 1} - \sqrt{2b - 6} = 2$
27. $\dfrac{3}{\sqrt{a}} + \sqrt{a} = \sqrt{a + 15}$
28. $\sqrt{a} + \sqrt{a - 3} = \dfrac{3}{\sqrt{a - 3}}$
29. $x^{2/3} - 2x^{1/3} - 3 = 0$
30. $2x^{2/3} + 9x^{1/3} - 5 = 0$
31. $y^{2/5} + 3y^{1/5} + 2 = 0$
32. $y^{2/5} + y^{1/5} - 6 = 0$

3-3 QUADRATIC INEQUALITIES

In Chapter 1 we addressed the problem of solving linear inequalities of the form $ax + b < cx + d$. We noted that solving linear inequalities differed in some respects from solving linear equalities.

First, if the replacement set for x is $\mathbb{R}$, then the solution set is generally an **interval of values.** For example, if $3x + 1 < 2x + 4$, then $x < 3$. The solution set is $\{x \mid x < 3\}$ or $(-\infty, 3)$.

Second, when an inequality is multiplied or divided by a negative number, the order of the inequality is reversed. Therefore, if $-3x < 9$, then $x > -3$, and $x \in (-3, \infty)$. These two properties of inequalities make the solving of quadratic inequalities distinctly different from that of quadratic equations.

Quadratic inequality

Definition: A quadratic inequality is a statement equivalent to $ax^2 + bx + c < 0$, $ax^2 + bx + c > 0$, $ax^2 + bx + c \leq 0$, or $ax^2 + bx + c \geq 0$ where a, b, and c are real numbers and $a \neq 0$.

The following theorem is a very important tool that is needed for solving quadratic inequalities:

Theorem: If a and b are any nonzero real numbers, then

(1) $ab > 0$ if and only if $a > 0$ and $b > 0$, or $a < 0$ and $b < 0$.

(2) $ab < 0$ if and only if $a > 0$ and $b < 0$, or $a < 0$ and $b > 0$.

The theorem states that the product of two real numbers is *positive* if and only if the two numbers are *alike in sign*. Or, the product of two real numbers is *negative* if and only if the two numbers are different in sign.

EXAMPLE 3–12 Solve $x^2 - 3x - 10 < 0$, and graph the solution set.

Solution: If $x^2 - 3x - 10 < 0$, then $(x - 5)(x + 2) < 0$. Using the above theorem, we find that $(x - 5)(x + 2) < 0$ if and only if

Set A		Set B
$x - 5 < 0$ and $x + 2 > 0$	or	$x - 5 > 0$ and $x + 2 < 0$
$x < 5$ and $x > -2$	or	$x > 3$ and $x < -2$

Thus, we have

$-2 < x < 5$	or	$\varnothing$

The solution set is $\{x \mid -2 < x < 5\}$ or $(-2, 5)$. The graph is Figure 3-1.

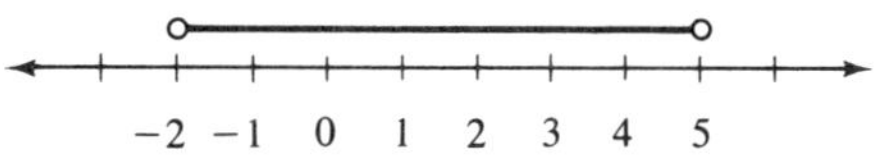

FIGURE 3-1 Solution set $(-2, 5)$

In Example 3–12 the solution set of the inequality is the **union** (or) of sets A and B. Set A consists of the **intersection** (and) of the sets of x's such that $x < 5$ and $x > -2$, that is, $-2 < x < 5$. Set B consists of the intersection of the sets of x's such that $x > 5$ and $x < -2$. (There are no real numbers that simultaneously satisfy both of these conditions, hence the empty set $\varnothing$.) The union of $(-2, 5)$ with $\varnothing$ is $(-2, 5)$.

Notice that the zeroes of the equations $x - 5 = 0$ and $x + 2 = 0$ are the endpoints of the interval of real numbers that is the solution set for the given inequality. In effect, therefore, the set of real numbers is divided into three disjoint subsets by the roots of the equations, $x = -2$ and $x = 5$: (a) $(-\infty, -2)$, (b) $(-2, 5)$, and (c) $(5, \infty)$. We can check the solution set by evaluating the given inequality with a specific value of x within each interval.

(a) $x \in (-\infty, -2)$, e.g., $x = -3$
$(-3)^2 - 3(-3) - 10 < 0$ (False)
(b) $x \in (-2, 5)$, e.g., $x = 0$
$0^2 - 3(0) - 10 < 0$ (True)
(c) $x \in (5, \infty)$, e.g., $x = 6$
$6^2 - 3(6) - 10 < 0$ (False)

Thus, $(-2, 5)$ contains the solution set of $x^2 - 3x - 10 < 0$.

Notice that only *one value of* x in a given interval is needed to determine whether or not *all values of* x in that interval are solutions of the given inequality. To demonstrate why this is true, consider the polynomial $P(x) = x^2 - 3x - 10$, where $x^2 - 3x - 10$ is the quadratic expression of Example 3–12. As noted, $P(-2) = P(5) = 0$. Furthermore, there are no other values of x for which $P(x) = 0$ and $P(x) = x^2 - 3x - 10$. *Therefore, $P(x)$ cannot change sign within* *any of the intervals* $(-\infty, -2)$, $(-2, 5)$, *and* $(5, \infty)$, *since* $P(x)$ *must equal* 0 *between any sign changes for* $P(x)$. That is, if $P(x)$ changes sign from $P(x) < 0$ to $P(x) > 0$ (or vice versa) on some interval of real numbers, then there must be at least one value of x on the interval for which $P(x) = 0$. Thus, if $x^2 - 3x - 10 < 0$ for any $x \in (-2, 5)$, then $x^2 - 3x - 10 < 0$ for all values of x within the interval, since there is no value of x on the interval for which $x^2 - 3x - 10 = 0$.

EXAMPLE 3–13 Solve and graph $3x^2 + x - 10 > 0$.

Solution: If $3x^2 + x - 10 > 0$, then $(3x - 5)(x + 2) > 0$. The roots of the equations $3x - 5 = 0$ and $x + 2 = 0$ are 5/3 and -2, respectively. The set of real numbers is therefore divided into the three disjoint subsets: $(-\infty, -2)$, $(-2, 5/3)$, and $(5/3, \infty)$.

(a) $x \in (-\infty, -2)$, e.g., $x = -3$
$3(-3)^2 + (-3) - 10 > 0$ (True)
(b) $x \in (-2, 5/3)$, e.g., $x = 0$
$3(0)^2 + 0 - 10 > 0$ (False)
(c) $x \in (5/3, \infty)$, e.g., $x = 2$
$3(2)^2 + 2 - 10 > 0$ (True)

The solution set is the union of the intervals of (a) and (c), that is, $(-\infty, -2) \cup (5/3, \infty)$, or $\{x \mid x < -2 \text{ or } x > 5/3\}$. The graph is Figure 3-2.

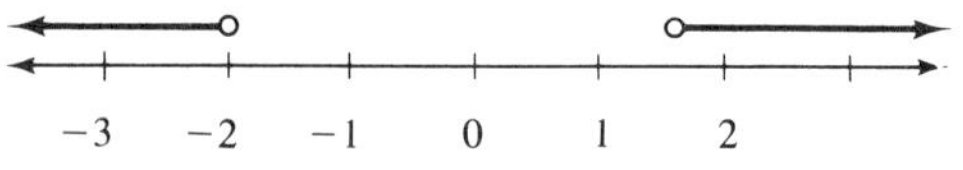

FIGURE 3-2 Solution set $(-\infty, -2) \cup (5/3, \infty)$

How do you decide what value of x to use in a given interval?

Any value of x in a given interval is either included or not included in the solution set. The values of x in each of the three intervals in Example 3–13 were arbitrarily chosen.

EXAMPLE 3–14 The areas of a set of photographs are between 220 and 532 square centimeters. The lengths are always 9 centimeters longer than the widths. What are the minimum and maximum dimensions of the photographs in the set?

Solution: If w is the width of any photograph, then $w + 9$ is the length of the same photograph. We then obtain the inequality

$$220 \leq w(w + 9) \leq 532$$

or

$220 \leq w^2 + 9w$	and	$w^2 + 9w \leq 532$
$0 \leq w^2 + 9w - 220$	and	$w^2 + 9w - 532 \leq 0$
$0 \leq (w + 20)(w - 11)$	and	$(w + 28)(w - 19) \leq 0$

Thus, $w \leq -20$ or $w \geq 11$, and $w \geq -28$ and $w \leq 19$.

The intersection of the two solution sets is $\{w \mid 11 \le w \le 19\}$. The widths of the given set of photographs are from 11 cm to 19 cm, and the corresponding lengths are from 20 cm to 28 cm. The minimum and maximum dimensions are therefore 11 by 20 cm and 19 by 28 cm. Notice that the interval of values $-28 \le w \le -20$ is rejected, since w cannot be negative.

Exercise Set 3–3

State and graph the solution sets of the following inequalities.

1. $(x + 2)(x - 2) > 0$
2. $x^2 \ge 4$
3. $-6(x - 3) < 0$
4. $(x - 3)(x - 2) \ge 0$
5. $x^2 - 25 \ge 0$
6. $2x^2 + x - 10 < 0$
7. $x^2 - 3x - 4 > 0$
8. $6x - 12x^2 > 0$
9. $2x^2 < -x + 15$
10. $3x^2 + 10x \ge -3$
11. $2x + 8 < x^2$
12. $3x^2 \le 2 - x$
13. $2x^2 - 3x > 2$
14. $6x^2 - 5 < -29x$
15. $x^2 - x \le 12$
16. $x^2 - 2x + 5 > x + 3$
17. $(x - 2)(x - 3) > 2$
18. $(2x - 1)(x - 1) < 3$
19. $x^2 - 6x + 9 \ge 0$
20. $x^2 + 4x + 4 < 0$
21. $-3x^2 + 12x - 12 > 0$
22. $-2x^2 - 16x - 32 \le 0$
23. A rocket is shot upward from the ground at a velocity of 96 ft/s. If the height S at time t is given by $S = -16t^2 + 96t$, find the time when the rocket is 128 feet or more above the ground.
24. A stone is thrown upward from a building that is 256 feet high. The height h at time t from the ground is given by $h = -16t^2 + 80t + 256$. At what time is the stone more than 320 feet above the ground?

MORE ON INEQUALITIES 3–4

In Section 3-3 we introduced the process of solving quadratic inequalities. We carefully chose all examples and exercises so that each quadratic expression was factorable over the set of integers. In Section 3-4 we consider inequalities that differ from those in Section 3-3.

EXAMPLE 3–15 Solve and graph $x^2 - 2 < 2x$.

Solution: If $x^2 - 2 < 2x$, then

$$x^2 - 2x - 2 < 0 \qquad \text{(subtraction of inequality)}$$

This quadratic expression cannot be factored over the set of integers. We therefore obtain the roots of the equation $x^2 - 2x - 2 = 0$ with the quadratic formula, and $x = 1 \pm \sqrt{3}$. Using a hand-held calculator with a square root key or Table 1 in the Appendix, we find that $\sqrt{3} \approx 1.73$. Therefore,

$$x \approx 2.73 \quad \text{or} \quad x \approx -0.73$$

The real numbers are divided into the three disjoint intervals: $(-\infty, -0.73)$, $(-0.73, 2.73)$, and $(2.73, \infty)$.

(a) $x \in (-\infty, -0.73)$, e.g., $x = -1$
$(-1)^2 - 2 < 2(-1)$, or $-1 < -2$ (False)
(b) $x \in (-0.73, 2.73)$, e.g., $x = 0$
$0^2 - 2 < 2(0)$, or $-2 < 0$ (True)
(c) $x \in (2.73, \infty)$, e.g., $x = 3$
$3^2 - 2 < 2(3)$, or $7 < 6$ (False)

The solution set is $(1 - \sqrt{3}, 1 + \sqrt{3})$ and the graph is Figure 3-3.

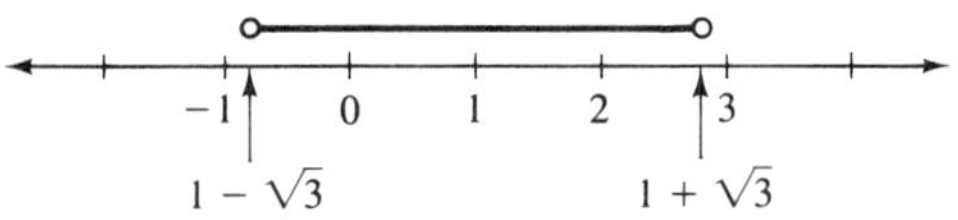

FIGURE 3-3 Solution set $(1 - \sqrt{3}, 1 + \sqrt{3})$

Example 3–15 illustrates why we need to approximate the roots of irrational numbers to estimate the roots of the quadratic equation that determine the intervals to be tested. Once the solution set is determined, the exact values of the endpoints of the intervals are used to name the solution set.

What if the quadratic formula gives us roots that are not real numbers?

If the quadratic formula gives us roots that are not real numbers, we always make one of two conclusions. The solution set is (a) the set of real numbers, $\mathcal{R}$, or (b) the empty set, $\varnothing$. Example 3–16 demonstrates how we can make the correct decision.

EXAMPLE 3–16 Solve and graph $4 > 2x - x^2$.

Solution: If $4 > 2x - x^2$, then

$$x^2 - 2x + 4 > 0 \qquad \text{(subtraction of inequality)}$$

and

$$x = 1 \pm \sqrt{-3} \qquad \text{(quadratic formula)}$$

Since $\sqrt{-3}$ is not a real number, the roots of the quadratic expression do not divide the set of real numbers into any disjoint intervals. Therefore, we check any real number in the original inequality. If $x = 0$, then

$$4 > 2(0) - 0^2 \qquad \text{(True)}$$

We therefore assert that the solution set is $\mathcal{R}$, the set of real numbers, and the graph is Figure 3-4. If $x = 0$ yielded a false inequality, we would conclude that the solution set is $\varnothing$, the empty set.

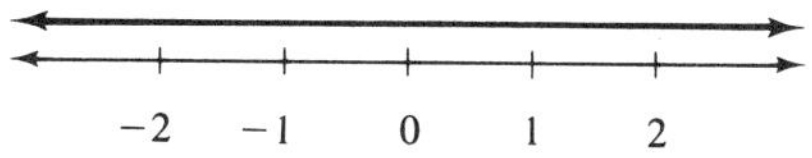

FIGURE 3-4 Solution set $\{x \mid x \in \mathcal{R}\}$

EXAMPLE 3–17 Solve and graph $\dfrac{2}{x-2} \le \dfrac{-3}{x+3}$ when $x \neq -3, 2$.

Solution: If $\dfrac{2}{x-2} \le \dfrac{-3}{x+3}$, then

$$\frac{2}{x-2} - \frac{-3}{x+3} \le 0 \qquad \text{(subtraction of inequality)}$$

$$\frac{2(x+3) - (-3)(x-2)}{(x-2)(x+3)} \le 0 \qquad \text{(subtraction of rational expressions)}$$

$$\frac{5x}{(x-2)(x+3)} \le 0$$

The roots of the three variable factors of the numerator and denominator of the indicated rational expression are -3, 0, and 2. The set of real numbers is divided into four disjoints sets: $(-\infty, -3)$, $(-3, 0)$, $(0, 2)$, and $(2, \infty)$.

(a) $x \in (-\infty, -3)$, e.g., $x = -4$

$$\frac{2}{-4-2} \le \frac{-3}{-4+3}, \text{ or } \frac{-1}{3} \le 3 \qquad \text{(True)}$$

(b) $x \in (-3, 0)$, e.g., $x = -1$

$$\frac{2}{-1-2} \le \frac{-3}{-1+3}, \text{ or } \frac{-2}{3} \le \frac{-3}{2} \qquad \text{(False)}$$

(c) $x \in (0, 2)$, e.g., $x = 1$

$$\frac{2}{1-2} \le \frac{-3}{1+3}, \text{ or } -2 \le \frac{-3}{4} \qquad \text{(True)}$$

(d) $x \in (2, \infty)$, e.g., $x = 3$

$$\frac{2}{3-2} \le \frac{-3}{3+3}, \text{ or } 2 \le \frac{-1}{2} \qquad \text{(False)}$$

Thus, the solution set is $(-\infty, -3) \cup [0, 2)$, that is, $\{x \mid x < -3 \text{ or } 0 \le x < 2\}$. The graph is given in Figure 3-5.

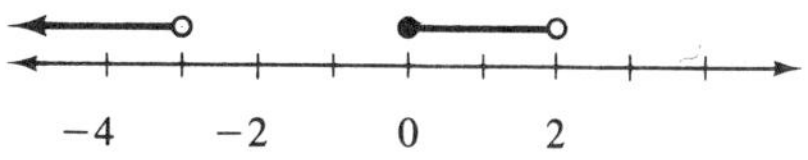

FIGURE 3-5 Solution set $(-\infty, 3) \cup [0, 2)$

Notice that since the equality relation is included in the given inequality, the root of the numerator factor is included in the solution set, because $x = 0$ makes the inequality true. However, the roots of the denominator factors are excluded from the replacement set of x.

It would be very tempting to clear the fraction by multiplying the inequality by $(x - 2)$ and $(x + 3)$, but we must resist that temptation. The reason is that the order of the inequality is reversed or not reversed as the value of $(x - 2)(x + 3)$ is negative or positive, respectively. For example, $(x - 2)(x + 3) > 0$ when $x = -4$. But when $x = -1$, $(x - 2)(x + 3) > 0$. Thus, *when we*

Why don't we just clear the fraction by multiplying the inequality by (x-2) and (x+3)?

multiply by a factor that contains the variable, we don't know whether we are multiplying by a positive or a negative number. Hence, we don't know whether or not the order of the inequality has been reversed. We therefore propose the following flow diagram.

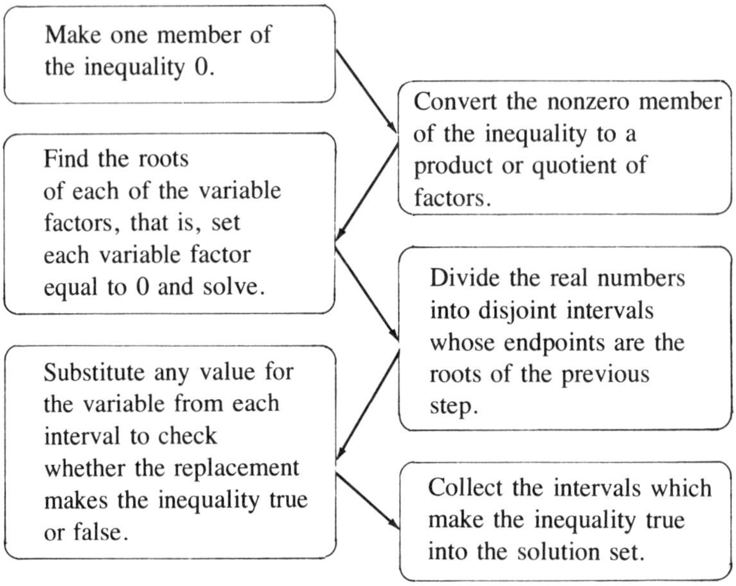

The following example illustrates the use of the flow diagram.

EXAMPLE 3–18 Solve and graph $\dfrac{x}{x+3} < \dfrac{2}{2x+1}$.

Solution: We follow the steps of the flow diagram:

$$\frac{x}{x+3} - \frac{2}{2x+1} < 0$$

$$\frac{x(2x+1) - 2(x+3)}{(x+3)(2x+1)} < 0$$

$$\frac{2x^2 - x - 6}{(x+3)(2x+1)} < 0$$

$$\frac{(2x+3)(x-2)}{(2x+1)(x+3)} < 0$$

The roots of the four factors are -3, $-3/2$, $-1/2$, and 2. The real numbers are divided into the following five disjoint intervals: $(-\infty, -3)$, $(-3, -3/2)$, $(-3/2, -1/2)$, $(-1/2, 2)$, and $(2, \infty)$.

(a) $x \in (-\infty, -3)$, e.g., $x = -4$

$$\frac{-4}{-4+3} < \frac{2}{2(-4)+1}, \text{ or } 4 < \frac{-2}{7} \qquad \text{(False)}$$

(b) $x \in (-3, -3/2)$, e.g., $x = -2$

$$\frac{-2}{-2+3} < \frac{2}{2(-2)+1}, \text{ or } -2 < \frac{-2}{3} \qquad \text{(True)}$$

(c) $x \in (-3/2, -1/2)$, e.g., $x = -1$

$$\frac{-1}{-1+3} < \frac{2}{2(-1)+1}, \text{ or } \frac{-1}{2} < -2 \qquad \text{(False)}$$

(d) $x \in (-1/2, 2)$, e.g., $x = 0$

$$\frac{0}{0+3} < \frac{2}{2(0)+1}, \text{ or } 0 < 2 \qquad \text{(True)}$$

(e) $x \in (2, \infty)$, e.g., $x = 3$

$$\frac{3}{3+3} < \frac{2}{2(3)+1}, \text{ or } \frac{1}{2} < \frac{2}{7} \qquad \text{(False)}$$

The solution set is the union of (b) and (d),

$$\{x \mid -3 < x < -3/2 \text{ or } -1/2 < x < 2\}$$

The graph is shown in Figure 3-6.

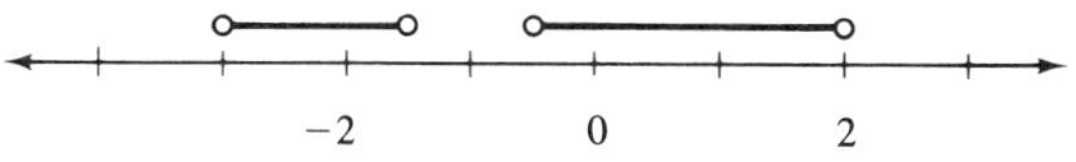

FIGURE 3-6 Solution set $(-3, -3/2) \cup (-1/2, 2)$

Exercise Set 3–4

State and graph the solution sets of the following inequalities.

1. $\dfrac{x-2}{x+1} < 0$
2. $\dfrac{x+1}{x-2} \geq 0$
3. $\dfrac{1}{x} \geq \dfrac{3}{2-x}$
4. $\dfrac{x}{x+1} > 1$
5. $\dfrac{5}{x+1} - \dfrac{6}{x-3} \geq 0$
6. $\dfrac{2x^2 - 3x - 20}{x+3} \geq 0$
7. $(x-1)(x-2)(x+3) < 0$
8. $(x+1)(x+2)(x+3) \geq 0$
9. $x^3 - x^2 - 2x < 0$
10. $(2x-3)^3(x+2)^2 \leq 0$
11. $\dfrac{(2x-3)^3}{(x+2)^2} \geq 0$
12. $\dfrac{x^3 - x}{x+2} > 0$
13. $2x^2 + x - 3 \leq 0$
14. $x^2 - 1 > 2x$
15. $x^2 - 6x + 7 > 0$
16. $x^2 - 2x + 11 \geq 0$
17. $x + \dfrac{1}{x} \geq 2$
18. $\dfrac{(x-1)^2}{(x+2)^2} < 0$
19. $(x^2 - 6)(x^2 - 3) \leq 0$
20. $\dfrac{x}{2} + \dfrac{2}{x} - 1 < x$

For each of the following inequalities, (a) determine the values of k for which the solution sets are the empty set or the set of real numbers and (b) determine whether the solution set is the set of real numbers or the empty set.

21. $x^2 - 4x + k < 0$
22. $3x^2 + 6x - k \geq 0$
23. $k + 4x - 2x^2 \leq 0$
24. $4k - 8x - x^2 \leq 0$
25. $x^2 + kx + 9 < 0$
26. $x^2 - 2kx + 25 \leq 0$
27. A number is at least 6 more than seven times its reciprocal. Find all such numbers.
28. The sum of a number and twice its reciprocal is at most 3. Find all such numbers.
29. The area of a rectangular plot of ground is to be between 1400 and 3300 square meters. If the width is always 5 meters less than the length, find the range of values for the length.
30. The volume of a rectangular solid with a square base is to be between 135 and 320 cubic meters. Find the range of values for an edge of the base if the height of the solid is always 5 times the length of an edge of the base.

3-5 ABSOLUTE VALUE EQUATIONS AND INEQUALITIES

Distance is a topic we frequently encounter in subjects such as engineering, science, and sports. Units for measuring distances range from the incredibly small picometer to the inconceivably large light-year. We find it necessary to make a distinction between *undirected* or *absolute distance* and *directed distance*.

Recall that we use the concepts of directed and undirected distance when we construct a real number line. The distance between points with consecutive integral coordinates is always the same, that is, it is the undirected unit length. For example, the **undirected distance** between points A and B with coordinates a and b, respectively, is 1, so long as a and b are consecutive integers. However, if $b > a$, then the **directed distance** from A to B is $+1$ (direction to the right is positive). But the directed distance from B to A is -1 (direction to the left is negative). Thus, *directed distances can be positive or negative, but undirected distances are always positive or zero.*

The operation of subtraction is used to compute the undirected distance between a pair of points on the number line. If A and B are two points on the number line with coordinates a and b, respectively, then the undirected or absolute distance between A and B, written $|AB|$, is given by the equation

$$|AB| = |a - b|$$

where $|a - b|$ is read "the absolute value of the difference between a and b." The following definition identifies the meaning of $|a - b|$.

Absolute value

Definition: If a and b are two real numbers, then

$$|a - b| = a - b \text{ if } a \geq b$$

and

$$|a - b| = b - a \text{ if } a < b$$

Thus, the undirected distance between points A and B with coordinates a and b, respectively, is computed by determining $|a - b|$, and $|a - b| \geq 0$. Furthermore, $|a - b|$ is represented geometrically as the distance between points A and B with coordinates a and b. The relationship between the absolute value of the difference between two numbers a and b and the undirected distance between a pair of points A and B will be valuable as we develop the concept further.

EXAMPLE 3–19 Compute $|AB|$ for the given values of a and b.

(a) $a = -8$ and $b = 4$ (b) $a = 4$ and $b = -8$

Solutions:

(a) With $a = -8$ and $b = 4$, then

$$|AB| = |-8 - 4| = 4 - (-8) = 12$$

(b) With $a = 4$ and $b = -8$, then

$$|AB| = |4 - (-8)| = 4 - (-8) = 12$$

Since $|a - b| \geq 0$ for any real numbers a and b, we can compute $a - b$ within the absolute value bars, then assert that the difference is nonnegative. For example, in the solution for Example 3–19,

(a) $|-8 - 4| = |-12| = 12$ (b) $|4 - (-8)| = |12| = 12$

> **Theorem:** If a and b are any real numbers, then
> $$|a - b| = |b - a|$$

If b is replaced with zero in the definition for $|a - b|$, then $|a - b| = |a - 0| = |a|$. Geometrically, $|a|$ represents the absolute distance that point A with coordinate a is from the origin of the number line.

> **Theorem:** For any real number a,
> $$|a| = a \text{ if } a \geq 0 \quad \text{and} \quad |a| = -a \text{ if } a < 0$$

EXAMPLE 3–20 Compute the following:

(a) $|17|$ (b) $|-10|$

Solutions:

(a) $17 > 0$; therefore, $|17| = 17$.

(b) $-10 < 0$; thus, $|-10| = -(-10) = 10$.

Notice that the absolute value of any real number is nonnegative.

If b and c are real numbers with $c > 0$, then the equation $|x - b| = c$ has a solution set $\{x_1, x_2\}$, where x_1 and x_2 are the coordinates of points X_1 and X_2. Points X_1 and X_2 are c units from point B with coordinate b, written $B(b)$. The following example illustrates the algebraic solution of such an equation.

EXAMPLE 3–21 Solve and graph $|x - 5| = 4$.

Solution: If $|x - 5| = 4$, then $b = 5$, $c = 4$, and

$$x - 5 = 4 \quad \text{or} \quad -(x - 5) = 4$$
$$x = 9 \quad \text{or} \quad x = 1$$

The solution set is $\{1, 9\}$. The graph is shown in Figure 3-7. Notice that points $X_1(1)$ and $X_2(9)$ are both 4 units from $B(5)$.

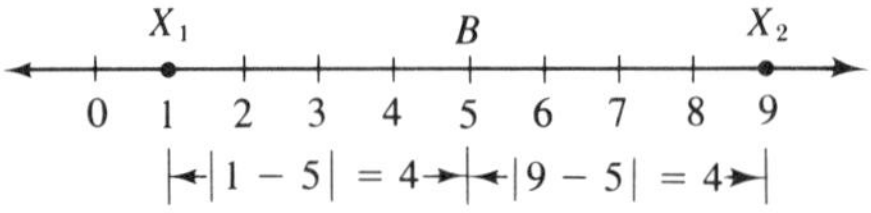

FIGURE 3-7 Solution set $\{1, 9\}$

If b and c are real numbers with $c > 0$, then the inequality $|x - b| < c$ has as a solution set $\{x \mid x_1 < x < x_2\}$, where x_1 and x_2 are endpoints of an open interval of real values *within c units of b*.

Notice that $|x - b| < c$ implies that

$$x - b < c \quad \text{and} \quad -(x - b) < c$$

But $-(x - b) < c$ is equivalent to the inequality $x - b > -c$, which can also be written as $-c < x - b$. Since $-c < x - b$ and $x - b < c$, we write simply

$$-c < x - b < c$$

Adding b to each member of the inequalities, we get

$$b - c < x < b + c$$

Thus, the values of x_1 and x_2, the endpoints of the open interval of real values that are the solution of the given inequality, are $b - c$ and $b + c$, respectively. Example 3–22 illustrates the algebraic solution of such an inequality.

EXAMPLE 3–22 Solve and graph $|2x + 3| \leq 9$.

Solution: If $|2x + 3| \leq 9$, then $b = -3$, $c = 9$, and

$$-9 \leq 2x + 3 \leq 9$$
$$-12 \leq 2x \leq 6$$
$$-6 \leq x \leq 3$$

The solution set is $\{x \mid -6 \leq x \leq 3\}$, and the graph is shown in Figure 3-8.

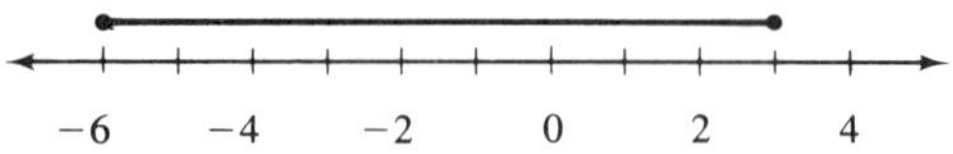

FIGURE 3-8 Solution set $[-6, 3]$

If b and c are real numbers with $c > 0$, then the inequality $|x - b| > c$ has a solution set $\{x \mid x < x_1 \text{ or } x > x_2\}$, where x_1 and x_2 are endpoints of disjoint open intervals of real values *more than c units from b*.

Notice that $|x - b| > c$ implies

$$x - b > c \quad \text{or} \quad -(x - b) > c$$

But $-(x - b) > c$ is equivalent to the inequality $x - b < -c$. Therefore,

$$x > b + c \quad \text{or} \quad x < b - c$$

and $x_1 = b - c$ and $x_2 = b + c$. The following example illustrates the algebraic solution of such an inequality.

EXAMPLE 3–23 Solve and graph $|x - 5| > 4$.

Solution: If $|x - 5| > 4$, then $b = 5$, $c = 4$, and

$$x - 5 < -4 \quad \text{or} \quad x - 5 > 4$$
$$x < 1 \quad \text{or} \quad x > 9$$

The solution set is $\{x \mid x < 1 \text{ or } x > 9\}$. The graph is given in Figure 3-9. Notice that the indicated intervals contain points which are more than 4 units from $B(5)$.

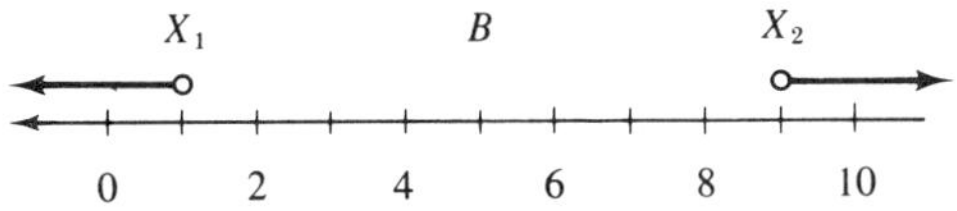

FIGURE 3-9 Solution set $(-\infty, 1) \cup (9, \infty)$

Exercise Set 3–5

Compute the following absolute values.

1. $|3|$
2. $|-4|$
3. $|6 - 8|$
4. $|-4 - 9|$
5. $|x^2|$
6. $|x^3|$
7. $|-x|$
8. $|x - y|$

Evaluate the following expressions for $x = 2$, $y = 3$, and $z = -1$.

9. $|2x + y|$
10. $|2y + 7z|$
11. $|5x - 3z|$
12. $|x^2 - 5y + 2z|$
13. $\left|\dfrac{x - 4y}{5z}\right|$
14. $\left|\dfrac{2z^3 - 2x}{y}\right|$

Solve the following absolute value equations. State solutions in set notation.

15. $|x - 1| = 3$
16. $|x + 1| = 5$
17. $|x| = 6$
18. $|x| = -6$
19. $|x| = x$
20. $|x| = -x$
21. $|2x + 3| = \dfrac{5}{2}$
22. $|3x + 2| = \dfrac{2}{5}$

23. $|x - 3| = x - 3$

24. $|x - 3| = 3 - x$

25. $|2 - x| = 3$

26. $|8 - 2x| = 4$

Solve and graph the following absolute value inequalities.

27. $|x - 3| < 2$

28. $|x + 1| \geq 1$

29. $|x + 3| > 2$

30. $|2x - 1| \leq 2$

31. $|3 - 4x| < 5$

32. $|6 - 5x| \geq 3$

33. $|x - 1| > -2$

34. $|x - 1| \leq -2$

35. $|x - 1| < 3$ and $|x - 4| < 1$

36. $|x - 2| \leq 5$ and $|x - 7| \geq 2$

37. $|x| \geq 4$ or $|x - 5| \leq 2$

38. $|x - 3| > 3$ or $|x + 1| > 5$

Write the following intervals of real numbers in absolute value notation.

39. $-2 \leq x \leq 2$

40. $x < -2$ or $x > 2$

41. $x \leq -3$ or $x \geq 5$

42. $0 < x < 8$

43. Use absolute value notation to identify the real numbers x that are the coordinates of points within 5 units of 2.

44. Use absolute value notation to identify the real numbers y that are the coordinates of points more than 7 units from -3.

Find the values of x and y for which the following open sentences are true.

45. $|xy| = |x| \cdot |y|$

46. $\left|\dfrac{x}{y}\right| = \dfrac{|x|}{|y|}$

47. $|x + y| = |x| + |y|$

48. $|x + y| < |x| + |y|$

49. $|x - y| \geq |x| - |y|$

50. $|x - y| \leq |x| + |y|$

51. Simplify the expression $\dfrac{|x|}{x}$

(a) if $x > 0$

(b) if $x < 0$

3-6 FACTOR THEOREM

The quadratic formula, as we have seen, provides us with a powerful tool for solving polynomial equations of degree two. There are, however, no simple formulas for finding the roots of polynomials of degree greater than two. But there are tools that do help us in solving such equations.

The first tool that we will consider is **synthetic division.** You may recall that synthetic division is used to short-cut the division of the polynomial $P(x)$ by the linear factor $x - c$. We use the following example to explain the process.

EXAMPLE 3–24 Divide $x^4 + 2x^3 - 5x - 10$ by $x - 2$.

Solution:

(1) Array the coefficients of the dividend. Insert a zero for any missing power of x.

$$\begin{array}{ccccc} 1 & 2 & 0 & -5 & -10 \end{array}$$

(2) Box off the value of c to the left of the array.

We are dividing by $x - 2$. Thus, $c = 2$, and

$$\begin{array}{c|ccccc} 2 & 1 & 2 & 0 & -5 & -10 \end{array}$$

(3) Skip a space, draw a line, then write below the line the first number in the array of coefficients.

$$\begin{array}{c|ccccc} 2 & 1 & 2 & 0 & -5 & -10 \\ & & & & & \\ \hline & 1 & & & & \end{array}$$

(4) Multiply this number by c and add the product to the second number in the array of coefficients.

$$\begin{array}{c|ccccc} 2 & 1 & 2 & 0 & -5 & -10 \\ & & 2 & & & \\ \hline & 1 & 4 & & & \end{array}$$

(5) Continue the process outlined in step (4), using the remaining numbers in the array of coefficients.

$$\begin{array}{c|ccccc} 2 & 1 & 2 & 0 & -5 & -10 \\ & & 2 & 8 & 16 & 22 \\ \hline & 1 & 4 & 8 & 11 & 12 \end{array}$$

The array obtained below the line contains the coefficients of the quotient, and the last value in the array is the remainder. Thus, we have

$$\frac{x^4 + 2x^3 - 5x - 10}{x - 2} = x^3 + 4x^2 + 8x + 11 + \frac{12}{x - 2}$$

The degree of the dividend is four (x^4 term) and the degree of the divisor is one (x term); thus, the quotient is the degree of the dividend reduced by one.

How did you know that the first term of the quotient was of degree three?

EXAMPLE 3–25 If $P(x) = x^4 + 2x^3 - 5x - 10$, find $P(2)$.

Solution: $P(2)$ is the value of $P(x)$ when x is everywhere replaced by 2. Thus,

$$P(2) = (2)^4 + 2(2)^3 - 5(2) - 10 = 12$$

Notice that *when* $x^4 + 2x^3 - 5x - 10$ *is divided by* $x - 2$, *the remainder is the value of* $P(x)$ *for* $x = 2$. In general, the remainder obtained when $P(x)$ is divided by $x - c$ is the value of $P(c)$. As a consequence, when the remainder is 0, then $P(c) = 0$, and c is a root of $P(x) = 0$. We state this important fact in the following theorem.

> **Factor Theorem:** If $P(x)$ is a polynomial and c is a real number, then $x - c$ is a factor of $P(x)$ if and only if c is a root of $P(x) = 0$, that is, $P(c) = 0$.

The value of the theorem is evident when we work with polynomials $P(x)$ of degree greater than two. When we obtain a root c of $P(x) = 0$, we divide $P(x)$ by

$x - c$ and obtain a polynomial $Q(x)$ of degree one less than $P(x)$. If possible, we continue the process on $Q(x)$ until a polynomial of degree two is obtained. We can then use the quadratic formula (if necessary) to obtain the remaining roots.

EXAMPLE 3–26 Solve $x^3 - 5x^2 - 17x + 21 = 0$ if 1 is a root.

Solution: If $P(x) = x^3 - 5x^2 - 17x + 21$ and 1 is a root of $P(x) = 0$, then we know that $P(1) = 0$ and $x - 1$ divides $P(x)$ evenly.

$$\begin{array}{r|rrrr} 1 & 1 & -5 & -17 & 21 \\ & & 1 & -4 & -21 \\ \hline & 1 & -4 & -21 & 0 \end{array}$$

The zero remainder verifies that 1 is a root of $P(x) = 0$. Furthermore,

$$\begin{aligned} P(x) &= (x - 1)(x^2 - 4x - 21) \\ &= (x - 1)(x - 7)(x + 3) \end{aligned}$$

and

$$x = 1 \quad \text{or} \quad x = 7 \quad \text{or} \quad x = -3$$

To check the values -3 and 7, we use synthetic division and the original equation.

$$\begin{array}{r|rrrr} -3 & 1 & -5 & -17 & 21 \\ & & -3 & 24 & -21 \\ \hline & 1 & -8 & 7 & 0 \end{array} \qquad \begin{array}{r|rrrr} 7 & 1 & -5 & -17 & 21 \\ & & 7 & 14 & -21 \\ \hline & 1 & 2 & -3 & 0 \end{array}$$

The 0 remainders verify that -3 and 7 are roots of the given equation, and the solution set is $\{-3, 1, 7\}$. In this section and the remaining sections of this chapter, we leave to you the responsibility to check the roots of the equations in all examples.

EXAMPLE 3–27 Find the value of k so that -1 is a root of

$$x^3 + 5x^2 - 2x + 3k = 0$$

Solution: For -1 to be a root of the given equation, the remainder must be 0 when we divide by -1.

$$\begin{array}{r|rrrr} -1 & 1 & 5 & -2 & 3k \\ & & -1 & -4 & 6 \\ \hline & 1 & 4 & -6 & 3k + 6 \end{array}$$

Thus, $3k + 6 = 0$ and $k = -2$. We conclude that for $P(x) = x^3 + 5x^2 - 2x - 6$, $P(-1) = 0$ and $x + 1$ is a factor of $P(x)$.

EXAMPLE 3–28 A particle is moving in a plane. The path of the particle is approximated with the polynomial equation $d = 2t^3 - 15t^2 + 28t - 15$, where d is the displacement of the particle relative to the horizontal axis and t is the time in seconds. The particle is observed on the t-axis at time $t = 1$ second.

(a) For what other values of t is $d = 0$?
(b) For what time intervals is the particle below the horizontal axis?

Solutions:
(a) If $d = 0$ when $t = 1$, then $P(1) = 0$.

1	2	−15	28	−15
		2	−13	15
	2	−13	15	0

Thus, we have

$$(t - 1)(2t^2 - 13t + 15) = d$$

and $d = 0$ for

$$2t^2 - 13t + 15 = 0 \quad \text{or} \quad (2t - 3)(t - 5) = 0$$

and $t = 3/2$ and 5. Therefore, when $t = 1$, $3/2$, and 5, the particle is on the horizontal axis.
(b) The particle is below the t-axis for values of t such that $P(t) < 0$. The zero points of part (a) divide the allowable values of t into four disjoint intervals: (0, 1), (1, 3/2), (3/2, 5) and (5, ∞).

(1) If $t \in (0, 1)$, e.g., $t = 0.5$, then

0.5	2	−15	28	−15
		1.0	−7	10.5
	2	−14	21	−4.5

and $P(t) < 0$.

(2) If $t \in (1, 3/2)$, e.g., $t = 1.1$, then

1.1	2	−15	28	−15
		2.2	−14.08	15.312
	2	−12.8	13.92	0.312

and $P(t) > 0$.

(3) If $t \in (3/2, 5)$, e.g., $t = 2$, then

2	2	−15	28	−15
		4	−22	12
	2	−11	6	−3

and $P(t) < 0$.

(4) If $t \in (5, \infty)$, e.g., $t = 6$, then

6	2	−15	28	−15
		12	−18	60
	2	−3	10	45

and $P(t) > 0$.

Therefore, for the time intervals (0, 1) and (3/2, 5), the particle is below the t-axis.

Exercise Set 3–6

If $P(x) = x^3 - 2x^2 - x + 2$, find the following.

1. $P(3)$
2. $P(1)$
3. $P(-1)$
4. $P(-3)$

Find all real roots of the following polynomial equations, given that the numbers in the braces are roots.

5. $x^3 - 2x^2 - x + 2 = 0$ $\{2\}$
6. $x^3 + x^2 - 4x - 4 = 0$ $\{-1\}$
7. $x^3 - 6x^2 + 25 = 0$ $\{5\}$
8. $x^3 + 3x^2 - 2x - 6 = 0$ $\{-3\}$
9. $x^3 + 3x - 4 = 0$ $\{1\}$
10. $x^3 - 9x^2 + 26x - 24 = 0$ $\{2\}$
11. $x^4 - x^3 - 5x^2 - x - 6 = 0$ $\{-2, 3\}$
12. $x^4 - 10x^2 + 9 = 0$ $\{1, 3\}$
13. $2x^4 - 5x^3 + 5x - 2 = 0$ $\{1, 2\}$
14. $x^4 + 10x^3 + 32x^2 + 37x + 10 = 0$ $\{-2, -5\}$

Some of the linear factors of a polynomial are given. Find the other linear factors of the polynomial.

15. $P(x) = x^3 - x^2 - 17x - 15$ $(x + 1)$
16. $P(x) = x^4 - 2x^3 - 3x^2 + 8x - 4$ $(x + 2), (x - 1)$
17. $P(x) = 2x^4 - 7x^3 - 47x^2 - 17x + 21$ $(2x - 1), (x + 3)$
18. $P(x) = x^4 - 20x + 64$ $(x + 4), (x - 4)$
19. $P(x) = x^4 - 2x^3 - 37x^2 + 4x + 70$ $(x + 5), (x - 7)$
20. $P(x) = 2x^3 - x^2 - 21x$ $(x + 3)$
21. Find the value of k so that 9 is a root of
$$P(x) = x^3 - 9x^2 - 7x + k$$
22. Find the value of k so that $x - 2$ is a factor of
$$P(x) = x^4 - 3x^3 + 2x^2 + kx - 2$$
23. Find the value of k so that 1 is a root of
$$P(x) = x^4 + 6x^3 + kx^2 - 8x + 2$$
24. Find the value of k so that $x - 2$ is a factor of
$$P(x) = x^3 + kx^2 + 11x + k$$
25. A particle moves in a plane such that the displacement d from the horizontal axis is given by
$$d = 2t^3 - 13t^2 + 22t - 8$$
where t is the time in seconds. The particle is observed on the t-axis at the time $t = 2$ seconds.
(a) For what values of t is the particle on the horizontal axis?
(b) For what time intervals is the particle above the horizontal axis?

26. The difference d between the selling price and the total cost per item of producing an article is determined for several values of n, the number of articles produced. If d is negative, then the cost of producing the article is greater than the revenue made by selling it. If d is positive, the revenue is greater than the cost. The relationship between d and n is described by the equation

$$d = n^3 - 70n^2 + 1000n$$

(a) For what values of n does $d = 0$?
(b) For what interval of values for n does the cost of producing the article exceed the revenue?

27. A company's profit statement p (in thousands of dollars) is plotted over a period of twelve months t, where t_0 = December 31, t_1 = January 31, t_2 = February 28, and so on, with t_{12} = December 31 again. The relationship between p and t is approximated with the equation

$$p = t^3 - 9t^2 + 11t + 21$$

(a) What was the company's profit statement at the beginning of the time period?
(b) Between what time interval did the company's profit become a loss?
(c) What was the company's profit statement at the end of the time period?

28. At a time $t = 0$, an object is moving with a velocity v of 10 meters per second. Without changing direction, the velocity is observed to change with time according to the equation

$$v = t^3 - 11t^2 + 24t + 10$$

The object is observed for time $t = 0$ to $t = 10$ seconds, that is, $t \in [0, 10]$.
(a) For what values of t is the object's velocity exactly 10 m/s?
(b) Over what time interval is the object's velocity less than 10 m/s?
(c) What is the object's velocity at the end of the observed time period?

INTEGRAL ROOT THEOREM 3-7

In Section 3-6 we were able to write a polynomial $P(x)$ as $Q(x)(x - c)$ when c is a root of $P(x) = 0$. Furthermore, if the degree of $P(x)$ is n, then the degree of $Q(x)$ is $n - 1$.

In this section we state the Integral Root Theorem, which enables us to find any integral roots that a given polynomial may have.

Integral Root Theorem: If $P(x)$ is a polynomial with integral coefficients and

$$P(x) = a_0x^n + a_1x^{n-1} + \cdots + a_n$$

and c is an integral root of $P(x) = 0$, then c is a factor of a_n.

We offer the following proof of the Integral Root Theorem. If c is a root of $P(x) = 0$, then $P(c) = 0$, and

$$0 = a_0c^n + a_1c^{n-1} + \cdots + a_{n-1}c + a_n$$
$$-a_n = a_0c^n + a_1c^{n-1} + \cdots + a_{n-1}c$$

Since c is a factor of every term in the right side of the equation, then c must also be a factor of the left side of the equation. Thus, c is a factor of a_n, the constant term of $P(x)$.

EXAMPLE 3–29 Identify the possible integral roots of

(a) $P(x) = x^5 + 9x^3 + 6$ (b) $P(x) = x^4 + 2x^2 - 8x$

Solutions:
(a) For $P(x) = x^5 + 9x^3 + 6$, the value of a_n is 6. Thus, the only possible integral roots of $P(x) = 0$ are $\{\pm 1, \pm 2, \pm 3, \pm 6\}$.
(b) In $P(x) = x^4 + 2x^2 - 8x$, $P(x)$ has no constant term; but

$$P(x) = x(x^3 + 2x - 8)$$
$$Q(x) = x^3 + 2x - 8$$

The roots of $Q(x)$ are also roots of $P(x)$; thus, the set of possible integral roots are factors of $a_n = 8$, or $\{\pm 1, \pm 2, \pm 4, \pm 8\}$.

EXAMPLE 3–30 Solve $x^3 - x^2 - 4x - 2 = 0$.

Solution: For the given polynomial equation, $a_n = 2$; thus, the only possible integral roots are ± 1 and ± 2.
If $x = 1$, then we have

$$\begin{array}{r|rrrr} 1 & 1 & -1 & -4 & -2 \\ & & 1 & 0 & -4 \\ \hline & 1 & 0 & -4 & -6 \end{array}$$

Since $P(1) = -6$, then 1 is not a root.
If $x = -1$, then

$$\begin{array}{r|rrrr} -1 & 1 & -1 & -4 & -2 \\ & & -1 & 2 & 2 \\ \hline & 1 & -2 & -2 & 0 \end{array}$$

Since $P(-1) = 0$, then -1 is a root, and

$$(x + 1)(x^2 - 2x - 2) = 0$$

With the quadratic formula and $x^2 - 2x - 2 = 0$, we get

$$x = \frac{2 \pm \sqrt{12}}{2} = 1 \pm \sqrt{3}$$

The solution set is $\{-1, 1 \pm \sqrt{3}\}$.

Is the "trial and error" method the only way to check possible integral roots?

The "trial and error" method is the only way by which we can actually check a given value in the set of possible integral roots. Also, the number of possible integral roots for a given equation can be quite extensive. For example, there are 12 possible integral roots for the equation $x^4 + x^3 + 7x^2 + 9x - 18 = 0$. Only 1 and -2 are integral roots as shown by the following divisions:

1	1	1	7	9	−18
		1	2	9	18
−2	1	2	9	18	0
		−2	0	−18	
	1	0	9	0	

Therefore,

$$x^4 + x^3 + 7x^2 + 9x - 18 = (x - 1)(x + 2)(x^2 + 9)$$

and the only *real roots* of the equation are 1 and −2. However, we can only obtain these roots through the process of elimination, using synthetic division.

The following theorems, stated without proof, state the conditions for an *upper boundary* (or limit) and *lower boundary* (or limit) for the real roots of a polynomial equation. Although a boundary has a rigorous definition in mathematics, we simply intend to use an intuitive idea of what a boundary is.

Therefore, the real number k, identified in the following theorem, is an **upper boundary** for the real roots of $P(x) = 0$ if and only if there are no real numbers r such that $r > k$ and $P(r) = 0$. That is, there are no real roots of the polynomial equation greater than k, and k is an upper boundary.

Similarly, the real number m is a **lower boundary** for the real roots of $P(x) = 0$ if and only if there are no real numbers s such that $s < m$ and $P(s) = 0$. That is, there are no real roots of the polynomial equation less than m, and m is a lower boundary.

Upper Boundary Theorem: Let $P(x) = 0$ be a polynomial equation with a positive leading coefficient, and let k be a nonnegative real number. If the coefficients of the quotient and remainder obtained when $P(x)$ is divided by $x - k$ are all positive numbers or zero, then k is an upper boundary for the real roots of the given equation.

Lower Boundary Theorem: Let $P(x) = 0$ be a polynomial equation with a positive leading coefficient, and let m be a nonpositive real number. If the coefficients of the quotient and remainder obtained when $P(x)$ is divided by $x - m$ are alternately nonnegative and nonpositive numbers, then m is a lower boundary for the real roots of the given equation.

In the Lower Boundary Theorem, a zero coefficient can be assigned either a positive or a negative sign to maintain the pattern of positive and negative terms.

Example 3–31 Solve $x^3 + 2x^2 - 3x - 6 = 0$.

Solution: The possible integral roots are factors of 6, that is, ±1, ±2, ±3, ±6.

If $x = 1$, then

1	1	2	−3	−6
		1	3	0
	1	3	0	−6

Thus, 1 is not a root.

If $x = 2$, then

2	1	2	−3	−6
		2	8	10
	1	4	5	4

Thus, 2 is not a root.

Notice that the final array contains only the positive numbers 1, 4, 5, and 4. Therefore, 2 is an upper boundary for the real roots of the given equation, and 3 and 6 are rejected as possible roots.

If $x = -1$, then

−1	1	2	−3	−6
		−1	−1	4
	1	1	−4	−2

Thus, −1 is not a root.
If $x = -2$, then

−2	1	2	−3	−6
		−2	0	6
	1	0	−3	0

Thus, −2 is a root.
Therefore, $(x + 2)(x^2 - 3) = 0$, and the solution set is $\{-2, -\sqrt{3}, \sqrt{3}\}$.

We now have the tools to analyze to some degree the roots of a polynomial equation. If $P(x) = 0$ has integral roots, we use the Integral Root Theorem to find them. Furthermore, if the equation has at most two nonintegral real roots, we can find all of them. Finally, we are also able to determine an upper and lower boundary for the real roots of the equation.

Exercise Set 3–7

Find all the real roots of the following polynomial equations and check.

1. $P(x) = x^3 - 3x^2 - x + 3$
2. $P(x) = x^3 + 3x^2 + 3x + 1$
3. $P(x) = x^3 - x^2 - 14x + 24$
4. $P(x) = 3x^4 + 4x^3 - 3x^2 - 4x$
5. $P(x) = x^3 + 10x^2 - 9x - 90$
6. $P(x) = x^3 - 8$
7. $P(x) = x^3 - x^2 - 4x + 4$
8. $P(x) = x^3 + 5x^2 - 7x + 1$
9. $P(x) = 6x^3 - 7x^2 - 7x + 6$
10. $P(x) = x^4 + 7x^3 - 25x^2 - 175x$
11. $P(x) = x^4 - 5x^3 + 8x^2 - 10x + 12$
12. $P(x) = x^4 - 14x^2 + 45$
13. $P(x) = 6x^5 + 23x^4 + 3x^3 - 32x^2 + 12x$
14. $P(x) = x^4 + x^3 + 3x^2 + 5x - 10$
15. $P(x) = x^4 - 4x^3 + 6x^2 - 4x + 1$
16. $P(x) = x^4 - 10x^3 + 37x^2 - 60x + 36$

17. $P(x) = x^4 + 5x^3 + 6x^2 - 4x - 8$

18. $P(x) = 9x^4 - 27x^3 + 2x^2 + 48x - 32$

19. $P(x) = x^5 + 5x^4 + 10x^3 + 10x^2 + 5x + 1$

20. $P(x) = x^5 - 15x^4 + 85x^3 - 225x^2 + 274x - 120$

Using only integers, determine the smallest intervals that contain the real roots of the following polynomial equations. Do not solve the equations.

21. $2x^3 - x^2 - 10x + 5 = 0$

22. $x^4 - x^2 - 6 = 0$

23. $x^4 - 4x^3 + 8x^2 + 2x - 1 = 0$

24. $x^4 - 2x^3 - 2x^2 + 4x + 2 = 0$

25. $x^4 - 4x^3 - 2x^2 + 12x - 3 = 0$

26. $x^4 - 6x^3 + 10x^2 - 18x + 7 = 0$

27. The product of three integers is 10. If one integer is 4 less than the first and another is 3 less than the first, find the three integers.

28. A box is to contain 72 cubic meters volume. The width is 2 m longer than the height, and the length is 7 m longer than the height. Find the dimensions of the box.

29. A box is to contain 144 cubic centimeters volume. The width is 3 cm longer than the height, and the length is 5 cm longer than the height. Find the dimensions of the box.

RATIONAL ROOT THEOREM 3–8

With the theorems of Section 3-7 we can determine the potential integral roots and an upper and lower boundary of the real roots of a polynomial equation. The theorem of this section enables us to determine the potential rational roots of a polynomial equation.

Rational Root Theorem: If p/q in lowest terms is a rational root of a polynomial equation $P(x) = 0$, with integral coefficients, $q \neq 0$, where

$$P(x) = a_0x^n + a_1x^{n-1} + \cdots + a_n$$

then p is a factor of a_n and q is a factor of a_0.

The proof of the Rational Root Theorem is as follows. Let p/q be a rational root of the polynomial equation $a_0x^n + a_1x^{n-1} + \cdots + a_n = 0$. Furthermore, recall p and q have no common factors other than 1. Then we have

$$a_0\left(\frac{p}{q}\right)^n + a_1\left(\frac{p}{q}\right)^{n-1} + \cdots + a_n = 0$$

Multiplying both sides of the equation by q^n, we get

$$a_0p^n + a_1qp^{n-1} + \cdots + a_nq^n = 0$$

We subtract a_nq^n from both sides of the equation and factor p from each term of the left side of the equation:

$$p(a_0p^{n-1} + a_1qp^{n-2} + \cdots + a_{n-1}q^{n-1}) = -a_nq^n$$

Since p is a factor of the left side of the equation, then p is also a factor of the right side of the equation. Since p and q have no common factors, then p must be a factor of a_n. A similar argument can show that q must be a factor of a_0.

EXAMPLE 3–32 Determine the possible rational roots of the equation $12x^3 - 16x^2 - 3x + 4 = 0$.

Solution: For the given equation, $a_0 = 12$ and $a_n = 4$. Therefore, the set of potential rational roots contains all possible ratios of the factors of 4 (that is, ± 1, ± 2, and ± 4) to the factors of 12 (that is, ± 1, ± 2, ± 3, ± 4, ± 6, and ± 12). The following set is the result:

$$\left\{\pm\frac{1}{12}, \pm\frac{1}{6}, \pm\frac{1}{4}, \pm\frac{1}{3}, \pm\frac{1}{2}, \pm 1, \pm\frac{2}{3}, \pm 2, \pm\frac{4}{3}, \pm 4\right\}$$

Aren't there more ratios possible than the ones listed?

There are 8 more ratios that can be written, but each of these ratios is equal to one that is already listed. For example, $\pm 2/2 = \pm 4/4 = \pm 1$, and only the ± 1 is listed. Notice also that the set of potential integral roots $\{\pm 1, \pm 2, \pm 4\}$ is a proper subset of the set of potential rational roots.

EXAMPLE 3–33 Solve $12x^3 - 16x^2 - 3x + 4 = 0$.

Solution: The set of possible rational roots is stated in Example 3–32. We check the integers of the set for roots of the equation. We must also watch the results of the division for boundaries of the real roots.

If $x = 1$, then

$$\begin{array}{r|rrrr} 1 & 12 & -16 & -3 & 4 \\ & & 12 & -4 & -7 \\ \hline & 12 & -4 & -7 & -3 \end{array}$$

and 1 is not a root or a boundary.

If $x = 2$, then

$$\begin{array}{r|rrrr} 2 & 12 & -16 & -3 & 4 \\ & & 24 & 16 & 26 \\ \hline & 12 & 8 & 13 & 30 \end{array}$$

Thus, 2 is not a root, but 2 is an upper boundary.

Furthermore, notice that $P(1) = -3$ and $P(2) = +30$. Since the polynomial changes sign between 1 and 2, there must be a real root r and $1 < r < 2$. From the set of rational roots, 4/3 is between 1 and 2.

If $x = 4/3$, then

$$\begin{array}{r|rrrr} 4/3 & 12 & -16 & -3 & 4 \\ & & 16 & 0 & -4 \\ \hline & 12 & 0 & -3 & 0 \end{array}$$

Thus, 4/3 is a root.

If $x = 4/3$, then $3x = 4$ and $3x - 4 = 0$. Therefore, if $P(x) = 12x^3 - 16x^2 - 3x + 4$, then $3x - 4$ is a factor of the polynomial. Then we have

$$(3x - 4)(4x^2 - 1) = 0$$

or

$$(3x - 4)(2x + 1)(2x - 1) = 0$$

The solution set is $\{-1/2, 1/2, 4/3\}$, and each root is a member of the set of potential rational roots.

EXAMPLE 3–34 Solve $3x^4 - 8x^3 + 9x^2 - 32x - 12 = 0$.

Solution: With $a_0 = 3$ and $a_n = 12$, the potential rational roots are $\pm 1/3$, $\pm 2/3$, ± 1, $\pm 4/3$, ± 2, ± 3, ± 4, ± 6, and ± 12. With synthetic division we reject 1 and 2 as roots and upper boundaries of the given equation.

If $x = 3$, then

$$\begin{array}{r|rrrrr} 3 & 3 & -8 & 9 & -32 & -12 \\ & & 9 & 3 & 36 & 12 \\ \hline & 3 & 1 & 12 & 4 & 0 \end{array}$$

Thus, 3 is a root and an upper boundary for the real roots. With synthetic division we reject 1/3, 2/3, and 4/3 as roots. We now work with the polynomial equation $3x^3 + x^2 + 12x + 4 = 0$, the quotient obtained above.

If $x = -1$, then

$$\begin{array}{r|rrrr} -1 & 3 & 1 & 12 & 4 \\ & & -3 & 2 & -14 \\ \hline & 3 & -2 & 14 & -10 \end{array}$$

Thus, -1 is not a root. Since the signs of the terms in the bottom array alternate in sign, -1 is a lower boundary for the real roots. We therefore check the rational numbers $-1/3$ and $-2/3$ from the set above.

If $x = -1/3$, then

$$\begin{array}{r|rrrr} -1/3 & 3 & 1 & 12 & 4 \\ & & -1 & 0 & -4 \\ \hline & 3 & 0 & 12 & 0 \end{array}$$

and $-1/3$ is a root. Thus, we have

$$(x - 3)(x + 1/3)(3x^2 + 12) = 0$$

or

$$(x - 3)(3x + 1)(x^2 + 4) = 0$$

Since $x^2 + 4$ has no real roots, the solution set for the given equation is $\{-1/3, 3\}$.

We are now ready to practice solving polynomial equations with the five tools we have developed in this chapter:

1. Factor Theorem
2. Integral Root Theorem
3. Upper Boundary Theorem
4. Lower Boundary Theorem
5. Rational Root Theorem

Using these theorems, attack the problems in the following exercise set.

Exercise Set 3–8

List the set of possible rational roots for the given equations.

1. $2x^3 - 7x^2 + 14x - 15 = 0$
2. $6x^3 - 8x^2 + 11x + 14 = 0$
3. $3x^5 - 19x^3 + 17x - 6 = 0$
4. $2x^5 - 4x^2 + 2x + 25 = 0$
5. $5x^5 + 7x^3 - 15 = 0$
6. $4x^5 + x^4 - 8x + 10 = 0$
7. $8x^7 + 3x^5 + x^3 - 1 = 0$
8. $20x^8 - 5x^6 + x^2 - 4 = 0$

Find the real roots of the following equations and check.

9. $4x^3 + 8x^2 - x - 2 = 0$
10. $x^3 - 3x^2 - 36x - 54 = 0$
11. $2x^3 - 5x^2 + 8x - 20 = 0$
12. $x^3 + 1 = 0$
13. $7x^3 - 15x^2 - 5x + 1 = 0$
14. $4x^3 - 23x^2 + 18x + 6 = 0$
15. $8x^3 - 27 = 0$
16. $4x^3 - 24x^2 + 35x - 12 = 0$
17. $8x^3 - 38x^2 - 11x + 5 = 0$
18. $24x^3 - 14x^2 - x + 1 = 0$
19. $x^3 - 2x^2 - 23x + 60 = 0$
20. $x^4 - 3x^2 + 2 = 0$
21. $6x^3 - 53x^2 + 82x - 35 = 0$
22. $8x^3 - 12x^2 + 6x - 1 = 0$
23. $2x^4 + x^3 - 11x^2 - 5x + 5 = 0$
24. $25x^4 + 200x^3 + 396x^2 - 32x - 64 = 0$
25. $6x^4 - 5x^3 - 17x^2 + 15x - 3 = 0$
26. $2x^4 - 11x^3 + 9x^2 + 2x - 2 = 0$
27. $x^5 - 2x^4 - 2x^3 + 4x^2 - 3x + 6 = 0$
28. $2x^5 + 4x^4 + 7x^3 + 14x^2 - 4x - 8 = 0$
29. $x^6 + x^5 - 9x^4 - 17x^3 - 8x^2 + 72x + 72 = 0$
30. $64x^5 + 272x^4 + 588x^3 + 675x^2 + 378x + 81 = 0$

SUMMARY

Quadratic Formula: The roots of any quadratic equation $ax^2 + bx + c = 0$ (where a, b, and c are real numbers and $a \neq 0$) are of the form

$$\frac{-b \pm \sqrt{b^2 - 4ac}}{2a}$$

Theorem: If r and s are roots of the quadratic equation $ax^2 + bx + c = 0$, then $r + s = -b/a$ and $rs = c/a$.

Theorem: If r and s are the roots of $ax^2 + bx + c = 0$, then

(1) r and s are real and unequal if $b^2 - 4ac > 0$.
(2) r and s are real and equal if $b^2 - 4ac = 0$.
(3) r and s are not real and unequal if $b^2 - 4ac < 0$.

Quadratic inequality

Definition: A quadratic inequality is a statement equivalent to $ax^2 + bx + c < 0$, $ax^2 + bx + c > 0$, $ax^2 + bx + c \leq 0$, or $ax^2 + bx + c \geq 0$, where a, b, and c are real numbers and $a \neq 0$.

Theorem: If a and b are any nonzero real numbers, then
(1) $ab > 0$ if and only if $a > 0$ and $b > 0$, or $a < 0$ and $b < 0$.
(2) $ab < 0$ if and only if $a > 0$ and $b < 0$, or $a < 0$ and $b > 0$.

Absolute value

Definition: If a and b are two real numbers, then $|a - b| = a - b$ if $a \geq b$, and $|a - b| = b - a$ if $a < b$.

Theorem: If a and b are any real numbers, then $|a - b| = |b - a|$.

Theorem: For any real number a, $|a| = a$ if $a \geq 0$, and $|a| = -a$ if $a < 0$.

Factor Theorem: If $P(x)$ is a polynomial and c is a real number, then $x - c$ is a factor of $P(x)$ if and only if c is a root of $P(x) = 0$.

Integral Root Theorem: If $P(x)$ is a polynomial with integral coefficients and

$$P(x) = a_0x^n + a_1x^{n-1} + \cdots + a_n$$

and c is an integral root of $P(x) = 0$, then c is a factor of a_n.

Upper Boundary Theorem: Let $P(x) = 0$ be a polynomial equation with a positive leading coefficient, and let k be a nonnegative real number. If the coefficients of the quotient and remainder obtained when $P(x)$ is divided by $x - k$ are all positive numbers or zero, then k is an upper boundary for the real roots of the given equation.

Lower Boundary Theorem: Let $P(x) = 0$ be a polynomial equation with a positive leading coefficient, and let m be a nonpositive real number. If the coefficients of the quotient and remainder obtained when $P(x)$ is divided by $x - m$ are alternately nonnegative and nonpositive numbers, then m is a lower boundary for the real roots of the given equation.

Rational Root Theorem: If p/q in lowest terms is a rational root of a polynomial equation $P(x) = 0$, with integral coefficients, $q \neq 0$, where

$$P(x) = a_0x^n + a_1x^{n-1} + \cdots + a_n$$

then p is a factor of a_n and q is a factor of a_0.

SELF-CHECKING EXERCISE

Find the real roots of the following equations. Check solutions with the roots theorem. If an equation has no real roots, write "none" for the solution set. **(3-1)**

1. $2x^2 + 11x = 21$
2. $2 - x - 4x^2 = 0$
3. $x^2 = 2x - 4$
4. $(x - 3)^2 = 2(5 - x)$

Determine the kinds of roots for the following equations. **(3-1)**

5. $9x^2 + 1 = 2x$
6. $5x^2 + 4/5 = 4x$

Determine a quadratic equation with the given roots as solutions. **(3-1)**

7. $4/5, -1/3$
8. $3 \pm \sqrt{2}$

Solve and check the following equations. **(3-2)**

9. $\dfrac{x}{x+3} + \dfrac{4}{2x-1} = 1$
10. $\dfrac{x^2}{4x^2 - 12x + 9} - \dfrac{3}{2x-3} = 7$
11. $\sqrt[3]{8-5x} + 3 = 0$
12. $\sqrt{x-2} = \sqrt{3x-2} - 2$
13. $3x^{1/3} = 2x^{2/3} - 5$
14. $3x^{1/2} - 14x^{1/4} + 8 = 0$

State and graph the solution sets of the following inequalities. **(3-3)**

15. $(2x-5)(3x+4) < 0$
16. $2x^2 + x \geq 6$
17. $\dfrac{x^2}{2} + \dfrac{31x}{8} \leq 1$
18. $2x^2 > 2x + 1$

State and graph the solution sets of the following inequalities. **(3-4)**

19. $x^3 - 2x^2 - 15x > 0$
20. $\dfrac{7x}{2x-3} \leq 2$
21. $\dfrac{x^2 - 9x + 20}{x+3} \leq 0$
22. $3x^2 > 2x - 1$

Evaluate the following expressions for $x = -1$, $y = -3$, and $z = 5$. **(3-5)**

23. $|3x - y + 2z|$
24. $|5x + z| \cdot |10y|$
25. $\left|\dfrac{7y}{z - 2x}\right|$
26. $|8x - y^2 - 3z|$

Solve the following absolute value equations. **(3-5)**

27. $|2x - 5| = 9$
28. $|5 + x| = 10$

Solve and graph the following absolute value inequalities. **(3-5)**

29. $|3x + 4| < 8$
30. $3|1 - 5x| \geq 12$

Let $P(x) = 3 - 2x + x^3$. **(3-6)**

31. Find $P(3)$
32. Find $P(-4)$

Find all real roots of the following polynomials, given that the numbers in braces are roots of the given polynomial. **(3-6)**

33. $2x^3 + 3x^2 - 14x + 6 = 0 \quad \{1/2\}$
34. $4x^4 - 25x^2 + 36 = 0 \quad \{-2, 3/2\}$

Completely factor the following polynomials with the given factors. **(3-6)**

35. $6x^3 - 31x^2 + 3x + 10 \quad (x - 5)$
36. $3x^4 - 7x^3 - 55x^2 + 79x - 20 \quad (x - 1), (x + 4)$

Solve and check the following polynomials. **(3-7)**

37. $x^3 - x^2 - 14x + 24 = 0$
38. $6x^4 - 5x^3 - 30x^2 + 20x + 24 = 0$
39. Find the value of k such that $x^3 + kx^2 - 8x + 12$ has $x - 2$ as a factor. **(3-7)**

List the set of possible rational roots of the following equations. **(3-8)**

40. $4x^3 - 2x + 3 = 0$

41. $6x^4 + 17x^2 - 10x = 0$

Find the real roots of the following equations and check. **(3-8)**

42. $12x^3 - 16x^2 - 5x + 3 = 0$

43. $24x^4 + 70x^3 + 21x^2 = 10x$

Answers to Self-Checking Exercise

1. $\{-7, 3/2\}$

2. $\left\{\frac{-1 \pm \sqrt{33}}{8}\right\}$

3. none

4. $\{2 \pm \sqrt{5}\}$

5. two unreal and unequal

6. two real and equal

7. $15x^2 - 7x - 4 = 0$

8. $x^2 - 6x + 7 = 0$

9. $\{15/2\}$

10. $\left\{\frac{13 \pm \sqrt{7}}{9}\right\}$

11. $\{7\}$

12. $\{2, 6\}$

13. $\{-1, 125/8\}$

14. $\{16/81, 256\}$

15. $(-4/3, 5/2)$

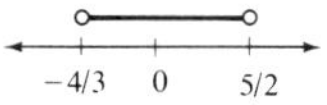

16. $(-\infty, -2] \cup [3/2, \infty)$

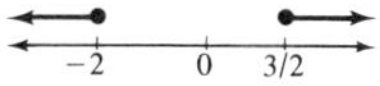

17. $[-8, 1/4]$

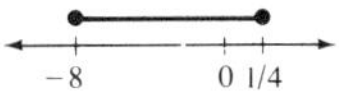

18. $\left(-\infty, \frac{1-\sqrt{3}}{2}\right) \cup \left(\frac{1+\sqrt{3}}{2}, \infty\right)$

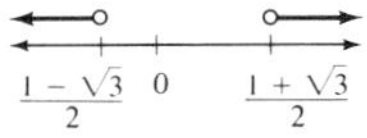

19. $(-3, 0) \cup (5, \infty)$

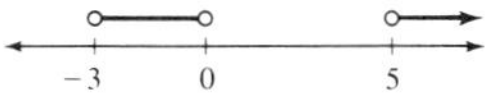

20. $[-2, 3/2)$

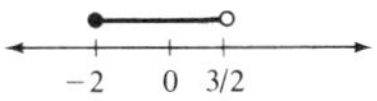

21. $(-\infty, -3) \cup [4, 5]$

22. $x \in$ reals

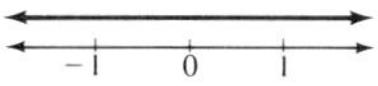

23. 10

24. 0

25. 3

26. 32

27. $\{-2, 7\}$

28. $\{-15, 5\}$

29. $(-4, 4/3)$

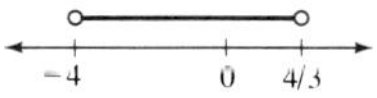

30. $(-\infty, -3/5] \cup [1, \infty)$

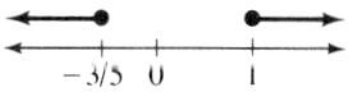

31. 24

32. -53

33. $\{1/2, -1 \pm \sqrt{7}\}$

34. $\{\pm 3/2, \pm 2\}$

35. $(x - 5)(3x - 2)(2x + 1)$

36. $(x - 5)(3x - 1)(x - 1)(x + 4)$

37. $\{-4, 2, 3\}$

38. $\{\pm 2, -2/3, 3/2\}$

39. $k = -1$

40. $\{\pm 1, \pm 1/2, \pm 1/4, \pm 3, \pm 3/2, \pm 3/4\}$

41. $\{\pm 1, \pm 1/2, \pm 1/3, \pm 1/6, \pm 2, \pm 2/3, \pm 5, \pm 5/2, \pm 5/3, \pm 5/6, \pm 10, \pm 10/3\}$

42. $\{-1/2, 1/3, 3/2\}$

43. $\{-5/2, -2/3, 0, 1/4\}$

Relations and Functions

4

An important subject in mathematics is relations and functions. In Chapter 1 the topic was introduced informally with polynomials. For example, the equation $P(x) = x^2 + 3$ implies that the value of $P(x)$ *is related to* or *is a function of* a replacement value for x.

In this chapter we consider a detailed analysis of relations and the terminology connected with relations. We review one of the coordinate systems used for graphing a relation, and we study some of the basic mathematical relations. Two sections on an operation that is unique to relations complete the chapter.

RELATIONS, FUNCTIONS, AND THE COORDINATE SYSTEM 4-1

Relations and Functions

In everyday conversation we often say that *pairing relationships* exist between two people, two animals, or two of almost anything. For example, the relationship "is a mother of" pairs a mother y with a child x. Or the relationship "is a husband of" pairs a husband y with his wife x. A pairing relationship between numbers is the intended meaning of a mathematical relation.

Relation

> **Definition:** A relation is a set of ordered pairs of numbers.

A common way of writing an ordered pair relationship is $\{(x, y)\}$, which is read "the set of ordered pairs x and y." Then y is the number that is paired with a given number x.

Domain

> **Definition:** The domain is the set of all first elements of a relation.

The variable that represents all the domain elements is called the **independent variable.**

Range

> **Definition:** The range is the set of all second elements of a relation.

The variable that represents all the range elements is called the **dependent variable.**

Function

> **Definition:** A function is a relation such that each domain element is paired with one and only one range element.

Therefore, if the relation $R = \{(x, y)\}$, then R is a function if for any x in the domain of R there is one and only one y in the range of R.

EXAMPLE **4–1** For the relations R and S below, (1) state the domain, (2) state the range, and (3) determine whether the relation is a function.
(a) $R = \{(0, 3), (1, 5), (2, 7), (3, 9)\}$
(b) $S = \{(-5, -2), (-1, 5), (0, 7), (-1, -3), (2, 6)\}$

Solutions:
(a) 1. The domain of $R = \{0, 1, 2, 3\}$.
2. The range of $R = \{3, 5, 7, 9\}$.
3. Since each element in the domain is paired with one and only one element in the range, R is a function.

(b) 1. The domain of $S = \{-5, -1, 0, 2\}$.
2. The range of $S = \{-3, -2, 5, 6, 7\}$.
3. Since -1 is paired with -3 and 5, S is not a function.

Is a listing of the ordered pairs the only way to write a relation?

Listing the ordered pairs of a relation is the least familiar way of writing a relation. Usually a **formula** is used to indicate the way the two variables are related to each other. For example, $R = \{(x, y) \mid y = 3x + 5\}$, which is read "the set of x and y's such that $y = 3x + 5$," identifies the elements of R with the equation (or formula) $y = 3x + 5$. Thus, each range element is 5 more than 3 times the corresponding domain element.

A third way of expressing a relation is with a graph of the relation. In this chapter we will practice expressing a given relation in all three ways. For example, we will graph relations that are identified by a formula. And, we will find a formula for a relation that is identified by a set of ordered pairs. In other words, we will see relations identified by each of the three methods given: (1) a set of ordered pairs, (2) a formula, and (3) a graph.

Rectangular Coordinate System

In Figure 4-1, two intersecting number lines are drawn to form a rectangular coordinate system. We can make the following observations relative to this system:

1. If a relation $R = \{(x, y)\}$ is graphed on the system, then the horizontal number line is labeled the x-axis and the vertical number line is the y-axis.

2. The point of intersection, O, of the number lines is the **origin** of the coordinate system and is labeled $O(0, 0)$.
3. If $P(x, y)$ is a point in the system, then x is the **horizontal displacement** and y is the **vertical displacement** from $O(0, 0)$ to the point P.
4. The system contains four **quadrants** labeled I through IV counterclockwise from the upper right quadrant.
5. If $P_1(x_1, y_1)$ is in I, then $x_1 > 0$ and $y_1 > 0$.
6. If $P_2(x_2, y_2)$ is in II, then $x_2 < 0$ and $y_2 > 0$.
7. If $P_3(x_3, y_3)$ is in III, then $x_3 < 0$ and $y_3 < 0$.
8. If $P_4(x_4, y_4)$ is in IV, then $x_4 > 0$ and $y_4 < 0$.
9. If $P_5(x_5, 0)$ is on the x-axis, then $x_5 \in$ reals and $y = 0$.
10. If $P_6(0, y_6)$ is on the y-axis, then $x = 0$ and $y_6 \in$ reals.

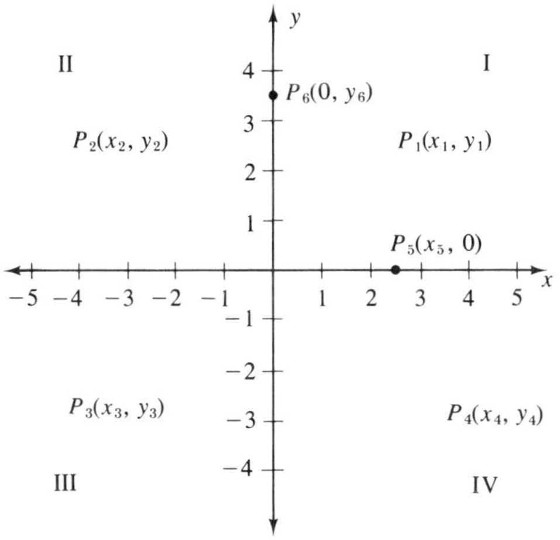

FIGURE 4-1 Rectangular coordinate system

The domain of a relation is shown in the **horizontal extent** of the graph. The range of a relation is shown in the **vertical extent** of the graph.

The **vertical line test** on the graph of a relation enables us to determine whether the graph is that of a function. *If no vertical line intersects the graph of a relation in more than one point, then the relation is a function.*

EXAMPLE 4–2 From the following graphs, (1) determine the domain, (2) determine the range, and (3) determine whether the relation is a function.

Solutions:

(a) The graph of R is shown in Figure 4-2.

1. The horizontal extent is from negative to positive infinity; thus, the domain of $R = \{x \mid x \in \text{reals}\}$.
2. The vertical extent is from negative to positive infinity; thus, the range of $R = \{y \mid y \in \text{reals}\}$.

3. No vertical line can intersect the graph of R in more than one point; thus, R is a function.

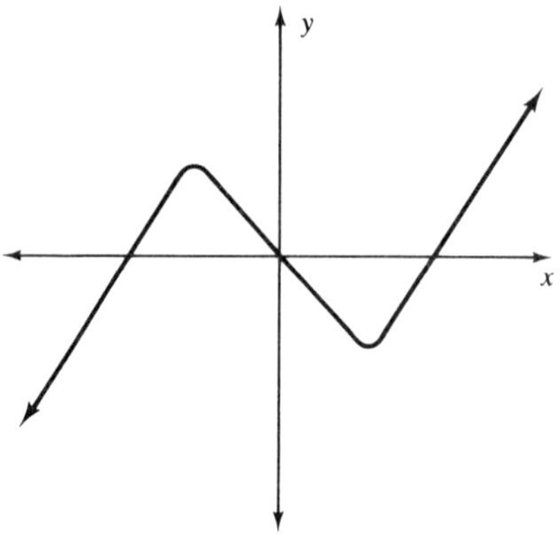

FIGURE 4-2 Graph of relation R

(b) The graph of S is shown in Figure 4-3.

1. The horizontal extent excludes the interval from $x = -3$ to $x = 3$; thus, the domain of $S = \{x \mid x \leq -3 \text{ or } x \geq 3\}$.
2. The vertical extent is from negative to positive infinity; thus, the range of $S = \{y \mid y \in \text{reals}\}$.
3. Any vertical line for $x > 3$ or $x < -3$ intersects the graph of S in two points; thus, S is not a function.

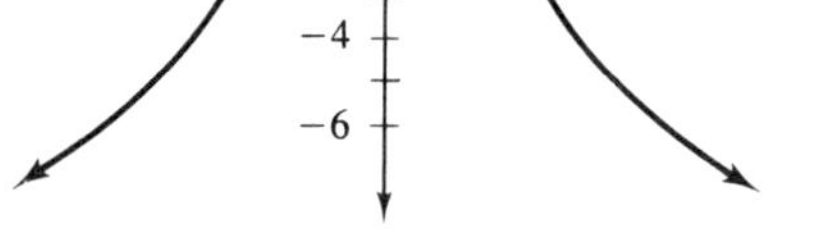

FIGURE 4-3 Graph of relation S

How do we determine the domain and range of a relation identified by a formula?

To determine the domain and range of a relation identified by a formula, we use the fact that *the variables in the formula can represent only real numbers.* Therefore, the independent variable is restricted to only those real numbers for which a real number is obtained for the dependent variable. The range is the set of values assumed by the dependent variable for every value of the independent variable in the domain.

EXAMPLE 4–3 If $R = \{(x, y) \mid y = x^2 - 6x\}$, (a) determine the domain of R, (b) determine the range of R, and (c) determine whether R is a function.

Solutions:

(a) Every real number value for x yields a real number for y; thus, the domain of $R = \{x \mid x \in \text{reals}\}$.

(b) If $y = x^2 - 6x$, then $x^2 - 6x - y = 0$. With $a = 1$, $b = -6$, and $c = -y$, then

$$x = \frac{6 \pm \sqrt{36 + 4y}}{2} = 3 \pm \sqrt{9 + y}$$

Now, x is a real number if and only if $9 + y \geq 0$; that is, $y \geq -9$. Thus, the range of $R = \{y \mid y \geq -9\}$.

(c) For any value of x in the domain of R, one and only one value of y is obtained; thus, R is a function.

> **Notation:** If y is a function of x, then we write $f(x)$ (read "f of x") for y.

The relation in Example 4–3 is a function; therefore,

$$R = \{(x, f(x)) \mid f(x) = x^2 - 6x\}$$

There are several advantages to using the symbol $f(x)$ rather than the variable y. First, the notation readily identifies the independent and dependent variables in a relationship. For example, if $f(x) = \dfrac{x + 1}{x - 1}$, then x is the independent variable and $f(x)$ is the dependent variable. In the relationship $f(t) = \sqrt{t^2 - 1}$, t is the independent variable and $f(t)$ is the dependent variable.

Second, a specific value of the function is readily paired with any value of the independent variable. For example, if $f(x) = x^2 - 6x$, then for any real number a in the domain of the function, $f(a)$ is the value of $f(x)$ when x is everywhere replaced with a. That is, $f(a) = a^2 - 6a$. In particular, if $x = 2$, then $f(2) = 2^2 - 6(2) = -8$. If $x = -5$, then $f(-5) = (-5)^2 - 6(-5) = 55$.

EXAMPLE 4–4 If $R = \left\{(a, f(a)) \mid f(a) = \dfrac{a}{a - 1}\right\}$, find

(a) $f(0)$ (b) $f(3)$ (c) $f(a + h)$

Solutions:

(a) $f(0) = \dfrac{0}{0 - 1} = 0$

(b) $f(3) = \dfrac{3}{3 - 1} = \dfrac{3}{2}$

(c) $f(a + h) = \dfrac{a + h}{a + h - 1}$

EXAMPLE 4–5 If $T = \left\{(x, y) \mid y = \dfrac{x - 1}{x + 1}\right\}$, (a) determine the domain of T, (b) determine the range of T, and (c) determine whether T is a function.

Solutions:
(a) T is a rational expression for which the denominator cannot be zero. If $x + 1 \neq 0$, then $x \neq -1$; and the domain of $T = \{x \mid x \in \text{reals}, x \neq -1\}$.
(b) If $y = \dfrac{x-1}{x+1}$, then

$$xy + y = x - 1 \quad \text{and} \quad y + 1 = x - xy$$

or

$$x(1 - y) = y + 1 \quad \text{and} \quad x = \frac{1+y}{1-y}$$

Since $1 - y \neq 0$, then $y \neq 1$; and the range of $T = \{y \mid y \in \text{reals}, y \neq 1\}$.
(c) Since any value of x yields one and only one value of y, T is a function.

In Example 4–5 we determined the range of T by solving the equation for x. However, we cannot always readily solve the equation of a function for the independent variable. In such a case we attempt to analyze the equation to determine whether there exist real values that the dependent variable cannot assume. For example, if $y = x^3 + x$, then we cannot easily solve the equation for x. But we can reason that for any real number y — both positive and negative — there does exist a real number x. Thus, the range of the function is the set of real numbers. If an analysis of the equation is too difficult, then you might try a sketch of the graph of the function.

Since polynomials are so frequently studied in mathematics, we state the following theorem:

Theorem: If P is a polynomial in x, then

(1) The domain of $P = \{x \mid x \in \text{reals}\}$.

(2) P is a function.

The range of P must be determined by analyzing the equation or by sketching a graph of the polynomial.

Exercise Set 4–1

For the following relations, (a) determine which relations are functions, and (b) find the domain and range of each set.

1. $\{(1, 2), (2, 3), (3, 4), (4, 0)\}$
2. $\{(-1, 2)\ (2, 4)\ (3, 5), (-1, 3)\}$
3. $\{(1/2, 3/4), (-1/2, 3/4), (2, 3/4), (0, 3/4)\}$
4. $\{(1, 1), (1, 2), (1, 3), (1\ ,4)\}$
5. $\{(-2, -2), (3, 3), (4, 4), (1, 1)\}$
6. $\{(-3, 1), (1, -3), (-2, -7), (-7, -2)\}$

Using the following graphs, (a) determine whether the graphs are functions, and (b) find the domain and range of each.

7.

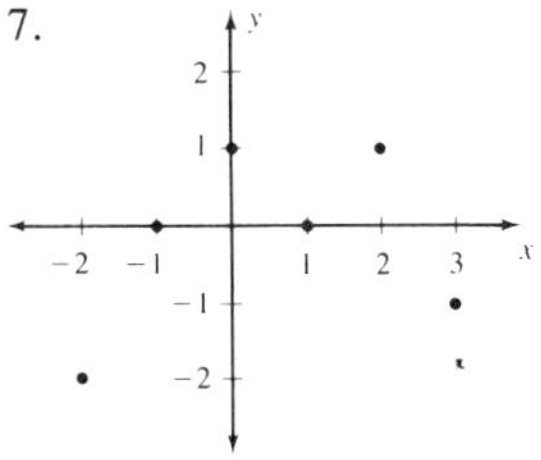

8.

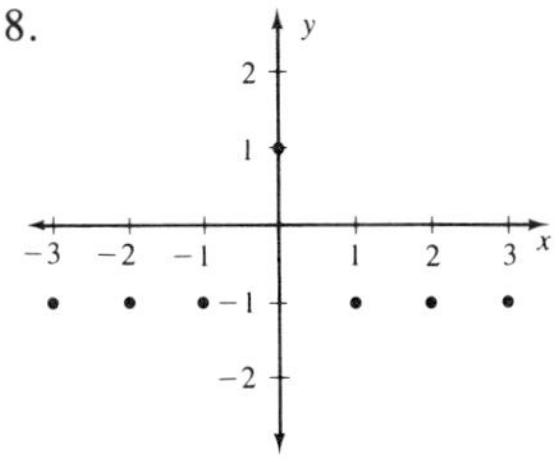

9.

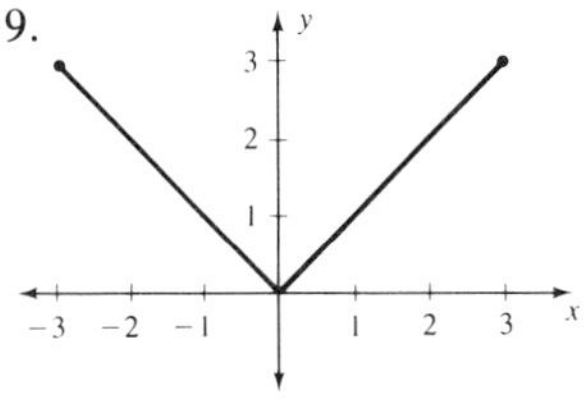

10.

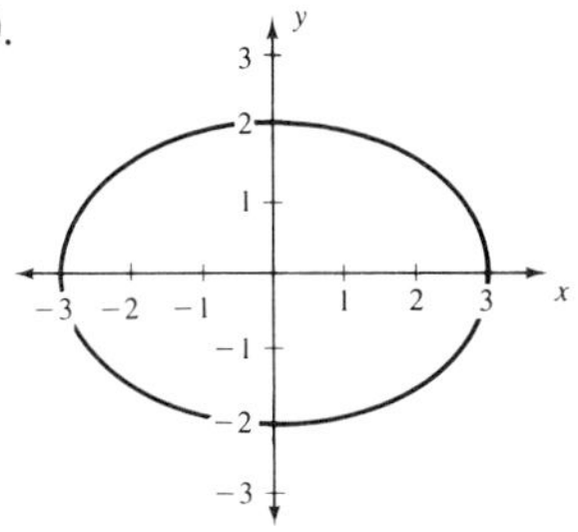

11.

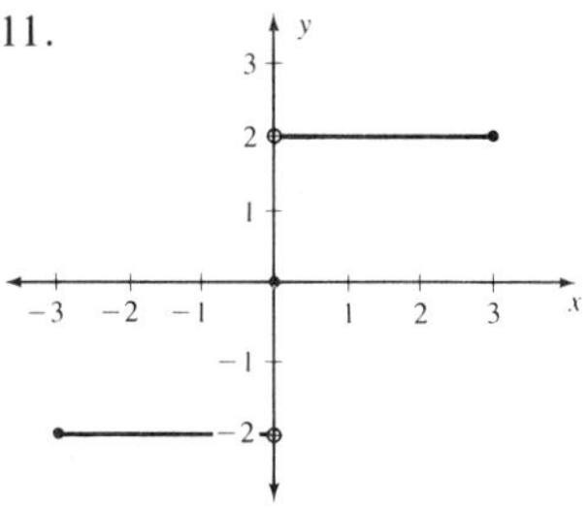

12.

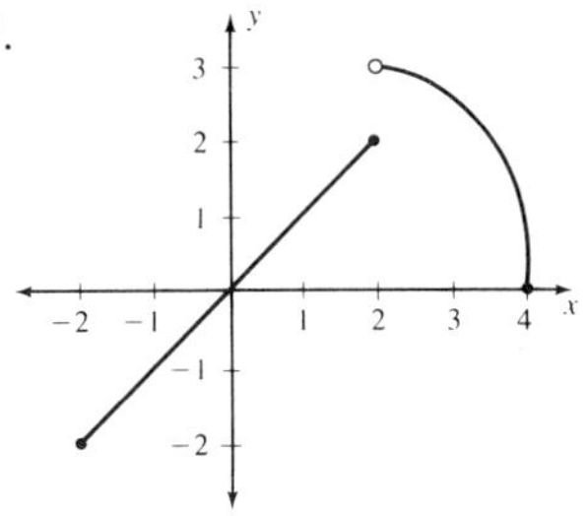

13.

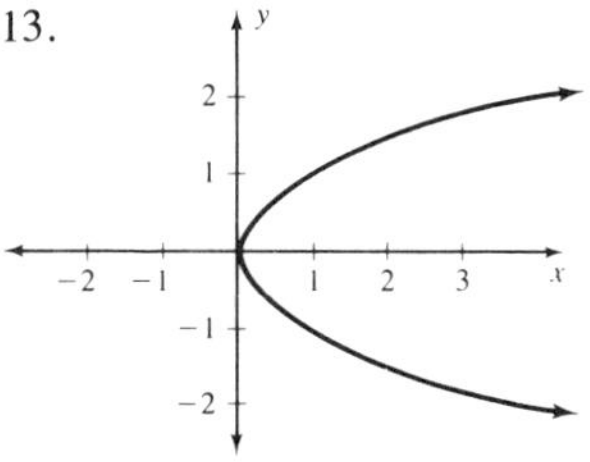

14.

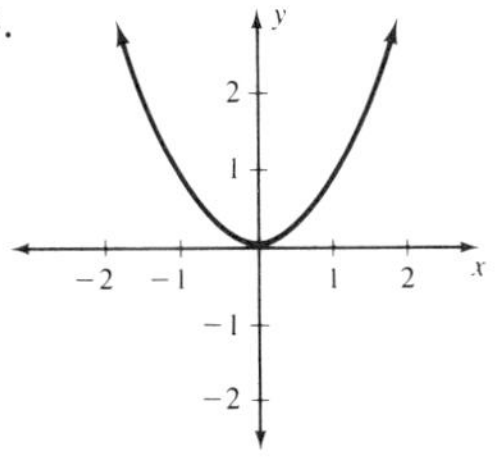

15.

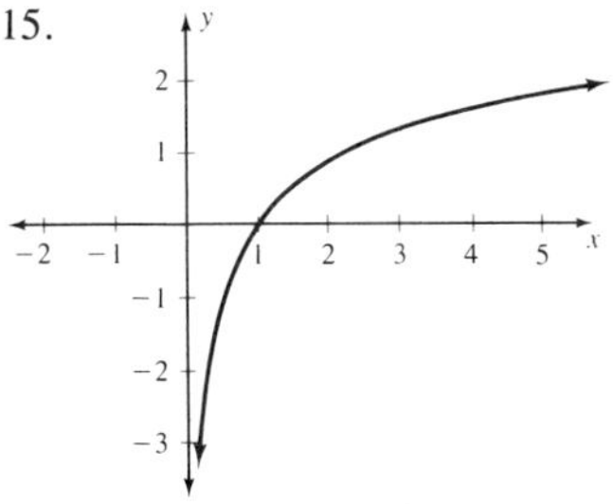

16.

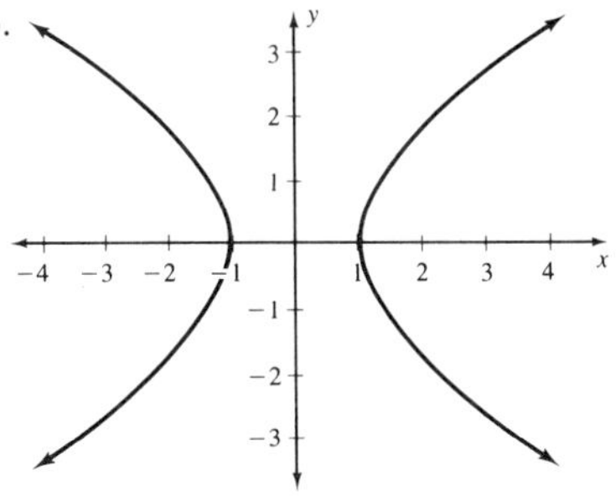

For the following relations, (a) determine the domain and range and (b) state whether or not the relation is a function.

17. $A = \{(x, y) \mid y = x^2\}$

18. $B = \{(x, y) \mid x = \sqrt{y}\}$

19. $C = \{(x, y) \mid x^2 - y^2 = 1\}$

20. $D = \left\{(x, y) \mid y = \dfrac{1}{x - 2}\right\}$

21. $E = \left\{(x, y) \mid y = \dfrac{x}{x^2 - 4}\right\}$

22. $F = \{(x, y) \mid x^2 + y^2 = 4\}$

23. $G = \left\{(x, y) \mid \dfrac{x^2}{4} + \dfrac{y^2}{9} = 1\right\}$

24. $H = \{(x, y) \mid y = (x - 1)^3\}$

25. $I = \left\{(x, y) \mid y = \dfrac{1}{x^2 + 1}\right\}$

26. $J = \{(x, y) \mid y = 2x - 5\}$

If $f(x) = \sqrt{\dfrac{x - 1}{x + 1}}$ and $g(x) = \dfrac{x^2}{2 - x}$, find

27. $f(2)$

28. $f(1)$

29. $f(-4)$

30. $g(0)$

31. $g(1)$

32. $g(3)$

33. $g(a)$

34. $g(a + h)$

35. $g(a + h) - g(a)$

36. $g(a^2)$

37. $g(2 - a)$

38. For the function $f(x) = \sqrt{x^2 - 3}$, determine the domain and range.

39. Determine the domain and the range of the function

$$h(x) = \sqrt{\frac{x + 2}{x - 4}}$$

4-2 CURVE SKETCHING

It has been said that a picture is sometimes worth a thousand words. The statement is very true when we work with relations and functions. A sketch of a relation is frequently a very valuable — and necessary — tool. We therefore set aside this section to discuss graphing.

A sketch of a graph implies a general picture that indicates the pattern of a relation's behavior. A detailed graph of the relation is not intended. A few details relevant to the graph of a relation are often enough to make the sketch. We will study each of these details separately and then illustrate them with an example. By considering all the keys discussed, we then analyze a relation and make a sketch of the relation.

x-intercepts

If a relation has one or more x-intercepts, then knowing these values will help us sketch the graph of the relation.

x-intercepts

> **Definition:** If $R = \{(x, y) \mid y = r(x)\}$, then the x-intercepts of R are the real values of x for which $r(x) = 0$.

We have called these real values of x the roots, or zeroes, of an equation $r(x) = 0$. For the relation R, if k is a real number such that $r(k) = 0$, then k is an x-intercept of the relation R, and $P(k, 0)$ is a point of the graph of R. Note that an x-intercept is a real number, not a point. But an x-intercept when paired with zero is a point on the graph.

Aren't these values of x the roots of an equation?

EXAMPLE 4–6 Determine the x-intercepts of the relations defined by the given equations.

(a) $y = \dfrac{1}{x - 1}$ (b) $y = \dfrac{x^2 - 1}{x}$

Solutions:

(a) If $y = 0$, then $\dfrac{1}{x - 1} = 0$. The solution set is empty, since a rational expression can only be 0 if the numerator is 0. If the denominator is 0, the expression is undefined. Thus, the relation defined by the given equation has no x-intercepts.

(b) If $y = 0$, then $\dfrac{x^2 - 1}{x} = 0$. The roots of $x^2 - 1 = 0$ are 1 and -1; thus, the x-intercepts of the relation are ± 1, and $P_1(1, 0)$ and $P_2(-1, 0)$ are points of the graph.

y-intercepts

The y-intercepts of a relation, if they exist, are also valuable tools for sketching the graph of the relation.

y-intercepts

> **Definition:** If $R = \{(x, y) \mid y = r(x)\}$, then the y-intercepts of R are the real values of y for which $x = 0$.

If k is a y-intercept of R, then $P(0, k)$ is a point of the graph of R.

EXAMPLE 4–7 Determine the y-intercepts of the relations defined by the following equations:

(a) $y = \dfrac{4 - x}{x - 2}$ (b) $y^2 = \dfrac{18}{x^2 + 2}$

Solutions:

(a) If $x = 0$, then $y = 4/-2 = -2$. Thus, the y-intercept is -2, and $P(0, -2)$ is a point of the graph of the relation.

(b) If $x = 0$, then $y^2 = 9$ and $y = \pm 3$. Thus, the y-intercepts are 3 and -3, and $P_1(0, 3)$ and $P_2(0, -3)$ are points of the graph of the relation.

Vertical Asymptotes

When the graph of a relation has one or more vertical asymptotes, locating these vertical lines is very important to sketching the graph of the relation.

Vertical asymptote

> The line $x = a$ is a vertical asymptote of the graph of a relation R if the values of $r(x)$ become infinitely large through positive values or infinitely small through negative values, or both, as the values of x approach a.

What do you mean by "the values of x approach a"?

To explain what we mean by "the values of x approach a," let us consider a relation

$$R = \left\{(x, y) \mid y = \frac{x}{x - 2}\right\}$$

The domain of $R = \{x \mid x \in \text{reals}, x \neq 2\}$. Thus, $x \neq 2$. But what happens to the values of y as x takes on values closer and closer to 2? The following table contains several values of y for x "near" 2.

x	1.000	1.500	1.900	1.990	1.999	3.000	2.500	2.100	2.010	2.001
y	−1	−3	−19	−199	−1999	3	5	21	201	2001

Notice that as the values of x approach 2 from the negative side of 2 (written $x \to 2^-$), the values of y become smaller and smaller through negative values. The closer the value of x is to 2 and on the negative side of 2, the smaller is the corresponding value of y. We say that *the values of y approach negative infinity as $x \to 2^-$*, and we write $y \to -\infty$.

As the values of x approach 2 from the positive side of 2 (written $x \to 2^+$), the values of y become larger and larger through positive values. The closer the value of x is to 2 and on the positive side of 2, the larger is the corresponding value of y. We say that *the values of y approach positive infinity as $x \to 2^+$*, and we write $y \to +\infty$.

Therefore, the vertical line $x = 2$ is a vertical asymptote for the graph of R. The phrase "x approaches a" means that the values of x get as close to a "as you please," but x never actually assumes the value of a.

To summarize, if R is a relation defined by the equation $y = r(x)$, then the graph of R has a vertical asymptote at $x = a$ if

$$r(x) \to +\infty \text{ as } x \to a$$

or

$$r(x) \to -\infty \text{ as } x \to a$$

or both of the above. To find the values of a for which $x = a$ is a vertical asymptote, we obtain values of x for which $r(x)$ is undefined or does not exist.

EXAMPLE 4–8 Determine the equation of any vertical asymptote for the graphs of the following relations:

(a) $R = \left\{(x, r(x)) \mid r(x) = \dfrac{x}{x^2 - 4}\right\}$

(b) $S = \left\{(x, s(x)) \mid s(x) = \dfrac{x^2 - 9}{x - 3}\right\}$

(c) $T = \left\{(x, t(x)) \mid t(x) = \sqrt{x - 1}\right\}$

Solutions:

(a) $r(x)$ is undefined when the denominator of $\frac{x}{x^2 - 4}$ is equal to 0. If $x^2 - 4 = 0$, then $(x + 2)(x - 2) = 0$ and $x = \pm 2$. The domain of $R = \{x | x \in \text{reals}, x \neq \pm 2\}$.

$$\text{As } x \to -2^-, \frac{x}{x^2 - 4} \to -\infty$$

$$\text{As } x \to -2^+, \frac{x}{x^2 - 4} \to +\infty$$

Thus, the line $x = -2$ is a vertical asymptote of the graph of R. A similar analysis shows that the line $x = 2$ is also an asymptote.

(b) $s(x)$ is undefined when the denominator of $\frac{x^2 - 9}{x - 3}$ is equal to 0. If $x - 3 \neq 0$, then $x \neq 3$ and the domain of $S = \{x \mid x \in \text{reals}, x \neq 3\}$. However, for $x \neq 3$,

$$\frac{x^2 - 9}{x - 3} = \frac{(x + 3)(x - 3)}{(x - 3)} = x + 3$$

As $x \to 3^-$ or $x \to 3^+$, $x + 3 \to 6$. Since the values of $s(x)$ approach a finite value as $x \to 3$, the graph of S does not have a vertical asymptote at $x = 3$.

(c) If $t(x) = \sqrt{x - 1}$, then $y \in$ reals for $x \geq 1$. However,

$$\sqrt{x - 1} \to 0 \text{ as } x \to 1^+$$

and

$$\sqrt{x - 1} \text{ does not exist as } x \to 1^-$$

Therefore, the graph of T does not have a vertical asymptote.

Horizontal Asymptotes

When the graph of a relation has one or more horizontal asymptotes, we can usually determine the location of the graph for very large or very small values of the independent variable.

Horizontal asymptote

> The line $y = h$ is a horizontal asymptote of the graph of a relation R if the values of $r(x)$ approach h as $|x|$ becomes infinitely large.

In symbols we write that the line $y = h$ is a horizontal asymptote if

$$r(x) \to h \text{ as } x \to +\infty$$

or

$$r(x) \to h \text{ as } x \to -\infty$$

or both of the above. To determine the horizontal asymptotes of a graph we need to study the behavior of $r(x)$ as x becomes positively or negatively infinite.

If $y = k/x$, where k is any nonzero real number, then the following table demonstrates the behavior of the values of y for larger and larger values of x.

x	1	10	100	1000	10,000
y	k	$\frac{k}{10}$	$\frac{k}{100}$	$\frac{k}{1000}$	$\frac{k}{10{,}000}$

Notice that as $x \to +\infty$, then $k/x \to 0$. Similarly, as $x \to -\infty$, then $k/x \to 0$. Furthermore, if $y = \frac{k}{x^n}$ and n is any positive real number, then $\frac{k}{x^n} \to 0$ as $x \to \infty$.

The following example illustrates the procedure for determining the horizontal asymptotes of certain functions. In the process we multiply numerator and denominator by $\frac{1}{x^n}$, where n is the degree of the denominator of the expression. Any term of the form $\frac{k}{x^n}$ in the function will approach 0 as we let $x \to \infty$.

EXAMPLE 4–9 Determine the equation of any horizontal asymptote of the graph of the following relations.

(a) $R = \left\{(x, r(x)) \mid r(x) = \frac{3x + 1}{5 - x}\right\}$

(b) $S = \left\{(x, s(x)) \mid s(x) = \frac{x + 2}{\sqrt{x}}\right\}$

(c) $T = \left\{(x, y) \mid y^2 = \frac{4x^2 - 1}{x^2}\right\}$

Solutions:

(a) The degree of the denominator is 1. Therefore, we multiply the numerator and denominator of $r(x)$ by $1/x$, and

$$r(x) = \frac{(3x + 1)\left(\frac{1}{x}\right)}{(5 - x)\left(\frac{1}{x}\right)} = \frac{3 + \frac{1}{x}}{\frac{5}{x} - 1}$$

As $x \to \infty$, $1/x \to 0$ and $5/x \to 0$, and

$$r(x) \to \frac{3 + 0}{0 - 1} \quad \text{or} \quad r(x) \to -3$$

The line $r(x) = -3$ is a horizontal asymptote of the graph of R.

(b) The degree of the denominator is 1/2. Therefore, we have

$$s(x) = \frac{(x + 2)\left(\frac{1}{\sqrt{x}}\right)}{\sqrt{x}\left(\frac{1}{\sqrt{x}}\right)} = \frac{\sqrt{x} + \frac{2}{\sqrt{x}}}{1}$$

As $x \to +\infty$, $2/\sqrt{x} \to 0$, but $\sqrt{x} \to +\infty$. Therefore, the graph of S has no horizontal asymptote.
(c) The degree of the denominator is 2. Therefore, we have

$$y^2 = 4 - \frac{1}{x^2}$$

As $x \to \infty$, $\left(4 - \frac{1}{x^2}\right) \to 4$. Thus, $y^2 \to 4$ and $y \to \pm 2$; and the lines $y = 2$ and $y = -2$ are horizontal asymptotes of the graph of T.

Symmetry

Although the graph of a relation can possess different kinds of symmetry, we shall consider only symmetry with respect to the vertical axis.

Symmetry with respect to the y-axis

> **Definition:** If R is a relation defined by the equation $y = r(x)$, and for every x in the domain of R it is true that $r(-x) = r(x)$, then the graph of R is symmetric with respect to the y-axis.

When the graph of a relation R is symmetric with respect to the y-axis, then the work of graphing R is cut in half. For example, if we determine that $P_1(a, r(a))$ is a point on the graph of R, and R is symmetric with respect to the y-axis, then we know that $P_2(-a, r(a))$ is also a point of the graph of R. Thus, every time we find one point on the graph of R, we automatically find a second point as a bonus.

EXAMPLE 4–10 Without graphing, determine whether the graphs of the following relations have symmetry with respect to the y-axis.
(a) $R = \{(x, y) \mid y = x^4 - 3x^2 + 1\}$ (b) $S = \{(x, y) \mid y = x^3 - 1\}$

Solutions:
(a) If $y = r(x) = x^4 - 3x^2 + 1$, then

$$\begin{aligned} r(-x) &= (-x)^4 - 3(-x)^2 + 1 \\ &= x^4 - 3x^2 + 1 = r(x) \end{aligned}$$

Since $r(-x) = r(x)$, the graph of R has symmetry with respect to the y-axis.
(b) If $y = s(x) = x^3 - 1$, then

$$\begin{aligned} s(-x) &= (-x)^3 - 1 \\ &= -x^3 - 1 \end{aligned}$$

Since $s(x) \neq s(-x)$, the graph of S does not have symmetry with respect to the y-axis.

EXAMPLE 4–11 Graph $R = \{(x, y) \mid y = x^3 - 4x\}$.

Solution:
(a) *x-intercepts.* If $y = 0$, then $x^3 - 4x = 0$ and $x(x + 2)(x - 2) = 0$. The

x-intercepts are 0, 2, and -2; thus, $P_1(0, 0)$, $P_2(2, 0)$, and $P_3(-2, 0)$ are points on the graph of R.
(b) *y-intercepts*. If $x = 0$, then $y = 0$. The y-intercept is 0, and $P_1(0, 0)$ is on the graph.
(c) *Vertical asymptotes*. Since the domain of $R = \{x \mid x \in \text{reals}\}$, the graph of R has no vertical asymptotes.
(d) *Horizontal asymptotes*. As $x \to \infty$, $(x^3 - 4x) \to \infty$, and the graph of R has no horizontal asymptote.
(e) *Symmetry*. If $y = r(x) = x^3 - 4x$, then

$$\begin{aligned} r(-x) &= (-x)^3 - 4(-x) \\ &= -x^3 + 4x \end{aligned}$$

Thus, the graph of R does not have symmetry with respect to the y-axis.
(f) *Selected values of* $r(x)$. We select a few values of x and y to locate generally the graph of R.

x	-3	-1	1	3
y	-15	3	-3	15

The graph of R is shown in Figure 4-4.

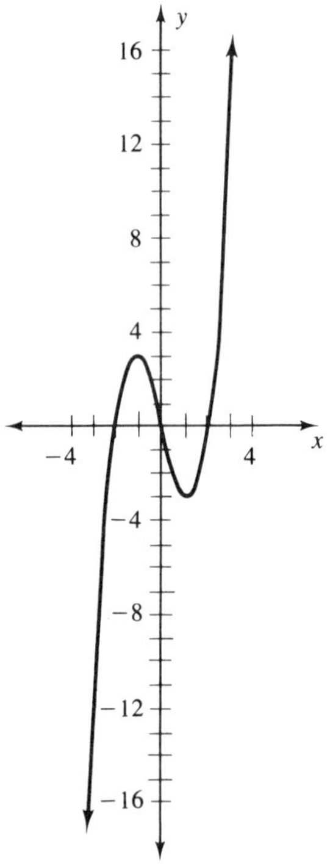

FIGURE 4-4 Graph of $y = x^3 - 4x$

EXAMPLE 4–12 Graph $S = \left\{(x, y) \,\middle|\, y = \dfrac{2x - 6}{3 + x}\right\}$.

Solution:

(a) *x-intercepts*. If $y = 0$, then $2x - 6 = 0$ and $x = 3$. Thus, $P_1(3, 0)$ is a point of the graph of S.

(b) *y-intercepts*. If $x = 0$, then $y = -6/3 = -2$. Thus, $P_2(0, -2)$ is a point of the graph of S.

(c) *Vertical asymptotes*. If $x = -3$, then $s(x)$ is undefined, and the line $x = -3$ is a vertical asymptote of the graph of S. Furthermore,

$$\text{as } x \to -3^-,\ y \to +\infty$$
$$\text{as } x \to -3^+,\ y \to -\infty$$

(d) *Horizontal asymptotes*. If $y = \dfrac{2x - 6}{3 + x}$, then

$$\frac{2 - \dfrac{6}{x}}{\dfrac{3}{x} + 1}$$

As $x \to \infty$, $y \to \dfrac{2 - 0}{0 + 1} = 2$. Also, as $x \to -\infty$, $y \to 2$. Thus, the line $y = 2$ is a horizontal asymptote.

(e) *Symmetry*. If $y = s(x)$, then

$$s(x) = \frac{2x - 6}{3 + x}$$

and

$$s(-x) = \frac{-2x - 6}{3 - x} \neq s(x)$$

The graph of S does not have symmetry with respect to the y-axis.

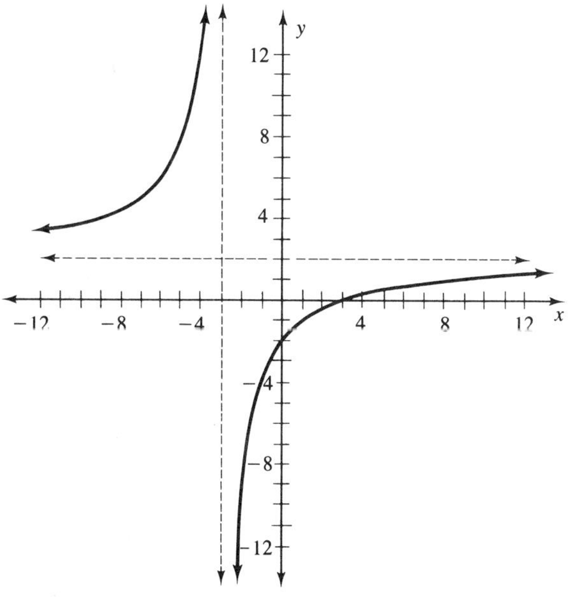

FIGURE 4-5 Graph of $y = \dfrac{2x - 6}{3 + x}$

(f) *Selected values of* $s(x)$.

x	-9	-6	-4	-2	9
y	4	6	14	-10	1

The graph is shown in Figure 4-5 with the asymptotes shown as dashed lines.

Exercise Set 4–2

Determine the x- and y-intercepts of the following relations, if they exist.

1. $y = \dfrac{x-2}{x+2}$
2. $y = \dfrac{x}{x+2}$
3. $y = x^3 - x$
4. $x^2 + y^2 = 4$
5. $y = \sqrt{4-x}$
6. $y^2 = \dfrac{16}{x^2+4}$

Determine the horizontal and vertical asymptotes of the graphs of the following relations, if they exist.

7. $y = \dfrac{2x-1}{x-2}$
8. $y = \dfrac{1-3x}{x+4}$
9. $y = \dfrac{1}{(x+1)^2}$
10. $y = \dfrac{x^3}{(x-3)^2}$
11. $y^2 = \dfrac{9x^2+1}{x^2}$
12. $y = \dfrac{4+x^2}{4-x^2}$

Without graphing, determine if the following relations have symmetry with respect to the y-axis.

13. $y = x^2$
14. $y = 2x^3 - 1$
15. $y^2 = x^3 + x^2 + 1$
16. $y^2 = x^4 + x^2 + 1$
17. $y = \sqrt{\dfrac{1-x^2}{1+x^2}}$
18. $y = \sqrt{4+x^2}$

Sketch graphs of the following relations, using intercepts, asymptotes, and symmetry.

19. $y = \dfrac{x+7}{x-1}$
20. $y = \dfrac{(x+7)^2}{x-1}$
21. $y = x^3$
22. $y = x^2$
23. $y = \dfrac{x^2}{x+2}$
24. $y = \dfrac{x^2+2x-8}{x^3}$
25. $y = \dfrac{(x-1)^2}{(x+1)^2}$
26. $y^2 = 4 - x^2$
27. $y = \dfrac{x-2}{x^2-2x-8}$
28. $y = \sqrt{9-x^2}$
29. $y = \dfrac{x^2+3}{x^2-4}$
30. $y = \dfrac{2x^2-1}{x^2+1}$

Find an equation of a function whose graph has the properties shown in the given sketch.

31.

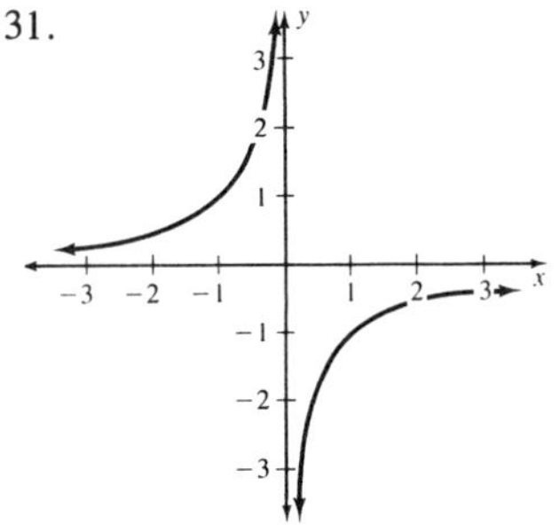

32.

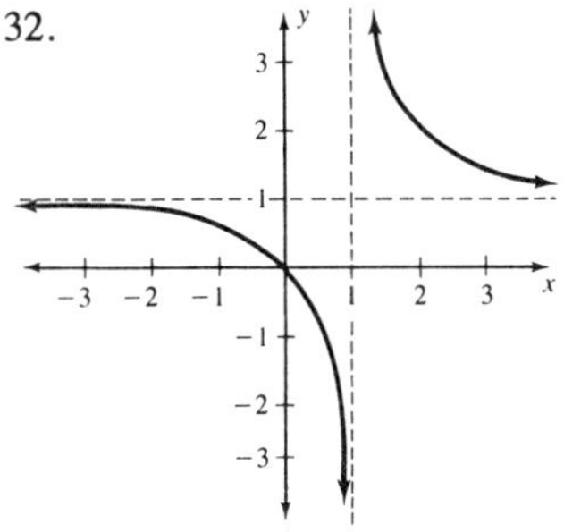

33.

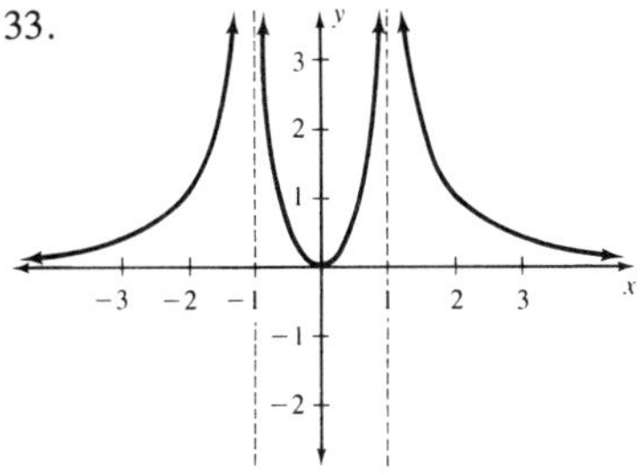

34.

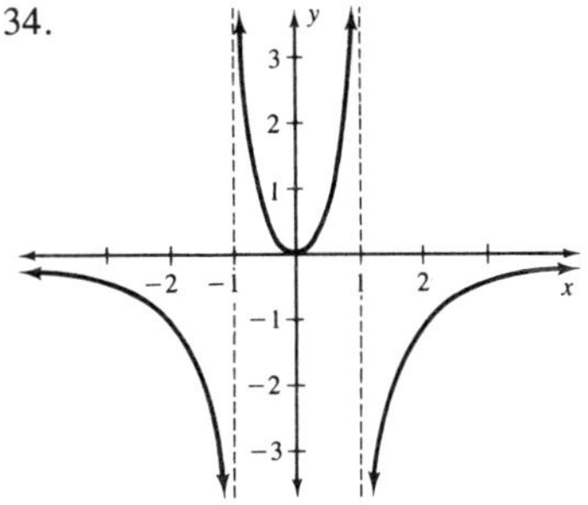

35.

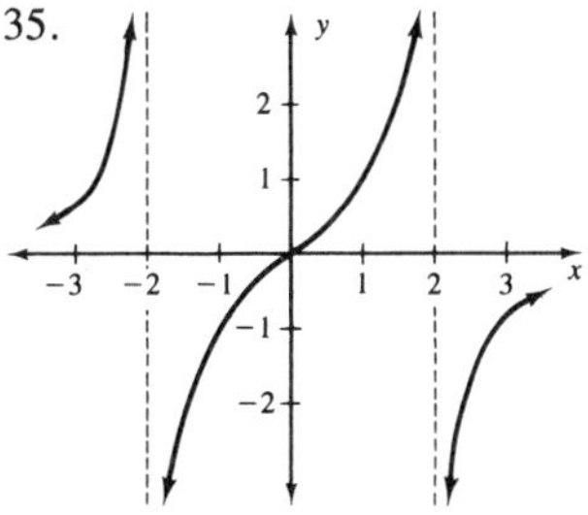

36.

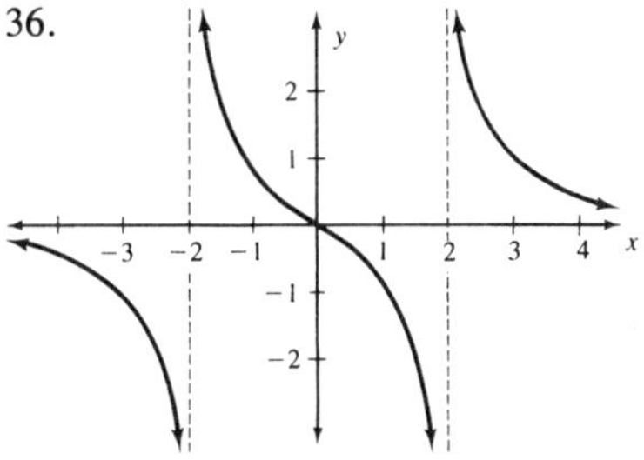

LINEAR RELATIONS 4–3

In Section 4-1 we defined a relation and a function, and we constructed the rectangular coordinate system for graphing relations. In this section we study the most widely encountered relation.

Linear relation

> **Definition:** A relation L that can be expressed in the form
>
> $$L = \{(x, y) \mid Ax + By + C = 0\}$$
>
> is a linear relation, where A, B, and C are real numbers but A and B are not both zero.

The equation $Ax + By + C = 0$ is the **standard linear equation.** Notice that L is called *linear* because the graph of L is a straight line — hence we say *linear* or *line equation*. Furthermore, any line in the coordinate plane has an equation of the form of L.

Basically, we have two goals to achieve in this section. First, we want to graph a linear relation with minimum effort. Second, we want to be able to determine the equation of a linear relation when certain information is given.

Two points determine a line; thus, any two solutions of the equation $Ax + By + C = 0$ determine the graph of the equation. If $P_1(x_1, y_1)$ and $P_2(x_2, y_2)$ are two points such that $Ax_1 + By_2 + C = Ax_2 + By_2 + C = 0$, then P_1 and P_2 determine the line that is the graph of the given equation.

EXAMPLE 4–13 Graph

(a) $L_1 = \{(x, y) \mid 3x + y = 6\}$ (b) $L_2 = \{(x, y) \mid x + 3 = 0\}$

Solutions:

(a) If $3x + y = 6$, then $(0, 6)$ and $(2, 0)$ are solutions of the equation. Thus, $P_1(0, 6)$ and $P_2(2, 0)$ determine the graph of L_1 in Figure 4-6.

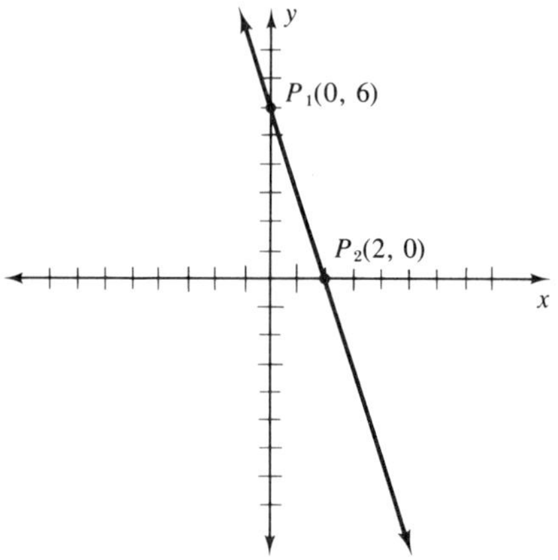

FIGURE 4-6 Graph of $3x + y = 6$

(b) If $x + 3 = 0$, then $(-3, 0)$ and $(-3, 4)$ are solutions of the equation. Therefore, $P_1(-3, 0)$ and $P_2(-3, 4)$ determine the graph of L_2 in Figure 4-7.

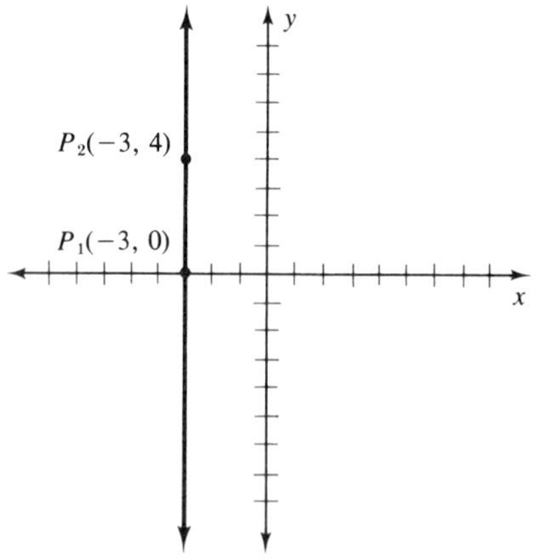

FIGURE 4-7 Graph of $x + 3 = 0$

Notice the difference between sets A and B defined below:

Why is the graph of $x+3=0$ a line and not just one point?

$$A = \{x \mid x + 3 = 0\}$$
$$B = \{(x, y) \mid x + 3 = 0\}$$

Set A contains but one element $\{-3\}$, and the graph is a dot on the number line. Set B contains an infinite number of ordered pairs, such as $\{\ldots, (-3, -2), (-3, \pi), (-3, 20), \ldots\}$. The graph of B is a vertical line, and the domain of the relation is $\{x \mid x = -3\}$.

y-intercept of L

> **Definition:** If $P(0, b)$ is a point of L, then b is the y-intercept of L.

x-intercept of L

> **Definition:** If $P(a, 0)$ is a point of L, then a is the x-intercept of L.

Slope

> **Definition:** If $P_1(x_1, y_1)$ and $P_2(x_2, y_2)$ are points on the graph of $Ax + By + C = 0$ and $B \neq 0$, then the slope of the graph of the relation is m, where
>
> $$m = \frac{y_2 - y_1}{x_2 - x_1} \quad \text{or} \quad m = \frac{y_1 - y_2}{x_1 - x_2}, \quad x_1 \neq x_2$$

We should note the following additional points concerning slope:

1. If $y_1 = y_2$, then $m = 0$ and the line is horizontal.
2. If $x_1 = x_2$, then m is undefined and the line is vertical.
3. We also refer to slope as

$$\frac{\text{rise}}{\text{run}} \quad \text{or} \quad \frac{\Delta y}{\Delta x}$$

where $\Delta y/\Delta x$ is read as "change in y over change in x."

We obtain an alternate form of the standard equation of a line by solving $Ax + By + C = 0$ for y. If $Ax + By + C = 0$, then

$$y = \frac{-A}{B}x + \frac{-C}{B}$$

Let $m = -A/B$ and $b = -C/B$, and we obtain the equation in the following definition.

Slope-intercept equation

> **Definition:** The slope-intercept equation of a line is of the form $y = mx + b$, where m is the slope and b is the y-intercept.

EXAMPLE 4–14 Graph $L = \{(x, y) \mid 5x + 2y = 8\}$.

Solution: If $5x + 2y = 8$, then

$$y = \frac{-5}{2}x + 4$$

The slope of L is $-5/2$ and the y-intercept is 4. We find $P_1(0, 4)$, then a vertical change of -5 units followed by a horizontal change of $+2$ units determines the second point of L, $P_2(2, -1)$. The line containing P_1 and P_2 is the graph of L in Figure 4-8.

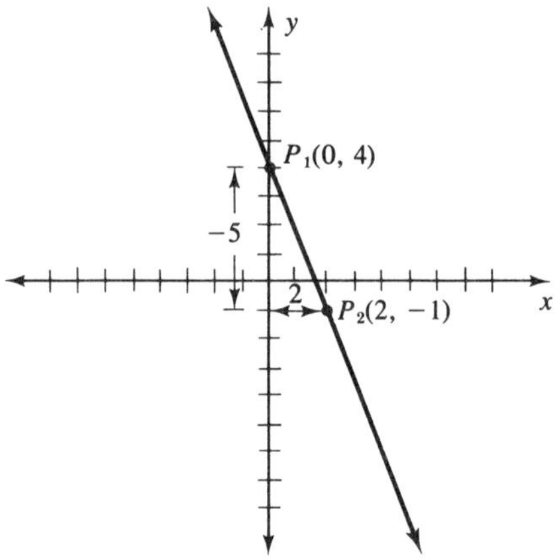

FIGURE 4-8 Graph of $5x + 2y = 8$

The following theorem on the domain and range of a linear relation completes the graphing part of this section.

> **Theorem:** If $L = \{(x, y) | Ax + By + C = 0\}$, then
>
> (1) The domain $= \{x | x \in \text{reals}\}$ and the range $= \{y | y \in \text{reals}\}$ for $A \neq 0$ and $B \neq 0$.
>
> (2) The domain $= \{x | x \in \text{reals}\}$ and the range $= \left\{y \mid y = \frac{-C}{B}\right\}$ for $A = 0$ and $B \neq 0$.
>
> (3) The domain $= \left\{x \mid x = \frac{-C}{A}\right\}$ and the range $= \{y | y \in \text{reals}\}$ for $A \neq 0$ and $B = 0$.

We use the point-slope equation of a line to determine the equation of a given line.

Point-slope equation

> **Definition:** The point-slope equation of a nonvertical line L with slope m and containing $P_1(x_1, y_1)$ is
>
> $$y - y_1 = m(x - x_1)$$

We call this equation the point-slope form because to use it we need to know both the slope of a line and a point on the line. Identifying the slope of the line or a point on the line is not enough to determine the equation of that line. There are infinitely many lines that have a given slope m. Similarly, there are infinitely many lines that contain a fixed point $P_1(x_1, y_1)$. But there is only one line through a fixed point with a given slope.

EXAMPLE 4–15 Find the equation of the line L that contains $P_1(3, 5)$ and $P_2(-2, 3)$. Then, graph L.

Solution: First we find the slope of L:

$$m = \frac{5 - 3}{3 - (-2)} = \frac{2}{5}$$

With the coordinates of $P_1(3, 5)$ and the point-slope equation, we get

$$y - 5 = \frac{2}{5}(x - 3)$$

$$5y - 25 = 2x - 6$$

$$2x - 5y = -19$$

To graph L we convert the standard form of the equation to the slope-intercept form:

$$y = \frac{2}{5}x + \frac{19}{5}$$

The graph is shown in Figure 4-9.

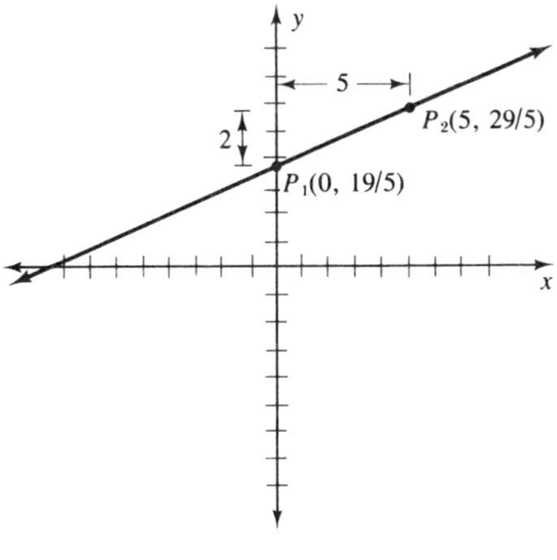

FIGURE 4-9 Graph of $2x - 5y = -19$

> **Theorem:** If L_1 and L_2 are nonvertical lines with slopes m_1 and m_2, respectively, then L_1 is **parallel** to L_2 (written $L_1 \parallel L_2$) if and only if $m_1 = m_2$.

> **Theorem:** If L_1 and L_2 are nonvertical lines with slopes m_1 and m_2, respectively, then L_1 is **perpendicular** to L_2 (written $L_1 \perp L_2$) if and only if $(m_1)(m_2) = -1$.

The proofs of the above theorems are included in the exercises.

EXAMPLE 4–16 Find the equation of the altitude to the side containing A and B of triangle ABC if $A(0, 0)$, $B(4, -1)$, and $C(2, 5)$ are the vertices.

Solution: By definition, an altitude to a side of a triangle contains the third vertex and is perpendicular to the indicated side. The slope of the line containing A and B is

$$m = \frac{0 - (-1)}{0 - 4} = \frac{-1}{4}$$

Therefore, the slope of the altitude = 4. (Note that $4(-1/4) = -1$.) Since the altitude contains $C(2, 5)$, the equation is

$$y - 5 = 4(x - 2)$$
$$4x - y = 3$$

In statistics, *inferences* are made by determining a mathematical equation for a set of data that have been collected. If the data consist of ordered pairs, and the graph of the ordered pairs approximates a straight line, then a regression line is used for the equation.

There are two ways of determining the equation of the regression line: (1) the **least-squares method** and (2) the **free-hand method.** The least-squares method requires a pair of equations and a considerable quantity of arithmetic. We will not use this method at this time, but if you are interested in knowing more about the subject, you can consult any book on elementary statistics.

The free-hand method is less accurate, but it is much simpler. When the **scatter diagram** of the set of data points is graphed, we simply draw a line that is estimated to be the best approximation of the relationship between the variables. By "best" we mean the line that comes closest to all the dots in the scatter diagram.

EXAMPLE 4–17 The following set of data points represents the scores of students on two math quizzes. Find the equation of the regression line that most closely describes the relationship between the scores. If x is the score on the first test and y is the score on the second test, then $\{(x, y)\} = \{(10, 11), (8, 10), (12, 10), (5, 7), (13, 15), (15, 14), (3, 8), (9, 8), (10, 9), (7, 7), (12, 13), (7, 10)\}$.

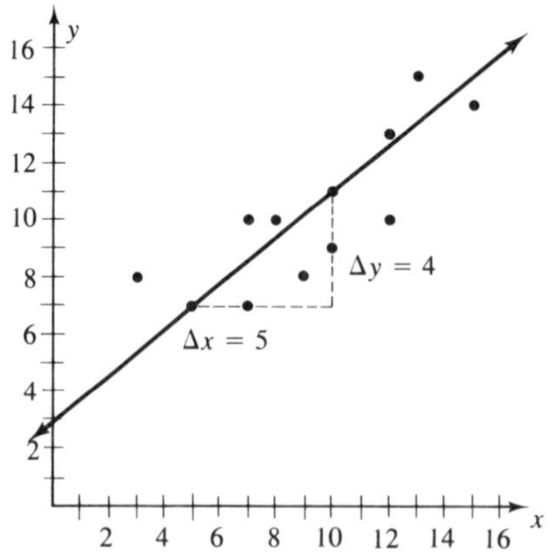

FIGURE 4-10 Graph of scatter diagram

Solution: The graph of the data points is given in Figure 4-10. The line is our "guess" for the regression line. We select two points on the line where the change in x and y can be determined.

From the diagram, $m = 4/5$ and $P(0, 3)$ are chosen. Thus, the equation of the regression line is

$$y - 3 = \frac{4}{5}(x - 0)$$

$$5y - 15 = 4x$$

$$4x - 5y + 15 = 0$$

Suppose a student is absent on the date of the second quiz, but had scored 10 on the first test. Use the regression line equation to estimate the score on the second quiz.

If $4x - 5y + 15 = 0$, and $x = 10$, then

$$4(10) - 5y + 15 = 0$$

$$y = 11$$

The student could be given a score of 11 on the second quiz based on the performance of the data collected on the class.

Exercise Set 4–3

Which of the following relations are linear?

1. $\{(x, y) | x^2 - y = 2\}$
2. $\{(x, y) | x = 2\}$
3. $\{(x, y) | y = 2\}$
4. $\{(x, y) | xy = 2\}$
5. $\{(x, y) | x + y = 2\}$
6. $\{(x, y) | 2y - x = 2\}$

Find the x- and y-intercepts of the following relations.

7. $2x + 3y = 12$
8. $x - y = 7$
9. $y = \frac{2}{3}x + 3$
10. $y = \frac{-3}{4}x - \frac{7}{3}$

Find the value of c such that (2, 3) belongs to each of the following relations.

11. $cx + y = 8$
12. $2x + 3y = c$
13. $cy = 3$
14. $c + 2x = 4$

Find the slope of the line which contains the following pairs of points.

15. (2, 1), (6, 5)
16. (3, −4), (3, 6)
17. (−1, −3), (2, −4)
18. (3, −3), (−2, 6)
19. (6, 5), (7, 5)
20. (a, b), $(2a, -b)$

For each of the following equations, find the slope and the y-intercept.

21. $y = \frac{3}{4}x + 2$
22. $2x + 3y = 6$
23. $x = \frac{7}{8}y + 3$
24. $y - \frac{4}{3} = \frac{2}{7}x$

Graph each of the following lines on a separate coordinate system.

25. $y = \frac{3}{4}x + 1$

26. $y = \frac{-4}{3}x + 2$

27. $4y - 3x = 12$

28. $3y - 4x = 4$

29. $2x + 3y = 12$

30. $3x - 12 = 2y$

31. If L_1 contains $P_1(2, 6)$ and $P_2(0, y)$, find y if $L_1 \parallel L_2$ and L_2 has slope 3/4.

32. If L_3 contains $P_3(2, -6)$ and $P_4(x, -7)$, find x if $L_3 \perp L_4$ and L_4 has slope 2/3.

In each of the following, find the equation of the line given the slope and the indicated point.

33. $m = 6$; (0, 1)

34. $m = -2/3$; (0, −2)

35. $m = 3/4$; (0, 4)

36. $m = 7/6$; (0, 2/3)

Find the equation of a line with the following properties.

37. Parallel to the line $y = 3x + 2$ and has a y-intercept of 4.

38. Perpendicular to the line $y = 2x - 5$ and has a y-intercept of −2.

Find the standard form of the equation of the line which contains the given points.

39. (−3, 2), (3, −2)

40. (−1, −6), (−2, −3)

41. (1, 6), (3, 6)

42. (a, b), $(2a, 3b)$

Find the standard form of the equation of the line satisfying the following conditions.

43. Perpendicular to the line $x = 2$ and contains the point (3, 6).

44. Parallel to the line $x = 2$ and contains the point (−3, −6).

45. Parallel to the line through the points (2, 3) and (−1, 6) and contains the point (1/2, 1/3).

46. Find the equations of the altitudes of triangle ABC with points $A(1, 2)$, $B(3, 4)$, and $C(2, -2)$.

47. The following set of ordered pairs represents the scores of golf players for two consecutive days on an 18-hole course. Find an equation of the regression line which approximates the relationship between scores on consecutive days, where x is the score on the first day and y is the score for the second day.

$$\{(x, y)\} = \{(88, 92), (68, 65), (74, 74), (73, 68), (66, 69), (74, 81), (81, 70), (90, 81), (95, 100), (84, 80)\}$$

48. If the following set of ordered pairs represents the scores of students on two consecutive spelling quizzes, with 10 points possible on each test, find an equation of a regression line which closely approximates the relationship.

$$\{(x, y)\} = \{(2, 3), (3, 5), (7, 8), (6, 2), (10, 10), (9, 9), (8, 7), (3, 7), (8, 5), (4, 4)\}$$

SYSTEMS OF LINEAR RELATIONS 4–4

In Section 4-3 we introduced and studied the linear relation. Recall that for $R = \{(x, y) | Ax + By = C\}$ there exists an infinite number of ordered pairs (x, y) that satisfy R.

In this section we consider simultaneously two or more linear relations and the intersection of their solution sets. That is, if R and S are two linear relations, what can be said about $R \cap S$? As we shall see, $R \cap S$ may be the null set, or $R \cap S$ may contain but one element, or $R \cap S$ may contain an infinite number of ordered pairs.

System of equations

> **Definition:** A system of equations is a set of two or more equations.

Solution of a system of two equations

> **Definition:** A solution of a system of equations in two unknowns is an ordered pair that satisfies each equation of the system.

Equivalent systems

> **Definition:** Two systems of equations are equivalent if and only if they have the same solution set.

We will study first a system of two equations in two unknowns. If

$$R = \{(x, y) | A_1x + B_1y = C_1\} \quad \text{with graph } L_1$$
$$S = \{(x, y) | A_2x + B_2y = C_2\} \quad \text{with graph } L_2$$

then the solution of the system of equations of R and S is related to lines L_1 and L_2.

How can the solution of a pair of equations be related to two lines?

Recall from geometry that if L_1 and L_2 are two lines that lie in the same plane, then L_1 and L_2 must (1) intersect in one and only one point, (2) be parallel, or (3) be the same line. Now keep in mind that the coordinates of every point on L_1 and L_2 are solutions to the equations of R and S.

Case 1. If L_1 and L_2 intersect, then the solution of the system of equations is the ordered pair that is the coordinate of the point of intersection. Such a system of equations is called **independent** and **consistent,** and

$$\frac{A_1}{A_2} \neq \frac{B_1}{B_2}$$

Case 2. If L_1 and L_2 are parallel, then the solution of the system of equations is the null set. The system of equations is called **inconsistent,** and

$$\frac{A_1}{A_2} = \frac{B_1}{B_2} \neq \frac{C_1}{C_2}$$

Case 3. If L_1 and L_2 are the same line, then the solution of the system of equations is an infinite set of ordered pairs. The system of equations is called **dependent,** and

$$\frac{A_1}{A_2} = \frac{B_1}{B_2} = \frac{C_1}{C_2}$$

One method of finding the solution of a pair of linear equations (one that we will not consider) is to graph the two lines and determine the solution set from the graph. There are also two algebraic methods commonly used: the **substitution method** and the **addition method.** We will demonstrate these two methods in the following examples.

EXAMPLE 4–18 Solve a system of equations with the substitution method.

Solution:

(1) Given two linear equations (a) $Ax + By = C$ (b) $Dx + Ey = F$	(1) Given the system (a) $5x - 3y = 7$ (b) $x + 2y = 4$
(2) Form equation (c) by solving (a) or (b) for x or y.	(2) Since $D = 1$, solve equation (b) for x, and (c) $x = 4 - 2y$
(3) Substitute the expression for the variable in (c) into the other equation.	(3) Substitute $4 - 2y$ for x in (a), and $5(4 - 2y) - 3y = 7$
(4) Solve this equation for the remaining variable.	(4) $20 - 10y - 3y = 7$ $y = 1$
(5) Use (c) to find the value of the other variable.	(5) If $y = 1$, then $x = 4 - 2(1) = 2$
(6) Check the values for x and y in (a) and (b).	(6) If $x = 2$ and $y = 1$, (a) $5(2) - 3(1) = 7$ (True) (b) $2 + 2(1) = 4$ (True)
(7) State the solution set for the system.	(7) $\{(2, 1)\}$ is the solution set.

EXAMPLE 4–19 Solve the following system with the substitution method:

$$\begin{aligned} &\text{(a)} \quad 5x + y = 4 \\ &\text{(b)} \quad 15x + 3y = 10 \end{aligned}$$

Solution: Solve (a) for y. Then for equation (c) we have

$$y = 4 - 5x$$

Now substitute $4 - 5x$ for y in (b), and

$$\begin{aligned} 15x + 3(4 - 5x) &= 10 \\ 15x + 12 - 15x &= 10 \\ 12 &= 10 \end{aligned}$$

Since $12 \neq 10$, the system is inconsistent and the solution set is null. Notice that $A_1/A_2 = B_1/B_2 = 1/3$, but $C_1/C_2 = 4/10$. We have therefore verified that the system is in fact inconsistent.

EXAMPLE 4–20 Solve a system of equations with the addition method.

Solution:

(1) Given two linear equations (a) $Ax + By = C$ (b) $Dx + Ey = F$	(1) Given the system (a) $5x + 4y = 8$ (b) $2x - 3y = -6$
(2) Convert (a) and (b) to equivalent equations (c) and (d) in which the coefficients of x or y are opposites, so that one variable will vanish when (c) and (d) are added.	(2) Multiply (a) by 3 and (b) by 4 to obtain opposite coefficients for the variable y. (Notice that 12 is the L.C.M. of 4 and -3.) (c) $15x + 12y = 24$ (d) $8x - 12y = -24$
(3) Add (c) and (d) to form (e), an equation in one variable.	(3) Add equations (c) and (d), and (e) $23x = 0$
(4) Solve (e) for the remaining variable.	(4) $x = 0$
(5) Substitute the root of (e) into (a) or (b) and solve.	(5) If $x = 0$, then in (a), $5(0) + 4y = 8$ $y = 2$
(6) Check the values for x and y in (a) and (b).	(6) If $x = 0$ and $y = 2$, (a) $5(0) + 4(2) = 8$ (True) (b) $2(0) - 3(2) = -6$ (True)
(7) State the solution set for the system.	(7) The solution set is $\{(0, 2)\}$.

EXAMPLE 4–21 Solve the following system with the addition method:

$$\text{(a)}\quad 2x + 5y = 10$$
$$\text{(b)}\quad 8x + 20y = 40$$

Solution: Multiply (a) by -4 to obtain opposite coefficients for the variable x. Then add (c) and (d) to form (e):

$$\begin{array}{lrcr} \text{(c)} & -8x - 20y & = & -40 \\ \text{(d)} & 8x + 20y & = & 40 \\ \hline \text{(e)} & 0 + \quad 0 & = & 0 \end{array}$$

Since $0 = 0$, the system is dependent and the solution set is

$$\{(x, y) \mid 2x + 5y = 10\}$$

Notice that $A_1/A_2 = B_1/B_2 = C_1/C_2 = 1/4$, and we have verified that the system is dependent.

EXAMPLE 4–22 Linda has a number of silver dimes and quarters. The face value of the coins is \$9.80, but the number of dimes she has is 8 more than twice the number of quarters. How many of each kind of coin does Linda have?

Solution: If x = the number of dimes, then $0.10x$ is the value of the dimes. If y = the number of quarters, then $0.25y$ is the value of the quarters. Therefore,

$$\text{(a) } x = 2y + 8$$
$$\text{(b) } 0.10x + 0.25y = 9.80$$

We use the substitution method to solve the system:

$$\begin{aligned} \text{(c) } 0.10(2y + 8) + 0.25y &= 9.80 \\ 0.45y &= 9.00 \\ y &= 20 \text{ quarters} \end{aligned}$$

Thus,

$$x = 2(20) + 8 = 48 \text{ dimes}$$

Check. The value of 20 quarters is \$5.00 and the value of 48 dimes is \$4.80. The total of \$9.80 checks with the given information.

EXAMPLE 4–23 Solve the system

$$\text{(a) } \frac{2}{x} + \frac{5}{y} = 9$$

$$\text{(b) } \frac{3}{x} - \frac{1}{y} = 5$$

Solution: Let $u = 1/x$ and $v = 1/y$. Then we have the system

$$\text{(a*) } 2u + 5v = 9$$
$$\text{(b*) } 3u - v = 5$$

From (b*), we get $v = 3u - 5$. Then substitute $3u - 5$ in (a*) for v:

$$\begin{aligned} \text{(c*) } 2u + 5(3u - 5) &= 9 \\ 17u &= 34 \\ u &= 2 \end{aligned}$$

With equation (a*),

$$\begin{aligned} 2(2) + 5v &= 9 \\ v &= 1 \end{aligned}$$

Since $u = 1/x$, then $x = 1/u = 1/2$. If $v = 1/y$ then $y = 1/1 = 1$. The solution set is $\{(1/2, 1)\}$. The check is omitted.

Equations (a), (b), and (c) below comprise a system of **three linear equations in three unknowns.** In the equations, A, B, C, and D are real numbers; and A, B, and C are not all zero at the same time in any of the three equations.

$$\begin{aligned}
&\text{(a)}\ A_1x + B_1y + C_1z = D_1\\
&\text{(b)}\ A_2x + B_2y + C_2z = D_2\\
&\text{(c)}\ A_3x + B_3y + C_3z = D_3
\end{aligned}$$

We can use the substitution and addition methods to solve such a system of equations. A recommended process for solving the system is outlined in the following steps:

1. Eliminate one of the variables from any pair of the equations, thereby forming (d), an equation in the remaining two variables.
2. Eliminate the *same variable* from a different pair of the equations, thereby forming (e), an equation in the same two variables as (d).
3. Equations (d) and (e) comprise a system of two equations in two unknowns and is solved by the addition or substitution method.
4. The solution of the system of equations (d) and (e) is substituted in (a), (b), or (c) to find the value of the third variable.

EXAMPLE 4–24 Solve the system

$$\begin{aligned}
&\text{(a)}\ 3x + 2y + z = 5\\
&\text{(b)}\ x - y + 3z = 9\\
&\text{(c)}\ 4x + 3y - 5z = 11
\end{aligned}$$

Solution:

(1) We arbitrarily elect to eliminate the variable z in forming (d) and (e). To form (d) we multiply (a) by 3 and (b) by -1 and add the two equations.

$$\begin{array}{llcll}
\text{(a)} & 3(3x + 2y + z = 5) & \rightarrow & \text{(a)} & 9x + 6y + 3z = 15\\
\text{(b)} & -1(x - y + 3z = 9) & \rightarrow & \text{(b)} & -x + y - 3z = -9\\
 & & & \text{(d)} & 8x + 7y \qquad = 6
\end{array}$$

(2) To form (e) we multiply (b) by 5 and (c) by 3 and add the two equations, thereby eliminating the variable z.

$$\begin{array}{llcll}
\text{(b)} & 5(x - y + 3z = 9) & \rightarrow & \text{(b)} & 5x - 5y + 15z = 45\\
\text{(c)} & 3(4x + 3y - 5z = 11) & \rightarrow & \text{(c)} & 12x + 9y - 15z = 33\\
 & & & \text{(e)} & 17x + 4y \qquad = 78
\end{array}$$

(3) To eliminate y in the system of equations (d) and (e), we multiply (d) by 4 and (e) by -7 and add the two equations.

$$\begin{array}{llcll}
\text{(d)} & 4(8x + 7y = 6) & \rightarrow & \text{(d)} & 32x + 28y = 24\\
\text{(e)} & -7(17x + 4y = 78) & \rightarrow & \text{(e)} & -119x - 28y = -546\\
 & & & & -87x \qquad = -522\\
 & & & & x \qquad = 6
\end{array}$$

(4) In (e), if $x = 6$, then $17(6) + 4y = 78$ and $y = -6$. In (a), if $x = 6$ and $y = -6$, then $3(6) + 2(-6) + z = 5$ and $z = -1$.

Check. With $x = 6$, $y = -6$, and $z = -1$,

$$\begin{array}{lrl} \text{(a)} & 3(6) + 2(-6) + (-1) = 5 & \text{(True)} \\ \text{(b)} & 6 - (-6) + 3(-1) = 9 & \text{(True)} \\ \text{(c)} & 4(6) + 3(-6) - 5(-1) = 11 & \text{(True)} \end{array}$$

The solution of the system is the **ordered triple** $\{(6, -6, -1)\}$.

Is there a geometric interpretation of this solution?

The graph of an equation of the form $Ax + By + Cz = D$, in which A, B, and C are not all zero at the same time, is a **plane** in a three-dimensional space. Thus, the ordered triple solution of the three equations in three unknowns in Example 4-24 is the intersection of three planes.

Exercise Set 4–4

Find the solution of the following systems of equations by any method.

1. $x + y = 10$
   $x - y = 8$
2. $x - y = 3$
   $2x - 2y = 4$
3. $x + 2y = 5$
   $2x + 4y = 10$
4. $4x + 2y = -1$
   $5x - 3y = 7$
5. $x + 5y = 2$
   $y = -3x$
6. $x + 3y = 7$
   $2x + 5y = 11$
7. $3y - 2x = 15$
   $y + 2x = 15$
8. $y + \dfrac{x}{4} = \dfrac{3}{8}$
   $\dfrac{5y}{12} + \dfrac{x}{6} = \dfrac{1}{4}$
9. $8x - 7 = y$
   $2x + 3y = 4$
10. $3x - 7y = 15$
    $3x + 2y = 15$
11. $4x + y = 10$
    $7x + 3y = 25$
12. $3x + y = 7$
    $2x - 5y = -1$
13. $\dfrac{x}{5} - \dfrac{2y}{3} = \dfrac{1}{2}$
    $\dfrac{x}{4} - \dfrac{y}{6} = \dfrac{3}{8}$
14. $8x = 7 - 2y$
    $3y + 12x = 5$
15. $5x = 6y + 4$
    $9y = 2x - 6$
16. $9x + 15y = 45$
    $20y + 12x = 60$
17. $4x + 3y = 5$
    $2x - y = 7$
18. $\dfrac{2x - 7y}{3} = \dfrac{x + 3y}{21}$
    $\dfrac{x + y}{3} = \dfrac{x - y + 4}{5}$
19. $5x - 7y = 0$
    $7x - 5y = 0$
20. $2x = 4y + 2$
    $2y = 3 - x$
21. $\dfrac{1}{x} + \dfrac{1}{y} = 5$
    $\dfrac{2}{x} + \dfrac{5}{y} = -4$
22. $\dfrac{1}{x} + \dfrac{1}{y} = \dfrac{2}{3}$
    $\dfrac{3}{x} - \dfrac{6}{y} = -10$

23. $\frac{3}{x} + \frac{7}{y} = 1$

$\frac{6}{x} + \frac{1}{y} = 28$

24. $\frac{5}{x} - \frac{4}{y} = 0$

$\frac{7}{x} + \frac{8}{y} = 17$

Solve the following systems of equations.

25. $x + y + z = 0$
$2x - 2y + z = 1$
$x + 3y - 5z = -1$

26. $2x - y - 3z = -4$
$x + y + 2z = 7$
$-3x - y + 4z = 7$

27. $x + y = 1$
$y - 3z = 0$
$2x - y + z = 18$

28. $x + z = 1$
$x + y = 1$
$2y - 7z = -25$

29. In a piggy bank with only nickels and pennies Steve finds 191 coins. If the value totals \$3.39, how many of each kind of coin does he have?

30. Find two integers whose sum is 34 and whose difference is 96.

31. The sum of Linda's and Gene's ages is 79 years. Four times Linda's age is 13 years less than three times Gene's age. How old are Linda and Gene?

32. The combined yearly wages of Lynn and Mike is \$25,500. If Lynn's salary is tripled and Mike's is doubled, then each of them would have the same yearly salary. How much does each of them gross per year?

33. The sum of 2 times an integer and 3 times a second integer is 5. Six times the reciprocal of the first added to 18 times the reciprocal of the second is -3. Find the two numbers.

34. The distance around a rectangular plot is 140 meters. The length of the plot is 2.5 times the width. Find the dimensions of the plot.

MATRICES AND DETERMINANTS 4–5

In Section 4-4 we solved systems of linear equations algebraically with the addition and substitution methods. In this section we consider the method of solving a system of equations using determinants and Cramer's Rule.

A study of determinants is frequently included in a study of matrix algebra. However, we feel that a thorough study of matrices (plural of *matrix*) is beyond the scope of our work at this time. We therefore study only those aspects of matrix theory that are necessary for an understanding of determinants.

Matrices

To begin this section, we offer the following definition of a matrix.

Matrix

> **Definition:** A matrix is a rectangular array of numbers arranged in rows and columns. The numbers are called *elements* of the matrix.

A pair of brackets is used to enclose the array of numbers to indicate the matrix.

Dimension of a matrix

> **Definition:** The dimension of a matrix with m rows and n columns is $m \times n$ (read "m by n").

EXAMPLE 4–25 State the dimensions of the following matrices:

(a) $A = \begin{bmatrix} 3 & 1 \\ 4 & 2 \\ 0 & 5 \end{bmatrix}$ (b) $B = \begin{bmatrix} 3 & 0 & -3 & 1 \\ 1 & 4 & 1 & 2 \end{bmatrix}$

Solutions:
(a) Matrix A has 3 rows and 2 columns; therefore the dimension of A is 3×2.
(b) Matrix B has 2 rows and 4 columns; thus, the dimension of B is 2×4.

Element of a matrix

> **Definition:** The element in the ith row and jth column of a matrix is the ijth element. For matrix A, the symbol for the ijth element is a_{ij}.

EXAMPLE 4–26 Identify the indicated element in matrix A, where

$$A = \begin{bmatrix} 3 & 0 & 1 \\ -2 & 2 & -1 \\ -3 & 4 & -4 \end{bmatrix}$$

(a) a_{31} (b) a_{22} (c) a_{13}

Solutions:
(a) a_{31} is the element in row three and column one of A; thus, $a_{31} = -3$.
(b) $a_{22} = 2$
(c) $a_{13} = 1$

Square matrix

> **Definition:** If A is an $n \times n$ matrix, then A is a square matrix of order n.

Determinants

The following definition asserts that there exists a real number that is related to each square matrix whose elements are real numbers.

Determinant of a square matrix

> **Definition:** If A is a square matrix of order $n \geq 2$ and every element of A is a real number, then there exists a real number D called the determinant of A.

We should note that there exists a *functional relationship* between the set of all square matrices and the set of real numbers. The domain of the function is the set of all square matrices with real number elements and the range is the set of real numbers.

If A is a square matrix whose elements are real numbers, then there exists one and only one real number that is the determinant of A. Thus, the relationship between square matrices and the determinants of the matrices is a function. The following definition states the method of computing the determinant of a 2×2 matrix.

Determinant of A

Definition: If

$$A = \begin{bmatrix} a_{11} & a_{12} \\ a_{21} & a_{22} \end{bmatrix}$$

then D, the determinant of A (also written $|A|$), is

$$D = \begin{vmatrix} a_{11} & a_{12} \\ a_{21} & a_{22} \end{vmatrix}$$

and

$$D = (a_{11})(a_{22}) - (a_{21})(a_{12})$$

EXAMPLE 4–27 Find

$$D = \begin{vmatrix} 3 & 1 \\ 4 & 7 \end{vmatrix}$$

Solution: Since $a_{11} = 3$, $a_{12} = 1$, $a_{21} = 4$, and $a_{22} = 7$, then $D = 3(7) - 4(1) = 17$. Thus, 17 is the value of the given determinant.

If A is a square matrix of order $n \geq 3$, we use the process of **expansion by minors** to find the value of the determinant. We will consider only determinants of 3×3 matrices, but the theory can be extended to any square matrix of order greater than three.

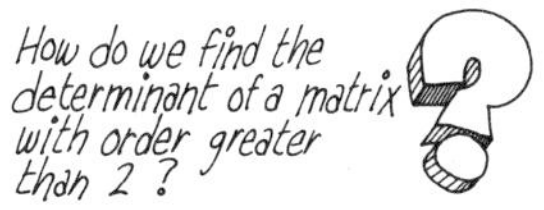

Minor of a_{ij}

Definition: The minor of the a_{ij} element in the determinant D is the determinant obtained by deleting the row and column containing the element a_{ij}.

EXAMPLE 4–28 Identify the minors of

(a) a_{21} (b) a_{32}

in the determinant

$$D = \begin{vmatrix} 4 & 2 & 0 \\ 1 & 3 & 5 \\ -2 & 0 & 2 \end{vmatrix}$$

Solutions:

(a) We cross out the row and column containing the element a_{21}. The remaining elements in D form the determinant that is the minor of a_{21}. That is,

$$\begin{vmatrix} 4 & 2 & 0 \\ 1 & 3 & 5 \\ -2 & 0 & 2 \end{vmatrix}$$

Thus, the minor of a_{21} is

$$\begin{vmatrix} 2 & 0 \\ 0 & 2 \end{vmatrix}$$

(b) We cross out the row and column containing the element a_{32}:

$$\begin{vmatrix} 4 & 2 & 0 \\ 1 & 3 & 5 \\ -2 & 0 & 2 \end{vmatrix}$$

The minor of a_{32} is

$$\begin{vmatrix} 4 & 0 \\ 1 & 5 \end{vmatrix}$$

If a_{ij} is an element in an $n \times n$ determinant, then the order of the minor of a_{ij} is $(n - 1) \times (n - 1)$. As seen in Example 4-28, for any element in a 3×3 determinant, the minor is a 2×2 determinant. The following theorem states the method of determining the value of a 3×3 determinant by the process called **expanding a determinant by minors.**

Theorem: If

$$D = \begin{vmatrix} a_{11} & a_{12} & a_{13} \\ a_{21} & a_{22} & a_{23} \\ a_{31} & a_{32} & a_{33} \end{vmatrix}$$

then the expansion of D by the elements in column one is

$$a_{11}\begin{vmatrix} a_{22} & a_{23} \\ a_{32} & a_{33} \end{vmatrix} - a_{21}\begin{vmatrix} a_{12} & a_{13} \\ a_{32} & a_{33} \end{vmatrix} + a_{31}\begin{vmatrix} a_{12} & a_{13} \\ a_{22} & a_{23} \end{vmatrix}$$

The following notes apply to the above theorem:

1. In the expansion of D, the minor of a_{11} is multiplied by a_{11}, the minor of a_{21} is multiplied by a_{21}, and the minor of a_{31} is multiplied by a_{31}.
2. We give the product of each element and its minor a positive or negative sign by using the following rule:

$$(-1)^{i+j} \text{ for the } ij\text{th element}$$

Thus, a positive sign is given to the products of a_{11} and its minor and a_{13} and its minor, since $(-1)^{1+1} = (-1)^{1+3} = +1$. Furthermore, a negative sign is given to the product of a_{21} and its minor, since $(-1)^{2+1} = -1$.

The value of D may be determined by expanding the determinant by the elements of any row or column. For example, for the determinant in the theorem, the value of D is also

Must we always expand D by using the elements of column 1?

$$-a_{21}\begin{vmatrix} a_{12} & a_{13} \\ a_{32} & a_{33} \end{vmatrix} + a_{22}\begin{vmatrix} a_{11} & a_{13} \\ a_{31} & a_{33} \end{vmatrix} - a_{23}\begin{vmatrix} a_{11} & a_{12} \\ a_{31} & a_{32} \end{vmatrix}$$

which is the expansion of D using the elements in row two.

EXAMPLE 4–29 Evaluate

$$D = \begin{vmatrix} 1 & 2 & -2 \\ 2 & -1 & 1 \\ 3 & -3 & 4 \end{vmatrix}$$

Solution: We arbitrarily choose to expand D by the elements in column one:

$$\begin{aligned} D &= 1\begin{vmatrix} -1 & 1 \\ -3 & 4 \end{vmatrix} - 2\begin{vmatrix} 2 & -2 \\ -3 & 4 \end{vmatrix} + 3\begin{vmatrix} 2 & -2 \\ -1 & 1 \end{vmatrix} \\ &= 1(-4+3) - 2(8-6) + 3(2-2) \\ &= -1 - 4 + 0 = -5 \end{aligned}$$

Cramer's Rule

We now turn our attention to a general system of two linear equations in two unknowns, using the following equations for demonstration:

$$\text{(a) } a_1x + b_1y = c_1$$
$$\text{(b) } a_2x + b_2y = c_2$$

We multiply (a) by b_2 and (b) by $(-b_1)$. Then we add the resulting equations and solve for x:

$$\begin{aligned} &\text{(a)} \quad a_1b_2x + b_1b_2y = b_2c_1 \\ &\text{(b)} \quad -a_2b_1x - b_1b_2y = -b_1c_2 \\ \hline &(a_1b_2 - a_2b_1)x = b_2c_1 - b_1c_2 \end{aligned}$$

and

$$x = \frac{b_2c_1 - b_1c_2}{a_1b_2 - a_2b_1}$$

The numerator and denominator of the ratio can be written as 2×2 determinants. That is,

$$x = \frac{\begin{vmatrix} c_1 & b_1 \\ c_2 & b_2 \end{vmatrix}}{\begin{vmatrix} a_1 & b_1 \\ a_2 & b_2 \end{vmatrix}} \tag{1}$$

The denominator in Equation (1) is the determinant of a matrix whose elements are the *coefficients of x and y* in the two given equations. The matrix

$$\begin{bmatrix} a_1 & b_1 \\ a_2 & b_2 \end{bmatrix}$$

is the **coefficient matrix,** and

$$D = \begin{vmatrix} a_1 & b_1 \\ a_2 & b_2 \end{vmatrix}$$

is called the **determinant of the coefficients.**

The numerator in Equation (1) is the determinant of a matrix whose elements are the *constants of the equations in place of the coefficients for x, together with the coefficients of y.* In symbols we write

$$A_x = \begin{bmatrix} c_1 & b_1 \\ c_2 & b_2 \end{bmatrix} \quad \text{and} \quad D_x = \begin{vmatrix} c_1 & b_1 \\ c_2 & b_2 \end{vmatrix}$$

The solution for x in Equation (1) can now be written $x = \dfrac{D_x}{D}$.

Can we solve the equations for y by using a similar pair of determinants?

When we replace the coefficients of y in D with the values of the constants, we obtain the determinant

$$D_y = \begin{vmatrix} a_1 & c_1 \\ a_2 & c_2 \end{vmatrix}$$

The value of y for the system of equations is obtained with the ratio

$$y = \frac{D_y}{D} \tag{2}$$

We have derived the set of equations known as **Cramer's Rule.**

Theorem: If

$$a_1x + b_1y = c_1 \quad \text{and} \quad a_2x + b_2y = c_2$$

and $a_1b_2 \neq a_2b_1$, then

$$x = \frac{D_x}{D} \quad \text{and} \quad y = \frac{D_y}{D}$$

where

$$D = \begin{vmatrix} a_1 & b_1 \\ a_2 & b_2 \end{vmatrix}$$

$$D_x = \begin{vmatrix} c_1 & b_1 \\ c_2 & b_2 \end{vmatrix}$$

$$D_y = \begin{vmatrix} a_1 & c_1 \\ a_2 & c_2 \end{vmatrix}$$

Notice that if $a_1b_2 = a_2b_1$, then $D = 0$, and the ratios for x and y are undefined. The system of equations is then either inconsistent or dependent.

EXAMPLE 4–30 Solve the system of equations

$$\begin{aligned} (1)\ & 2x - 3y = 5 \\ (2)\ & 4x + y = 3 \end{aligned}$$

Solution: For the given set of equations,

$$D = \begin{vmatrix} 2 & -3 \\ 4 & 1 \end{vmatrix} = 2(1) - 4(-3) = 14$$

$$D_x = \begin{vmatrix} 5 & -3 \\ 3 & 1 \end{vmatrix} = 5(1) - 3(-3) = 14$$

$$D_y = \begin{vmatrix} 2 & 5 \\ 4 & 3 \end{vmatrix} = 2(3) - 4(5) = -14$$

Thus, $x = 14/14 = 1$ and $y = -14/14 = -1$. The solution set for the system of equations is $\{(1, -1)\}$.

Check. If $(x, y) = (1, -1)$, then in Equation (1),

$$2(1) - 3(-1) = 5 \qquad \text{(True)}$$

and in Equation (2),

$$4(1) + (-1) = 3 \qquad \text{(True)}$$

EXAMPLE 4–31 Solve the system of equations

$$\begin{aligned} &(1)\quad 5x + 2y = 10 \\ &(2)\quad 15x + 6y = 35 \end{aligned}$$

Solution: For the given set of equations,

$$D = \begin{vmatrix} 5 & 2 \\ 15 & 6 \end{vmatrix} = 5(6) - 15(2) = 0$$

Since $D = 0$, the system is either inconsistent or dependent.

$$D_x = \begin{vmatrix} 10 & 2 \\ 35 & 6 \end{vmatrix} = 10(6) - 35(2) = -10$$

Since $D_x \neq 0$, the system is inconsistent and the solution set is $\varnothing$. Notice that if $D = 0$, but $D_x \neq 0$ or $D_y \neq 0$, then the given system is inconsistent. The graphs of the given equations are parallel lines. However, if $D = 0$ and $D_x = D_y = 0$, then the system is dependent. The graphs of the equations of such a system are the same line.

Cramer's Rule can be extended to systems of n equations in n unknowns. The rule is extended to the system

$$\begin{aligned} &\text{(a)}\quad a_1x + b_1y + c_1z = d_1 \\ &\text{(b)}\quad a_2x + b_2y + c_2z = d_2 \\ &\text{(c)}\quad a_3x + b_3y + c_3z = d_3 \end{aligned}$$

with the following equations:

$$x = \frac{D_x}{D}, \qquad y = \frac{D_y}{D}, \qquad z = \frac{D_z}{D}$$

where

$$D = \begin{vmatrix} a_1 & b_1 & c_1 \\ a_2 & b_2 & c_2 \\ a_3 & b_3 & c_3 \end{vmatrix} \qquad D_x = \begin{vmatrix} d_1 & b_1 & c_1 \\ d_2 & b_2 & c_2 \\ d_3 & b_3 & c_3 \end{vmatrix}$$

$$D_y = \begin{vmatrix} a_1 & d_1 & c_1 \\ a_2 & d_2 & c_2 \\ a_3 & d_3 & c_3 \end{vmatrix} \qquad D_z = \begin{vmatrix} a_1 & b_1 & d_1 \\ a_2 & b_2 & d_2 \\ a_3 & b_3 & d_3 \end{vmatrix}$$

EXAMPLE 4–32 Solve the following system of equations:

$$\begin{aligned} &\text{(a)} \quad x - 2y + 3z = 1 \\ &\text{(b)} \quad 2x + y - 2z = 3 \\ &\text{(c)} \quad -2x + 2y + z = -1 \end{aligned}$$

Solution: For the given set of equations,

$$D = \begin{vmatrix} 1 & -2 & 3 \\ 2 & 1 & -2 \\ -2 & 2 & 1 \end{vmatrix} = 1\begin{vmatrix} 1 & -2 \\ 2 & 1 \end{vmatrix} - 2\begin{vmatrix} -2 & 3 \\ 2 & 1 \end{vmatrix} + (-2)\begin{vmatrix} -2 & 3 \\ 1 & -2 \end{vmatrix}$$

$$= 5 + 16 - 2 = 19$$

$$D_x = \begin{vmatrix} 1 & -2 & 3 \\ 3 & 1 & -2 \\ -1 & 2 & 1 \end{vmatrix} = 1\begin{vmatrix} 1 & -2 \\ 2 & 1 \end{vmatrix} - 3\begin{vmatrix} -2 & 3 \\ 2 & 1 \end{vmatrix} + (-1)\begin{vmatrix} -2 & 3 \\ 1 & -2 \end{vmatrix}$$

$$= 5 + 24 - 1 = 28$$

$$D_y = \begin{vmatrix} 1 & 1 & 3 \\ 2 & 3 & -2 \\ -2 & -1 & 1 \end{vmatrix} = 1\begin{vmatrix} 3 & -2 \\ -1 & 1 \end{vmatrix} - 2\begin{vmatrix} 1 & 3 \\ -1 & 1 \end{vmatrix} + (-2)\begin{vmatrix} 1 & 3 \\ 3 & -2 \end{vmatrix}$$

$$= 1 - 8 + 22 = 15$$

$$D_z = \begin{vmatrix} 1 & -2 & 1 \\ 2 & 1 & 3 \\ -2 & 2 & -1 \end{vmatrix} = 1\begin{vmatrix} 1 & 3 \\ 2 & -1 \end{vmatrix} - 2\begin{vmatrix} -2 & 1 \\ 2 & -1 \end{vmatrix} + (-2)\begin{vmatrix} -2 & 1 \\ 1 & 3 \end{vmatrix}$$

$$= -7 + 0 + 14 = 7$$

Thus, $x = 28/19$, $y = 15/19$, and $z = 7/19$, and the solution set is $\{(28/19, 15/19, 7/19)\}$. The check is left for the reader.

Exercise Set 4–5

State the dimensions of the following matrices.

1. $\begin{bmatrix} 3 & 0 & 5 \\ 0 & 4 & 2 \end{bmatrix}$

2. $\begin{bmatrix} 2 & 8 & 0 \\ 9 & -1 & 4 \\ 0 & 3 & -5 \\ 7 & 4 & 1 \end{bmatrix}$

3. $\begin{bmatrix} 3 & 2 \\ 7 & -5 \\ 0 & 1 \end{bmatrix}$

4. $\begin{bmatrix} 1 & 4 & 6 & 0 & 3 \\ 0 & 9 & -4 & 2 & -1 \\ 5 & -6 & 0 & 0 & 2 \end{bmatrix}$

5. $\begin{bmatrix} 2 & 5 \\ 1 & -8 \\ -3 & 3 \\ 0 & 7 \end{bmatrix}$

6. $\begin{bmatrix} 0 & 1 & 2 \\ -3 & 2 & 5 \\ 1 & 9 & -4 \end{bmatrix}$

Identify the indicated element in matrix A.

$$A = \begin{bmatrix} 0 & 4 & -2 & 1 \\ 3 & -3 & -1 & 9 \\ 7 & 5 & -6 & 8 \\ 2 & 6 & -4 & -5 \end{bmatrix}$$

7. a_{33}
8. a_{42}
9. a_{24}
10. a_{13}
11. a_{22}
12. a_{31}

Evaluate the following determinants.

13. $\begin{vmatrix} 2 & -3 \\ 3 & 9 \end{vmatrix}$

14. $\begin{vmatrix} -1 & 0 \\ 9 & 6 \end{vmatrix}$

15. $\begin{vmatrix} 7 & -2 \\ -1 & 3 \end{vmatrix}$

16. $\begin{vmatrix} -1 & 1 \\ 1 & 6 \end{vmatrix}$

17. $\begin{vmatrix} 3 & 5 \\ 21 & 35 \end{vmatrix}$

18. $\begin{vmatrix} 2 & -16 \\ -7 & 56 \end{vmatrix}$

19. $\begin{vmatrix} 2 & 2 & 2 \\ 0 & 1 & 3 \\ 0 & -1 & 2 \end{vmatrix}$

20. $\begin{vmatrix} 0 & 0 & 8 \\ 1 & -3 & -29 \\ 2 & 5 & 38 \end{vmatrix}$

21. $\begin{vmatrix} 2 & -2 & 1 \\ 0 & 1 & 1 \\ 3 & -2 & 0 \end{vmatrix}$

22. $\begin{vmatrix} 6 & 1 & 2 \\ 1 & 2 & 1 \\ 0 & 1 & 2 \end{vmatrix}$

23. $\begin{vmatrix} 3 & 1 & -2 \\ 1 & 2 & 3 \\ 3 & 1 & -2 \end{vmatrix}$

24. $\begin{vmatrix} -2 & 3 & 1 \\ 1 & 2 & 3 \\ 10 & -15 & -5 \end{vmatrix}$

Solve the following systems of equations using Cramer's Rule.

25. $3x + 4y = 7$
$5x + 6y = 8$

26. $7x + 3y = -2$
$3x - 7y = -5$

27. $2x - y - 1 = 0$
$3x - 4y - 5 = 0$

28. $x + y + 3 = 0$
$2x + 3y - 1 = 0$

29. $10y = x - 8$
$2x = 4y + 5$

30. $x = -y$
$5x + 2y = 2$

31. $3x = 1 - 6y$
$-15x = 30y - 7$

32. $y - 5x + 7 = 0$
$20x = 28 + 4y$

33. $x = 4$
$4y - 5x = 20$

34. $3y = 5$
$2y - 6x = 9$

35. $\begin{aligned} 3x + 2y + 7z &= -1 \\ x + z &= 2 \\ 5x + 3y + 11z &= 1 \end{aligned}$

36. $\begin{aligned} x + y + z &= 5 \\ x - y + 2z &= 8 \\ x + 2y &= 0 \end{aligned}$

37. $\begin{aligned} 3x + 6y + 6z &= 1 \\ x - z &= 1 \\ 3x - 12y &= -8 \end{aligned}$

38. $\begin{aligned} x + y + z &= 4 \\ 2x - y - 2z &= -1 \\ x - 2y - z &= 1 \end{aligned}$

39. $\begin{aligned} 2x - 3y + 11z &= -3 \\ -2x + 4y - 13z &= 4 \\ x - 2y + 7z &= -2 \end{aligned}$

40. $\begin{aligned} 2x + 4y + 2z &= 3 \\ 6x + 2y - 2z &= 0 \\ x - y - z &= 1 \end{aligned}$

4-6 LINEAR INEQUALITIES

Recall that when we solved inequalities in one variable we used the property that a point on a line divides the line into three disjoint subsets of points: the point itself, the half-line to the left of the point, and the half-line to the right of the point. In a similar way, a line divides a plane into three disjoint subsets of points: the points on the line itself, the points of a half-plane "above" the line, and the points of a half-plane "below" the line. We use this property of a line to graph the solution sets of linear inequalities.

Therefore, if L is a line in a plane and the equation of L is $Ax + By = C$, then the set of points in the two half-planes have coordinates for which $Ax + By > C$ or $Ax + By < C$. The half-plane that contains the points whose coordinates satisfy a given inequality depends on the values of A, B, and C. Suppose, for example, that the values of A, B, and C are such that the graph in Figure 4-11 illustrates the three disjoint subsets of points in the plane.

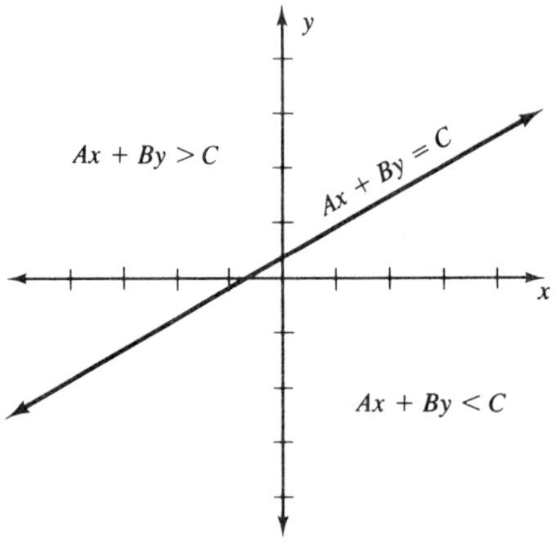

FIGURE 4-11

The first step in solving a linear inequality is to graph the line that is the boundary between the two half-planes. That is, we convert the given inequality $Ax + By < C$ or $Ax + By > C$ into the equality $Ax + By = C$. We then graph the line. To determine which half-plane contains the solution set to the given inequality, we select a point *not on the line*. The coordinates of the point are then substituted in the original inequality. If the substitution yields a true statement, the solution set is the set of points in that half-plane. If the substitution yields a false statement, the solution set is the set of points in the other half-plane.

EXAMPLE 4–33 Graph the solution set of $3x - 4y < 12$.

Solution: Graph the line $3x - 4y = 12$. The line is shown in Figure 4-12 as a dashed line, since the line is not included in the solution set.

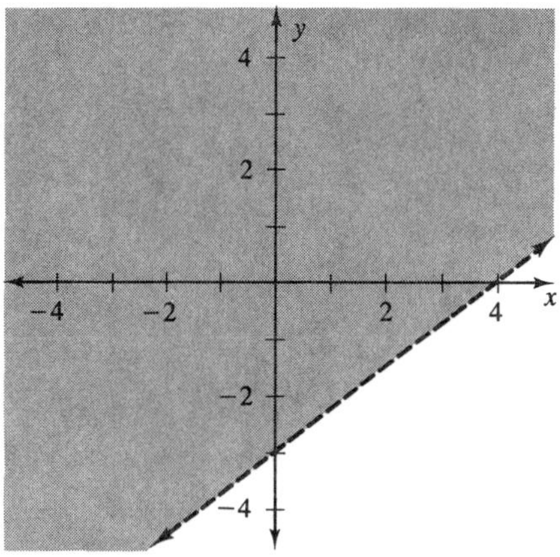

FIGURE 4-12 Graph of $3x - 4y < 12$

We arbitrarily select the origin $O(0, 0)$ and substitute the coordinates in the original inequality. If $(x, y) = (0, 0)$, then

$$3(0) - 4(0) < 12 \qquad \text{(True)}$$

Since a true statement is obtained, the half-plane containing the origin is shaded. Notice that if the resulting statement were false, the other half-plane would be shaded.

The shaded half-plane contains all the points in the plane whose coordinates make the original inequality true.

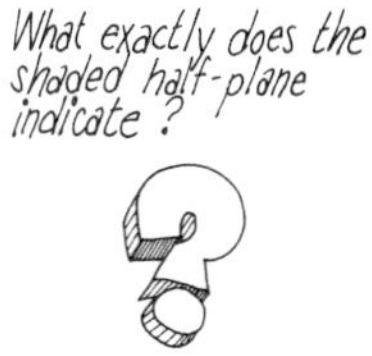

EXAMPLE 4–34 Graph $\{(x, y) | -6 < x + 2y < 8\}$.

Solution: The given inequality is compound and may be written

$$\{(x, y) | -6 < x + 2y\} \cap \{(x, y) | x + 2y < 8\}$$

We graph the two lines $-6 = x + 2y$ and $x + 2y = 8$. That is,

$$y = \frac{-1}{2}x - 3 \quad \text{and} \quad y = \frac{-1}{2}x + 4$$

Notice that the two lines have equal slopes and are therefore parallel. The two dashed lines are shown in Figure 4-13. Now we select three points, one point in each of the indicated disjoint regions, to determine the solution set of the given inequality.

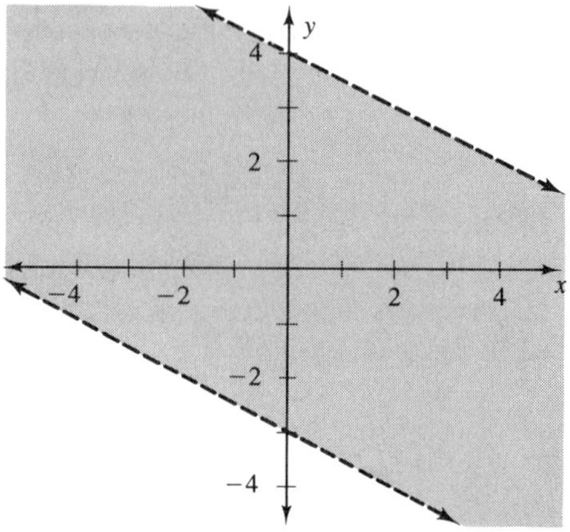

FIGURE 4-13 Graph of $-6 < x + 2y < 8$

If $(x, y) = (3, 4)$, then

$$-6 < 3 + 2(4) < 8 \quad \text{or} \quad -6 < 11 < 8 \qquad \text{(False)}$$

since $11 \nless 8$. If $(x, y) = (0, 0)$, then

$$-6 < 0 + 2(0) < 8 \quad \text{or} \quad -6 < 0 < 8 \qquad \text{(True)}$$

If $(x, y) = (0, -4)$, then

$$-6 < 0 + 2(-4) < 8 \quad \text{or} \quad -6 < -8 < 8 \qquad \text{(False)}$$

since $-6 \nless -8$. Thus, the solution set is the region between the parallel lines and is correspondingly shaded.

EXAMPLE 4–35 Graph $\{(x, y) | 2x + y < 3\} \cap \{(x, y) | 5x - 2y \geq -4\}$.

Solution: Let L_1 and L_2 be the boundary lines of $2x + y < 3$ and $5x - 2y > -4$, respectively. For L_1,

$$2x + y = 3$$
$$y = -2x + 3$$

For L_2,

$$5x - 2y = -4$$
$$y = \frac{5}{2}x + 2$$

L_1 is shown as a dashed line in Figure 4-14, but L_2 is a solid line because the given inequality includes the equality relationship. (The distinction between dashed and solid lines is similar to the distinction between hollow and solid dots for inequalities of one variable.)

The plane is divided into four regions, and a point in each region is selected to determine the solution set. Notice that the solution set is an **intersection of inequalities;** therefore, *both* inequalities must be satisfied by a substitution of the coordinates of a point.

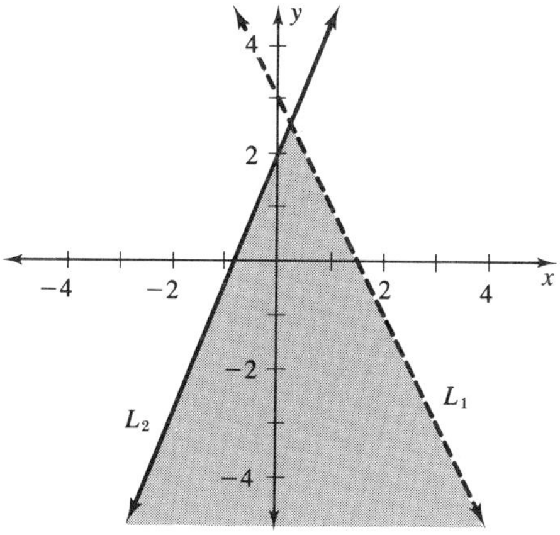

FIGURE 4-14 Graph of $\{(x, y) \mid 2x + y < 3\} \cap \{(x, y) \mid 5x - 2y \geq -4\}$

(a) If $(x, y) = (3, 0)$, then

$$2(3) + 0 < 3 \quad \text{and} \quad 5(3) - 2(0) \geq -4$$
$$6 < 3 \quad \text{and} \quad 15 \geq -4 \qquad \text{(False)}$$

since $6 \not< 3$.

(b) If $(x, y) = (0, 4)$, then

$$2(0) + 4 < 3 \quad \text{and} \quad 5(0) - 2(4) \geq -4$$
$$4 < 3 \quad \text{and} \quad -8 \geq -4 \qquad \text{(False)}$$

for both inequalities.

(c) If $(x, y) = (-3, 0)$, then

$$2(-3) + 0 < 3 \quad \text{and} \quad 5(-3) - 2(0) \geq -4$$
$$-6 < 3 \quad \text{and} \quad -15 \geq -4 \qquad \text{(False)}$$

since $-15 \not\geq -4$.

(d) If $(x, y) = (0, 0)$, then

$$2(0) + 0 < 3 \quad \text{and} \quad 5(0) - 2(0) \geq -4$$
$$0 < 3 \quad \text{and} \quad 0 \geq -4 \qquad \text{(True)}$$

We therefore shade the region containing $O(0, 0)$ for the solution set of the given inequality.

Exercise Set 4–6

Graph the solution set of the following inequalities on a coordinate plane.

1. $x + y > 3$
2. $y - x \leq 2$
3. $2x \leq 3y - 6$
4. $y < -x$
5. $x - 6y \geq 2$
6. $2x + y < 4$
7. $x \leq -3$
8. $y > 4$

9. $y - 5 < 0$
10. $2x - 3 \geq 4$
11. $2 < x < 4$
12. $-3 \leq y \leq -1$
13. $-2 \leq x + 2y \leq 2$
14. $1 \geq y - 3x \geq -1$
15. $0 < x + 3y < 3$
16. $4 < x - 2y < 6$
17. $y \leq x$ or $y \geq 3x$
18. $x + y \leq 2$ or $x - y \leq 2$
19. $2x + 3y \leq 12$ and $3x - 2y \leq 6$
20. $y > x - 9$ and $2x + y > 3$
21. $y \leq 2$ and $y \leq -x$
22. $x \leq 3$ and $x - y > -1$
23. $x > 0$ and $y > 0$ and $x + y < 6$
24. $x - 4y \leq 1$ and $3x + 4y \leq 7$ and $x - y + 5 \geq 0$

4-7 ABSOLUTE VALUE FUNCTION

In Chapter 3 we defined the absolute value of a number. Recall that for any real number x,

$$|x| = x \quad \text{if } x \geq 0$$
$$|x| = -x \quad \text{if } x < 0$$

We also solved absolute value equations and inequalities in one variable.

In this section we define the absolute value function. The function is analyzed so that a quick method for graphing the function can be developed.

Absolute value function

> **Definition:** If a, b, and c are real numbers and $a \neq 0$, then the absolute value function is F, where
>
> $$F = \{(x, y) | y = a|x - b| + c\}$$

The domain of F, unless otherwise restricted, is the set of real numbers. That is, the domain of $F = \{x | x \in \text{ reals}\}$. The range of $F = \{y | y \geq c, \text{ for } a > 0\}$ or $\{y | y \leq c, \text{ for } a < 0\}$.

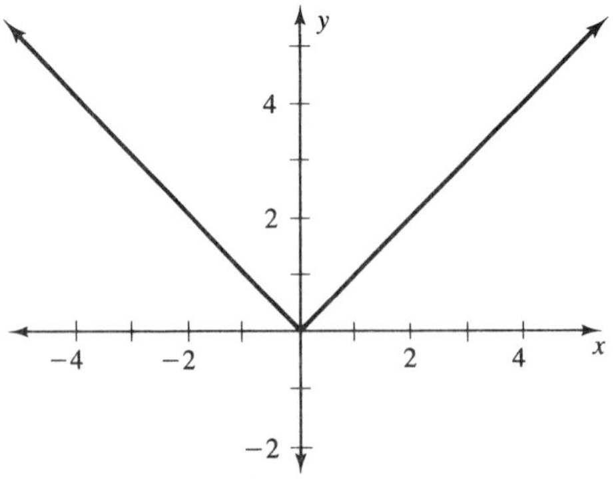

FIGURE 4-15 Graph of $y = |x|$

If $a = 1$ and $b = c = 0$, then $F = \{(x, y)|y = |x|\}$. Recall that $|x| = x$ for $x \geq 0$, and $|x| = -x$ for $x < 0$. Thus, $y = |x|$ is a compound equation such that

$$(1)\ y = x \text{ and } x \geq 0 \quad \text{or} \quad (2)\ y = -x \text{ and } x < 0$$

The graph of (1) is a half-line with slope 1 and y-intercept 0. The graph of (2) is a half-line with slope -1 and y-intercept 0. The union of the two half-lines is the graph in Figure 4-15. The V-shaped graph is standard for all absolute value functions where the domain is the set of real numbers.

The effect of different values for a, b, and c only vary the shape and position of the V, but the general shape is always retained. The point of intersection of the two half-lines is the **vertex** V of the graph.

Do you mean that for different values of a, b, and c we still get a V-shaped graph?

EXAMPLE 4–36 Graph the following functions:
(a) $y = 2|x|$ (b) $y = -3|x|$

Solutions:
(a) If $x \geq 0$, then Equation (1) is $y = 2x$. The graph of Equation (1) is a half-line with slope 2 and y-intercept 0. If $x < 0$, then Equation (2) is $y = -2x$. The graph of Equation (2) is a half-line with slope -2 and y-intercept 0. The union of the two half-lines is the graph in Figure 4-16. Note that the slopes of the two half-lines are ± 2 and the vertex is $V(0, 0)$. For comparison, the graph of $y = |x|$ is shown as a dashed line.

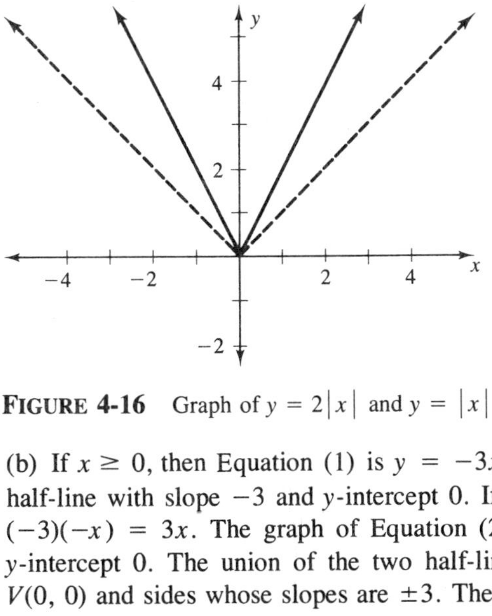

FIGURE 4-16 Graph of $y = 2|x|$ and $y = |x|$ with dashed lines

(b) If $x \geq 0$, then Equation (1) is $y = -3x$. The graph of Equation (1) is a half-line with slope -3 and y-intercept 0. If $x < 0$, then Equation (2) is $y = (-3)(-x) = 3x$. The graph of Equation (2) is a half-line with slope 3 and y-intercept 0. The union of the two half-lines is the inverted V with vertex $V(0, 0)$ and sides whose slopes are ± 3. The graph is shown in Figure 4-17.

> We can summarize the effect of the constant a on the graph of the equation $y = a|x|$, $a \neq 0$.
>
> **Case 1.** If $a > 0$, the V opens upward and the slopes of the half-lines are $\pm a$.
>
> **Case 2.** If $a < 0$, the V opens downward and the slopes of the half-lines are $\pm a$.

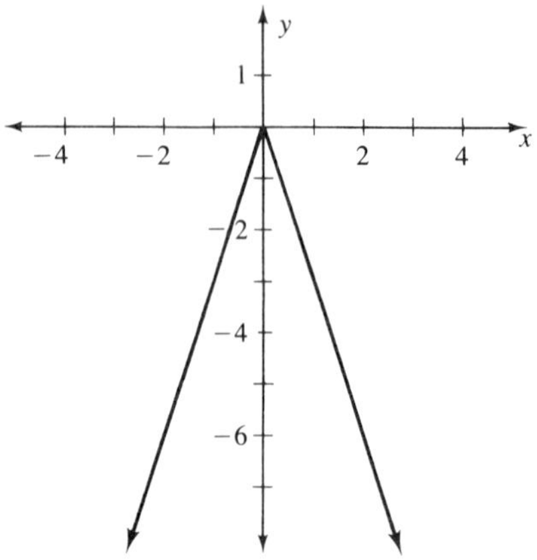

FIGURE 4-17 Graph of $y = -3|x|$

EXAMPLE 4–37 Graph the following functions:
(a) $y = |x - 3|$ (b) $y = |2x + 2|$

Solutions:
(a) If $x - 3 \geq 0$ (that is, $x \geq 3$), then Equation (1) is $y = x - 3$. The graph of Equation (1) is a line with slope 1 and y-intercept -3. However, since $y \geq 0$ for all values of x (the absolute value of any number is nonnegative), we draw only the half-line above the x-axis. If $x - 3 < 0$ (that is, $x < 3$), then Equation (2) is $y = -(x - 3)$, or $y = -x + 3$. The graph of Equation (2) is a line with slope -1 and y-intercept 3. Again, we draw only the half-line above the x-axis. The graph, shown in Figure 4-18, is a V with vertex $V(3, 0)$ and sides whose slopes are ± 1. (Note that the value of a for the given equation is 1.)

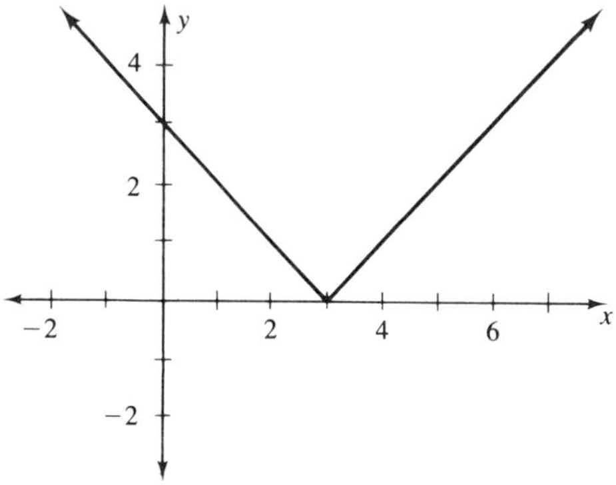

FIGURE 4-18 Graph of $y = |x - 3|$

(b) If $y = |2x + 2|$, then $y = 2|x + 1| = 2|x - (-1)|$. Therefore, for the given equation, $a = 2$ and $b = -1$. The slopes of the sides of the V are ± 2, and $V(-1, 0)$ is the vertex. The graph is shown in Figure 4-19.

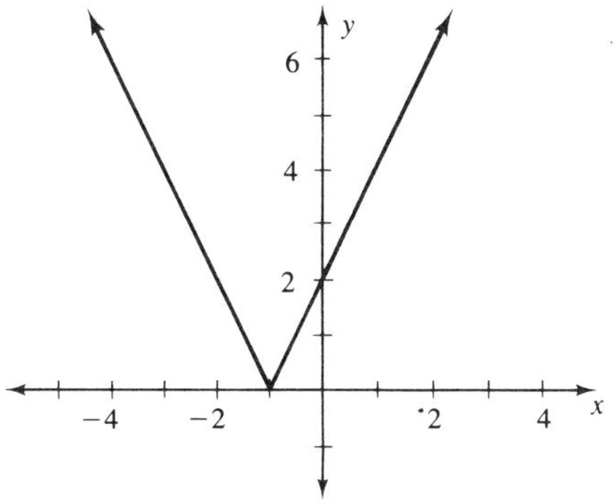

FIGURE 4-19 Graph of $y = 2|x + 1|$

> We can summarize the effect of the constant b on the graph of the equation $y = |x - b|$:
>
> The vertex of the graph is the point $V(b, 0)$.

Notice that, as written, the value of b requires that the coefficient of x be 1. If the coefficient of x is not 1, then we simply factor it out from within the absolute value sign.

EXAMPLE 4–38 Graph the following functions:
(a) $y = |x| - 3$ (b) $y = |x - 2| + 1$

Solutions:
(a) With $a = 1$, the slopes of the sides of the V are ± 1. If $x \geq 0$, then $y = x - 3$; and if $x < 0$, then $y = -x - 3$. With $c = -3$, the vertex is the point $V(0, -3)$. The graph is shown in Figure 4-20. Notice that since $b = 0$, the vertex has not been shifted to the left or right.

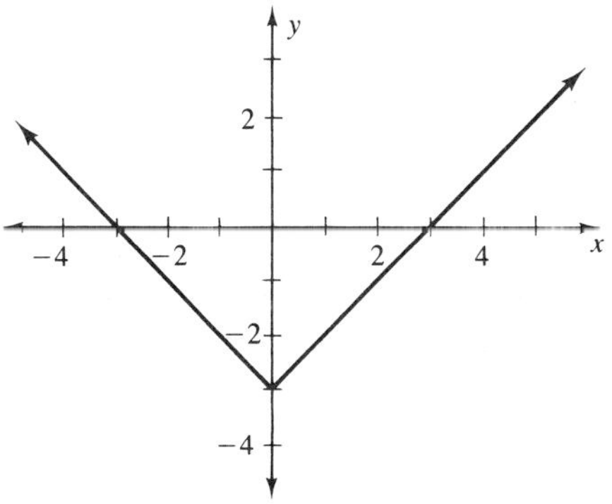

FIGURE 4-20 Graph of $y = |x| - 3$

(b) With $a = 1$, the slopes of the sides of the V are ± 1. Since $b = 2$, the vertex is shifted 2 units to the right. With $c = 1$, the vertex is shifted up 1 unit. The graph with vertex $V(2, 1)$ is shown in Figure 4-21.

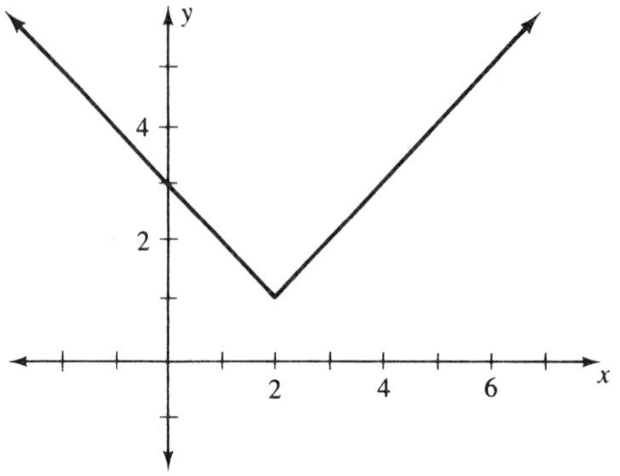

FIGURE 4-21 Graph of $y = |x - 2| + 1$

> In summary, the graph of the function
>
> $$F = \{(x, y) \mid y = a|x - b| + c\}$$
>
> with domain the set of real numbers, has a V shape for which the slopes of the sides are $\pm a$. The V opens upward for $a > 0$ and downward for $a < 0$. The vertex of the V is the point $V(b, c)$.

If the domain of F is the set of real numbers, then the function divides the points of the plane into the following disjoint subsets:

1. The points of the V.
2. The points "inside" the V.
3. The points "outside" the V.

The graph of the absolute value relation

$$R = \{(x, y) \mid y < a|x - b| + c\}$$

or

$$R = \{(x, y) \mid y > a|x - b| + c\}$$

is the region inside the V or outside the V. The approach to graphing a given absolute value relation is similar to the one we used for graphing linear inequalities. That is, we graph the boundary V between regions, then we test the coordinates of an arbitrarily selected point to establish the solution set of the given inequality.

EXAMPLE 4–39 Graph the solution set of $y > 2|x + 1| - 2$.

Solution: We first graph the function $y = 2|x + 1| - 2$. With $a = 2$, $b = -1$, and $c = -2$, the vertex is the point $V(-1, -2)$ and the slopes of the sides of the V are ± 2. The original inequality is strictly "greater than"; thus, the graph is shown as a dashed V in Figure 4-22.

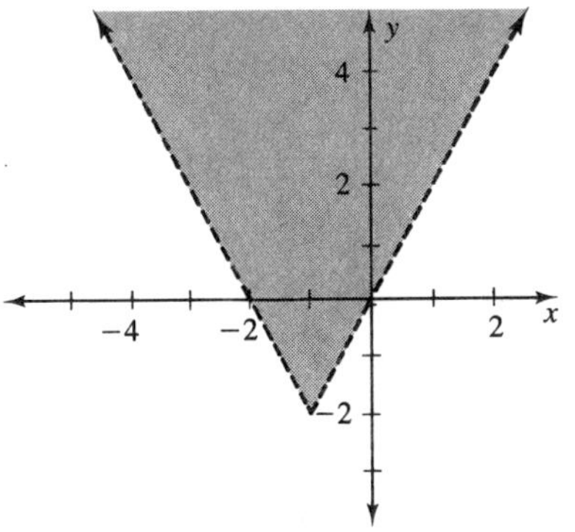

FIGURE 4-22 Graph of $y > 2|x + 1| - 2$

The point $P(-1, 0)$ is in the interior of the V. If $(x, y) = (-1, 0)$, then $0 > 2|-1 + 1| - 2$, or $0 > -2$. This inequality is true; therefore, the solution set is the interior of the V and is accordingly shaded.

EXAMPLE 4–40 Solve the inequality $|x - 2| \leq |2x + 4|$.

Solution: Let $y_1 = |x - 2|$ and $y_2 = |2x + 4| = 2|x + 2|$. The graphs of the two functions are shown in Figure 4-23. Since the solution set of the given inequality is the set of x's such that $|x - 2| \leq |2x + 4|$, we want those values of x for which $y_1 \leq y_2$. From Figure 4-23, we see that the graph of y_1 is below the graph of y_2 everywhere except on the interval between x_1 (approximately -6) and x_2 (approximately $-1/2$).

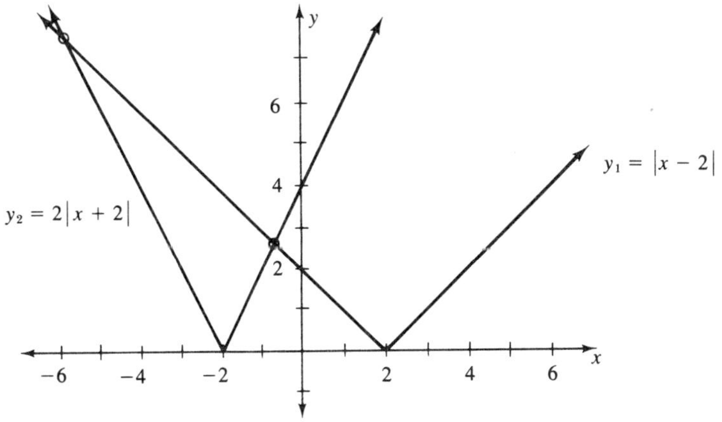

FIGURE 4-23 Graph of $y_1 = |x - 2|$ and $y_2 = 2|x + 2|$

The exact values of x_1 and x_2, the x-coordinates of the points of intersection of y_1 and y_2, are determined by the following procedure. For $x_1 \approx -6$,

$$|x - 2| = -(x - 2) = -x + 2$$

and

$$|2x + 4| = -(2x + 4) = -2x - 4$$

If $|x - 2| = |2x + 4|$, then $-x + 2 = -2x - 4$, and $x = -6$ exactly. For $x_2 \approx -1/2$,

$$|x - 2| = -(x - 2) = -x + 2$$

and

$$|2x + 4| = 2x + 4$$

If $|x - 2| = |2x + 4|$, then $-x + 2 = 2x + 4$, and $x = -2/3$ exactly.

Since the open interval from -6 to $-2/3$ is excluded from the solution set, the solution set is $(-\infty, -6] \cup [-2/3, \infty)$ or $\{x \mid x \leq -6 \text{ or } x \geq -2/3\}$. The graph of the solution set is shown in Figure 4-24.

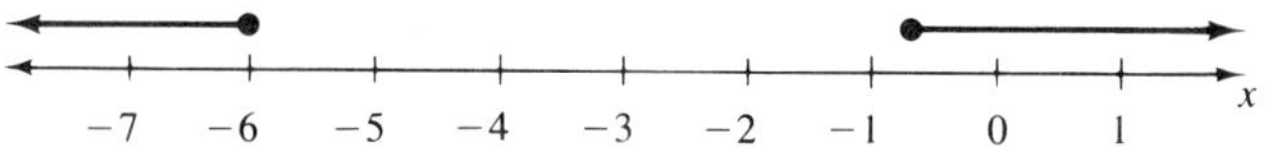

FIGURE 4-24 Graph of $|x - 2| \leq |2x + 4|$

Exercise Set 4–7

Sketch a graph of the following absolute value functions.

1. $y = 3|x|$
2. $y = \frac{1}{4}|x|$
3. $y = -2|x|$
4. $y = \frac{-1}{2}|x|$
5. $y = |x - 1|$
6. $y = |x + 1|$
7. $y = -|x + 5|$
8. $y = -\left|x - \frac{5}{3}\right|$
9. $y = |x| + 1$
10. $y = |x| - 3$
11. $y = -|x| + 2$
12. $y = -|x| - 3$
13. $y = 2|x - 1|$
14. $y = -3|x + 1|$
15. $y = -2|x| - 1$
16. $y = \frac{3}{4}|x| + 3$
17. $y = -3|x - 2| + 5$
18. $y = \frac{1}{3}|x + 4| - 2$
19. $y = 5|x + 3| - 5$
20. $y = \frac{2}{3}\left|x + \frac{1}{2}\right| - \frac{3}{4}$

Graph the following absolute inequalities.

21. $y < |x|$
22. $y \geq |x - 2|$
23. $y > 2|x - 1| + 3$
24. $y < |x - 2| - 2$

25. $y \leq -|x| + 4$

26. $y > 4|x + 2| - 3$

When the positions of x and y are interchanged, a relation is formed whose graph is a V that opens to the left or to the right.

27. Graph $x = |y|$.

28. Graph $x = |y - 2|$.

29. Graph $x = |y| - 2$.

30. Graph $x = -|y| + 1$.

Solve the following equations.

31. $|x - 5| = x - 5$

32. $|x + 2| = -x - 2$

33. $|x - 3| = |2x - 1|$

34. $|2x - 3| = |4x - 7|$

Solve the following inequalities.

35. $|x + 5| < |x + 1|$

36. $|x - 7| \geq |3x + 1|$

37. $|x - 6| \geq |x + 4|$

38. $|2x - 4| < |3x - 9|$

COMPOSITE FUNCTIONS 4–8

In Chapter 1 we performed the rational operations on two polynomials $P(x)$ and $Q(x)$. Recall that polynomials are functions for which the domain is the set of real numbers. In this section we extend the rational operations to any two functions f and g.

Sum, difference, product, and quotient of functions f and g

Definition: If

$$f = \{(x, y) \mid y = f(x)\} \quad \text{and} \quad g = \{(x, y) \mid y = g(x)\}$$

then

(1) $f + g = \{(x, y) \mid y = (f + g)(x) = f(x) + g(x)\}$
(2) $f - g = \{(x, y) \mid y = (f - g)(x) = f(x) - g(x)\}$
(3) $(f)(g) = \{(x, y) \mid y = ((f)(g))(x) = (f(x))(g(x))\}$
(4) $\dfrac{f}{g} = \left\{(x, y) \mid y = \dfrac{f}{g}(x) = \dfrac{f(x)}{g(x)},\ g(x) \neq 0\right\}$

The domain of the function h that is the sum, difference, product, or quotient of f and g is the intersection of the domains of f and g, with the exception that in the division of f and g, $g(x) \neq 0$. That is, only those values of x that are in both the domain of f and the domain of g are in the domain of h.

EXAMPLE 4–41 If $f = \{(0, 0), (1, 1), (2, 4), (3, 9)\}$ and $g = \{(0, 2), (1, 1), (3, 3)\ (4, -1)\}$, find
(a) the domain of the function that is the sum, difference, product, or quotient of f and g
(b) $f + g$ (c) $f - g$ (d) $(f)(g)$ (e) f/g

Solutions:
(a) The domain of $f = \{0, 1, 2, 3\}$ and the domain of $g = \{0, 1, 3, 4\}$. Thus, the domain of the function that is the sum, difference, product, or quotient of f and g is the set $\{0, 1, 3\}$.

Notice that since $g(x) \neq 0$ for any value of x, there is no additional restriction necessary for the domain of f/g.

For each of the elements in the domain above, we find $f(x)$ and $g(x)$:

$$\text{If } x = 0, \text{ then } f(0) = 0 \text{ and } g(0) = 2.$$
$$\text{If } x = 1, \text{ then } f(1) = 1 \text{ and } g(1) = 1.$$
$$\text{If } x = 3, \text{ then } f(3) = 9 \text{ and } g(3) = 3.$$

(b) By definition, $f + g = \{(x, y) \mid y = f(x) + g(x)\}$. Thus,

$$\text{for } x = 0, \quad y = f(0) + g(0) = 0 + 2 = 2$$
$$\text{for } x = 1, \quad y = f(1) + g(1) = 1 + 1 = 2$$
$$\text{for } x = 3, \quad y = f(3) + g(3) = 9 + 3 = 12$$

Therefore, $f + g = \{(0, 2), (1, 2), (3, 12)\}$.
(c) By definition, $f - g = \{(x, y) \mid y = f(x) - g(x)\}$. Thus, $f - g = \{(0, -2), (1, 0), (3, 6)\}$.
(d) By definition, $(f)(g) = \{(x, y) \mid y = (f(x))(g(x)\}$. Thus, $(f)(g) = \{(0, 0), (1, 1), (3, 27)\}$.
(e) By definition, $\dfrac{f}{g} = \left\{(x, y) \mid y = \dfrac{f(x)}{g(x)}, g(x) \neq 0\right\}$. Thus, $\dfrac{f}{g} = \{(0, 0), (1, 1), (3, 3)\}$.

EXAMPLE 4–42 If

$$f = \{(x, f(x)) \mid f(x) = x^2 + 1\} \quad \text{and} \quad g = \{(x, g(x)) \mid g(x) = 1/x\}$$

find
(a) the domain of $f + g$, $f - g$, fg, and f/g
(b) $f + g$ (c) $f - g$ (d) fg (e) f/g

Solutions:
(a) The domain of $f = \{\text{reals}\}$ and the domain of $g = \{\text{reals}, x \neq 0\}$. Therefore, the domain of the sum, difference, product, and quotient of f and g is $\{\text{reals}, x \neq 0\}$.

Notice that $g(x) \neq 0$; thus, no additional restriction is necessary for the domain of f/g.

(b) $(f + g)(x) = (x^2 + 1) + \dfrac{1}{x} = \dfrac{x^3 + x + 1}{x}$

(c) $(f - g)(x) = (x^2 + 1) - \dfrac{1}{x} = \dfrac{x^3 + x - 1}{x}$

(d) $((f)(g))(x) = (x^2 + 1)\left(\dfrac{1}{x}\right) = \dfrac{x^2 + 1}{x}$

(e) $\dfrac{f}{g}(x) = \dfrac{x^2 + 1}{\dfrac{1}{x}} = x^3 + x$

We now define an operation for functions that has no counterpart in the operations on real numbers.

Composition of functions f and g

> **Definition:** If f and g are functions, then the composition of f and g, written $f \circ g$, is given by the equation
>
> $$f \circ g = \{(x, y) \mid y = f(g(x))\}$$
>
> The domain of $f \circ g$ is the set of x's in the domain of g such that $g(x)$ is in the domain of f.

The symbol $f(g(x))$ is read "f of g of x." Recall that if $f = \{(x, y) \mid y = f(x)\}$, then for any value b in the domain of f, $f(b)$ is the value of y when x is everywhere replaced with b. Thus, $f(g(x))$ is the expression that is obtained when x in $f(x)$ is everywhere replaced with $g(x)$. Furthermore, the only values of x in the domain of $f \circ g$ are those values of x in the domain of g for which $g(x)$ is in the domain of f.

EXAMPLE 4–43 If $f = \{(0, 0), (1, 1), (2, 4), (3, 9)\}$ and $g = \{(0, 2), (1, 1), (3, 3), (4, -1)\}$, find
(a) $f \circ g$ (b) $g \circ f$

Solutions:
(a) The domain of $g = \{0, 1, 3, 4\}$.
For $x = 0$, $g(0) = 2$ and $f(2) = 4$; thus, $(0, 4) \in f \circ g$.
For $x = 1$, $g(1) = 1$ and $f(1) = 1$; thus, $(1, 1) \in f \circ g$.
For $x = 3$, $g(3) = 3$ and $f(3) = 9$; thus, $(3, 9) \in f \circ g$.
For $x = 4$, $g(4) = -1$, but $f(-1)$ does not exist. Thus, 4 is not in the domain of $f \circ g$.
Therefore, $f \circ g = \{(0, 4), (1, 1), (3, 9)\}$, and the domain of $f \circ g$ is $\{0, 1, 3\}$.
(b) The domain of $f = \{0, 1, 2, 3\}$.
For $x = 0$, $f(0) = 0$ and $g(0) = 2$; thus, $(0, 2) \in g \circ f$.
For $x = 1$, $f(1) = 1$ and $g(1) = 1$; thus, $(1, 1) \in g \circ f$.
For $x = 2$, $f(2) = 4$ and $g(4) = -1$; thus, $(2, -1) \in g \circ f$.
For $x = 3$, $f(3) = 9$, but $g(9)$ does not exist. Thus, 3 is not in the domain of $g \circ f$.
Therefore, $g \circ f = \{(0, 2), (1, 1), (2, -1)\}$, and the domain of $g \circ f$ is $\{0, 1, 2\}$.

In general, $f \circ g \neq g \circ f$; therefore, *the operation of composition is not commutative*. Notice also in the above example that the domains of $f \circ g$ and $g \circ f$ are different. In general, this will also be the case for the composition of functions f and g.

EXAMPLE 4–44 If

$$f = \{(x, f(x)) \mid f(x) = x^2 + 1\} \quad \text{and} \quad g = \{(x, g(x)) \mid g(x) = 1/x\}$$

find
(a) $(f \circ g)(x)$ (b) the domain of $f \circ g$
(c) $(g \circ f)(x)$ (d) the domain of $g \circ f$

Solutions:

(a) $(f \circ g)(x) = f(g(x)) = \left(\frac{1}{x}\right)^2 + 1 = \frac{1 + x^2}{x^2}$

(b) The domain of g = {reals, $x \neq 0$}; thus, the range of g = {reals, $g(x) \neq 0$}. Since the domain of f = {reals}, then any element in the range of g will give an element in the domain of f; and the domain of $f \circ g$ = {reals, $x \neq 0$}.

(c) $(g \circ f)(x) = g(f(x)) = \frac{1}{x^2 + 1}$

(d) The domain of f = {reals}, and the range of $f = \{f(x) \mid f(x) \geq 1\}$. The domain of g = {reals, $x \neq 0$}, and every element in the range of f is also an element in the domain of g; therefore, the domain of $g \circ f$ = {reals}.

Recall that for the undefined operations on the real numbers — addition and multiplication — there exist unique real numbers that are the **identity elements** for the operations. That is, if $a \in$ real numbers, then

$$a + 0 = 0 + a = a \quad \text{and} \quad a(1) = 1(a) = a$$

Thus, 0 and 1 are the identity elements, respectively, for addition and multiplication. The following theorem identifies the **identity function** for the operation composition.

> **Theorem:** The function $I = \{(x, I(x)) \mid I(x) = x\}$ is the identity function for the operation composition, such that for any function f,
>
> $$(f \circ I)(x) = (I \circ f)(x) = f(x)$$

Figure 4-25 shows the graph of the identity function. We also refer to the identity function as "the $y = x$ line" and "the 45° line" because the graph of $I(x) = x$ makes a 45° angle with the x-axis.

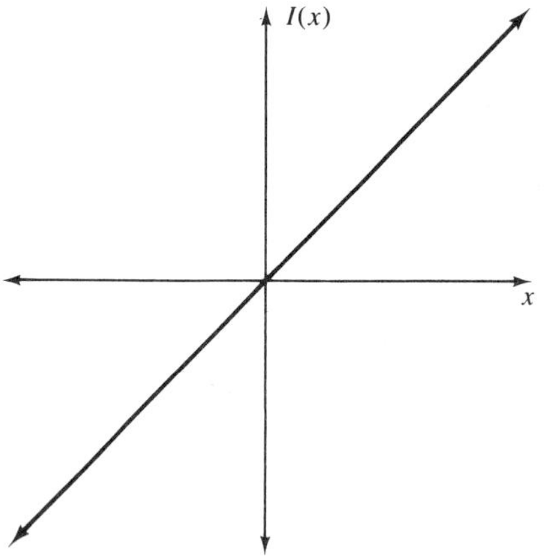

FIGURE 4-25 Graph of identity function $I(x) = x$

EXAMPLE 4–45 If $f = \{(x, f(x)) \mid f(x) = x^3 - 5x\}$ and I is the identity function, find

(a) $(f \circ I)(x)$ (b) $(I \circ f)(x)$

Solutions:
(a) $(f \circ I)(x) = f(I(x)) = x^3 - 5x = f(x)$
(b) $(I \circ f)(x) = I(f(x)) = x^3 - 5x = f(x)$

Notice that $(f \circ I)(x)$ requires that we replace x everywhere in $f(x)$ with x. Since we are replacing x with itself, obviously we will always get $f(x)$ for the result. Furthermore, the domain of $f \circ I$ and $I \circ f$ is the domain of f, for any function f.

Exercise Set 4–8

For each of the following, find
(a) the domain of the sum, difference, product, and quotient of the given functions f and g
(b) $(f + g)(x)$
(c) $(f - g)(x)$
(d) $(f \cdot g)(x)$
(e) $(f/g)(x)$

1. $f(x) = 3x - 1, \quad g(x) = 4 - x$
2. $f(x) = x^3 + 2, \quad g(x) = x^2 - 1$
3. $f(x) = \dfrac{2x}{x^2 - 4}, \quad g(x) = \dfrac{1}{x}$
4. $f(x) = \dfrac{1}{x^2 + 5x}, \quad g(x) = \dfrac{1}{x^2 - 3x}$
5. $f(x) = \dfrac{1}{x^2 - 1}, \quad g(x) = \dfrac{1}{x^2 - 3x + 2}$
6. $f(x) = \dfrac{x}{x + 1}, \quad g(x) = \dfrac{x + 1}{x + 2}$
7. $f(x) = \sqrt{x + 1}, \quad g(x) = \dfrac{1}{\sqrt{x}}$
8. $f(x) = x + 2, \quad g(x) = (x + 2)^2$
9. $f(x) = \dfrac{1}{\sqrt{x + 1}}, \quad g(x) = \dfrac{1}{\sqrt{x - 1}}$
10. $f(x) = \sqrt{x^2 + 4}, \quad g(x) = \sqrt{x^2 - 4}$

For each of the following pairs of functions, find
(a) $(g \circ f)(x)$
(b) the domain of $g \circ f$

11. $f(x) = \sqrt{x}, \quad g(x) = x^2$
12. $f(x) = x^2, \quad g(x) = 1/x^3$
13. $f(x) = \dfrac{x - 1}{x}, \quad g(x) = x^2$
14. $f(x) = 3x^2 - 2x, \quad g(x) = 1/x$
15. $f(x) = x - 1, \quad g(x) = x + 1$
16. $f(x) = 3x - 2, \quad g(x) = \dfrac{x + 2}{3}$

17–22. For each of the pairs of functions for exercises 11–16, find
(a) $(g \circ f)(x)$
(b) the domain of $g \circ f$

23. If $f(x) = x^2 + \dfrac{2}{x}$ and I is the identity function, find
(a) $(f \circ I)(x)$
(b) $(I \circ f)(x)$

24. If $g(x) = x^2 - 1$ and I is the identity function, find
(a) $(g \circ I)(x)$
(b) $(I \circ g)(x)$

4-9 INVERSE FUNCTIONS

In Chapter 1 we noted that for any real number a there exists a unique real number $-a$, called the **additive inverse** of a, such that

$$a + (-a) = (-a) + a = 0$$

Notice that the sum of a and its additive inverse is the *identity element for addition*.

Furthermore, if $a \neq 0$, then there exists a unique real number $1/a$, called the **multiplicative inverse** of a, such that

$$a\left(\frac{1}{a}\right) = \left(\frac{1}{a}\right)a = 1$$

Notice that the product of a and its multiplicative inverse is the *identity element for multiplication*.

In the last section we defined an operation **composition** for functions and identified the identity function I, such that for any function f,

$$f \circ I = I \circ f = f$$

In this section we study the **inverse** of a function f, (written f^{-1}), such that

$$f \circ f^{-1} = f^{-1} \circ f = I$$

Doesn't the negative exponent mean that we write f^{-1} as $\frac{1}{f}$?

Remember that a number with a negative exponent means the reciprocal of the number. But f is a function, and f^{-1} symbolizes the inverse of the function, not the reciprocal. More importantly though, we need to establish that *not all functions have inverse functions*. Therefore, we define the set of functions that always have inverse functions.

One-to-one function

> **Definition:** A function f is one-to-one if and only if for each element in the range of f, there is only one element in the domain of f.

Recall that a relation is a function if and only if every element in the domain is paired with only one element in the range. But a function is one-to-one if and only if every element in the range is paired with only one element in the domain. Thus, $R = \{(1, 2), (2, 2)\}$ is a function, but the function is not one-to-one.

EXAMPLE 4–46 Determine whether the stated functions are one-to-one.
(a) $f = \{(5, 3), (4, 1), (-1, 4), (8, 2)\}$
(b) $g = \{(3, 7), (0, 2), (-4, 5), (-9, 2)\}$

Solutions:
(a) f is a one-to-one function because each of the range elements $\{3, 1, 4, 2\}$ is paired with only one element in the domain of f.
(b) g is not a one-to-one function because the range element 2 is paired with two domain elements, namely 0 and -9.

Example 4–47 Determine whether the following graphs are of one-to-one functions.

(a) Figure 4-26 (b) Figure 4-27

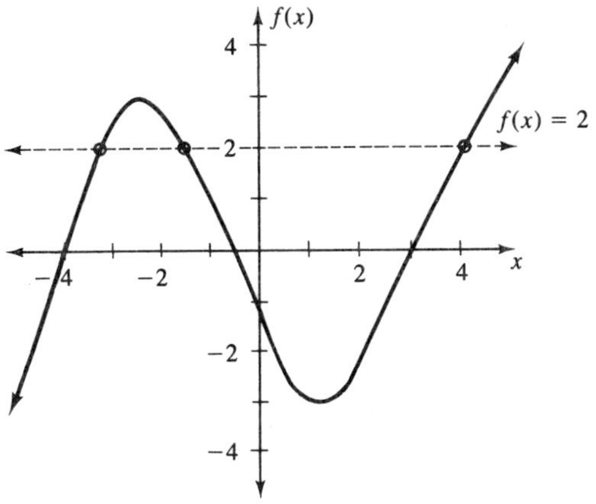

Figure 4-26 Graph of a function, f

Solutions:

(a) The horizontal line $f(x) = 2$ intersects the graph of f in three points as indicated. Thus, the range element 2 is paired with three domain elements, and f is not a one-to-one function.

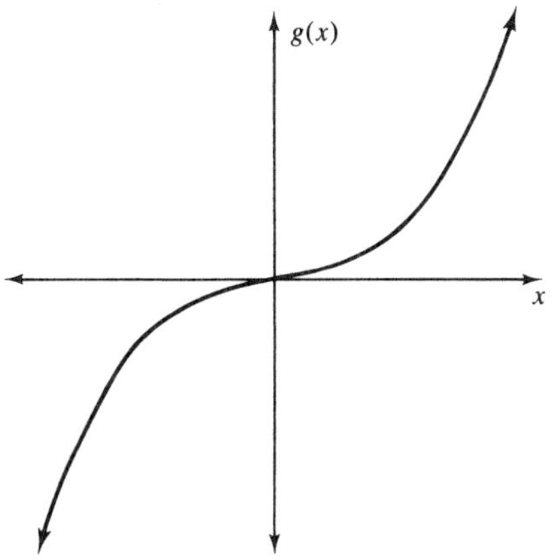

Figure 4-27 Graph of a function, g

(b) No horizontal line can intersect the graph of g in more than one point; therefore, every range element in g is paired with exactly one domain element of g, and the function is one-to-one.

The above example illustrates the use of the horizontal line test on the graph of a function to determine whether the function is one-to-one.

> **Horizontal Line Test:** If any horizontal line can intersect the graph of a function in *at most one point,* then the function is one-to-one.

EXAMPLE 4–48 Determine whether the following functions are one-to-one.
(a) $f = \{(x, y) \mid y = 3x + 5\}$ (b) $g = \{(x, y) \mid y = x^2 + 2x\}$

Solutions:

(a) If $y = 3x + 5$, then $x = \dfrac{y - 5}{3}$. Since every replacement for y with an element in the range of f yields only one value for x, f is one-to-one.
(b) If $y = x^2 + 2x$, then

$$x^2 + 2x - y = 0$$
$$x = \frac{-2 \pm \sqrt{4 + 4y}}{2}$$
$$= -1 \pm \sqrt{1 + y}$$

Every replacement for y, such that $y \geq -1$, yields two values of x, that is, $-1 + \sqrt{1 + y}$ and $-1 - \sqrt{1 + y}$. Therefore, g is not a one-to-one function.

The following definition identifies the inverse function.

Inverse function

> **Definition:** If f is a one-to-one function, then f^{-1} is the inverse function of f if and only if
>
> $$f^{-1} = \{(b, a) \mid (a, b) \in f\}$$

The definition states that the numbers in each ordered pair of f and f^{-1} are the same, but the order of the numbers is reversed for f and f^{-1}. For example, if $f = \{(1, 2), (2, 7), (3, 12), (4, 17)\}$, then $f^{-1} = \{(2, 1), (7, 2), (12, 3), (17, 4)\}$.

Then the domains and ranges of f and f^{-1} must be related.

The domains and ranges of f and f^{-1} are related, and the following equations identify the relationship:

The domain of f = the range of f^{-1}.

The range of f = the domain of f^{-1}.

For example, if $f = \{(x, y) \mid y = \sqrt{x - 2}\}$, the domain of $f = \{x \mid x \geq 2\}$ and the range of $f = \{y \mid y \geq 0\}$. Therefore, the domain of $f^{-1} = \{x \mid x \geq 0\}$, and the range of $f^{-1} = \{y \mid y \geq 2\}$.

The following equations state an important relationship between f and f^{-1}:

> If f is a one-to-one function with inverse f^{-1}, then
>
> $$f \circ f^{-1} = f^{-1} \circ f = I$$

These equations state that the composition of f with its inverse function is the *identity function for composition*.

The following flowchart and example outline the process for determining the inverse of a one-to-one function. Notice that the process "works" because of the definition of f^{-1}. That is, by interchanging the variables in the equation for f, we in effect are interchanging the order of every ordered pair in f and obtaining the ordered pairs in f^{-1}.

Given f is a one-to-one function such that

$$f = \{(x, y) \mid y = f(x)\},$$

↓

form $f^{-1} = \{(x, y) \mid x = f(y)\}$. That is, replace x with y and y with x everywhere in the equation for f.

↓

If possible, express f^{-1} as a function of x by solving $x = f(y)$ for y.

↓

Verify that f^{-1} is the inverse of f by computing $f \circ f^{-1}$ and $f^{-1} \circ f$ to obtain $I = \{(x, y) \mid y = x\}$.

EXAMPLE 4–49 If $f = \{(x, y) \mid y = 2x + 5\}$, find f^{-1}.

Solution: If $f = \{(x, y) \mid y = 2x + 5\}$, then $f^{-1} = \{(x, y) \mid x = 2y + 5\}$.

If $x = 2y + 5$, then $y = \dfrac{x - 5}{2}$ and

$$f^{-1} = \left\{(x, y) \mid y = \frac{x - 5}{2}\right\}.$$

Check.

(a) $(f \circ f^{-1})(x) = f(f^{-1}(x))$

$$= 2\left(\frac{x - 5}{2}\right) + 5$$

$$= x - 5 + 5 = x$$

(b) $(f^{-1} \circ f)(x) = f^{-1}(f(x))$

$$= \frac{(2x + 5) - 5}{2}$$

$$= \frac{2x}{2} = x$$

EXAMPLE 4–50 If $f = \{(x, y) \mid y = x^3 + 2\}$,
(a) Find f^{-1}.
(b) Graph f, f^{-1}, and I on the same set of axes.

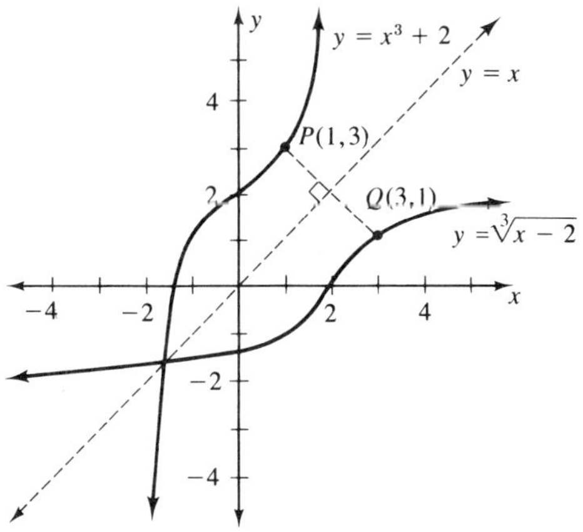

FIGURE 4-28 Graphs of f, f^{-1}, and I

Solutions:
(a) If $f = \{(x, y) \mid y = x^3 + 2\}$, then $f^{-1} = \{(x, y) \mid x = y^3 + 2\}$. If $x = y^3 + 2$, then

$$y = \sqrt[3]{x - 2} \quad \text{and} \quad f^{-1} = \{(x, y) \mid y = \sqrt[3]{x - 2}\}$$

(b) The graphs of f, f^{-1}, and I are shown in Figure 4-28.

On the basis of the graphs of f, f^{-1}, and I in Figure 4-28, we make the following observations regarding the graphs of any one-to-one function, its inverse function, and the identity function:

1. For every point $P(a, b)$ on f, there is a point $Q(b, a)$ on f^{-1} such that the graph of I is the perpendicular bisector of line segment $\overline{PQ}$. For example, in Figure 4-28, $P(1, 3)$ is on the graph of f, $Q(3, 1)$ is on the graph of f^{-1}, and the line $y = x$ is the perpendicular bisector of line segment $\overline{PQ}$.
2. The graphs of f and f^{-1} are *symmetric* with respect to the graph of the identity function. That is, when the plane containing the graphs of f and f^{-1} is folded along the line $y = x$, the two graphs coincide.
3. If the graphs of f and f^{-1} intersect, then their intersection is a point on the graph of the identity function.

If a function is not one-to-one, then it doesn't have an inverse?

We can form the inverse of any function by the method outlined in this section. However, *if f is not a one-to-one function, then the inverse of f is a relation and not a function.*

For example, if $f = \{(x, y) \mid y = x^2\}$, then the inverse of $f = \{(x, y) \mid y = \pm\sqrt{x}\}$; and the inverse of f is a relation that is not a function.

We sometimes use a procedure to convert a function f that is not one-to-one to a function that is one-to-one. The process requires that we restrict the natural domain of f. For example, if $f = \{(x, y) \mid y = x^2\}$, the domain of $f = \{x \mid x \in \text{reals}\}$, the range $= \{y \mid y \geq 0\}$, and the inverse of f is not a function. However, if we restrict the domain of f to $\{x \mid x \geq 0\}$, then the inverse of f is a function. The graphs of f, f^{-1}, and I for the restricted domain are shown in Figure 4-29.

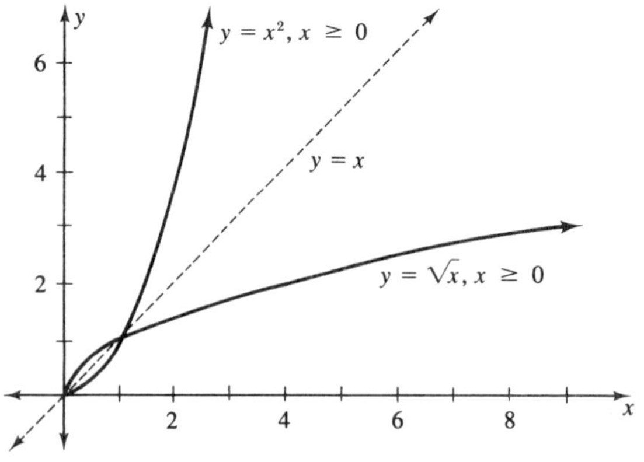

FIGURE 4-29 Graphs of f, f^{-1}, and I

Exercise Set 4–9

Determine whether the following functions are one-to-one.

1. $f = \{(3, 1), (4, 2), (5, 3), (2, -1)\}$
2. $g = \{(1, 7), (2, 6), (3, 7), (4, 2)\}$
3. $h = \{(-1, -1), (2, 2), (-3, -3), (1, -1)\}$
4. $j = \{(1/2, 1/3), (7/8, 3/4), (-3/4, 1/4), (-9/8, 3/5)\}$

Determine whether the following functions are one-to-one.

5.

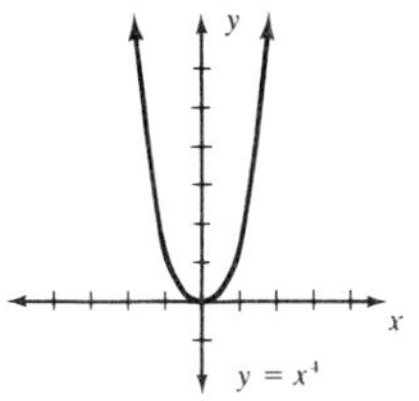

6.

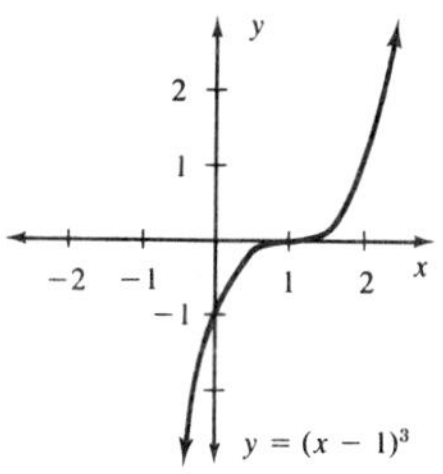

7.

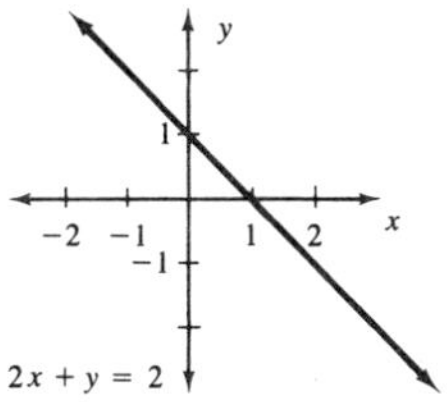

8.

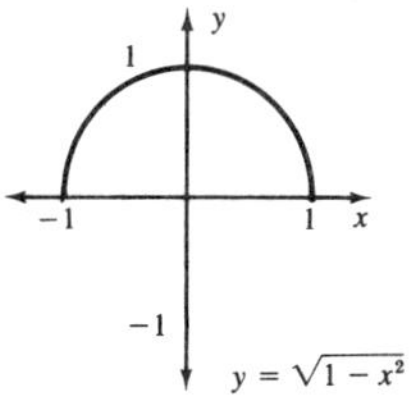

Determine whether the following functions are one-to-one.

9. $f = \{(x, y) \mid y = -2x + 3\}$
10. $h = \{(x, y) \mid y = x^2 - 4x + 4\}$
11. $k = \{(x, y) \mid y = (x + 1)^3\}$
12. $m = \{(x, y) \mid y = \sqrt{1 + x^2}\}$

For the following one-to-one functions, find f^{-1} and verify that $f \circ f^{-1}$ and $f^{-1} \circ f$ both equal I.

13. $f = \{(x, y) \mid y = 2x + 1\}$
14. $f = \{(x, y) \mid y = 2x^3\}$
15. $f = \left\{(x, y) \mid y = \dfrac{1}{x}\right\}$
16. $f = \left\{(x, y) \mid y = \sqrt{x - 1}\right\}$
17. $f = \left\{(x, y) \mid y = \dfrac{2x}{3x + 1}\right\}$
18. $f = \left\{(x, y) \mid y = \dfrac{x^2}{1 + x}, x > 0\right\}$
19. If $f = \{(x, y) \mid y = x^2 + 2, x \geq 0\}$,
    (a) Find f^{-1}.
    (b) Graph f and f^{-1} on the same set of axes.
20. If $g = \{(x, y) \mid y = 3x - 2\}$,
    (a) Find g^{-1}.
    (b) Graph g and g^{-1} on the same set of axes.

21. If $h = \{(x, y) \mid y = \sqrt{1 - x^2},\ 0 \le x \le 1\}$,
(a) Find h^{-1}.
(b) Graph h and h^{-1} on the same set of axes.

22. If $k = \{(x, y) \mid y = x^3\}$,
(a) Find k^{-1}.
(b) Graph k and k^{-1} on the same set of axes.

SUMMARY

Relation — **Definition:** A relation is a set of ordered pairs of numbers.

Domain — **Definition:** The domain of a relation is the set of all first elements.

Range — **Definition:** The range of a relation is the set of all second elements.

Function — **Definition:** A function is a relation such that each domain element is paired with one and only one range element.

Theorem: If P is a polynomial in x, then

(1) The domain of $P = \{x \mid x \in \text{reals}\}$.
(2) P is a function.

x-intercepts — **Definition:** If $R = \{(x, y) \mid y = r(x)\}$, then the x-intercepts of R are the real values of x for which $r(x) = 0$.

y-intercepts — **Definition:** If $R = \{(x, y) \mid y = r(x)\}$, then the y-intercepts of R are the real values of y for which $x = 0$.

Vertical asymptote — The line $x = a$ is a vertical asymptote of the graph of a relation R if the values of $r(x)$ become infinitely large through positive values or infinitely small through negative values, or both, as the values of x approach a.

Horizontal asymptote — The line $y = h$ is a horizontal asymptote of the graph of a relation R if the values of $r(x)$ approach h as $|x|$ becomes infinitely large.

Symmetry with respect to the y-axis — **Definition:** If R is a relation defined by the equation $y = r(x)$, and for every x in the domain of R it is true that $r(-x) = r(x)$, then the graph of R is symmetric with respect to the y-axis.

Linear relation — **Definition:** A relation L that can be expressed in the form

$$L = \{(x, y) \mid Ax + By + C = 0\}$$

is a linear relation, where A, B, and C are real numbers but A and B are not both zero.

y-intercept of L — **Definition:** If $P(0, b)$ is a point on L, then b is the y-intercept of L.

x-intercept of L — **Definition:** If $P(a, 0)$ is a point of L, then a is the x-intercept of L.

Slope — **Definition:** If $P_1(x_1, y_1)$ and $P_2(x_2, y_2)$ are points on the graph of $Ax + By + C = 0$ and $B \neq 0$, then the slope of the graph of the relation is m, where

$$m = \frac{y_2 - y_1}{x_2 - x_1} \quad \text{or} \quad m = \frac{y_1 - y_2}{x_1 - x_2}, \quad x_1 \neq x_2$$

Definition: The slope-intercept equation of a line is of the form $y = mx + b$, where m is the slope and b is the y-intercept. **Slope-intercept equation**

Definition: The point-slope equation of a nonvertical line L with slope m and containing $P_1(x_1, y_1)$ is $y - y_1 = m(x - x_1)$. **Point-slope equation**

Theorem: If L_1 and L_2 are nonvertical lines in the same plane with slopes m_1 and m_2, respectively, then L_1 is parallel to L_2 if and only if $m_1 = m_2$.

Theorem: If L_1 and L_2 are nonvertical lines in the same plane with slopes m_1 and m_2, respectively, then L_1 is perpendicular to L_2 if and only if $(m_1)(m_2) = -1$.

Definition: A system of equations is a set of two or more equations. **System of equations**

Definition: A solution of a system of two equations in two unknowns is an ordered pair that satisfies each equation of the system. **Solution of a system of two equations**

Definition: Two systems of equations are equavalent if and only if they have the same solution set. **Equivalent systems**

Definition: A matrix is a rectangular array of numbers arranged in rows and columns. The numbers are called *elements* of the matrix. **Matrix**

Definition: The dimension of a matrix with m rows and n columns is $m \times n$ (read "m by n"). **Dimension of a matrix**

Definition: The element in the ith row and jth column of a matrix is the ijth element. For matrix A, the symbol for the ijth element is a_{ij}. **Element of a matrix**

Definition: If A is an $n \times n$ matrix, then A is a square matrix of order n. **Square matrix**

Definition: If A is a square matrix of order $n \geq 2$, and every element of A is a real number, then there exists a real number D called the determinant of A. **Determinant of a square matrix**

Definition: If **Determinant of A**

$$A = \begin{bmatrix} a_{11} & a_{12} \\ a_{21} & a_{22} \end{bmatrix}$$

then the determinant of A is

$$D = \begin{vmatrix} a_{11} & a_{12} \\ a_{21} & a_{22} \end{vmatrix}$$

and $D = (a_{11})(a_{22}) - (a_{21})(a_{12})$.

Definition: The minor of the a_{ij} element in the determinant D is the determinant obtained by deleting the row and column containing the element a_{ij}. **Minor of a_{ij}**

Theorem: If

$$D = \begin{vmatrix} a_{11} & a_{12} & a_{13} \\ a_{21} & a_{22} & a_{23} \\ a_{31} & a_{32} & a_{33} \end{vmatrix}$$

then the expansion of D by the elements in column one is

$$a_{11}\begin{vmatrix} a_{22} & a_{23} \\ a_{32} & a_{33} \end{vmatrix} - a_{21}\begin{vmatrix} a_{12} & a_{13} \\ a_{32} & a_{33} \end{vmatrix} + a_{31}\begin{vmatrix} a_{12} & a_{13} \\ a_{22} & a_{23} \end{vmatrix}$$

Theorem: If

$$a_1x + b_1y = c_1 \quad \text{and} \quad a_2x + b_2y = c_2$$

and $a_1b_2 \neq a_2b_1$, then

$$x = \frac{D_x}{D} \quad \text{and} \quad y = \frac{D_y}{D}$$

where

$$D = \begin{vmatrix} a_1 & b_1 \\ a_2 & b_2 \end{vmatrix}, \quad D_x = \begin{vmatrix} c_1 & b_1 \\ c_2 & b_2 \end{vmatrix}, \quad D_y = \begin{vmatrix} a_1 & c_1 \\ a_2 & c_2 \end{vmatrix}$$

Absolute value function

Definition: If a, b, and c are real numbers, and $a \neq 0$, then the absolute value function is F, where $F = \{(x, y) \mid y = a|x - b| + c\}$.

Sum, difference, product, and quotient of functions *f* and *g*

Definition: If $f = \{(x, y) \mid y = f(x)\}$ and $g = \{(x, y) \mid y = g(x)\}$, then

(1) $f + g = \{(x, y) \mid y = f(x) + g(x)\}$

(2) $f - g = \{(x, y) \mid y = f(x) - g(x)\}$

(3) $(f)(g) = \{(x, y) \mid y = (f(x))(g(x))\}$

(4) $\dfrac{f}{g} = \left\{(x, y) \mid y = \dfrac{f(x)}{g(x)}, g(x) \neq 0\right\}$

Composition of functions *f* and *g*

Definition: If f and g are functions, then the composition of f and g, written $f \circ g$, is given by the equation $f \circ g = \{(x, y) \mid y = f(g(x))\}$.

Theorem: The function $I = \{(x, I(x)) \mid I(x) = x\}$ is the identity function for the operation composition, such that for any function f, $(f \circ I)(x) = (I \circ f)(x) = f(x)$.

One-to-one function

Definition: A function f is one-to-one if and only if for each element in the range of f, there is one and only one element in the domain of f.

Inverse function

Definition: If f is a one-to-one function, then f^{-1} is the inverse function of f if and only if $f^{-1} = \{(b, a) \mid (a, b) \in f\}$.

SELF-CHECKING EXERCISE

For each of the following relations, (a) determine which are functions and (b) find the domain and range of each set. **(4-1)**

1. $\{(2, -3), (3, -3), (7, 2), (-4, 1)\}$
2. $\{(5, 0), (5, -10), (5, \sqrt{3}), (5, 192)\}$
3. $\{(-15, -2), (-3, -2), (\sqrt{2}, -2), (35/2, -2), (99, -2)\}$
4. $\{(\text{aunt } A, \text{nephew of } A)\}$

For each of the following relations, (a) determine the domain and range and (b) state whether or not the relation is a function. **(4-1)**

5. $R = \left\{(x, y) \mid y = \dfrac{x + 1}{x - 1}\right\}$

6. $S = \{(x, y) \mid y = 4 - x^2\}$

If $f(x) = \dfrac{x + 1}{x}$, find the following. **(4-1)**

7. $f(-1)$

8. $f(3/2)$

9. $f(a + h) - f(a)$

10. $\dfrac{f(a + h) - f(a)}{h}$

Determine the x- and y-intercepts of the following relations, if they exist. **(4-2)**

11. $y = x^2 + 4x - 12$

12. $y = \dfrac{x^2 - 16}{x + 2}$

Determine the horizontal and vertical asymptotes of the graphs of the following relations, if they exist. **(4-2)**

13. $y = x^3 - 4x + 1$

14. $y = \dfrac{x + 1}{x^2 - 9}$

Without graphing, determine whether the following relations have symmetry with respect to the y-axis. **(4-2)**

15. $y = \dfrac{4}{x^4 + x^3}$

16. $y = \dfrac{\sqrt{36 - 9x^2}}{2}$

Sketch graphs of the following relations. **(4-2)**

17. $y = \dfrac{3x - 6}{x + 2}$

18. $y = \dfrac{x + 9}{x^2 - 9}$

For the following linear relations, find (a) the y-intercept and (b) the slope. **(4-3)**

19. $3x + y - 5 = 0$

20. $4y - 5x = 12$

Find the slope of the line that contains the following pairs of points. **(4-3)**

21. $(3, -2), (1, -8)$

22. $(8, -4), (0, 4)$

23. Find the standard form of the equation of the line containing the points $(-3, 4)$ and $(6, -3)$. **(4-3)**

24. Find the equation of the line perpendicular to the line $2x + 3y - 7 = 0$ and containing the point $(-1, -4)$. **(4-3)**

Solve the following systems of equations. **(4-4)**

25. $$\begin{aligned} 3x + 2y &= 5 \\ x - 3y &= -2 \end{aligned}$$

26. $$\begin{aligned} 6x - 6y + 1 &= 0 \\ 3y &= 5 - 6x \end{aligned}$$

27. $$\begin{aligned} \frac{1}{x} + \frac{2}{y} + 8 &= 0 \\ \frac{4}{x} - \frac{3}{y} + 10 &= 0 \end{aligned}$$

28. $$\begin{aligned} 2x + y \phantom{{}- z} &= 3 \\ 3y - z &= 4 \\ x - y + 2z &= -2 \end{aligned}$$

State the dimensions of the following matrices. **(4-5)**

29. $$\begin{bmatrix} 2 & 4 \\ 1 & 0 \\ -5 & 6 \end{bmatrix}$$

30. $$\begin{bmatrix} -1 & 0 & 2 & 0 & 5 \\ 3 & 1 & -4 & 2 & 0 \\ -2 & 6 & 0 & -3 & 1 \\ 0 & 4 & 9 & 2 & 7 \end{bmatrix}$$

Identify the indicated element in matrix B. **(4-5)**

$$B = \begin{bmatrix} 0 & 4 & -1 & -4 \\ 2 & 1 & -2 & 6 \\ -3 & 5 & 3 & -5 \end{bmatrix}$$

31. b_{34} 32. b_{23} 33. b_{12} 34. b_{31}

Evaluate the following determinants. **(4-5)**

35. $\begin{vmatrix} 3 & 2 \\ 4 & -1 \end{vmatrix}$

36. $\begin{vmatrix} -2 & 5 \\ -3 & 6 \end{vmatrix}$

37. $\begin{vmatrix} 1 & -2 & 3 \\ 0 & 5 & -10 \\ -3 & 4 & -1 \end{vmatrix}$

38. $\begin{vmatrix} 1 & 2 & -3 \\ 5 & 3 & -6 \\ 3 & -1 & 0 \end{vmatrix}$

Use Cramer's Rule to solve the following systems of equations. **(4-5)**

39. $2x + 3y + 4 = 0$
$7y + 5x + 8 = 0$

40. $2x + 3y + 6z = 1$
$x + y - z = 0$
$5x + 2y + z = 2$

Graph the solution set of the following inequalities on a coordinate plane. **(4-6)**

41. $3x - y < 2$

42. $y \leq 2x + 1$ and $2y - 2 \geq x$

Graph the following absolute value equations and inequalities. For each graph, label the coordinates of the vertex. **(4-7)**

43. $y = |3x + 6|$

44. $y = \dfrac{-|x - 1|}{2} + 3$

45. $y > 2|x| - 4$

46. $y \leq -|x + 2| + 1$

If $f = \{(x, f(x)) \mid f(x) = x^2\}$ and $g = \{(x, g(x)) \mid g(x) = 1/x\}$, find the following. **(4-8)**

47. $(f + g)(x)$ 48. $(f - g)(x)$ 49. $(f \cdot g)(x)$

50. $(f/g)(x)$ 51. $(f \circ g)(x)$ 52. $(g \circ f)(x)$

53. The domain of the sum, difference, product, and quotient of f and g.

54. The domain of $f \circ g$.

55. $(f \circ I)(x)$

Determine whether the following functions are one-to-one. **(4-9)**

56. $f = \left\{(x, y) \mid y = \sqrt{\dfrac{3x + 2}{1 - x}}\right\}$

57. $g = \{(x, y) \mid y^3 - 1 = 2x\}$

For the following one-to-one functions, (a) find the inverse function, (b) state the domain and range of the inverse function, and (c) verify that the composition of the function and its inverse is I. **(4-9)**

58. $f = \{(x, y) \mid y = x^3 + 2\}$

59. $g = \left\{(x, y) \mid y = \dfrac{x}{x + 1}\right\}$

Answers to Self-Checking Exercise

1. (a) Yes (b) Domain = $\{-4, 2, 3, 7\}$; range = $\{-3, 1, 2\}$
2. (a) No (b) Domain = $\{5\}$; range = $\{-10, 0, \sqrt{3}, 192\}$

3. (a) Yes (b) Domain = $\{-15, -3, \sqrt{2}, 35/2, 99\}$; range = $\{-2\}$

4. (a) No, a given nephew may have more than one aunt.
   (b) Domain = {all aunts in the world}; range = {all nephews in the world}

5. (a) Domain = $\{x \mid x \in \text{reals}, x \neq 1\}$; range = $\{y \mid y \in \text{reals}, y \neq 1\}$
   (b) Yes

6. (a) Domain = $\{x \mid x \in \text{reals}\}$; range = $\{y \mid y \leq 4\}$ (b) Yes

7. 0

8. 5/3

9. $\dfrac{-h}{a(a+h)}$

10. $\dfrac{-1}{a(a+h)}$

11. x-intercepts -6, 2
    y-intercept -12

12. x-intercepts -4, 4
    y-intercept -8

13. No horizontal or vertical asymptote

14. Horizontal asymptote $y = 0$; vertical asymptotes $x = -3$ and $x = 3$

15. No

16. Yes

17.

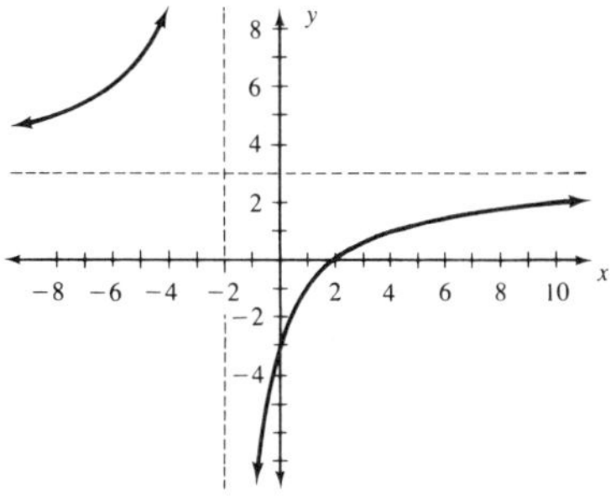

18.

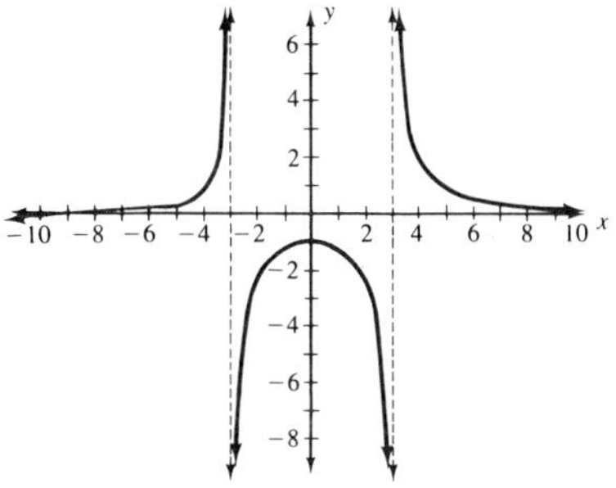

19. (a) y-intercept = 5
    (b) Slope = -3

20. (a) y-intercept = 3
    (b) Slope = 5/4

21. 3

22. -1

23. $7x + 9y = 15$

24. $3x - 2y = 5$

25. (1, 1)

26. (1/2, 2/3)

27. $(-1/4, -1/2)$

28. $(1, 1, -1)$

29. 3×2

30. 4×5

31. -5

32. -2

33. 4

34. -3

35. -11

36. 3

37. 20

38. 0

39. $(4, -4)$

40. $(1/2, -1/3, 1/6)$

41. 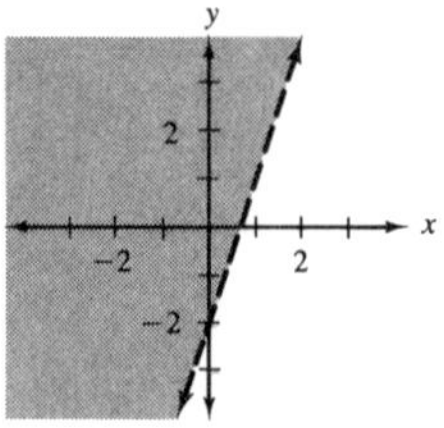

42. 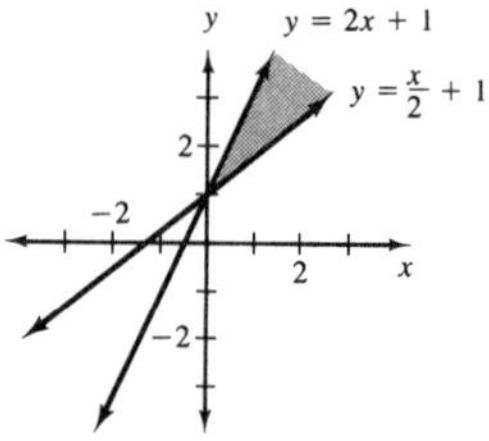

43. 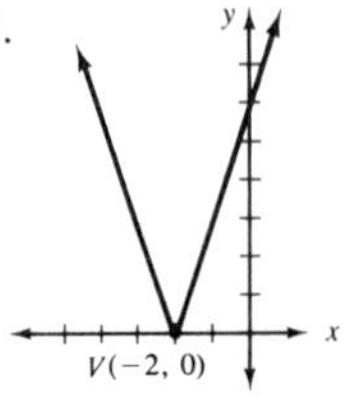

44. 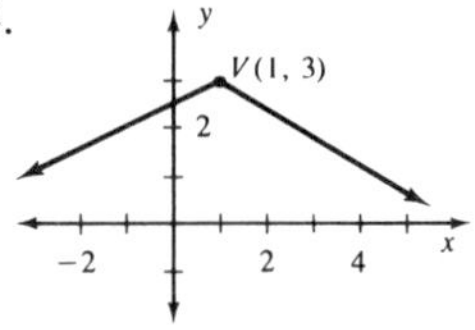

45. 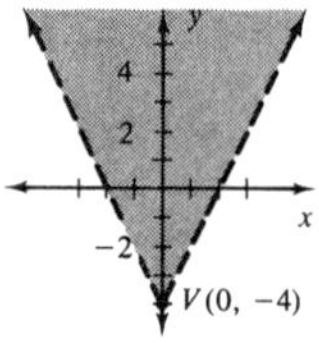

46. 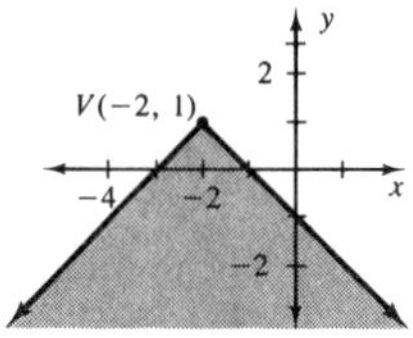

47. $\dfrac{x^3 + 1}{x}$

48. $\dfrac{x^3 - 1}{x}$

49. x

50. x^3

51. $1/x^2$

52. $1/x^2$

53. $\{x \mid x \in \text{reals}, x \neq 0\}$

54. $\{x \mid x \in \text{reals}, x \neq 0\}$

55. x^2

56. Yes

57. Yes

58. (a) $f^{-1} = \{(x, y) \mid y = \sqrt[3]{x - 2}\}$

(b) Domain = $\{x \mid x \in \text{reals}\}$; range = $\{y \mid y \in \text{reals}\}$

59. (a) $g^{-1} = \left\{(x, y) \mid y = \dfrac{x}{1 - x}\right\}$

(b) Domain = $\{x \mid x \in \text{reals}, x \neq 1\}$; range = $\{y \mid y \in \text{reals}, y \neq -1\}$

5 Exponential and Logarithmic Functions

We have dedicated many pages and much effort to the study of relations and functions. Our mathematical vocabulary contains terms such as polynomials, quadratic relations, domain and range, roots of functions, solutions of systems of relations, and more. In this chapter we add to the list new terms and two more functions.

Recall that we have used integers and rational numbers as exponents. However, the exponents have always been a constant. One of the functions introduced in this chapter uses a *variable* as an exponent and is called an *exponential function*. The domain of this function is the set of real numbers. The inverse of the exponential function is a function that is defined in terms of *logarithms*. As we shall see, the domain of this function is the set of positive real numbers.

EXPONENTIAL FUNCTION 5-1

To introduce the exponential function, we restrict the independent variable to the set of integers.

Exponential function

> **Definition:** If b is a real number, $b > 0$, and $b \neq 1$, then
>
> $$f = \{(x, f(x)) \mid f(x) = b^x\}$$
>
> is an exponential function. The domain of f = {integers} and the range of f = {positive real numbers}.

To graph the exponential function we will use the variable y rather than $f(x)$, since we are more accustomed to graphing ordered pairs (x, y) than $(x, f(x))$.

EXAMPLE 5–1 If $x \in$ {integers}, graph the following functions:

(a) $y = 2^x$ (b) $y = (1/2)^x$

Solutions:
(a) The set $\{(-3, 1/8), (-2, 1/4), (-1, 1/2), (0, 1), (1, 2), (2, 4), (3, 8)\}$ contains elements of the given function. The graph is shown in Figure 5-1.
(b) The set $\{(-3, 8), (-2, 4), (-1, 2), (0, 1), (1, 1/2), (2, 1/4), (3, 1/8)\}$ contains elements of the given function. The graph is shown in Figure 5-2.

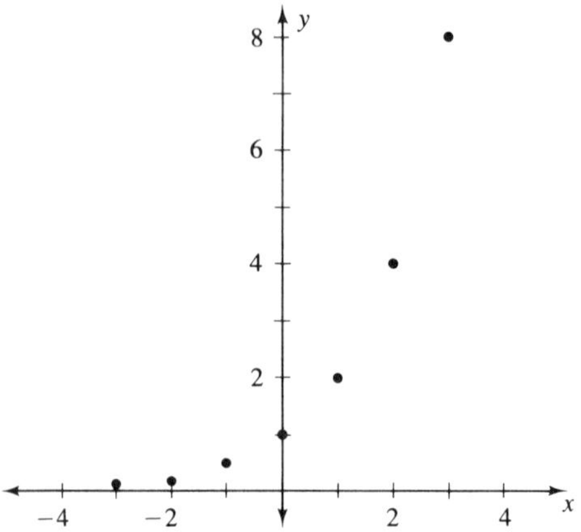

FIGURE 5-1 Graph of $y = 2^x$, $x \in \{\text{integers}\}$

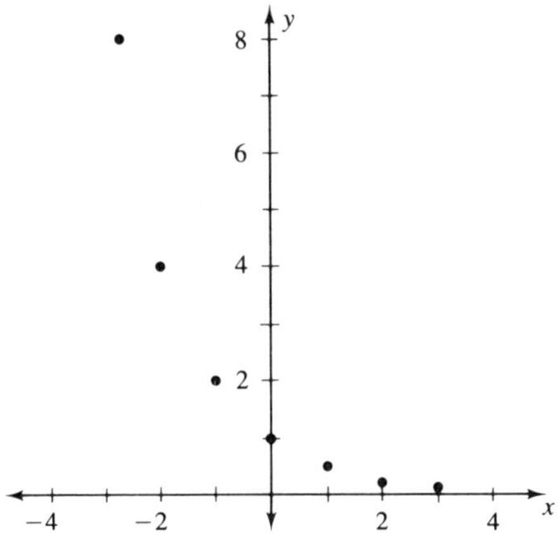

FIGURE 5-2 Graph of $y = \left(\frac{1}{2}\right)^x$, $x \in \{\text{integers}\}$

How did you get those values for y when x is less than 0?

Recall that $a^{-n} = 1/a^n$ for $a \neq 0$ and $n > 0$. Therefore, $y = 2^x$ and $x = -1$ implies $y = 2^{-1} = 1/2$; and for $x = -2$, $y = 1/4$. Recall also that $a^0 = 1$ for any real number not equal to zero; thus, $y = 2^0 = 1$ for $x = 0$.

We can make the following observations from the graphs in Figures 5-1 and 5-2:

(a) $y = 2^x$

1. For $x > 0$, $y > 1$.
2. For $x < 0$, $0 < y < 1$.

(b) $y = (1/2)^x$

1. For $x > 0$, $0 < y < 1$.
2. For $x < 0$, $y > 1$.

3. As x increases from $-\infty$ to $+\infty$, y increases from 0 to $+\infty$.
4. The negative x-axis is an asymptote of the graph.

3. As x increases from $-\infty$ to $+\infty$, y decreases from $+\infty$ to 0.
4. The positive x-axis is an asymptote of the graph.

We now extend the domain of the exponential function to include real numbers that are not integers. We are tempted to simply "connect the dots" to form a smooth curve for the functions shown in Figures 5-1 and 5-2. We can argue that since $f(x)$ increases as x increases for $f(x) = 2^x$, then $2^0 < 2^{1/2} < 2^1$, since $0 < 1/2 < 1$. Similarly, for any rational number p/q and integers N and $N + 1$, we assert that $2^N \leq 2^{p/q} < 2^{N+1}$ whenever $N \leq p/q < N + 1$.

Furthermore, if x is an irrational number such as $\sqrt{3}$, we can argue that for $f(x) = 2^{\sqrt{3}}$ that $2^1 < 2^{\sqrt{3}} < 2^2$, since $1 < \sqrt{3} < 2$. In general, we can argue that for any real number x — rational or irrational — there exists a real value $f(x)$, and $f(x) = 2^x$. The proof of such an assumption is beyond the scope of this text; therefore, we will accept the truth of this statement without formal proof. The following theorem summarizes this discussion.

Theorem: If b is a positive real number and $b \neq 1$, then for any real number x there exists a positive real number $f(x)$, such that $f(x) = b^x$.

We have therefore extended the domain of the exponential function to the set of real numbers. The range of the function is the set of positive real numbers.

EXAMPLE 5–2 If $x \in \{\text{real numbers}\}$, sketch the graphs of
(a) $y = 4^x$ (b) $y = (1/4)^x$

Solutions:
(a) If $y = 4^x$, then $\{(-2, 1/16), (-1, 1/4), (0, 1), (1, 4), (2, 16)\}$ are elements of the function. A smooth curve is drawn through these points and is shown in Figure 5-3.

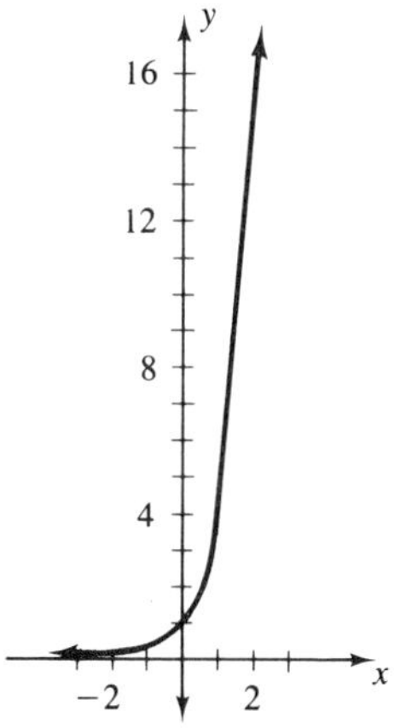

FIGURE 5-3 Graph of $y = 4^x$, $x \in \{\text{reals}\}$

(b) If $y = (1/4)^x$, then $\{(-2, 16), (-1, 4), (0, 1), (1, 1/4), (2, 1/16)\}$ are elements of the function. A smooth curve is drawn through these points and is shown in Figure 5-4.

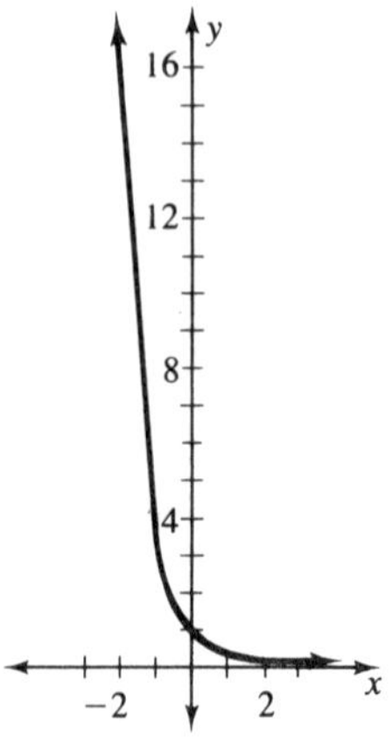

FIGURE 5-4 Graph of $y = \left(\frac{1}{4}\right)^x$, $x \in \{\text{reals}\}$

Any positive real number different from 1 can be used as the value for b. Two real numbers that appear frequently for b are 10 and the irrational number e. Ten is used as a base since the base of our number system is 10. The number e, with approximate value 2.71828, is important in the study of calculus.

The general shape of the graph of the exponential function differs slightly from the graphs of $y = 4^x$ and $y = (1/4)^x$ in Figures 5-3 and 5-4. When the base b is increased for $b > 1$ or decreased for $0 < b < 1$, the ''steepness'' of the curve is increased. But the basic shape of the curve is unaltered. We can now summarize a few of the more significant details regarding the exponential function.

If $f(x) = b^x$, $x \in \{\text{real numbers}\}$, and

(a) $b > 1$	(b) $0 < b < 1$
then	
1. $f(x)$ increases as x increases.	1. $f(x)$ decreases as x increases.
2. $f(x)$ decreases as x decreases.	2. $f(x)$ increases as x decreases.
3. $f(x)$ is always positive.	3. $f(x)$ is always positive.
4. $f(x) = 1$ for $x = 0$.	4. $f(x) = 1$ for $x = 0$.
5. f is a one-to-one function.	5. f is a one-to-one function.

EXAMPLE 5–3 The amount of money A obtained for A_0 dollars deposited at a rate of r percent compound interest for t interest periods is given by the equation $A = A_0(1 + r)^t$. Compute A for $A_0 = \$1{,}000$, $r = 5$ percent yearly interest compounded quarterly, and the money is deposited for two years.

Solution: If interest is compounded quarterly, then two years contain eight interest periods, and $t = 8$. Furthermore, 5 percent yearly interest computes at

1¼ percent quarterly. Therefore, $A = \$1{,}000(1 + 0.0125)^8 = \$1{,}104.50$. Thus, the initial \$1,000 earned \$104.50 interest.

An equation that contains at least one term with a variable in the exponent is called an exponential equation. For example $3^x = 1/81$ and $2^{x^2-4} = 8^x$ are exponential equations. Such an equation can be solved with the following theorem:

> **Theorem:** If $b > 0$ and $b \neq 1$, then $b^n = b^m$ if and only if $n = m$.

Therefore, if $3^x = 1/81$, then $3^x = 3^{-4}$ implies that $x = -4$. Furthermore, if $2^{x^2-4} = 2^{3x}$, then $x^2 - 4 = 3x$, and $x = 4$ or -1.

EXAMPLE 5–4 Solve for x:

$$(\sqrt{5})^{3x+1} = 125^{x-2}$$

Solution: We convert both sides of the equation to equivalent expressions with the same base b. Since $\sqrt{5} = 5^{1/2}$ and $125 = 5^3$, then

$$(5^{1/2})^{3x+1} = (5^3)^{x-2}$$
$$5^{(3x+1)/2} = 5^{3x-6}$$

If $b^n = b^m$, then $n = m$. Therefore,

$$\frac{3x+1}{2} = 3x - 6$$
$$x = 13/3$$

The laws of exponents also apply to irrational exponents, and we use the next theorem to extend the laws of exponents to include *all real number exponents.*

Do the laws of exponents also apply to irrational exponents?

> **Theorem:** If a and b are positive real numbers and x and y are any real numbers, then
>
> (1) $(b^x)(b^y) = b^{x+y}$
> (2) $(b^x)^y = b^{xy}$
> (3) $\dfrac{b^x}{b^y} = b^{x-y}$
> (4) $(ab)^x = (a^x)(b^x)$
> (5) $\left(\dfrac{a}{b}\right)^x = \dfrac{a^x}{b^x}$

Exercise Set 5–1

Sketch a graph of the following exponential functions, where the domain is the set of real numbers.

1. $y = 3^x$
2. $y = (1/3)^x$
3. $y = (4/3)^x$

4. $y = 2^{-x}$ 5. $y = 2^x + 1$ 6. $y = 2^{x+1}$

Simplify the following expressions.

7. $(3^{\sqrt{5}})(3^{-\sqrt{5}})$
8. $(5^{(\sqrt{2}-1)})(5^{(2-\sqrt{2})})$
9. $(2^{\sqrt{2}})^{\sqrt{8}}$
10. $((\sqrt{2})^{\sqrt{2}})^{\sqrt{2}}$
11. $\dfrac{5^{\sqrt{27}}}{5^{\sqrt{3}}}$
12. $\dfrac{4^{\sqrt{27}}}{2^{\sqrt{27}}}$
13. $(2^{1/\pi})^{-3\pi}$
14. $\dfrac{\left(\frac{2}{3}\right)^{\sqrt{2}}}{\left(\frac{3}{2}\right)^{-\sqrt{2}}}$

Solve the following exponential equations.

15. $2^x = 2^{2x-3}$
16. $3^{3x-1} = 3^{x+2}$
17. $3^x = 9^{x+1}$
18. $2^{(x^2)} = 8^3$
19. $125^{2x-3} = 5^x$
20. $8^x = (1/2)^{x-1}$
21. $49^{x^2} = (1/7)^{3/4}$
22. $(\sqrt{2})^{x+3} = 2^{x^2}$
23. If \$200 is invested in a savings and loan company at a rate of 6 percent compounded semiannually, compute the amount in the account at the end of 2 years. (See Example 5-3.)
24. The population of a certain country is found to be determined by the equation $P = P_0 2^{kt}$, where P_0 is the current population, k is the rate of growth, and t is the time. Find the population in the year 2000 if the population in 1975 is 2 million and the population grows at the rate of 6 percent a year.

5-2 LOGARITHM FUNCTION

One of the properties we noted for the exponential function $f = \{(x, y) \mid y = b^x, b > 0 \text{ and } b \neq 1\}$ is that f is one-to-one. That is, for every positive real number y, there is one and only one real number x. Therefore, *the inverse of f is also a function.*

In this section we study the inverse of the exponential function, called the *logarithm function*. We should note that the logarithm has been studied as a tool for doing involved computations. However, with the availability of the hand-held calculator, the need for logarithms as such a tool has diminished.

We still use logarithms in the study of mathematics for other purposes. We will use **logs** (an abbreviation for logarithms) in calculus where their existence is essential. Therefore, we need to study this important function carefully to know and appreciate its unique properties.

By definition, if $f = \{(x, y) \mid y = b^x, b > 0, b \neq 1\}$, then $f^{-1} = \{(x, y) \mid x = b^y, b > 0, b \neq 1\}$. The domain of $f^{-1} = \{x \mid x > 0\}$, and the range of

$f^{-1} = \{y \mid y \in \text{reals}\}$. To solve the equation $x = b^y$ for x, the team *logarithm* is used.

$\log_b x$

> **Definition:** $x = b^y$ if and only if $y = \log_b x$.

In the equation $x = b^y$, y is an exponent. Therefore, in the equation $y = \log_b x$, y is also an exponent. It is significant to remember that the logarithm of a number x to a base b is an exponent to which b is raised to get the number x.

EXAMPLE 5–5 Write the following in terms of logarithms.

(a) $16 = 2^4$ (b) $1/36 = 6^{-2}$
(c) $27 = 9^{3/2}$ (d) $1 = 7^0$

Solutions:

(a) Since $16 = 2^4$, then $\log_2 16 = 4$. Notice that 4 is the exponent (y) to which 2, the base (b), is raised to get 16, the number (x).
(b) Since $1/36 = 6^{-2}$, then $\log_6(1/36) = -2$.
(c) $27 = 9^{3/2}$ is equivalent to $\log_9 27 = 3/2$.
(d) $1 = 7^0$ is equivalent to $\log_7 1 = 0$.

EXAMPLE 5–6 Write the following in terms of exponents.

(a) $\log_4 2 = 1/2$ (b) $\log_2(1/16) = -4$
(c) $\log_{10} 1 = 0$ (d) $\log_{10} 10 = 1$

Solutions:

(a) $\log_4 2 = 1/2$ is equivalent to $2 = 4^{1/2}$. Notice that 4 is the base of the logarithm and also the base of the exponent 1/2.
(b) $\log_2(1/16) = -4$ is equivalent to $1/16 = 2^{-4}$.
(c) $\log_{10} 1 = 0$ is equivalent to $1 = 10^0$.
(d) $\log_{10} 10 = 1$ is equivalent to $10 = 10^1$.

For any positive real number $b (b \neq 1)$, equations (c) and (d) of Example 5–6 are true. We state this fact in the following theorem.

Wouldn't parts (c) and (d) of Example 5-6 be true for any base of a logarithm?

> **Theorem:** If b is any positive real number ($b \neq 1$), then
>
> $$\log_b 1 = 0 \qquad (1)$$
> $$\log_b b = 1 \qquad (2)$$

Equation (1) is true, since for any $b \neq 0$, $b^0 = 1$; therefore, $\log_b 1 = 0$. Equation (2) is true, since for any real number b, $b^1 = b$; thus, $\log_b b = 1$.

In Figure 5-5 the graphs of $f = \{(x, y) \mid y = 2^x\}$ and $f^{-1} = \{(x, y) \mid y = \log_2 x\}$ are shown. The identity function $I = \{(x, y) \mid y = x\}$ is shown as a dashed line. Notice that f and f^{-1} are symmetric with respect to the line $y = x$. The following observations are made about the two graphs:

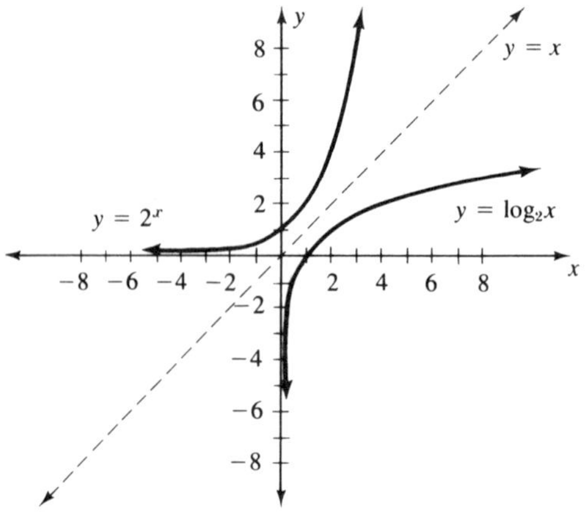

FIGURE 5-5 Graphs of $y = 2^x$, $y = \log_2 x$, and $y = x$

(a) $y = 2^x$

1. The domain = {reals}.
2. The range = {positive reals}.
3. As x increases, y increases.
4. f is a one-to-one function.

(b) $y = \log_2 x$

1. The range = {reals}.
2. The domain = {positive reals}.
3. As y increases, x increases.
4. f^{-1} is a one-to-one function.

Notice that since the domain of f^{-1} = {positive reals}, the logarithm of a negative number does not exist. For example, if $\log_2(-4) = y$, then $2^y = -4$. But there does not exist a real number y such that $2^y \leq 0$. Therefore, *the logarithm of any nonpositive number does not exist.*

EXAMPLE 5–7 Solve for x:

(a) $\log_5 x = 2$ (b) $\log_x 64 = 6$ (c) $\log_{25} 125 = x$

Solutions:

(a) If $\log_5 x = 2$, then $x = 5^2$ and $x = 25$.

(b) If $\log_x 64 = 6$, then $64 = x^6$. But $64 = 2^6$; therefore, $2^6 = x^6$, and $x = 2$.

(c) If $\log_{25} 125 = x$, then $125 = 25^x$. But $125 = 5^3$ and $25^x = (5^2)^x = 5^{2x}$. Therefore, $5^3 = 5^{2x}$ and $2x = 3$, or $x = 3/2$.

Recall from the section on inverse functions that for any one-to-one function f and its inverse function f^{-1},

$$(f \circ f^{-1})(x) = f(f^{-1}(x)) = x$$
$$(f^{-1} \circ f)(x) = f^{-1}(f(x)) = x$$

When these equations are applied to the exponential and logarithmic functions, we obtain the equations in the following theorem.

Theorem: If b is a positive real number, $b \neq 1$, and $x \in \{\text{reals}\}$, then

$$\log_b b^x = x \qquad (3)$$

$$b^{\log_b x} = x, \quad x > 0 \qquad (4)$$

Equations (3) and (4) are obtained directly by applying the equation for $f \circ f^{-1}$ and $f^{-1} \circ f$. For example, if $f(x) = 3^x$ and $f^{-1}(x) = \log_3 x$, then in Equation (3),

$$f^{-1}(f(x)) = \log_3 3^x = x$$

That is, $\log_3 3^x$ is an exponent to which we raise the base 3 to get the number 3^x. Obviously, we must raise 3 to the x-power to get the number 3^x for any real number x. A similar analysis can be used to explain Equation (4).

Exercise Set 5–2

Rewrite the following exponential expressions as logarithmic expressions.

1. $9 = 3^2$
2. $1/7 = 7^{-1}$
3. $1 = 11^0$
4. $1/16 = 64^{-2/3}$
5. $9 = 27^{2/3}$
6. $4 = (1/2)^{-2}$

Rewrite the following logarithmic expressions as exponential expressions.

7. $\log_5 125 = 3$
8. $\log_2(1/8) = -3$
9. $\log_3 3 = 1$
10. $\log_{1/2} 8 = -3$
11. $\log_2 32 = 5$
12. $\log_8 4 = 2/3$

Solve the following equations for x.

13. $\log_{10} 100 = x$
14. $\log_4 1 = x$
15. $\log_8 16 = x$
16. $\log_{1/5} 25 = x$
17. $\log_7 343 = x$
18. $\log_{1/2} 64 = x$
19. $\log_{81} 3 = x$
20. $\log_8 x = -2/3$
21. $\log_x 25 = 4$
22. $\log_{10} x = -3$
23. $\log_x 4 = 2/3$
24. $\log_x 9 = 2$
25. $\log_x 0.01 = 2$
26. $\log_{10}(x^2 + 64) = 2$
27. Graph on the same set of axes $y = 5^x$ and $y = \log_5 x$.
28. Graph on the same set of axes $y = (1/5)^x$ and $y = \log_{1/5} x$.

5-3 LOGARITHMIC OPERATIONS

In Sections 5-1 and 5-2 we established a functional relationship between the exponential and logarithmic functions. In Chapter 2 we established theorems related to operations with exponents. In this section we now establish a set of theorems that govern operations with logarithms. Because of the relationship between the exponential and logarithmic functions, we will use exponential laws to prove the theorems on logarithms. The examples and exercises are designed to give us practice using the theorems given in this section so that we can become as proficient in manipulating logarithmic expressions as we are at manipulating exponential expressions.

Theorem: If x, y, and b are positive real numbers, $b \neq 1$, and k is a positive integer, then

(1) $\log_b xy = \log_b x + \log_b y$

(2 $\log_b \left(\frac{x}{y}\right) = \log_b x - \log_b y$

(3) $\log_b x^y = y \log_b x$

(4) $\log_b \sqrt[k]{x} = \frac{1}{k} \log_b x$

Proof.
(1) If $M = \log_b x$ and $N = \log_b y$, then by definition $b^M = x$ and $b^N = y$, and

$$(b^M)(b^N) = xy$$
$$b^{M+N} = xy$$

Then we have

$$M + N = \log_b xy$$
$$\log_b x + \log_b y = \log_b xy$$

Equation (1) in the theorem states that *the logarithm of a product is equal to the sum of the logarithms of the factors of the product*. For example, $50 = 5(10)$; therefore, $\log_b 50 = \log_b 5(10) = \log_b 5 + \log_b 10$.
(2) The proof of Equation (2) is part of the exercise set for this section.
(3) Recall that for any $x > 0$, $x = b^{\log_b x}$. Thus, for any real number y, $x^y = (b^{\log_b x})^y$. Since $\log_b x$ is an exponent and the exponent law states that $(a^m)^n = a^{mn}$, then

$$x^y = b^{y \log_b x}$$
$$\log_b x^y = \log_b b^{y \log_b x}$$
$$\log_b x^y = y \log_b x$$

(4) The proof of Equation (4) is part of the exercise set.

EXAMPLE 5–8 Rewrite the following as logarithms of prime numbers:
(a) $\log_b 24$ (b) $\log_b 0.25$

Solutions:

(a) $24 = (2)(2)(2)(3) = 2^3(3)$. Therefore,

$$\begin{aligned}\log_b 24 &= \log_b 2^3(3)\\ &= 3\log_b 2 + \log_b 3\end{aligned}$$

(b) $0.25 = 1/4 = 2^{-2}$. Therefore,

$$\begin{aligned}\log_b 0.25 &= \log_b 2^{-2}\\ &= -2\log_b 2\end{aligned}$$

EXAMPLE 5–9 Rewrite as a single logarithm.

(a) $\frac{1}{2}\log_b 3 + \log_b 2$ (b) $\frac{1}{3}(\log_b x - \log_b 3)$

Solutions:

(a) $$\begin{aligned}\frac{1}{2}\log_b 3 + \log_b 2 &= \log_b 3^{1/2} + \log_b 2\\ &= \log_b 2(3^{1/2})\\ &= \log_b 2\sqrt{3}\end{aligned}$$

(b) $$\begin{aligned}\frac{1}{3}(\log_b x - \log_b 3) &= \frac{1}{3}\log_b\left(\frac{x}{3}\right)\\ &= \log_b\sqrt[3]{\frac{x}{3}}\end{aligned}$$

There should be a restriction on the variable x, since $\log_b x$ is valid only when $x > 0$. Rather than repeating this restriction whenever we take the logarithm of a variable, let us agree that the variable always represents only positive real numbers.

Shouldn't there be a restriction placed on the variable x?

EXAMPLE 5–10 If $\log_b 2 = j$, $\log_b 3 = k$, and $\log_b 5 = m$, express the following in terms of j, k, and m.

(a) $\log_b 30$ (b) $\log_b\sqrt{2.5}$

(c) $\dfrac{\log_b 2}{\log_b 3}$ (d) $\log_b 2^{\log_b 3}$

Solutions:

(a) $30 = 2(3)(5)$; therefore,

$$\begin{aligned}\log_b 30 &= \log_b 2(3)(5)\\ &= \log_b 2 + \log_b 3 + \log_b 5\\ &= j + k + m\end{aligned}$$

(b) $2.5 = 5/2$; therefore,

$$\begin{aligned}\log_b\sqrt{2.5} &= \log_b\left(\frac{5}{2}\right)^{1/2}\\ &= \frac{1}{2}(\log_b 5 - \log_b 2)\\ &= \frac{1}{2}(m - j)\end{aligned}$$

(c) $\dfrac{\log_b 2}{\log_b 3} = \dfrac{j}{k}$ Note: $\log_b\left(\dfrac{2}{3}\right) \neq \dfrac{\log_b 2}{\log_b 3}$

(d) $$\begin{aligned}\log_b 2^{\log_b 3} &= (\log_b 3)(\log_b 2)\\ &= kj \quad \text{or} \quad jk\end{aligned}$$

The following theorem enables us to convert the logarithm of a number with an acceptable base b to the logarithm of the number to some standard base k.

Theorem: If $a > 0$, $b > 0$, $k > 0$, $b \neq 1$, and $k \neq 1$, then

$$\log_b a = \frac{\log_k a}{\log_k b}$$

Proof. Let $N = \log_b a$. Then $b^N = a$, and

$$\begin{aligned}\log_k b^N &= \log_k a\\ N \log_k b &= \log_k a\\ N &= \frac{\log_k a}{\log_k b}\end{aligned}$$

On substitution, $\log_b a = \dfrac{\log_k a}{\log_k b}$.

EXAMPLE 5–11 Express the following as a logarithm with base 10.
(a) $\log_3 15$ (b) $\log_e 10$ (c) $\log_5 3^2$

Solutions:
(a) With $k = 10$, $a = 15$, and $b = 3$, then

$$\log_3 15 = \frac{\log_{10} 15}{\log_{10} 3}$$

(b) $\log_e 10 = \dfrac{\log_{10} 10}{\log_{10} e} = \dfrac{1}{\log_{10} e}$

(c) $\log_5 3^2 = 2 \log_5 3 = 2\left(\dfrac{\log_{10} 3}{\log_{10} 5}\right)$

EXAMPLE 5–12 Solve $\log_3(x + 1) + \log_3 2 = \log_3 4 + \log_3 5$.

Solution: Use the laws of logarithms to rewrite the equation. If $\log_3(x + 1) + \log_3 2 = \log_3 4 + \log_3 5$, then

$$\begin{aligned}\log_3 2(x + 1) &= \log_3 20\\ 2(x + 1) &= 20\\ x &= 9\end{aligned}$$

Check.

$$\log_3(9 + 1) + \log_3 2 = \log_3 4 + \log_3 5$$

$$\log_3 20 = \log_3 20 \qquad \text{(True)}$$

Thus, the solution set is $\{9\}$.

It is always good mathematical practice to check the roots of an equation. However, for a log equation the check is imperative. A value obtained for the variable could be a restricted value for the given equation. For example, if a term in the equation is $\log_b(1 - x)$ and we obtain a value of $x = 3$ in solving the equation, then $\log_b(1 - x) = \log_b(-2)$. Therefore, the potential solution of 3 is rejected from the solution set because $1 - x < 0$.

Exercise Set 5–3

Use the rules of logarithms to rewrite the following as logarithms of only *prime integers*.

1. $\log_b 6$
2. $\log_b \frac{35}{11}$
3. $\log_b 30$
4. $\log_b 9$
5. $\log_b \sqrt[3]{77}$
6. $\log_b \sqrt[4]{\frac{5}{7}}$

Use the rules of logarithms to rewrite the following as a single logarithm.

7. $\log_b 7 + \log_b 2$
8. $\log_b 7 - \log_b 2$
9. $\log_b 3 - \log_b 7 + \log_b 2$
10. $2(\log_b 5 - \log_b 5^2)$
11. $\frac{1}{3}(\log_b 8 - \log_b 27)$
12. $\frac{-3}{4}(\log_b 16 - \log_b 25)$

If $\log_b 2 = j$, $\log_b 3 = k$, and $\log_b 5 = m$, express the following in terms of j, k, and m.

13. $\log_b 25$
14. $\log_b 12$
15. $\log_b(\log_b 9)$
16. $\log_b(3^{\log_b 2})$
17. $\log_b(b^{\log_b 3})$
18. $\log_b \sqrt{1.5}$

Express the following logarithms as a logarithm with the base 10.

19. $\log_4 7$
20. $\log_2 7^3$
21. $\log_3 5^2$
22. $\log_2 \sqrt{3}$

Solve the following equations for x.

23. $\log_3 x = 3 \log_3 5$
24. $\log_7 x + \log_7 4 = 1$
25. $\log_3 x - \log_3 4 - \log_3 20$
26. $\log_5 x = \log_5 18 - \log_5 3$
27. $\log_3 x = \log_3 17 + \log_3 4$
28. $\log_5 2 = \log_5 x - \log_5 3 + \log_5 4$
29. $\log_3 28 = \log_3 7 + 2 \log_3 x$
30. $\log_3(x + 1) + \log_3(x - 1) = 1$
31. Prove $\log_b\left(\frac{x}{y}\right) = \log_b x - \log_b y$, where x, y, and b are positive real numbers and $b \neq 1$.
32. Prove $\log_b \sqrt[k]{x} = \frac{1}{k} \log_b x$, where x and b are positive real numbers, k is a positive integer, and $b \neq 1$.

5-4 COMMON AND NATURAL LOGARITHMS

We have restricted the base of the logarithm function to any positive real number b, $b \neq 1$. Therefore, for a given positive real number a, there are an infinite number of logarithms of a to the different bases b. If we were to attempt to construct a table of values for the logarithm of all positive real numbers to any base b, we would need an infinite number of tables.

Fortunately, the last theorem of Section 5-3 reduces the required number of tables to one. However, there are two real numbers, 10 and e, for which logarithm tables are provided in the Appendix. In this section we discuss each base and do some computations using both bases.

Common Logarithms

The logarithm of a positive number a to the base 10 is called a *common logarithm*.

Common logarithm

Definition: A common logarithm has a base 10, and we write

$$\log_{10} a = \log a, \quad \text{for } a > 0$$

Theorem: If a is a positive real number, and in scientific notation $a = x(10^k)$, then

$$\log a = k + \log x$$

Note that $\log a$ and $\log x$ imply that the base is 10. The term k is called the **characteristic** of $\log a$. The term $\log x$ is called the **mantissa.** Recall that $1 \leq x < 10$ for a number written in scientific notation.

Table 2 in the Appendix contains the common logarithms of selected values of x. Notice that the table values are approximated to four decimal places, since the logarithms of most numbers are irrational numbers.

EXAMPLE 5–13 Use Table 2 to find the following logarithms.
(a) log 5,630,000 (b) log 0.00575

Solutions:
(a) In scientific notation, $5{,}630{,}000 = 5.63 \times 10^6$. Then

$$\begin{aligned}\log 5{,}630{,}000 &= \log(5.63 \times 10^6)\\ &= 6 + \log 5.63\end{aligned}$$

From Table 2, $\log 5.63 = 0.7505$. Therefore, $\log(5.63 \times 10^6) = 6 + 0.7505 = 6.7505$.
(b) In scientific notation, $0.00575 = 5.75 \times 10^{-3}$. Then

$$\begin{aligned}\log 0.00575 &= \log(5.75 \times 10^{-3})\\ &= -3 + 0.7597 = -2.2403\end{aligned}$$

Note that although $\log 0.00575 = -2.2403$, we frequently leave the answer as

0.7597 − 3. In this form, the mantissa is positive and the negative characteristic indicates that x is a number between 0 and 1, that is, $0 < x < 1$.

Recall that for any suitable base b, $\log_b b^a = a$. Therefore, for $b = 10$,

$$\begin{aligned} \log a &= \log x(10^k) \\ &= \log x + \log 10^k \\ &= \log x + k \end{aligned}$$

Why is the characteristic of the logarithm the exponent on 10?

and the exponent on 10 is the characteristic of log a.

Many hand-held calculators have a key labeled LOG. Such a calculator will give the logarithm of any positive number to several decimal places.

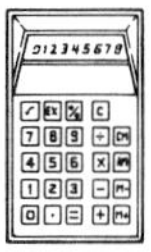

Example 5–14 Find the following logarithms with a hand-held calculator.
(a) log 721,000 (b) log 0.000368

Solutions:
(a) Enter the number 721,000. Depress key labeled LOG. Display reads 5.85794 to five decimal places.
(b) Enter the number 0.000368. Depress LOG key. Display reads −3.43415 to five decimal places.

Antilogarithm

> **Definition:** If c is the logarithm of a number and a is a positive real number, then
>
> antilog $c = a$ if and only if $\log a = c$

Example 5–15 Find
(a) antilog 3.7543 (b) antilog(−1.2248)

Solutions:
(a) If a = antilog 3.7543, then $\log a = 3.7543$. Therefore, $k = 3$ and $\log x = 0.7543$. From Table 2, log 5.68 = 0.7543; therefore, $x = 5.68$ and $a = 5.68 \times 10^3 = 5680$.
(b) If a = antilog(−1.2248), then $\log a = -1.2248$. To obtain a positive mantissa, we add and subtract 2 from −1.2248, and we get $0.7752 - 2 = \log a$. (Note that 2 is the smallest integer that yields a positive mantissa.) From Table 2, log 5.96 = 0.7752; therefore, $a = 5.96 \times 10^{-2} = 0.0596$.

The entries in Table 2 are based on numbers between 1 and 10; therefore, all the mantissas are positive. We must convert a negative logarithm to an equivalent logarithm with a positive mantissa to use Table 2. However, a hand-held calculator with a 10^x key gives us the antilogarithm of any real number, positive or negative.

Why did you have to change that last antilog to a positive mantissa?

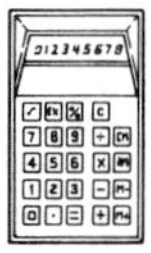

Example 5–16 Use a hand-held calculator to find the following.
(a) antilog 2.6324 (b) antilog(−1.4773)

Solutions:
(a) Enter 2.6324. Depress 10^x key. Display reads 428.94 to two decimal places. Therefore, antilog 2.6324 is approximately 429.
(b) Enter -1.4773. Depress 10^x key. Display reads 0.03332 to five decimal places. Therefore, antilog(-1.4773) is approximately 0.0333.

EXAMPLE 5–17 Find $560(\sqrt[4]{0.035})$.

Solution: Let $N = 560\ (\sqrt[4]{0.035})$. Then

$$\log N = \log 560 + \frac{1}{4}\log 0.035$$

$$= 2.7482 + \frac{1}{4}(0.5441 - 2)$$

Since both terms of log 0.035 must be multiplied by 1/4, we add and subtract 2 so that the resulting characteristic is an integer:

$$\log N = 2.7482 + \frac{1}{4}(2.5441 - 4)$$

$$= 2.7482 + 0.6360 - 1 = 2.3842$$

Therefore, N = antilog 2.3842, or N is approximately 242.

Natural Logarithms

We identified e as an irrational number with approximate value 2.71828. The following definition identifies e as the base of a *natural logarithm*.

Natural logarithm

> **Definition:** A natural logarithm has base e, and we write
>
> $$\log_e a = \ln a, \quad \text{for } a > 0$$

Therefore, the symbol ln is used to indicate a natural logarithm. The values of $\ln a$ for $1.0 \le a \le 10.0$ are given in Table 3 to four decimal places. Notice that since the rules of logarithms are true for any $b > 0$, they are also valid for base e.

How does it happen that an irrational number is used as the base of a logarithm table?

In calculus a relationship is established between the rate of change of a function and the slope of a tangent to the graph of the function. The exact details of the relationship cannot be explained in a few words, so we will omit them at this time. However, for the function $f = \{(x, y) \mid y = e^x\}$ and a value of $x = x_1$, the value of $y_1 = e^{x_1}$ and the slope of the tangent to the graph of f at $P(x_1, y_1)$ are the same. This relationship is unique for the exponential functon with base e and is very important in the study of calculus. Therefore, e is the number used as the base of the natural logarithm. The following example illustrates the use of Table 3 to find $\ln a$ for any $a > 0$.

EXAMPLE 5–18 With Table 3, find
(a) ln 580 (b) ln 0.0057

Solutions:
(a) We know that $580 = 5.8 \times 10^2$; thus,

$$\begin{aligned}\ln 580 &= \ln(5.8 \times 10^2)\\ &= \ln 5.8 + 2 \ln 10\end{aligned}$$

From Table 3, $\ln 5.8 = 1.7579$, and $\ln 10 = 2.3026$. Therefore, $\ln 580 = 1.7579 + 2(2.3026) = 6.3630$. Thus, $e^{6.3630}$ is approximately equal to 580.
(b) We know that $0.0057 = 5.7 \times 10^{-3}$; thus, $\ln 0.0057 = \ln(5.7 \times 10^{-3})$. From Table 3, $\ln 5.7 = 1.7405$ and

$$\ln 0.0057 = 1.7405 - 3(2.3026) = -5.1673$$

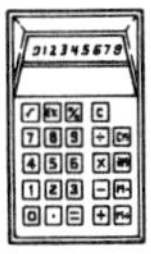

EXAMPLE 5–19 With a calculator, find the following.
(a) ln 36 (b) ln 0.058

Solutions:
(a) Enter 36. Depress key labeled LN. Display reads 3.5835 to 4 decimal places.
(b) Enter 0.058. Depress key labeled LN. Display reads −2.8473.

EXAMPLE 5–20 With Table 3, find x if
(a) $\ln x = 6.4457$ (b) $\ln x = -3.0576$

Solutions:
(a) Since $0.0000 \le \ln x \le 2.3026$ in Table 3 for $1.0 \le x \le 10.0$, we must subtract multiples of $\ln 10 = 2.3026$ from 6.4457 to obtain the logarithm of a number in the range of the table values:

$$\begin{aligned}6.4457 - 2.3026 &= 4.1431\\ 4.1431 - 2.3026 &= 1.8405\\ \ln 6.3 &= 1.8405\\ \ln 630 &= 6.4457 \quad \text{or} \quad x = 630\end{aligned}$$

Notice that 6.3 is multiplied by 10^2, since two values of ln 10 are subtracted from x to yield a number in the range of Table 3. If three values of ln 10 are necessary, then the table value of the logarithm is multiplied by 10^3, and so on.
(b) We must add multiples of 2.3026 to the given x to obtain a value in the range of Table 3.

$$\begin{aligned}-3.0576 + 2.3026 &= -0.7550\\ -0.7550 + 2.3026 &= 1.5476\\ \ln 4.7 &= 1.5476\\ \ln 0.047 &= -3.0576 \quad \text{or} \quad x = 0.047\end{aligned}$$

Notice that 4.7 is multiplied by 10^{-2}, since two values of ln 10 are added to x to yield a number in the range of Table 3.

If a calculator has a key labeled e^x, we simply enter the value of $\ln x$, depress the e^x key, and read the display. For example, if $\ln x = 8.5691$, we enter the value in the calculator, depress the e^x key, and read 5266.39 to two decimal places. If $\ln x = -10.0492$, then $x = 0.000043$ to six decimal places.

Can a calculator be used to find the value of x in Example 5-20?

Exercise Set 5–4

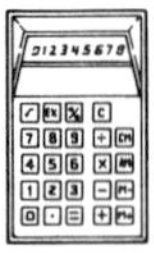

To find the following, (a) use Table 2 and (b) use a hand-held calculator.

1. log 19.6
2. log 196
3. log 0.0196
4. log 19,600
5. log 52,300,000
6. log 0.0964
7. log 673
8. log 79,100
9. log 7,000,000
10. log 0.0999

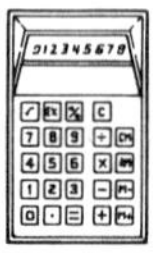

To find x, (a) use Table 2 and (b) use a hand-held calculator.

11. $\log x = 0.6551$
12. $\log x = 2.6551$
13. $\log x = -3 + 0.6551$
14. $\log x = -1.3449$
15. $\log x = 5.3118$
16. $\log x = -1 + 0.9552$
17. $\log x = -3.0009$
18. $\log x = 1.0170$
19. $\log x = 7.6010$
20. $\log x = 7.3222 - 10$

To find the following, (a) use Table 3 and (b) use a hand-held calculator.

21. ln 8.3
22. ln 83
23. ln 0.83
24. ln 830
25. ln 7200
26. ln 0.0091

To find x, (a) use Table 3 and (b) use a hand-held calculator.

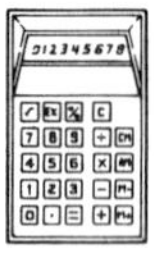

27. $\ln x = 0.1823$
28. $\ln x = 2.2083$
29. $\ln x = 2.4849$
30. $\ln x = 4.5109$
31. $\ln x = -2.1203$
32. $\ln x = 11.5029$

33. Sketch a graph of $y = \ln 2x$ and $y = \frac{1}{2}e^x$ on the same coordinate axes.
34. Sketch a graph of $y = 3 \ln x$ and $y = \ln 3x$ on the same coordinate axes.
35. Graph $y = 2e^x$ and $y = e^{2x}$ on the same coordinate axes.

Use the rules of natural logarithms to rewrite each of the following as a single natural logarithm.

36. $\ln 7 + \ln 3$
37. $\ln 15 - \ln 3$
38. $2(\ln 5 - \ln 5^2)$
39. $\frac{1}{3} \ln 8$
40. $2(\ln 2e + \ln 3e)$
41. $3 \ln e^2 - 2 \ln e$

Calculate to three digits, using logarithms.

42. $(46.8)(1.94)$
43. $\dfrac{(13.4)}{0.000349}$
44. $(1.98)^5$
45. $\sqrt[5]{47.6}$
46. $(0.419)^{-3}$
47. $(0.00164)^{2/5}$

48. If \$3000 is invested for 30 years at 7 percent interest, how much is the investment worth at the end of 30 years? Use the equation $A = A_0(1 + r)^t$, where A_0 is the original amount invested, t is the time, and r is the interest rate.

49. Thirty-two discs of various sizes are placed on one of three pegs so that the largest disc to the smallest disc are arranged from bottom to top. The discs are moved from one peg to a second peg by moving one disc at a time and by never placing a larger disc on top of a smaller disc. The number of moves necessary to move the discs is N, and $N = 2^{32} - 1$.
(a) How many moves are necessary?
(b) At one move per second, how long does it take to move all 32 discs?

50. A culture of yeast grows according to the equation $y = y_0e^{kx}$, where x is the time the culture grows in hours, y_0 is the initial quantity of culture used, k is a constant, and y is the amount of culture present at any time x. If $y_0 = 10$, and $y = 100$ when $x = 1$ hour, find the amount of yeast when $x = 2$ hours.

51. For the yeast culture of Exercise 50, find k if $y_0 = 100$ and 200 units of culture are observed when $x = 1$ hour.

52. For the yeast culture of Exercise 50, find the amount of culture present at the end of 3 hours if $y_0 = 20$ and 100 units of culture are observed when $x = 1$ hour.

EXPONENTIAL EQUATIONS 5–5

Exponential equations frequently occur in physical problems. In this section we develop the procedure for solving such equations. Recall that an equation is exponential if it contains at least one term with a variable as an exponent. First, we consider two examples to develop the procedure for solving such an equation, then we do two physical applications.

EXAMPLE 5–21 Solve $3^{2x+1} = 8$.

Solution: Notice that we are looking for a number x such that 3^{2x+1} is 8. Since $3^1 = 3$ and $3^2 = 9$, we can expect that $1 < 2x + 1 < 2$. If $3^{2x+1} = 8$, then

$$\begin{aligned} \log 3^{2x+1} &= \log 8 \\ (2x + 1)\log 3 &= \log 8 \\ 2x + 1 &= \frac{\log 8}{\log 3} \\ x &= \frac{1}{2}\left(\frac{\log 8}{\log 3} - 1\right) \end{aligned}$$

which is the *exact answer*. To find an approximate value for x, we use Table 2 or a calculator to evaluate log 8 and log 3. Thus,

$$x = \frac{1}{2}\left(\frac{0.9031}{0.4771} - 1\right) = 0.4464$$

to four decimal places. Notice that $\frac{\log 8}{\log 3} = 1.8928$, which is a value within the expected range of values for $2x + 1$.

EXAMPLE 5–22 Find x correct to three significant digits:

$$40^{x-1} = 2^{3x+2}$$

Solution: If $40^{x-1} = 2^{3x+2}$, then $(x - 1)\log 40 = (3x + 2)\log 2$. But $\log 40 = 1.6021$ and $\log 2 = 0.3010$. Therefore,

$$\begin{aligned}(x - 1)(1.6021) &= (3x + 2)(0.3010)\\ 1.6021x - 1.6021 &= 0.9030x + 0.602\\ 0.6991x &= 2.2041\\ x &= \frac{2.2041}{0.6991}\\ &= 3.15\end{aligned}$$

EXAMPLE 5–23 The formula used to compute compound interest is $A = P(1 + r)^N$, where A is the amount of money on deposit, r is the interest rate, P is the original principal, and N is the number of conversion periods. Find the length of time needed to double a given initial principal when the interest is 5 percent compounded annually.

Solution: If P is the initial principal, we want to find N such that $A = 2P$. Therefore, if $r = 0.05$, then

$$\begin{aligned}2P &= P(1 + 0.05)^N\\ 2 &= 1.05^N\\ \log 2 &= N \log 1.05\\ N &= \frac{\log 2}{\log 1.05}\\ &= \frac{0.3010}{0.0212}\\ &= 14.2 \text{ years to the nearest tenth}\end{aligned}$$

EXAMPLE 5–24 If P is the population of a group of people at a given time t, then P_0 is the initial population for time $t = 0$. If k is a constant determining the rate of growth, then $P = P_0e^{kt}$ represents the relationship between the variables. The formula is used to predict future populations or is used to determine the number of years required to achieve a certain population, if the current rate of growth is known. For example, if the population of a given country is 10 million and grows at a rate of 3 percent, how long will it take for the population to double?

Solution: From the given information, $P = 20$ million, $P_0 = 10$ million, and $k = 0.03$. Solving the equation $P = P_0e^{kt}$ for t, we get

$$\begin{aligned} 20 &= 10e^{0.03t} \\ 2 &= e^{0.03t} \\ \ln 2 &= \ln e^{0.03t} \\ \ln 2 &= 0.03t \end{aligned}$$

since $\ln e = 1$. We know that $\ln 2.0 = 0.6931$; therefore,

$$t = \frac{0.6931}{0.03} = 23.1 \text{ years}$$

We should note that the computed value for t is independent of the initial population size. Thus, it would take *any population* 23.1 years to double its population when the rate of annual growth is 3 percent.

Would the 23.1 years be the value of t for a different value of P_0?

EXAMPLE 5–25 Find x if $3e^{x+2} = 20$.

Solution: If $3e^{x+2} = 20$, then

$$\begin{aligned} \ln 3e^{x+2} &= \ln 20 \\ \ln 3 + (x + 2)\ln e &= \ln 20 \\ x + 2 &= \ln 20 - \ln 3 \\ x &= \ln 20 - \ln 3 - 2 \end{aligned}$$

or $x = -0.1029$ to four decimal places.

Exercise Set 5–5

Solve the following for x. State the *exact answer* and *approximate answer* correct to three significant digits.

1. $2^x = 5$
2. $3^{x+1} = 10$
3. $4^{x-1} = 151$
4. $5^{x^2} = 625$
5. $(3.5)^{2x+1} = 100$
6. $\left(\frac{7}{3}\right)^{x+1} = 7$
7. $27^{x-4} = 2^{x+1}$
8. $10^{3x-1} = 119$
9. $(8.11)^x = 0.567$
10. $(1.91)^x(1.62)^{2x+1} = 348$
11. $e^x = 2.36$
12. $e^{2x}(0.191)^x = 13$
13. If \$500 is invested in a bank and interest is compounded annually at 6 percent, how long will it take for the investment to be worth \$1000?
14. If an initial investment is doubled in 15 years, at what interest rate is the investment made if interest is compounded annually?
15. If \$3000 is invested in a bank at 4 percent interest, how long will it take for the investment to be worth \$4000 if interest is compounded annually?

16. At the end of 25 years an investment is worth \$10,000. What was the initial investment if the interest rate was 3.5 percent and interest is compounded annually?

17. How long does it take \$2000 compounded semiannually at 5 percent to be worth \$3000?

18. If the present world population is 4 billion and grows at the rate of 4 percent per year, when will the population be 5 billion?

19. If the population of a country increases from 10 million to 30 million in 30 years, what was the rate of growth?

20. In a certain city the population was 10,000 in the year 1970. When will the population reach 25,000 if the rate of growth is 8 percent?

21. The half-life of a certain substance is 2000 years. Thus, one gram of the substance reduces to one-half gram in 2000 years. Find the decay constant k so that $P = P_0e^{-kt}$ is the formula for the rate of decay.

22. Using the constant of Exercise 21, determine how long it takes the substance to decay from 10 grams to 1 gram.

SUMMARY

Exponential function

Definition: If b is a real number, $b > 0$, and $b \neq 1$, then $f = \{(x, f(x)) \mid f(x) = b^x\}$ is an exponential function. The domain of $f = \{\text{reals}\}$, and the range of $f = \{\text{positive real numbers}\}$.

Theorem: If b is a positive real number and $b \neq 1$, then for any real number x there exists a positive real number $f(x)$, such that $f(x) = b^x$.

Theorem: If $b > 0$ and $b \neq 1$, then $b^n = b^m$ if and only if $n = m$.

Theorem: If a and b are positive real numbers and x and y are any real numbers, then

(1) $(b^x)(b^y) = b^{x+y}$

(2) $(b^x)^y = b^{xy}$

(3) $\dfrac{b^x}{b^y} = b^{x-y}$

(4) $(ab)^x = (a^x)(b^x)$

(5) $\left(\dfrac{a}{b}\right)^x = \dfrac{a^x}{b^x}$

$\log_b x$

Definition: $x = b^y$ if and only if $y = \log_b x$.

Theorem: If b is any positive real number ($b \neq 1$), then $\log_b 1 = 0$ and $\log_b b = 1$.

Theorem: If b is a positive real number, $b \neq 1$, and $x \in \{\text{reals}\}$, then $\log_b b^x = x$ and $b^{\log_b x} = x, x > 0$.

Theorem: If x, y, and b are positive real numbers, $b \neq 1$, and k is a positive integer, then

(1) $\log_b xy = \log_b x + \log_b y$

(2) $\log_b \frac{x}{y} = \log_b x - \log_b y$

(3) $\log_b x^y = y \log_b x$

(4) $\log_b \sqrt[k]{x} = \frac{1}{k} \log_b x$

Theorem: If $a > 0$, $b > 0$, $k > 0$, $b \neq 1$, and $k \neq 1$, then $\log_b a = \frac{\log_k a}{\log_k b}$.

Common logarithm

Definition: A common logarithm has a base 10, and we write $\log_{10} a = \log a$, for $a > 0$.

Theorem: If a is a positive real number, and in scientific notation $a = x(10^k)$, then $\log a = k + \log x$.

Antilogarithm

Definition: If c is the logarithm of a number and a is a positive real number, then antilog $c = a$ if and only if $\log a = c$.

Natural logarithm

Definition: A natural logarithm has base e and we write $\log_e a = \ln a$, for $a > 0$.

SELF-CHECKING EXERCISE

Graph the following equations. **(5-1)**

1. $y = 7^x$
2. $y = (1/7)^x$

Simplify the following expressions. **(5-1)**

3. $(3^{\sqrt{2}})(3^{\sqrt{8}})$
4. $(7^{\sqrt{2}})^{\sqrt{2}}$
5. $\left(\frac{5^{2\sqrt{3}}}{5^{\sqrt{3}}}\right)^{-\sqrt{3}}$
6. $(2^{\sqrt{3}} \cdot 3^{\sqrt{2}})(2^{\sqrt{3}} \cdot 3^{\sqrt{2}})$

Solve for x. **(5-1)**

7. $16^x = (1/2)^{2-x}$
8. $5^{x^2-5} = (1/625)^x$

Express in logarithmic form. **(5-2)**

9. $64^{2/3} = 16$
10. $e^{3/4} = 2.117$

Express in exponential form. **(5-2)**

11. $\log_{10} 0.001 = -3$
12. $\log_6 7776 = 5$

Solve for x. **(5-2)**

13. $\log_x 125 = 3$
14. $\log_{49} 343 = x$

Graph on the same coordinate axes. **(5-2)**

15. $y = 3^x$ and $y = \log_3 x$
16. $y = e^x$ and $y = \ln x$

If $\log_b 2 = k$, $\log_b 3 = m$, and $\log_b 5 = n$, express the following in terms of k, m, or n. **(5-3)**

17. $\log_b 30$
18. $\log_b 32$
19. $\log_b 12.5$
20. $\log_b \sqrt[3]{15}$
21. $\log_2 3$
22. $\log_3 2$

Solve for x. **(5-3)**

23. $2 \log_5 x = \log_5(x + 1) - \log_5 2$
24. $\log_3 3x = \log_3(x + 2) + \log_3(x - 2)$

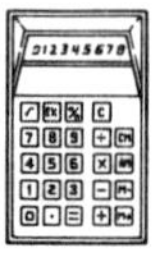

To find the following, use Table 2 or a hand calculator. **(5-4)**

25. log 92.6
26. log 0.0458

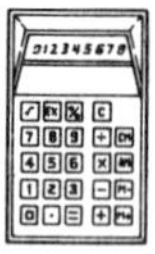

To find the following, use Table 2 or a hand calculator. **(5-4)**

27. antilog 3.7033
28. antilog(−1.1198)

Calculate the following, using logarithms, and round each answer to three significant digits. **(5-4)**

29. $(219)(-0.0164)$
30. $\sqrt{\dfrac{0.236}{18.7}}$
31. $(1.07)^{15}$
32. $\log_4 70$
33. If $1000 is invested at an interest rate compounded annually at 4 percent, how long will it take for the investment to be worth $1500? **(5-4)**
34. A city with a current population of 13,000 grows at the rate of 8 percent per year. When will the population reach 100,000? **(5-4)**

Solve for x. **(5-5)**

35. $3^{2x-1} = 13$
36. $e^{2x-1} = 2.96$

Answers to Self-Checking Exercise

1. 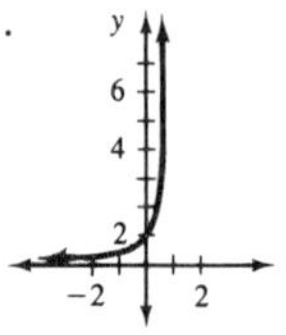

2. 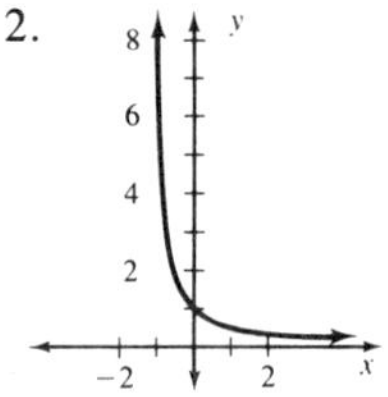

3. $3^{3\sqrt{2}}$
4. 49
5. 1/125
6. $2^{2\sqrt{3}} \cdot 3^{2\sqrt{2}}$ or $(2^{\sqrt{3}} \cdot 3^{\sqrt{2}})^2$
7. $\{-2/3\}$
8. $\{-5, 1\}$
9. $\log_{64} 16 = 2/3$
10. $\ln 2.117 = 3/4$
11. $10^{-3} = 0.001$
12. $6^5 = 7776$
13. $\{5\}$
14. $\{3/2\}$

15.

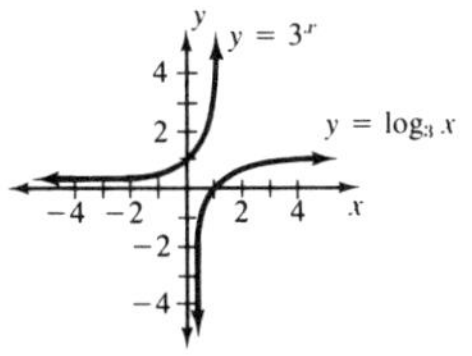

16.

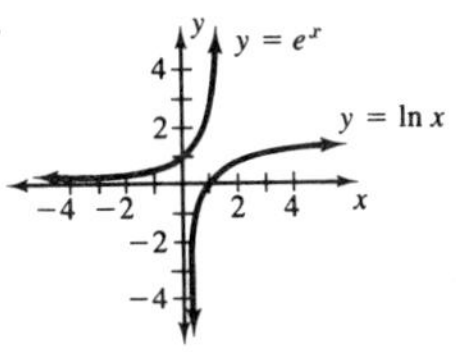

17. $k + m + n$ 18. $5k$ 19. $2n - k$

20. $\frac{1}{3}(m + n)$ 21. m/k 22. k/m

23. $\{1\}$ 24. $\{4\}$ 25. 1.9666

26. $0.6609 - 2$ 27. 5,050 28. 0.0759

29. -3.59 30. 0.112 31. 2.76

32. 3.06 33. 10.3 years 34. 25.5 years

35. $\frac{1}{2}\left(\frac{\log 13}{\log 3} + 1\right)$ or 1.67 36. $\frac{1}{2}(\ln 2.96 + 1)$ or 1.04

6 Analytic Geometry

In this chapter we study a new family of mathematical relations. Each relation in the family is a member of C, where

$$C = \{(x, y) \mid Ax^2 + Bxy + Cy^2 + Dx + Ey + F = 0\}$$

and A, B, C, D, E, and F are real numbers. In this chapter B is always 0, although there are equations for which $B \neq 0$.

The graphs of each relation discussed in this chapter can be obtained by intersecting a cone with a plane. That is, when the cone in Figure 6-1 is intersected with a plane, the points of the intersection form a graph of one of the relations in C. The details of the association are given when each relation is introduced. However, the relationship between the cone and the relations from C is the reason why the graphs are sometimes called **conic sections,** or simply *conics*.

FIGURE 6-1 Cone

6-1 DISTANCE FORMULA AND MIDPOINT FORMULA

Some of the relations in C have equations that can be interpreted geometrically as describing distances related to two points, or distances related to a point and a line. Therefore, in this section we introduce two equations for computing the distance between $P_1(x_1, y_1)$ and $P_2(x_2, y_2)$ and the distance between $P_1(x_1, y_1)$ and $L = \{(x, y) \mid Ax + By + C = 0\}$. In the second part of this section, we discuss the equation for computing the midpoint of line segment $\overline{P_1P_2}$.

Distance Formula

Recall that for points $P_1(x_1, y_1)$ and $P_2(x_2, y_2)$ in the coordinate plane, the quantities $y_2 - y_1$ and $x_2 - x_1$ are used to determine the slope m of the line through points P_1 and P_2. That is,

$$m = \frac{y_2 - y_1}{x_2 - x_1}, \quad x_1 \neq x_2$$

We now use the quantities $y_2 - y_1$ and $x_2 - x_1$ to calculate the distance between the points P_1 and P_2.

Distance formula

Definition: If $P_1(x_1, y_1)$ and $P_2(x_2, y_2)$ are any two points in the coordinate plane, then the distance between P_1 and P_2, written $|P_1P_2|$, is given by the equation

$$|P_1P_2| = \sqrt{(x_2 - x_1)^2 + (y_2 - y_1)^2}$$

In Figure 6-2 a horizontal line through P_1 intersects a vertical line through P_2 at the point $P_3(x_2, y_1)$. Since triangle $P_1P_2P_3$ is a right triangle, we can apply the Pythagorean Theorem to the lengths of the sides of the triangle. Furthermore, with P_1 and P_3 on a horizontal line, then

$$|P_1P_3| = |x_2 - x_1|$$

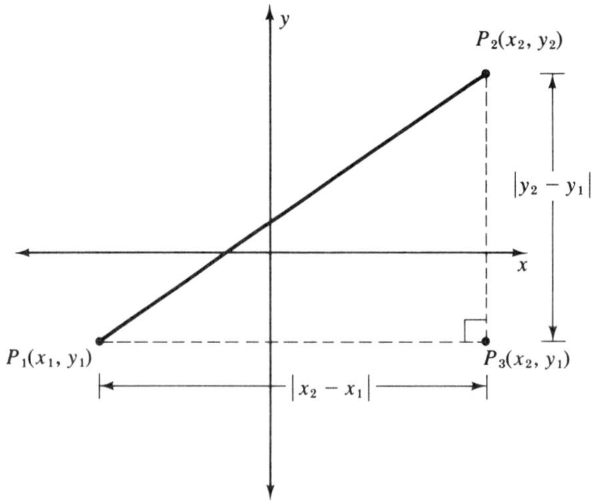

FIGURE 6-2 $|P_1P_2| = \sqrt{(x_2 - x_1)^2 + (y_2 - y_1)^2}$

which is the equation used to determine the distance between points on a number line. With P_2 and P_3 on a vertical line, we have $|P_2P_3| = |y_2 - y_1|$. Therefore,

$$|P_1P_2|^2 = |x_2 - x_1|^2 + |y_2 - y_1|^2$$
$$|P_1P_2| = \sqrt{(x_2 - x_1)^2 + (y_2 - y_1)^2}$$

The square of any real number is positive; therefore, the absolute value symbols on the terms in the radicand are unnecessary. Furthermore, since $|P_1P_2|$ is a distance, only the *positive square root* of the radical expression is accepted.

EXAMPLE 6–1 Find $|P_1P_2|$ for $P_1(-2, -1)$ and $P_2(1, 3)$.

Solution: With $x_1 = -2$, $y_1 = -1$, $x_2 = 1$, and $y_2 = 3$, we have

$$|P_1P_2| = \sqrt{(1 - (-2))^2 + (3 - (-1))^2}$$
$$= \sqrt{9 + 16} = \sqrt{25} = 5$$

EXAMPLE 6–2 Determine whether the triangle with vertices $A(-4, 2)$, $B(-1, 5)$, and $C(-5, 6)$ is isosceles.

Solution: Triangle ABC is isosceles if and only if two sides of the triangle are equal in length.

$$|AB| = \sqrt{(-1 - (-4))^2 + (5 - 2)^2} = \sqrt{18} = 3\sqrt{2}$$
$$|AC| = \sqrt{(-5 - (-4))^2 + (6 - 2)^2} = \sqrt{17}$$
$$|BC| = \sqrt{(-5 - (-1))^2 + (6 - 5)^2} = \sqrt{17}$$

Since $|AC| = |BC|$, triangle ABC is isosceles.

Frequently, we need to find the distance d from a point $P(x_1, y_1)$ to a line $L = \{(x, y) \mid Ax + By + C = 0\}$. Unless otherwise stated, d is measured along the line through P *perpendicular* to L; thus, d is the *shortest distance* from P to L.

> **Theorem:** The distance d from point $P(x_1, y_1)$ to line $L = \{(x, y) \mid Ax + By + C = 0\}$ is computed with the equation
>
> $$d = \frac{|Ax_1 + By_1 + c|}{\sqrt{A^2 + B^2}}$$

EXAMPLE 6–3 Find the distance from $P(3, 4)$ to

$$L = \{(x, y) \mid 5x + 12y - 24 = 0\}$$

Solution: With $A = 5$, $B = 12$, $C = -24$, $x_1 = 3$, and $y_1 = 4$, we have

$$d = \frac{|5(3) + 12(4) - 24|}{\sqrt{5^2 + 12^2}}$$
$$= \frac{|39|}{13} = 3$$

Thus, $P(3, 4)$ is 3 units from L. Notice that 3 is the shortest distance from P to L.

Midpoint Formula

Whenever the coordinates of the endpoints of a line segment are known, then the coordinates of the midpoint of the segment can be determined.

Midpoint of a line segment

> **Definition:** Point M is the midpoint of line segment $\overline{P_1P_2}$ if and only if
>
> (1) M is colinear with $P_1(x_1, y_1)$ and $P_2(x_2, y_2)$.
>
> (2) $|P_1M| = |MP_2|$

Can't $|P_1M| = |MP_2|$ and not have M as the midpoint of segment $\overline{P_1P_2}$?

There are infinitely many points that are equally distant from two points and in the coordinate plane that contains them. But there is *only one point M* that is equally distant from P_1 and P_2 and colinear with them.

> **Theorem:** If $P_1(x_1, y_1)$ and $P_2(x_2, y_2)$ are any two points in the coordinate plane, then
>
> $$M\left(\frac{x_1 + x_2}{2}, \frac{y_1 + y_2}{2}\right)$$
>
> is the midpoint of $\overline{P_1P_2}$.

Proof. In Figure 6-3, $P_1(x_1, y_1)$ and $P_2(x_2, y_2)$ are shown with a horizontal line through P_1 and a vertical line through P_2 intersecting in point $P_3(x_2, y_1)$. Without loss of generality, we can assume that P_1 is to the left of P_3; thus, $x_1 < x_2$ and $x_2 - x_1 > 0$.

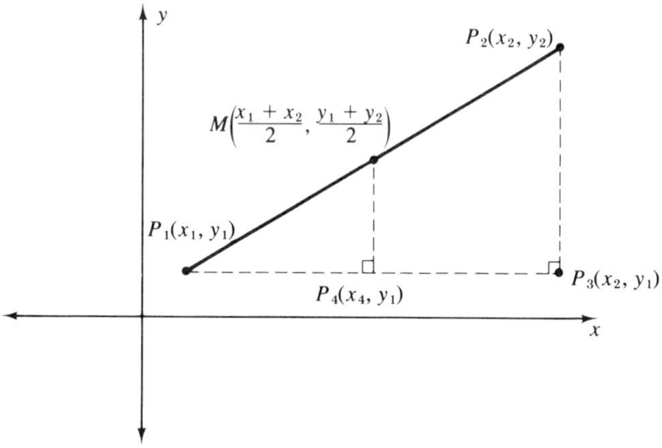

FIGURE 6-3 M, the midpoint of line segment P_1P_2

Now let $P_4(x_4, y_1)$ be the midpoint of $\overline{P_1P_3}$; therefore, $|P_1P_4| = |P_4P_3|$. Notice that since P_4 is between P_1 and P_3, $x_1 < x_4 < x_2$, and $x_4 - x_1 > 0$. If $|P_1P_4| = |P_4P_3|$, then

$$x_4 - x_1 = x_2 - x_4$$

$$2x_4 = x_1 + x_2$$

$$x_4 = \frac{x_1 + x_2}{2}$$

The vertical segment through P_4 intersects $\overline{P_1P_2}$ in a point M. The resulting right triangles P_1P_4M and $P_1P_3P_2$ are similar, since they have the common acute angle P_1. The ratio of the lengths of the corresponding sides are therefore equal, and

$$\frac{|P_1M|}{|P_1P_2|} = \frac{|P_1P_4|}{|P_1P_3|} = \frac{1}{2}$$

and

$$|P_1M| = \frac{1}{2}|P_1P_2|$$

Therefore, $|P_1M| = |MP_2|$, and point M with x-coordinate $\dfrac{x_1 + x_2}{2}$ is the midpoint of $\overline{P_1P_2}$. In a similar way we can show that the y-coordinate of point M is $\dfrac{y_1 + y_2}{2}$.

Example 6–4 Find the equation of the perpendicular bisector of $\overline{P_1P_2}$ with endpoints $P_1(4, -6)$ and $P_2(6, 2)$.

Solution: By definition, *the perpendicular bisector of a line segment is a line perpendicular to the segment and containing the midpoint of the segment.*

The midpoint of $\overline{P_1P_2}$ is $\left(\dfrac{4+6}{2}, \dfrac{-6+2}{2}\right) = (5, -2)$. The slope of $\overline{P_1P_2} = \dfrac{2-(-6)}{6-4} = \dfrac{8}{2} = 4$. Therefore, the slope of the line perpendicular to $\overline{P_1P_2} = -1/4$. The equation of the perpendicular bisector of $\overline{P_1P_2}$ is

$$y - (-2) = \frac{-1}{4}(x - 5)$$

$$4y + 8 = -x + 5$$

$$x + 4y + 3 = 0$$

If L is the perpendicular bisector of $\overline{P_1P_2}$, then $L = \{(x, y) \mid x + 4y + 3 = 0\}$. Therefore, any point equally distant from P_1 and P_2 has coordinates that belong to the solution set of L. Thus, if $A(x_1, y_1)$ is on L, then $x_1 + 4y_1 + 3 = 0$ is a true statement.

Exercise Set 6–1

Find the length of the line segment joining the following pair of points.

1. $(1, 2), (3, 1)$
2. $(2, -1), (-3, 5)$
3. $(-2, -5), (-7, 8)$
4. $(3, 0), (2, 4)$
5. $(1, \sqrt{3}), (3, -2\sqrt{3})$
6. $(1 + \sqrt{2}, 5), (2 + \sqrt{2}, -1)$
7. $(a, b), (2a, -3b)$
8. $(0, b), (b, 0)$

Find the coordinates of the midpoint of the line segment joining the following points.

9. (7, 3), (−3, 5)
10. (1, 5), (−1, −7)
11. (2, 9), (3, 6)
12. (−8, −14), (3, 11)
13. (6, −7), (−2, 11)
14. (−17, 33), (−19, 0)
15. $(\sqrt{3} + 1, \sqrt{2})$, $(1 - \sqrt{3}, -\sqrt{2})$
16. $(3a, 2b)$, $(5a, 8b)$
17. Given $C(2, -6)$, find the coordinates of point D such that (4, 7) is the midpoint of $\overline{CD}$.
18. Given $B(-3, -7)$, find the coordinates of point A such that (−5, 8) is the midpoint of $\overline{AB}$.

In each of the following problems the vertices of a triangle are given. Find the length of each side of the triangle and then (a) determine if the triangle is isosceles, (b) determine if the triangle is equilateral (all three sides are equal in length), (c) determine if the triangle is a right triangle, and (d) find the perimeter.

19. (5, 0), (1, 2), (5, 2)
20. (0, 0), (10, 0), $(5, 5\sqrt{3})$
21. (3, −1), (2, −7), (4, −7)
22. (1, 1), (3, 3) (−1, −3)
23. Find the equation of the perpendicular bisector of the segment joining the points (−3, 7) and (1, 3).
24. Given that $A(1, 2)$, $B(3, 4)$, and $C(5, -6)$ are vertices of triangle ABC, find the equations of the three medians of the triangle. A *median* is a line containing a vertex and the midpoint of the opposite side of the triangle.
25. Using the distance formula, show that (1, −1), (3, 1), and (6, 4) are colinear points.
26. Let (0, 0), $(a, 0)$, and $(0, b)$ be three vertices of a rectangle. Find the fourth vertex. Show that the diagonals of the rectangle bisect each other.

Find the distance from point P to the line with the given equation.

27. $P(7, 3)$; $5x + 12y + 7 = 0$
28. $P(-1, 2)$; $4x - 3y + 5 = 0$
29. $P(5, -4)$; $15y - 8x + 15 = 0$
30. $P(-5, -2)$; $7x - 24y + 17 = 0$
31. $P(0, 0)$; $3y + 7x - 8 = 0$
32. $P(0, 5)$; $4x + 5y = 10$

6-2 CIRCLES

With the distance formula we can determine the distance between any two points in a plane. In this section we determine the equation of the set of points in a plane that are a fixed distance from a fixed point. We then analyze the equation so that the graph of the equation can be readily sketched.

Circle

> **Definition:** A circle is a set of points in a plane that are a fixed distance from a fixed point.

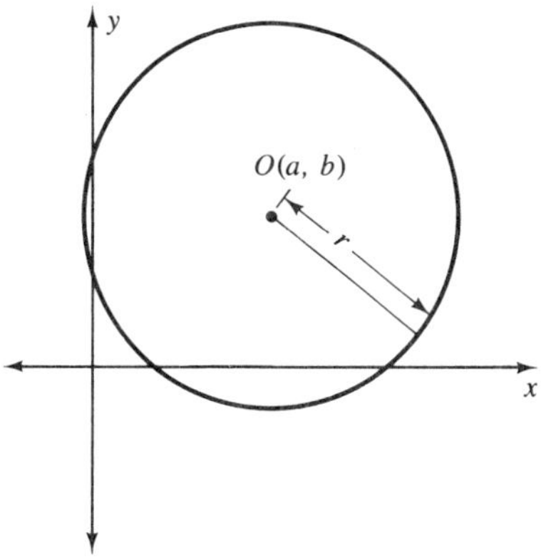

FIGURE 6-4 Graph of circle with center $O(a, b)$ and radius r

The fixed point is the **center** of the circle, and the fixed distance is the **radius.** In Figure 6-4, a circle C is drawn with center $O(a, b)$ and radius r. Any point $P(x, y)$ in the plane of C and r units from O is an element of C. With the distance formula, we get

$$r^2 = (x - a)^2 + (y - b)^2$$

Therefore, the set of all points — and only those points — whose coordinates satisfy the stated equation are elements of C.

> **Theorem:** The **center-radius equation** of the circle C with center $O(a, b)$ and radius r $(r > 0)$ is of the form
>
> $$C = \{(x, y) \mid (x - a)^2 + (y - b)^2 = r^2\}$$

The domain of C contains values of x such that the difference between x and a is less than or equal to r. That is, $|x - a| \leq r$, or

$$-r \leq x - a \leq r \quad \text{and} \quad a - r \leq x \leq a + r$$

Therefore, the domain and range of C are $\{x \mid a - r \leq x \leq a + r\}$ and $\{y \mid b - r \leq y \leq b + r\}$, respectively. Recall that the domain and range of a relation are the **horizontal** and **vertical extents** of the graphs of the relation, respectively.

EXAMPLE 6–5 Sketch the graph of $C = \{(x, y) \mid (x - 2)^2 + (y + 1)^2 = 25\}$.

Solution: The key to graphing a circle with a minimum amount of effort is to identify the center and radius of the circle. When we compare the equation of C with the center-radius equation of the above theorem, we determine that $a = 2$, $b = -1$, and $r = \sqrt{25} = 5$. Therefore, the center of C is $O(2, -1)$ and the radius is 5. The graph is shown in Figure 6-5.

EXAMPLE 6–6 Sketch the graph of

$$C = \{(x, y) \mid x^2 + y^2 + 8x - 4y - 16 = 0\}$$

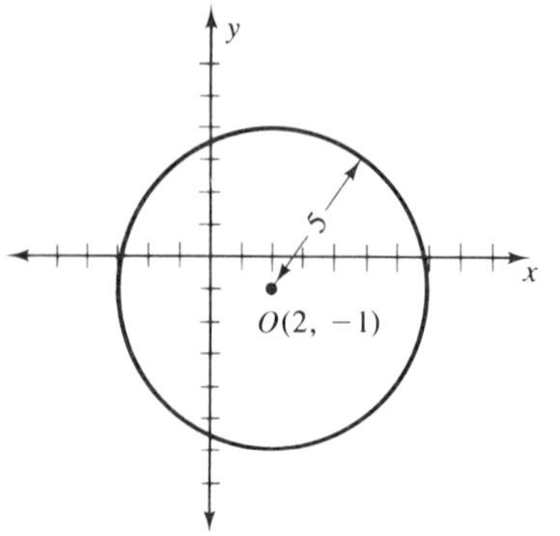

FIGURE 6-5 Graph of $C = \{(x, y) \mid (x - 2)^2 + (y + 1)^2 = 25\}$

Solution: To graph C, we first convert the given equation to the center-radius form. To eliminate the x and y terms in the equation we complete the square in x and y as follows:

$$x^2 + 8x \qquad + y^2 - 4y \qquad = 16 \qquad \text{(add 16 to both sides)}$$

$$(x^2 + 8x + 16) + (y^2 - 4y + 4) = 16 + 16 + 4 \qquad \text{(add the squares of 1/2 the coefficients of } x \text{ and } y \text{ to both sides)}$$

$$(x + 4)^2 + (y - 2)^2 \qquad = 36$$

Thus, $O(-4, 2)$ is the center of C and the radius is 6. The graph is shown in Figure 6-6.

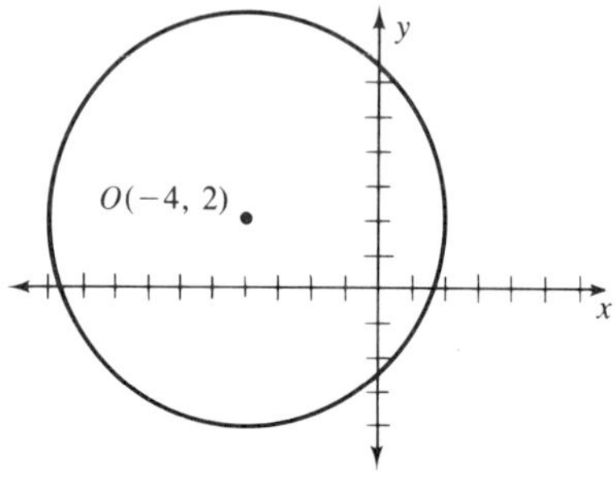

FIGURE 6-6 Graph of $C = \{(x, y) \mid x^2 + y^2 + 8x - 4y - 16 = 0\}$

How did you know that the graph of C is a circle?

The graph of any second-degree equation in both x and y in which the coefficients of x^2 and y^2 are equal can be a circle (provided there is no xy term). That is, the graph of an equation of the form $Ax^2 + Cy^2 + Dx + Ey + F = 0$ can be a circle if $A = C$.

When we convert the given equation to the center-radius form of the circle equation, we get $(x - a)^2 + (y - b)^2 = r^2$.

(1) If $r^2 > 0$, then the graph is a circle.

(2) If $r^2 = 0$, then the graph is a point, sometimes called a *point-circle*.

(3) If $r^2 < 0$, then the graph does not exist in the real coordinate plane.

EXAMPLE 6–7 Determine the equation of the circle C with center $(5, -1)$ and radius 3.

Solution: With $(a, b) = (5, -1)$, $r = 3$, and the center-radius equation of a circle, we get

$$\begin{aligned} C &= \{(x, y) \mid (x - 5)^2 + (y + 1)^2 = 3^2\} \\ &= \{(x, y) \mid x^2 + y^2 - 10x + 2y + 17 = 0\} \end{aligned}$$

EXAMPLE 6–8 Determine the equation of the circle C that contains the point $P(0, 0)$ and has center where $L_1 = \{(x, y) \mid 3x - y = 13\}$ and $L_2 = \{(x, y) \mid x + 5y = -17\}$ intersect.

Solution: We solve the equations for L_1 and L_2 simultaneously with the substitution method. For L_2, $x = -5y - 17$. In L_1,

$$\begin{aligned} 3(-5y - 17) - y &= 13 \\ -16y &= 64 \\ y &= -4 \end{aligned}$$

Then $x = -5(-4) - 17 = 3$. Thus, $(3, -4)$ is the center of the circle C. If C contains $P(0, 0)$, then

$$r^2 = (0 - 3)^2 + (0 - (-4))^2 = 25$$

Therefore,

$$\begin{aligned} C &= \{(x, y) \mid (x - 3)^2 + (y + 4)^2 = 25\} \\ &= \{(x, y) \mid x^2 + y^2 - 6x + 8y = 0\} \end{aligned}$$

EXAMPLE 6–9 Determine the equation of circle C that contains $P_1(4, -6)$, $P_2(-3, 1)$, and $P_3(-2, 2)$.

Solution: For $C = \{(x, y) \mid Ax^2 + Cy^2 + Dx + Ey + F = 0\}$ that contains P_1, P_2, and P_3, the coordinates of the three points must each satisfy the general equation for C. Since for any circle $A = C$, then

$$Ax^2 + Cy^2 + Dx + Ey + F = x^2 + y^2 + \frac{D}{A}x + \frac{E}{A}y + \frac{F}{A}$$

Now let $a = D/A$, $b = E/A$, and $c = F/A$. Then for $x = 4, -3$, and -2, and $y = -6, 1$, and 2, we obtain the following system of equations:

$$\begin{aligned} &(1)\quad 16 + 36 + 4a - 6b + c = 0 \\ &(2)\quad 9 + 1 - 3a + b + c = 0 \\ &(3)\quad 4 + 4 - 2a + 2b + c = 0 \end{aligned}$$

The solution of the three equations yields $a = -2$, $b = 4$, and $c = -20$. Therefore, $C = \{(x, y) \mid x^2 + y^2 - 2x + 4y - 20 = 0\}$.

How can the cone in Figure 6-1 be intersected with a plane to get a circle?

When we intersect the cone with the plane held perpendicular to the **axis** of the cone (the line through the center of the cone), the intersection is a circle. However, the intersection can be a point (that is, a point-circle) if the plane intersects the vertex of the cone (see Figure 6-7).

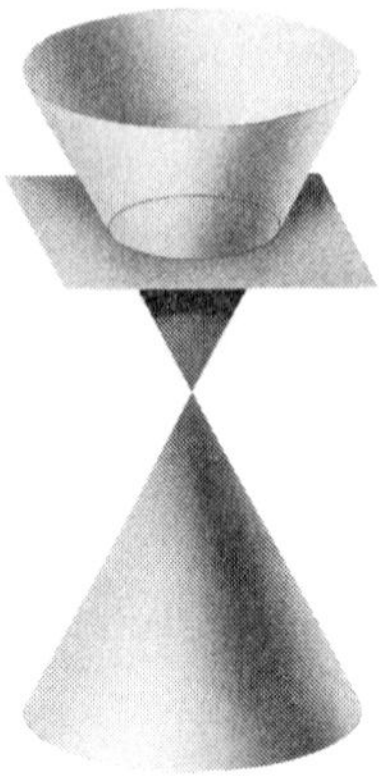

FIGURE 6-7 Circle

Exercise Set 6–2

Find the equation of the circle with the given center and radius.

1. $(2, 3)$; 5
2. $(-1,4)$; 7
3. $(-5, -2)$; 3
4. $(4, -8)$; $\sqrt{5}$
5. $(1/2, 2/3)$; $3/4$
6. $(-3/4, -1/2)$; 2
7. $(a, 3a)$; $\sqrt{3}$
8. $(5b, 3b)$; $\sqrt{10}$

(a) Find the center and radius, and (b) sketch a graph of the following circles. When necessary, approximate the radius to one decimal place to sketch the graph of the circle.

9. $x^2 + y^2 = 36$
10. $x^2 + y^2 - 8x - 8y = 32$
11. $x^2 + y^2 - 2x - 4y - 11 = 0$
12. $x^2 + y^2 + 4x + 10y - 71 = 0$
13. $x^2 + y^2 + 14y = 0$
14. $x^2 - 8x + y^2 = 9$
15. $x^2 + x + y^2 + y - 1 = 0$
16. $x^2 - 3x + y^2 - 2y = 7$
17. $4x^2 + 4y^2 - 10x - 8y = 0$
18. $4x^2 = 28x + 8y - 4y^2 - 37$

For the following problems, find the equation of the circle satisfying the given conditions.

19. Center $(-1, 2)$; radius 18
20. Center at the origin and containing the point $(5, 12)$
21. Center at the point $(2, 3)$ and containing the point $(3, 4)$
22. Tangent to the y-axis with center $(-3, 2)$

23. Ends of a diameter at (1, 1) and (−3, −5)

24. Tangent to both the x and y axes, with center in quadrant IV and radius of 3

If $C = \{(x, y) \mid Ax^2 + Ay^2 + Dx + Ey + F = 0\}$ is a circle with center $O(a, b)$ and radius r, then the graph of the relation $Ax^2 + Ay^2 + Dx + Ey + F < 0$ is the set of points in the *interior* of the circle with center $O(a, b)$ and radius r. Furthermore, the graph of $Ax^2 + Ay^2 + Dx + Ey + F > 0$ is the set of points in the *exterior* of the same circle.

Sketch a graph of the given equations or inequalities.

25. $x^2 + y^2 < 9$

26. $(x + 2)^2 + y^2 > 4$

27. $x^2 + y^2 - 4x + 6y - 51 < 0$

28. $x^2 + y^2 + 4x - 4y - 1 \geq 0$

29. $y = \sqrt{16 - x^2}$ (*Hint:* Note $y \geq 0$. Now square both sides of the equation.)

30. $x = -\sqrt{25 - y^2}$ (*Hint:* Note $x \leq 0$. Now square both sides of the equation.)

31. $y \leq \sqrt{36 - x^2}$

32. $x > \sqrt{64 - y^2}$

Find the equation of the circle containing the following points.

33. $A(0, 0)$; $B(6, 0)$; $C(0, 8)$

34. $A(1, 1)$; $B(2, 5)$; $C(-2, 5)$

35. $A(2, 1)$; $B(3, 0)$; $C(1, 4)$

36. $A(-1, 2)$; $B(-2, -2)$; $C(0, 0)$

ELLIPSES 6–3

The planets of the solar system orbit the sun with paths that are similar to a circle. However, the path of each planet is not a fixed distance from one point, but the distance is a constant when measured from two fixed points. The figure that describes such a path is an ellipse. In this section we study the equation of the ellipse and develop a procedure for graphing the ellipse with a minimum amount of work.

Ellipse

> **Definition:** An ellipse is a set of points in a plane the sum of whose distances from two fixed points is a constant.

Each fixed point is a **focus** of the ellipse. The plural of focus is *foci* (pronounced "fo-sigh"). In this section we will study only ellipses with foci on one of the coordinate axes. We call these standard ellipses.

Standard ellipse

> **Definition:** A standard ellipse has foci that are either
>
> $$F_1(c, 0) \quad \text{and} \quad F_2(-c, 0)$$
>
> or
>
> $$F_1(0, c) \quad \text{and} \quad F_2(0, -c)$$
>
> where $c \in \{\text{reals}\}$ and $c > 0$.

Theorem:
Case 1. If E is an ellipse with foci $F_1(c, 0)$ and $F_2(-c, 0)$, then

$$E = \left\{(x, y) \mid \frac{x^2}{a^2} + \frac{y^2}{b^2} = 1\right\} \quad \text{and} \quad a > b > 0$$

Case 2. If E is an ellipse with foci $F_1(0, c)$ and $F_2(0, -c)$, then

$$E = \left\{(x, y) \mid \frac{y^2}{a^2} + \frac{x^2}{b^2} = 1\right\} \quad \text{and} \quad a > b > 0$$

Notice that since $a > b > 0$, then $a^2 > b^2$. Therefore, the foci are on the coordinate axis of the variable with the larger denominator. For example, the foci are on the x-axis of the ellipse with equation $\frac{x^2}{3} + \frac{y^2}{2} = 1$. But the foci are on the y-axis of the ellipse with equation $\frac{y^2}{3} + \frac{x^2}{2} = 1$.

Are a and b somehow related to the coordinate c of the foci?

We can show that $a^2 - b^2 = c^2$. In Figure 6-8 a standard ellipse E with foci $F_1(c, 0)$ and $F_2(-c, 0)$ is drawn. Therefore, the equation of E is $\frac{x^2}{a^2} + \frac{y^2}{b^2} = 1$. (Notice that $a > c$.) Now, if $y = 0$, then $x^2/a^2 = 1$, $x^2 = a^2$, and $x = \pm a$. Thus, a and $-a$ are x-intercepts of E. If $x = 0$, then $y^2/b^2 = 1$ and $y = \pm b$. Therefore, b and $-b$ are the y-intercepts of E.

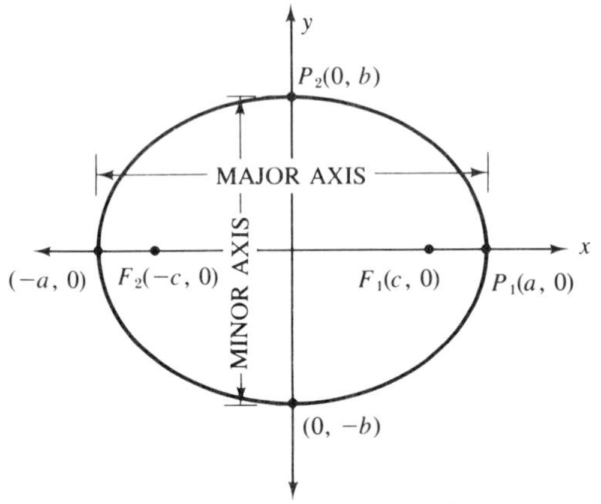

FIGURE 6-8 Graph of $E = \left\{(x, y) \,\middle|\, \frac{x^2}{a^2} + \frac{y^2}{b^2} = 1\right\}$

Recall that the definition of an ellipse states that the sum of the distances between any point of E and the foci is a constant. In particular, for $P_1(a, 0)$ and $P_2(0, b)$, then

$$|P_1F_1| + |P_1F_2| = k, \quad k > 0$$
$$|P_2F_1| + |P_2F_2| = k$$
$$|P_1F_1| + |P_1F_2| = |P_2F_1| + |P_2F_2|$$

With the distance formula, we have

$$\sqrt{(a-c)^2} + \sqrt{(a-(-c))^2} = \sqrt{c^2+(-b)^2} + \sqrt{(-c)^2+(-b)^2}$$

Since $a > c$, $a - c > 0$, and $\sqrt{(a-c)^2} = a - c$ and $\sqrt{(a+c)^2} = a + c$. Therefore,

$$(a-c) + (a+c) = \sqrt{c^2+b^2} + \sqrt{c^2+b^2}$$
$$2a = 2\sqrt{c^2+b^2}$$
$$a^2 = c^2 + b^2$$
$$a^2 - b^2 = c^2$$

The following observations are made for E from the graph in Figure 6-8.

1. The points $V_1(a, 0)$ and $V_2(-a, 0)$ are the *vertices* of the ellipse.
2. The *domain* of $E = \{x \mid -a \le x \le a\}$.
3. The *range* of $E = \{y \mid -b \le y \le b\}$.
4. Since any vertical line drawn between the vertices of E intersects the graph in two points, *E does not define a function.*
5. The coordinate axes are *lines of symmetry* for E. Therefore, if $P_1(x, y)$ is a point of E, then $P_2(x, -y)$, $P_3(-x, y)$, and $P_4(-x, -y)$ are also points of E.
6. The line segment with endpoints $(a, 0)$ and $(-a, 0)$ is called the *major axis,* and the line segment with endpoints $(0, b)$ and $(0, -b)$ is called the *minor axis* of the ellipse.

In Figure 6-9 the graph of a standard ellipse E with equation $\frac{y^2}{a^2} + \frac{x^2}{b^2} = 1$ is drawn. The points $V_1(0, a)$ and $V_2(0, -a)$ are the vertices. The domain and range of E are $\{x \mid -b \le x \le b\}$ and $\{y \mid -a \le y \le a\}$, respectively. E does not define a function, and the coordinate axes are lines of symmetry for the

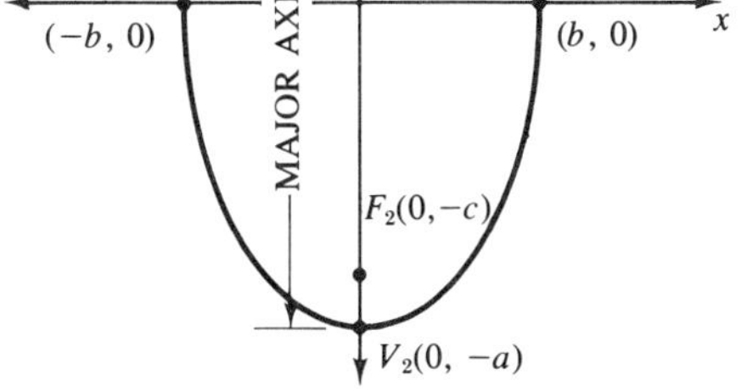

FIGURE 6-9 Graph of $E = \left\{(x, y) \,\middle|\, \frac{y^2}{a^2} + \frac{x^2}{b^2} = 1\right\}$

graph of E. The major axis is the line segment $\overline{V_1V_2}$, and the minor axis is the line segment with endpoints $(b, 0)$ and $(-b, 0)$. Finally, the relationship $a^2 - b^2 = c^2$ is valid for the equation of E.

Example 6–10 Sketch the graph of $E = \{(x, y) \mid 4x^2 + 9y^2 = 36\}$.

Solution: Convert the equation of E to standard form. If $4x^2 + 9y^2 = 36$, then

$$\frac{x^2}{9} + \frac{y^2}{4} = 1$$

Since $9 > 4$, the foci are located on the x-axis. Thus, $a^2 = 9$, and $V_1(3, 0)$ and $V_2(-3, 0)$ are vertices of E. With $b^2 = 4$, $(0, 2)$ and $(0, -2)$ are points of E. To obtain additional points of E, we let $x = 2$. Then

$$4(2)^2 + 9y^2 = 36$$

$$y = \pm 1.5 \text{ to one decimal point}$$

Thus, by symmetry, $P_1(2, 1.5)$, $P_2(2, -1.5)$, $P_3(-2, 1.5)$, and $P_4(-2, -1.5)$ are all points of E. The graph is shown in Figure 6-10.

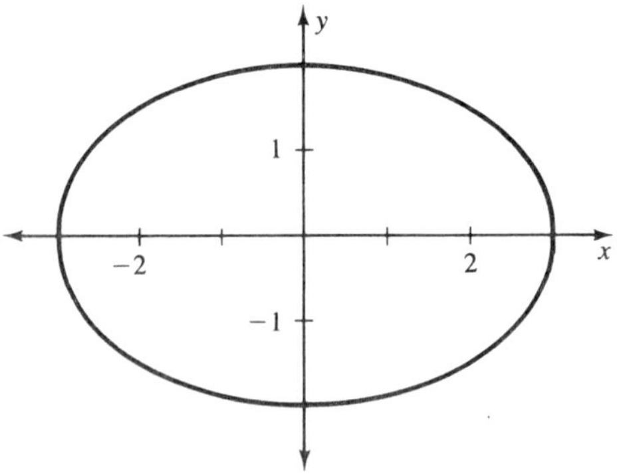

Figure 6-10 Graph of $E = \{(x, y) \mid 4x^2 + 9y^2 = 36\}$

How did you know that the graph of E was an ellipse?

Recall the general equation of the relations $C = \{(x, y) \mid Ax^2 + Cy^2 + Dx + Ey + F = 0\}$. If $D = E = 0$ and $F < 0$, then the graph of C is an ellipse whenever $A > 0$ and $C > 0$. (If $AC > 0$ and $A = C$, then the graph is a circle.) Notice that if $F = 0$, then the graph is a point; and if $F > 0$ while $A > 0$ and $C > 0$, then there is no graph of C in the real coordinate plane. For example, $4x^2 + 9y^2 + 36 = 0$ does not have a graph in the real plane.

Example 6–11 Graph the equation $E = \left\{(x, y) \mid y = 4\sqrt{1 - x^2}\right\}$.

Solution: Notice that the range of $E = \{y \mid y \geq 0\}$. Keeping this restriction in mind, we now square both sides of the given equation:

$$y^2 = 16(1 - x^2)$$

$$y^2 + 16x^2 = 16$$

$$\frac{y^2}{16} + \frac{x^2}{1} = 1$$

Since $y \geq 0$, the graph of E is half of an ellipse, or **semi-ellipse,** with a focus on the y-axis (since $16 > 1$). With $x = 0.5$, $y = 4\sqrt{1 - (0.5)^2}$, or $y = 3.5$ to one decimal point. Thus, $P_1(0.5, 3.5)$ and $P_2(-0.5, 3.5)$ are points on the graph of E shown in Figure 6-11.

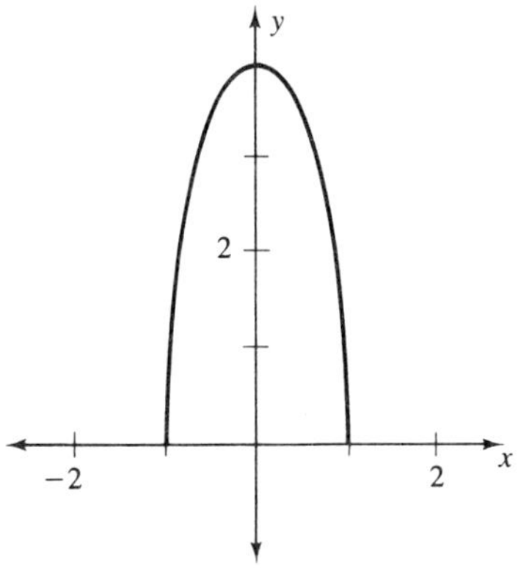

FIGURE 6-11 Graph of $E = \{(x, y) \mid y = 4\sqrt{1 - x^2}\}$

EXAMPLE 6–12 Determine the equation of the ellipse with foci $F_1(3, 0)$ and $F_2(-3, 0)$ and vertices $V_1(5, 0)$ and $V_2(-5, 0)$.

Solution: Since the foci are points on the x-axis, the equation is of the form $\frac{x^2}{a^2} + \frac{y^2}{b^2} = 1$. Since $F_1(3, 0)$ is a focus, $c = 3$. Furthermore, with $V_1(5, 0)$ a vertex, $a = 5$. Therefore, we have

$$\begin{aligned} a^2 - b^2 &= c^2 \\ 25 - b^2 &= 9 \\ b^2 &= 16 \end{aligned}$$

The equation of the ellipse is therefore

$$E = \left\{(x, y) \mid \frac{x^2}{25} + \frac{y^2}{16} = 1\right\}$$

or

$$E = \{(x, y) \mid 16x^2 + 25y^2 - 400 = 0\}$$

Notice that $A = 16$, $C = 25$, and $F < 0$; thus, the conditions for an ellipse equation are satisfied.

In Figure 6-12 the cone is intersected with a plane that is not parallel to either the axis or the edge of the cone. The result of the intersection is an ellipse. However, we can obtain a point as the intersection if we cut the cone at its vertex. In this case, as noted previously, $F = 0$.

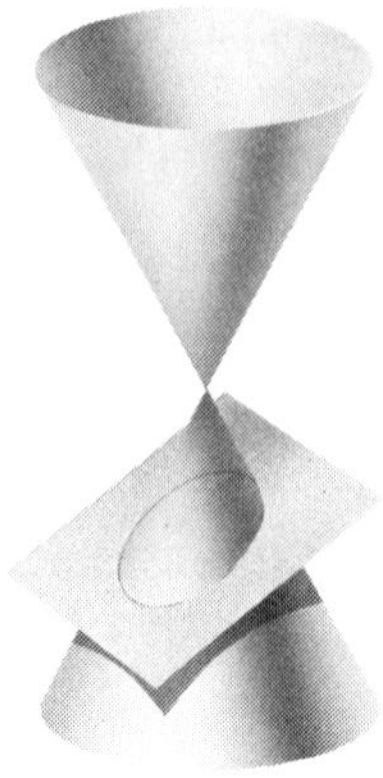

FIGURE 6-12 Ellipse

Exercise Set 6–3

Sketch a graph of the following standard ellipses showing all intercepts. State the coordinates of the foci.

1. $x^2 + 4y^2 = 16$
2. $9x^2 + 25y^2 = 225$
3. $5x^2 + 2y^2 = 50$
4. $9x^2 + y^2 = 9$
5. $4x^2 + 25y^2 = 100$
6. $x^2 + 2y^2 = 2$
7. $25x^2 + 16y^2 = 400$
8. $16x^2 + 9y^2 = 36$
9. $x^2 + 36y^2 = 36$
10. $49x^2 + 9y^2 = 441$

Find the equation of the following standard ellipses with the given information.

11. Vertices: $V_1(6, 0)$, $V_2(-6, 0)$; Foci: $F_1(\sqrt{33}, 0)$, $F_2(-\sqrt{33}, 0)$
12. Vertices: $V_1(0, 5)$, $V_2(0, -5)$; Foci: $F_1(0, \sqrt{21})$, $F_2(0, -\sqrt{21})$
13. Vertices: $V_1(7, 0)$, $V_2(-7, 0)$; Intercepts: $I_1(0, 3)$, $I_2(0, -3)$
14. Vertices: $V_1(0, 8)$, $V_2(0, -8)$; Intercepts: $I_1(4, 0)$, $I_2(-4, 0)$
15. Foci: $F_1(4, 0)$, $F_2(-4, 0)$; Sum of distances from foci, $2a$, is 10
16. Foci: $F_1(0, 5)$, $F_2(0, -5)$; Sum of distances from foci, $2a$, is 12

If $E = \{(x, y) \mid Ax^2 + Cy^2 + F = 0\}$ is a standard ellipse, then the graph of the relation $Ax^2 + Cy^2 + F < 0$ is the set of points in the interior of the ellipse. Furthermore, the graph of the relation $Ax^2 + Cy^2 + F > 0$ is the set of points in the exterior of the ellipse.

Sketch a graph of the following relations.

17. $x^2 + 4y^2 \leq 16$
18. $49x^2 + 4y^2 > 196$
19. $25x^2 + y^2 \geq 25$
20. $4x^2 + 9y^2 < 36$
21. $y = 3\sqrt{1 - x^2}$ (*Hint:* Note that $y \geq 0$. Then square both sides of the equation.)

22. $x \geq 4\sqrt{1 - y^2}$ (*Hint:* Note that $x \geq 0$. Then square both sides of the equation.)

23. $9 \leq x^2 + 9y^2 \leq 36$

24. $25 \leq y^2 + 25x^2 \leq 100$

25. $x^2 + 4y^2 < 4$ or $x^2 + 4y^2 > 36$

26. $9y^2 + 16x^2 < 144$ or $9y^2 + 16x^2 > 576$

27. In designing a playhouse, a father decides to make an elliptical opening in one wall. The wall measures 6 feet by 6 feet, and he wants to leave a border around the opening of 1 foot on the top and bottom and 1.5 feet on each side. Where should he locate the foci so that he can draw the ellipse? What is the equation of the ellipse that is centered on the wall?

28. Pete wants to build a coffee tabletop in the shape of an ellipse. The top is to be cut from a piece of wood 4 feet by 2 feet and Pete wants to use the full length and width of the board. Where should he locate the foci so that with pencil and string he can outline the table on the piece of wood?

PARABOLAS 6-4

In an earlier chapter we solved quadratic equations of the form $ax^2 + bx + c = 0$, with $a \neq 0$. In this section we will study quadratic functions of the form

$$P = \{(x, y) \mid y = ax^2 + bx + c,\ a \neq 0\}$$

The graph of P is called a parabola. The quadratic function is used extensively in mathematical studies, and many physical relationships are described by this function. Notice that if $y = 0$, then we obtain the quadratic equation $ax^2 + bx + c = 0$. If the roots of the equation are real, then the roots are the x-intercepts of the function.

P is a member of C, where $C = \{(x, y) \mid Ax^2 + Cy^2 + Dx + Ey + F = 0\}$. When $C = 0$, we obtain the equation $Ax^2 + Dx + Ey + F = 0$. This equation can be converted to the form of the quadratic function above. Furthermore, when $A = 0$, then

Is the quadratic function a member of set C ?

$$Cy^2 + Dx + Ey + F = 0$$

This is the equation of the **quadratic relation** discussed later in this section.

Parabola

> **Definition:** A parabola is a set of points in the plane that are equally distant from a fixed point and a fixed line.

Notice that the parabola, like the circle and ellipse, is defined in terms of distances between points in the plane. We call the fixed point the **focus** of the parabola and the fixed line the **directrix.** In this section we will restrict *directrices* (plural of *directrix*) to only horizontal or vertical lines.

In Figure 6-13 the graph of a parabola with focus $F(0, c)$ and horizontal directrix $y = -c$ is shown. Notice that we have selected a parabola with the focus on the positive y-axis and containing the origin of the coordinate system.

Therefore, since the distance from $F(0, c)$ to $O(0, 0)$ is equal to the distance from $O(0, 0)$ to the directrix, then $P(0, -c)$ must be a point of the directrix. We conclude that the equation of the directrix is $y = -c$ for the given parabola.

By definition, for any point $P(x, y)$ on the parabola, $|PF| = |PD|$, where $D(x, -c)$ is the point of intersection of a vertical line through P and the directrix.

Vertex

> **Definition:** The vertex is the point of the parabola closest to the directrix.

Axis of summetry

> **Definition:** The axis of symmetry of a parabola is the line containing the focus and perpendicular to the directrix.

In Figure 6-13, $O(0, 0)$ is the vertex and the y-axis is the axis of symmetry. For parabolas with horizontal directrices, the vertex is the *maximum or minimum point* on the graph of the parabola. For parabolas with vertical directrices, *the vertex is the greatest displacement to the right or to the left of the graph.* As we shall see, the location of the vertex and axis of symmetry of a parabola are key factors in sketching the graph of the parabola.

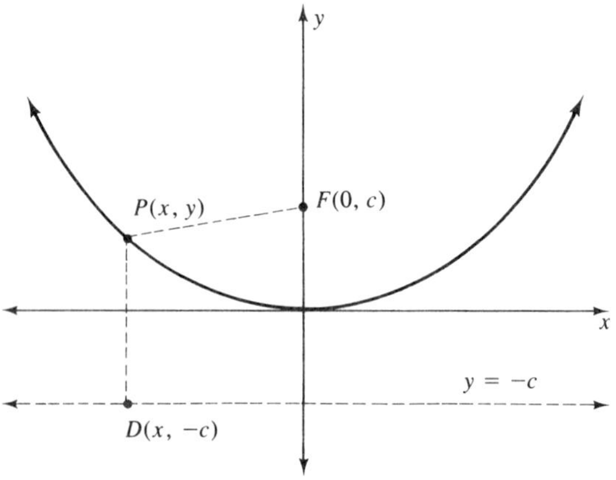

FIGURE 6-13 Graph of $P = \{(x, y) \mid y = ax^2, a > 0\}$

> **Theorem:** When the equation of the quadratic function $P = \{(x, y) \mid y = ax^2 + bx + c, a \neq 0\}$ is written in the form
>
> $$P = \{(x, y) \mid y = a(x - h)^2 + k\},$$
>
> then the axis of symmetry is the line $x = h$, and the vertex is the point $V(h, k)$.

When the equation of P is written in the alternate form, we can state that the domain of $P = \{x \mid x \in \text{reals}\}$ and the range $= \{y \mid y \geq k, a > 0\}$ and range $= \{y \mid y \leq k, a < 0\}$.

Why is the range of P in two parts?

As we shall see, the graph of P opens "upward" for $a > 0$; thus, $V(h, k)$ is a **minimum point.** For $a < 0$, the graph of P opens "downward" and $V(h, k)$ is a **maximum point.** The constants a, h, and k alter the graph of P in a way that is very similar to the effect of the constants a, b, and c on the graph of the absolute value function with equation $y = a|x - b| + c$. As a result, in the following example we will include graphs of both functions for similar values of the constants.

EXAMPLE 6–13 Sketch the graph of the following parabolas.
(a) $y = 2x^2$ (b) $y = -3x^2$

Solutions:
(a) Comparing the equation $y = 2x^2$ with the equation $y = a(x - h)^2 + k$, we see that $a = 2$, $h = 0$, and $k = 0$. Thus, $V(0, 0)$ is the vertex and $x = 0$ (the y-axis) is the axis of symmetry. The points $(1, 2)$, $(-1, 2)$ $(2, 8)$, and $(-2, 8)$ are used to graph the parabola. For the sake of comparison, the absolute value function $y = 2|x|$ is graphed with a dashed line. The graphs are shown in Figure 6-14.

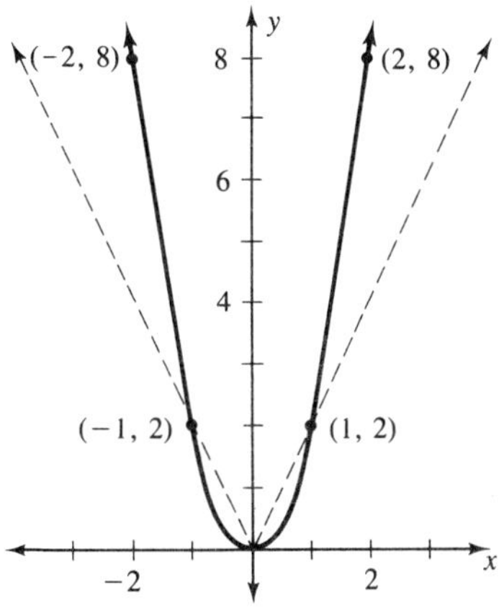

FIGURE 6-14 Graphs of $y = 2x^2$ and $y = 2|x|$

(b) For the equation $y = -3x^2$, $a = -3$ and $h = k = 0$. The points $(1, -3)$, $(-1, -3)$, $(2, -12)$, and $(-2, -12)$ are used to graph the parabola. This graph and the graph of the absolute value function $y = -3|x|$ are shown in Figure 6-15.

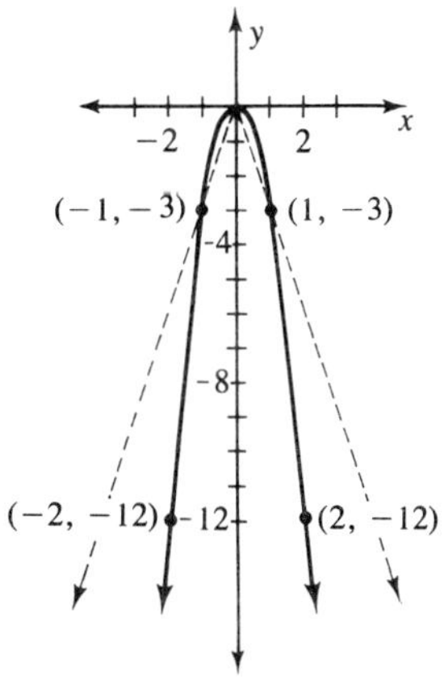

FIGURE 6-15 Graphs of $y = -3x^2$ and $y = -3|x|$

From the graphs in Example 6-13 we see that the graph of P opens "upward" for $a > 0$ and opens "downward" for $a < 0$.

EXAMPLE 6-14 Sketch the graph of the parabola $y = (x - 2)^2$.

Solution: Comparing the equation $y = (x - 2)^2$ with the equation $y = a(x - h)^2 + k$, we see that $a = 1, h = 2$, and $k = 0$. Thus, $V(2, 0)$ is the vertex and $x = 2$ is the axis of symmetry. The points (0, 4), (1, 1), (3, 1), and (4, 4) are used to graph the parabola. The graph is shown in Figure 6-16.

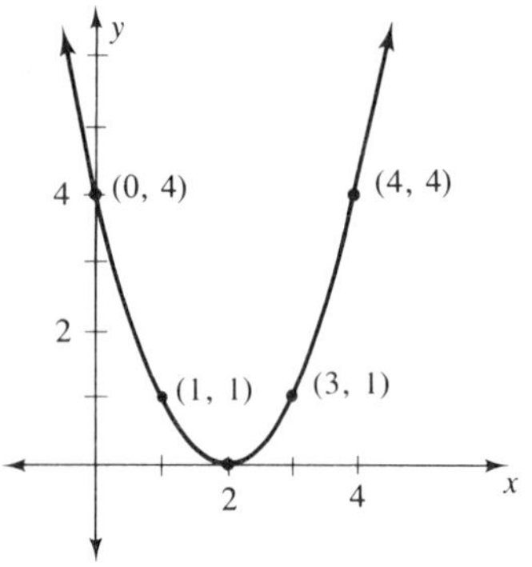

FIGURE 6-16 Graph of $y = (x - 2)^2$

From the graph in Example 6-14 we see that the graph of P is shifted h units to the right of the origin for $h > 0$. If $h < 0$, the graph of the parabola is shifted h units to the left of the origin.

The effect of different values of k on the graph of the parabola is the same as the effect of similar values of k on the absolute value function. Thus, for the parabola $P = \{(x, y) \mid y = ax^2 + k\}$, the vertex is the point $V(0, k)$.

> To summarize, if $P = \{(x, y) \mid y = a(x - h)^2 + k\}$, then
>
> (1) The parabola opens upward for $a > 0$ and downward for $a < 0$.
>
> (2) The vertex is the point $V(h, k)$.
>
> (3) The axis of symmetry is the vertical line $x = h$.

EXAMPLE 6–15 Graph the parabola $P = \{(x, y) \mid y = 2(x - 1)^2 - 3\}$.

Solution: For the given equation, $a = 2$, $h = 1$, and $k = -3$. Therefore, the parabola opens upward and has vertex $V(1, -3)$ and axis of symmetry $x = 1$. The points $(-1, 5)$, $(0, -1)$, $(2, -1)$, and $(3, 5)$ are used to sketch P in Figure 6-17.

When the equation of P is given in the form $y = ax^2 + bx + c$, then we *complete the square in x* to convert the given equation to the form $y = a(x - h)^2 + k$.

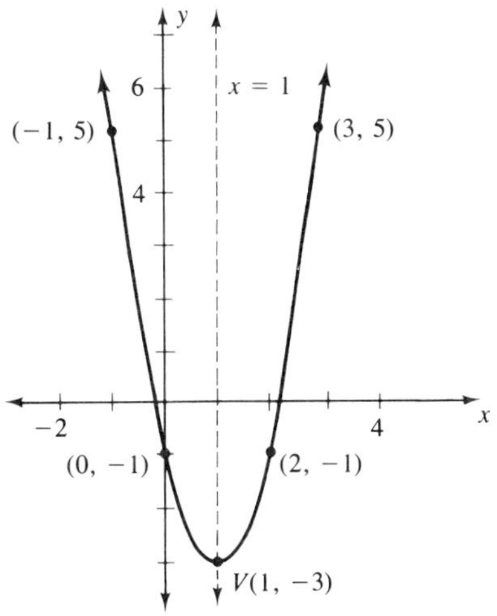

FIGURE 6-17 Graph of $y = 2(x - 1)^2 - 3$

EXAMPLE 6–16 Sketch the graph of $P = \{(x, y) \mid y = x^2 - 6x + 8\}$.

Solution: Convert the equation of P to the form $y = a(x - h)^2 + k$.

$y = x^2 - 6x + 8$ (notice that $a = 1$)

$y - 8 = x^2 - 6x$ (subtract 8 from each side of the equation)

$y - 8 + 9 = x^2 - 6x + 9$ (add the square of one-half the coefficient of x to each side)

$y + 1 = (x - 3)^2$ (write the right side of the equation as a perfect square)

$y = (x - 3)^2 - 1$ (subtract 1 from each side of the equation)

Thus, $a = 1, h = 3$, and $k = -1$. The vertex is $V(3, -1)$, the axis of symmetry is $x = 3$, and the points (1, 3), (2, 0), (4, 0), and (5, 3) are used to graph P in Figure 6-18.

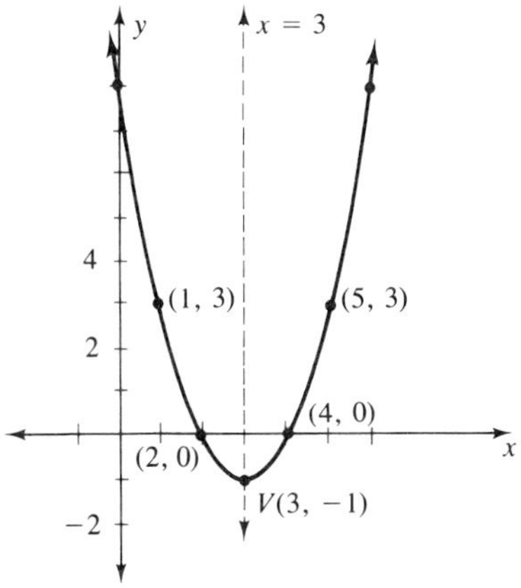

FIGURE 6-18 Graph of $y = x^2 - 6x + 8$

EXAMPLE 6–17 Determine the equation of the parabola P with vertex $V(-1/2, 2)$ that contains the point $P_1(1, 11)$.

Solution: From the given information, $h = -1/2$ and $k = 2$. Therefore, $P = \{(x, y) \mid y = a(x + 1/2)^2 + 2\}$. To determine the value of a we use the coordinates of P_1:

$$11 = a(1 + 1/2)^2 + 2$$

$$\frac{9}{4}a = 9$$

$$a = 4$$

Thus, $P = \{(x, y) \mid y = 4(x + 1/2)^2 + 2\}$ or

$$P = \{(x, y) \mid y = 4x^2 + 4x + 3\}$$

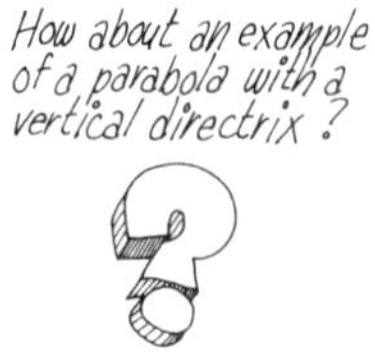

The equation of a parabola with a vertical directrix has an equation of the form $P = \{(x, y) \mid x = ay^2 + by + c\}$ or $P = \{(x, y) \mid x = a(y - h)^2 + k\}$. The graph of P has a vertex at point $V(k, h)$ and horizontal axis of symmetry $y = h$. If $a > 0$, the parabola opens to the right, and if $a < 0$, the parabola opens to the left.

EXAMPLE 6–18 Sketch the graph of $P = \{(x, y) \mid x = y^2 - 6y + 7\}$.

Solution:

$x = y^2 - 6y + 7$ (notice that $a = 1$)

$x - 7 = y^2 - 6y$ (subtract 7 from each side of the equation)

$x - 7 + 9 = y^2 - 6y + 9$ (add the square of one-half the coefficient of y to each side)

$x + 2 = (y - 3)^2$ (write the right side of the equation as a perfect square)

$x = (y - 3)^2 - 2$ (subtract 2 from each side of the equation)

Thus, $V(-2, 3)$ is the vertex of P and the axis of symmetry is the line $y = 3$. The points $(2, 5)$, $(-1, 4)$, $(-1, 2)$, and $(2, 1)$ are used to graph P in Figure 6-19.

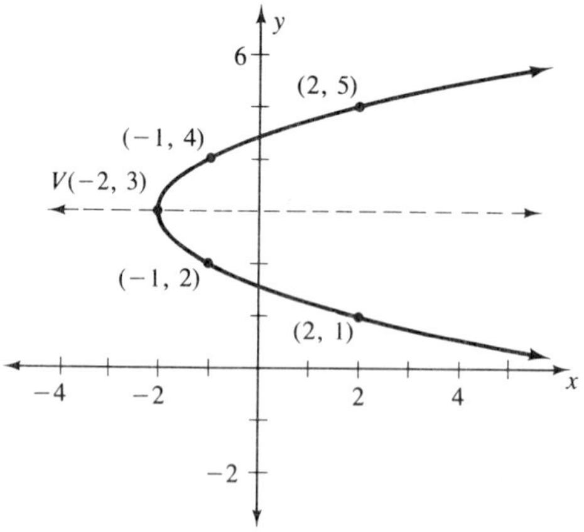

FIGURE 6-19 Graph of $x = y^2 - 6y + 7$

FIGURE 6-20 Parabola

In Figure 6-20 the plane is held parallel to an edge of the cone. The intersection of the cone and plane is then a parabola.

How do you cut the cone with a plane to get the parabola ?

Exercise Set 6–4

Sketch a graph of the following parabolas. Indicate the coordinates of the vertex on the graph.

1. $y = 3x^2$
2. $y = \frac{-1}{3}x^2$
3. $y = (x + 2)^2$
4. $y = -(x - 3)^2$

Graph the following.

5. $y = 2(x - 1)^2 + 1$
6. $y = -3(x + 1)^2 + 2$
7. $y = x^2 - 8x + 12$
8. $y = 5 - 2x - x^2$
9. $2y = x^2 - 8x + 20$
10. $3y = 5 - 4x - x^2$
11. $2y = 3x^2 - 6x - 1$
12. $3y + 2x^2 + 12x = 0$
13. $x = y^2$
14. $x = (y - 1)^2$
15. $x = y^2 + 2$
16. $x = 2y^2 + y + 2$
17. $x = 1 - 2y - y^2$
18. $x = y^2 + 8y - 6$

If $P = \{(x, y) \mid y = ax^2 + bx + c\}$, then the graph of the inequality $y < ax^2 + bx + c$ is either the interior or the exterior of the parabola, depending on the sign of a.

Graph the following quadratic inequalities.

19. $y < x^2 - 4x$
20. $y \geq 4x^2 + 4x + 1$
21. $y \leq 6x - x^2$
22. $y > 1 + 4x - x^2$
23. $2y > x^2 - 8x$
24. $3y < 12x - 2x^2 - 3$

Find the equation of the following parabolas with the information given.

25. Vertex $V(0, 0)$ and containing $P(2, 8)$
26. Vertex $V(1, 2)$ and containing $P(3, 3)$
27. Vertex $V(0, 3)$ and containing $P(1, -1)$
28. Vertex $V(4, 2)$ and containing $P(0, -6)$
29. Vertex $V(0, 0)$ and directrix $y = -2$ (*Hint:* Use the definition of a parabola and the distance formula.)
30. Vertex $V(2, 2)$ and directrix $y = -4$ (See hint in Exercise 29.)
31. Focus $F(1, 2)$ and directrix $y = -2$ (See hint in Exercise 29.)
32. Focus $F(0, 6)$ and directrix $y = 10$ (See hint in Exercise 29.)

6-5 HYPERBOLAS

In Section 6-3 we defined an ellipse as the set of points in a plane such that the *sum* of the distances between each point in the set and two fixed points is a constant. In this section we study the set of points in a plane the *difference* of whose distances from two fixed points is a constant.

Hyperbola

> **Definition:** A hyperbola is the set of points in a plane such that the difference in the distances between each point in the set and two fixed points is a constant.

The two fixed points are the foci of the hyperbola.

Standard hyperbola

> **Definition:** A standard hyperbola has foci that are either $F_1(c, 0)$ and $F_2(-c, 0)$ or $F_1(0, c)$ and $F_2(0, -c)$, where $c \in \{\text{reals}\}$ and $c > 0$.

Notice that in each case the foci are located the same distance from — but on opposite sides of — the origin on a coordinate axis.

> **Theorem:**
> *Case 1.* If H is a hyperbola with foci $F_1(c, 0)$ and $F_2(-c, 0)$, then
>
> $$H = \left\{(x, y) \,\middle|\, \frac{x^2}{a^2} - \frac{y^2}{b^2} = 1\right\}$$
>
> *Case 2.* If H is a hyperbola with foci $F_1(0, c)$ and $F_2(0, -c)$, then
>
> $$H = \left\{(x, y) \,\middle|\, \frac{y^2}{a^2} - \frac{x^2}{b^2} = 1\right\}$$

Notice that the foci are located on the axis of the variable for which the denominator is a^2. For example, if the equation of H is $\frac{x^2}{4} - \frac{y^2}{9} = 1$, then the foci are contained on the x-axis. But, if the equation of H is $\frac{y^2}{9} - \frac{x^2}{4} = 1$, then the foci are contained on the y-axis.

The constants a, b, and c are related with the equation $a^2 + b^2 = c^2$ for the hyperbola. (Recall that $a^2 = b^2 + c^2$ for the ellipse.) In Figure 6-21 a standard hyperbola H is drawn with foci on the x-axis. The equation of H is therefore of the form $\frac{x^2}{a^2} - \frac{y^2}{b^2} = 1$. If $y = 0$, then $x^2 = a^2$ and $x = \pm a$. Thus, a and $-a$ are the x-intercepts of H, and $V_1(a, 0)$ and $V_2(-a, 0)$ are vertices of the graph of H. Notice that $c > a$; therefore, the foci of the standard hyperbola are farther from the origin than the vertex located on the same side of the axis. The line segment with endpoints V_1 and V_2 is called the **transverse axis** of the hyperbola.

If $x = 0$ in the above equation, then $y^2 = -b^2$. There are no real values of y such that $y^2 < 0$; therefore, H does not have any y-intercepts. However, the line segment with endpoints $(0, b)$ and $(0, -b)$ is called the **conjugate axis** of the hyperbola.

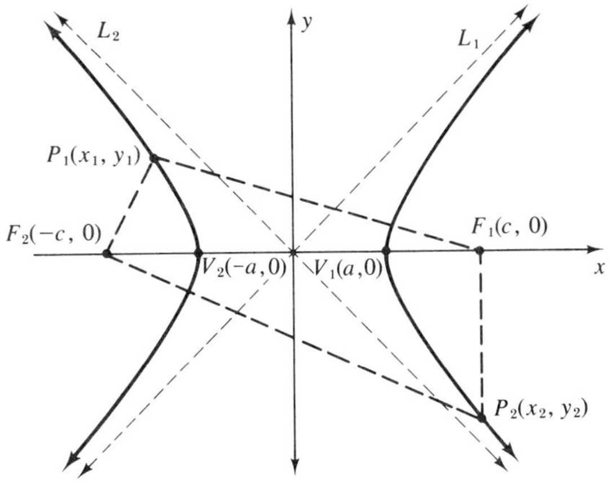

FIGURE 6-21 Graph of $H = \left\{(x, y) \;\middle|\; \frac{x^2}{a^2} - \frac{y^2}{b^2} = 1\right\}$

We make the following observations for H from the graph in Figure 6-21.

1. The graph of H consists of two disjoint curves, or *branches*. Therefore, $P_1(x_1, y_1)$ is on the graph of H if and only if

$$|P_1F_1| - |P_1F_2| = k, \quad k > 0$$

and $P_2(x_2, y_2)$ is on the graph of H if and only if

$$|P_2F_2| - |P_2F_1| = k, \quad k > 0$$

2. The *domain* of $H = \{x \mid |x| \geq a\}$.
3. The *range* of $H = \{y \mid y \in \{\text{reals}\}\}$.

4. The graph of H is *symmetric* with respect to both coordinate axes. Thus, if $P_1(x, y)$ is a point on the graph of H, then $P_2(x, -y)$, $P_3(-x, y)$, and $P_4(-x, -y)$ are also points of H.
5. The dashed lines L_1 and L_2 are *asymptotes* of the hyperbola.

How do we find the asymptotes for the graph of a hyperbola?

The equations of the lines that are the asymptotes of hyperbola H with equation $\frac{x^2}{a^2} - \frac{y^2}{b^2} = 1$ are obtained by solving the equation for y. That is,

$$y^2 = \frac{b^2}{a^2}(x^2 - a^2)$$

$$y = \pm\frac{b}{a}\sqrt{x^2 - a^2}$$

But a is a constant; therefore, $\sqrt{x^2 - a^2}$ approaches the value of x as $|x|$ becomes infinitely large. That is, the value of a^2 becomes insignificant for large values of x. Therefore, as $|x|$ becomes infinitely large,

$$y \approx \pm\frac{b}{a}x$$

and the lines L_1 and L_2 with equations $y = \frac{b}{a}x$ and $y = \frac{-b}{a}x$, respectively, are asymptotes for a standard hyperbola with foci on the x-axis.

EXAMPLE 6–19 Sketch the graph of $H = \{(x, y) \mid 4x^2 - 9y^2 = 36\}$.

Solution: If $4x^2 - 9y^2 = 36$, then

$$\frac{x^2}{9} - \frac{y^2}{4} = 1$$

and H is a standard hyperbola with foci on the x-axis. Furthermore, $a^2 = 9$ and $a = \pm 3$; therefore, the vertices of the hyperbola are $V_1(3, 0)$ and $V_2(-3, 0)$. With $b^2 = 4$ and $b = \pm 2$, the equations of the asymptotes are

$$y = \pm\frac{2}{3}x$$

The graph of H is shown in Figure 6-22.

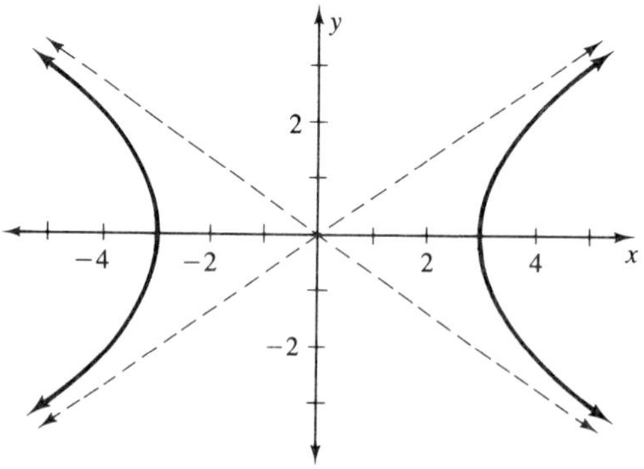

FIGURE 6-22 Graph of $H = \{(x, y) \mid 4x^2 - 9y^2 = 36\}$

We can determine that the graph of a quadratic relation is a hyperbola by knowing the values of A, C, and F. Recall the general equation of the quadratic relations, $C = \{(x, y) \mid Ax^2 + Cy^2 + Dx + Ey + F = 0\}$. If $D = E = 0$ and $F \neq 0$, then *the graph of* C *is a hyperbola whenever* $AC < 0$. If $F = 0$, then the graph consists of intersecting lines.

Did you know that the graph of H was a hyperbola before you changed the equation to standard form?

In Figure 6-23 the graph of a standard hyperbola with foci on the y-axis is shown. We make the following observation for H from the graph.

1. The y-intercepts of H are $\pm a$; thus, $V_1(0, a)$ and $V_2(0, -a)$ are vertices of H.
2. H does not have any x-intercepts.
3. The *domain* of $H = \{x \mid x \in \{\text{reals}\}\}$.
4. The *range* of $H = \{y \mid |y| \geq a\}$.
5. The graph of H is *symmetric* with respect to the x and y axes.
6. The lines L_1 and L_2 with equations $y = \frac{a}{b}x$ and $y = \frac{-a}{b}x$, respectively, are the *asymptotes* of the graph of H.

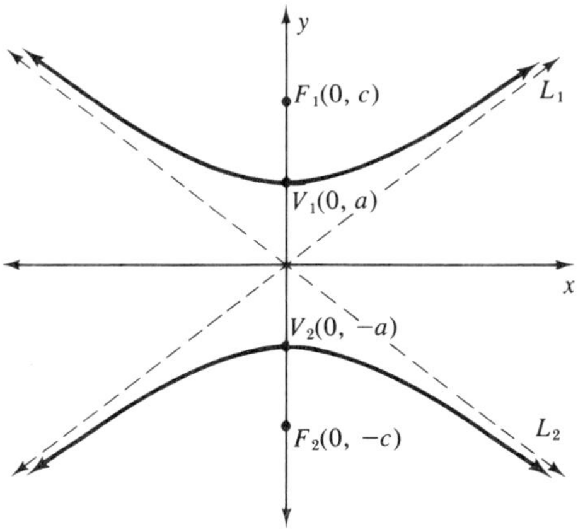

FIGURE 6-23 Graph of $H = \left\{(x, y) \;\middle|\; \frac{y^2}{a^2} - \frac{x^2}{b^2} = 1\right\}$

EXAMPLE 6–20 Sketch the graph of $H = \{(x, y) \mid y^2 - 4x^2 = 4\}$.

Solution: If $y^2 - 4x^2 = 4$, then

$$\frac{y^2}{4} - \frac{x^2}{1} = 1$$

Thus, the graph of H is a standard hyperbola with foci on the y-axis. Furthermore, $a^2 = 4$, and $a = \pm 2$. The vertices of H are $V_1(0, 2)$ and $V_2(0, -2)$. With $b^2 = 1$ and $b = \pm 1$, the equations of the asymptotes are $y = 2x$ and $y = -2x$. The graph of H is shown in Figure 6-24.

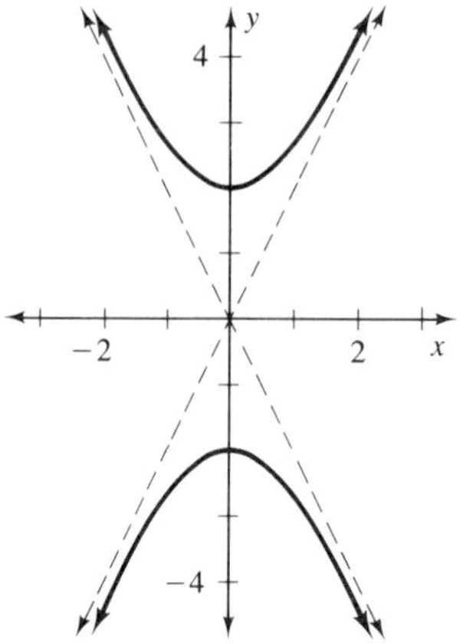

FIGURE 6-24 Graph of $H = \{(x, y) \mid y^2 - 4x^2 = 4\}$

EXAMPLE 6–21 Determine the equation of the standard hyperbola with focus $F(5, 0)$ and $b = 2$.

Solution: Since $F(5, 0)$ is a focus, then $c = 5$. With $b = 2$ and the equation $a^2 + b^2 = c^2$, we have

$$a^2 + 4 = 25$$
$$a^2 = 21$$

The equation of the hyperbola is therefore

$$\frac{x^2}{21} - \frac{y^2}{4} = 1 \quad \text{or} \quad 4x^2 - 21y^2 = 84$$

How do we cut the cone with the plane so that the intersection is a hyperbola?

In Figure 6-25, the cone is intersected with the plane held parallel to the axis of the cone. Notice that if the plane also contains the axis of the cone, then the graph is two intersecting lines.

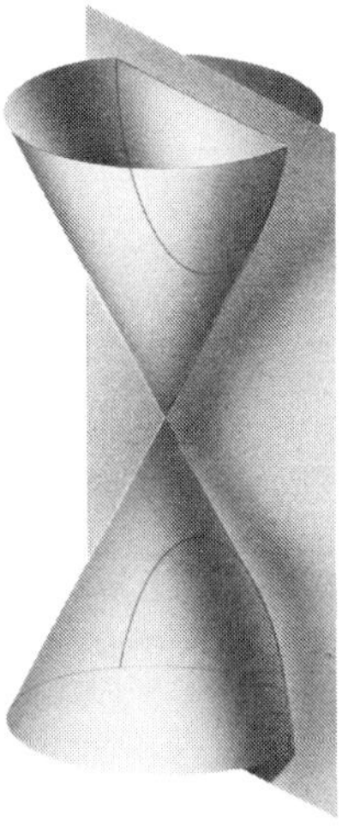

FIGURE 6-25 Hyperbola

Exercise Set 6–5

Graph the following hyperbolas. Label the vertices, and state the equations of the asymptotes.

1. $x^2 - y^2 = 4$
2. $25x^2 - 9y^2 = 225$
3. $x^2 - 4y^2 = 16$
4. $9x^2 - y^2 = 9$
5. $4y^2 - 25x^2 = 100$
6. $27x^2 - 48y^2 = 432$
7. $4x^2 - 4y^2 = 1$
8. $\frac{y^2}{9} - \frac{x^2}{4} = -1$

Determine an equation of a standard hyperbola with the given information.

9. Focus $(3, 0)$; x-intercept $= 2$
10. Focus $(-4, 0)$; x-intercept $= -1$
11. Focus $(0, 4)$; y-intercept $= 3$
12. Focus $(0, -6)$; y-intercept $= -2$
13. x-intercept $= 5$; $b = 4$
14. x-intercept $= -6$; $b = 8$
15. y-intercept $= 2$; $b = 3$
16. y-intercept $= -5$; $b = 2$
17. Focus $(5, 0)$; $b = 3$
18. Focus $(0, -3)$; $b = 3/2$

Graph the following.

19. $25x^2 + 4y^2 = 100$
20. $25x^2 - 4y^2 = 100$
21. $25x^2 + 25y^2 = 100$
22. $25x^2 - 25y^2 = 100$
23. $x = \sqrt{16 - 4y^2}$
24. $x = \sqrt{16 + 4y^2}$
25. $y = \frac{-1}{3}\sqrt{4x^2 - 1}$
26. $y = \frac{-1}{3}\sqrt{1 - 4x^2}$

Graph each of the following pairs of quadratic equations on the same coordinate axes and estimate the points of intersection from the graphs.

27. $y = x$
    $x^2 + y^2 = 8$
28. $x^2 + y^2 = 4$
    $x^2 - y^2 = 4$
29. $x^2 - y^2 = 5$
    $3y^2 - x^2 = 3$
30. $4x^2 + y^2 = 4$
    $y = x^2 - 3$

TRANSLATION OF AXES 6–6

Suppose we are instructed to graph set T, where

$$T = \{(x, y) \mid 4x^2 + 9y^2 - 16x - 18y - 11 = 0\}$$

The given quadratic equation identifies T as a member of set C, where

$$C = \{(x, y) \mid Ax^2 + Cy^2 + Dx + Ey + F = 0\}$$

with $A = 4$, $C = 9$, $D = -16$, $E = -18$, and $F = -11$. Since $AC = (4)(9) = 36$, and $36 > 0$, there is a possibility that the graph of T is an ellipse. However, $D \neq 0$ and $E \neq 0$; therefore, the ellipse (if it exists) is *not standard*. That is, the foci are not contained on either the x or the y coordinate axis.

Are we then going to have to compute a lot of values for x and y to graph T?

Because we have a method of graphing T that is easier, and also more accurate, we will not have to compute many values for x and y. To demonstrate the process, let us first complete the square in x and y as follows. If $4x^2 + 9y^2 - 16x - 18y - 11 = 0$, then make the x^2 and y^2 coefficients 1 with factoring and add the constant term to the right side of the equation:

$$4(x^2 - 4x \qquad) + 9(y^2 - 2y \qquad) = 11$$

Now add 4 times the square of one-half the x-coefficient and 9 times the square of one-half the y-coefficient to both sides of the equation:

$$4(x^2 - 4x + 4) + 9(y^2 - 2y + 1) = 11 + 16 + 9$$

Write the two trinomials as perfect squares:

$$4(x - 2)^2 + 9(y - 1)^2 = 36$$

Convert the constant to 1 by dividing both sides of the equation by 36:

$$\frac{(x-2)^2}{9} + \frac{(y-1)^2}{4} = 1$$

Let $x^* = x - 2$ and $y^* = y - 1$. (Read x^* as "x-star".) Then

$$\frac{(x^*)^2}{9} + \frac{(y^*)^2}{4} = 1$$

Thus, T defines a standard ellipse in an x^*y^*-coordinate system in which the x^*-intercepts are ± 3, the vertices are $V_1^*(3, 0)$ and $V_2^*(-3, 0)$, and y^*-intercepts are ± 2. The graph is shown in Figure 6-26.

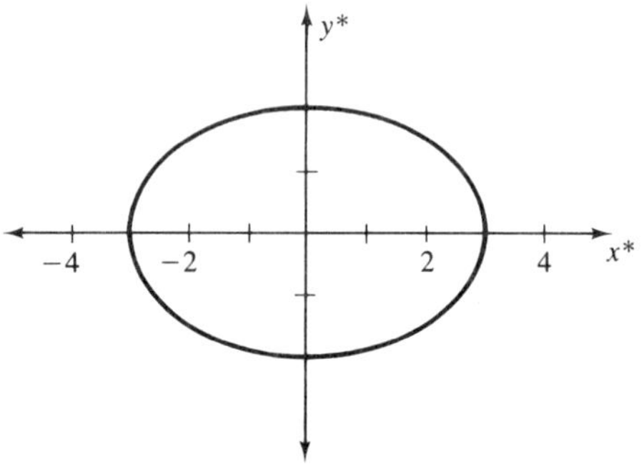

FIGURE 6-26 Graph of $\frac{(x^*)^2}{9} + \frac{(y^*)^2}{4} = 1$

Does that mean we will just ignore the xy-coordinate system?

Since the given equation for T is stated in terms of the variables x and y, we will not ignore the xy-coordinate system. We simply used the substitution of x^* and y^* for the quantities $(x - 2)$ and $(y - 1)$, respectively, to obtain a coordinate system in which the given ellipse is standard. The substitution we employed translated the equation of T from one in the variables x and y into one in the variables x^* and y^*. But the two coordinate systems exist simultaneously in the same plane.

To study the relationship between the two systems, recall the equations we used to relate the two pairs of variables. That is,

$$x^* = x - 2 \quad \text{and} \quad y^* = y - 1 \tag{1}$$

or

$$x = x^* + 2 \quad \text{and} \quad y = y^* + 1 \tag{2}$$

If $(x^*, y^*) = O^*(0, 0)$, then from Equation (2), $O^*(0, 0) = P(2, 1)$. That is, the origin of the x^*y^*-coordinate system is the point $P(2, 1)$ in the xy-coordinate system. The x and x^* axes are parallel to each other, as are the y and y^* axes. The graph of T and both coordinate systems are displayed in Figure 6-27, with the x^* and y^* axes shown as dashed lines.

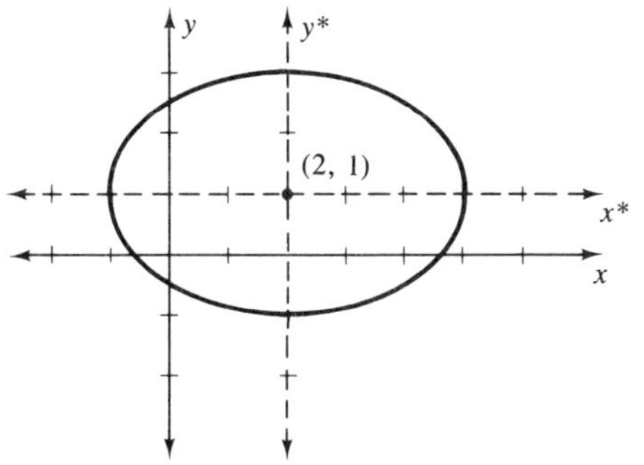

FIGURE 6-27 Graph of $\dfrac{(x - 2)^2}{9} + \dfrac{(y - 1)^2}{4} = 1$

We now generalize the results of the above example in the following theorem. We use the diagram in Figure 6-28 to verify the equations stated in the theorem.

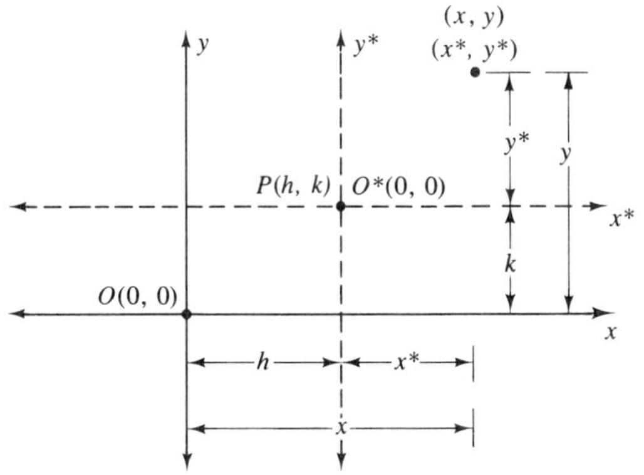

FIGURE 6-28 Two coordinate systems, xy and x^*y^*

Theorem: If the point $P(h, k)$ in an xy-coordinate system is the origin of an x^*y^*-coordinate system with similar units, then any point in the plane has two ordered pairs (x, y) and (x^*, y^*) such that

$$x^* = x - h \quad \text{and} \quad y^* = y - k \tag{1}$$

and

$$x = x^* + h \quad \text{and} \quad y = y^* + k \tag{2}$$

Notice that we have previously used the equations in (1) to assist us in graphing a circle. Recall that the form of the equation of a circle with center $O(a, b)$ and radius $r > 0$ is

$$(x - a)^2 + (y - b)^2 = r^2$$

If we now let $x^* = x - a$ and $y^* = y - b$, then

$$(x^*)^2 + (y^*)^2 = r^2$$

and the center of the circle is now the origin of the x^*y^*-coordinate system. As we shall see in a later example, a similar analysis can be made for a parabola with vertex at point $P(h, k)$.

EXAMPLE 6–22
(a) Sketch the graph of $R = \{(x, y) \mid 4x^2 - 9y^2 - 16x - 54y - 101 = 0\}$.
(b) Determine the domain and range of R.

Solutions:
(a) Since $A = 4$ and $C = -9$, and $4(-9) < 0$, the graph of R is potentially a hyperbola. We now complete the square in x and y:

$$\begin{aligned} 4x^2 - 16x \quad - 9y^2 - 54y \quad &= 101 \\ 4(x^2 - 4x + 4) - 9(y^2 + 6y + 9) &= 101 + 16 - 81 \\ \frac{(x-2)^2}{9} - \frac{(y+3)^2}{4} &= 1 \end{aligned}$$

The graph of R is a standard hyperbola in the x^*y^*-coordinate system for which $x^* = x - 2$ and $y^* = y + 3$. For the equation

$$\frac{(x^*)^2}{9} - \frac{(y^*)^2}{4} = 1$$

the x^*-intercepts $= \pm 3$, and the equations of the asymptotes are $y^* = \dfrac{\pm b}{a}x^*$ and $y^* = \dfrac{\pm 2}{3}x^*$. Furthermore, if $x^* = x - 2$, then $x = x^* + 2$; and for $x^* = 0$, $x = 2$. If $y^* = y + 3$, then $y = y^* - 3$; and for $y^* = 0$, $y = -3$. Therefore, $O^*(0, 0) = P(2, -3)$, and the graph of R is shown in Figure 6-29.
(b) The domain of $R = \{x^* \mid x^* \leq -3 \text{ or } x^* \geq 3\} = \{x \mid x \leq -1 \text{ or } x \geq 5\}$. The range of $R = \{y^* \mid y^* \in \{\text{reals}\}\} = \{y \mid y \in \{\text{reals}\}\}$.

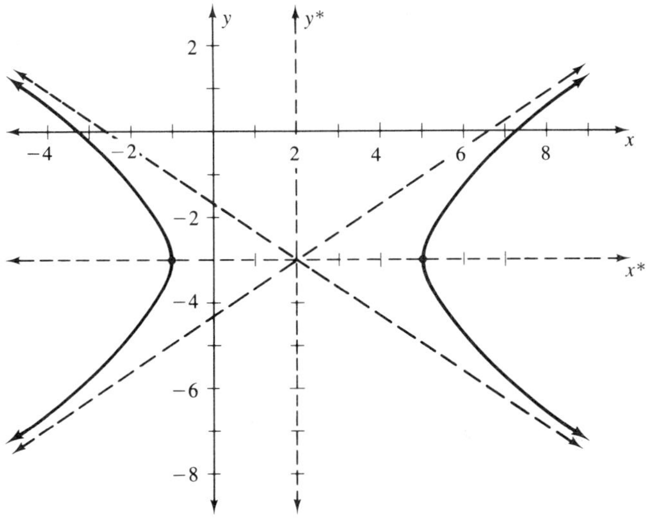

FIGURE 6-29 Graph of $R = \{(x, y) \mid 4x^2 - 9y^2 - 16x - 54y - 101 = 0\}$

We should appreciate the method of translating axes as a tool for graphing quadratic relations. In Example 6-22 we obtained an accurate graph of a very complex equation. We would need many solutions of quadratic equations to obtain a similar graph by using the variables x and y and the original equation.

EXAMPLE 6–23
(a) Sketch $R = \{(x, y) \mid 4y^2 - 2x - 12y + 3 = 0\}$.
(b) Determine the domain and range of R.

Solutions:
(a) Since $A = 0$ and $C = 4$, the graph of R is a parabola that opens to the right. If $4y^2 - 2x - 12y + 3 = 0$, then we first separate the y terms from the x term and constant:

$$2x - 3 = 4y^2 - 12y$$

Now factor the right side to make the y^2 coefficient 1:

$$2x - 3 = 4(y^2 - 3y \qquad)$$

Add 4 times the square of $1/2(-3)$ to both sides of the equation:

$$\begin{aligned} 2x - 3 + 9 &= 4(y^2 - 3y + 9/4) \\ 2x + 6 &= 4(y - 3/2)^2 \\ x + 3 &= 2(y - 3/2)^2 \end{aligned}$$

Let $x^* = x + 3$ (and $x = x^* - 3$), and $y^* = y - 3/2$ (and $y = y^* + 3/2$). The equation is then written

$$x^* = 2(y^*)^2$$

The graph of R is a parabola with vertex $O^*(0, 0)$ and axis of symmetry $y^* = 0$. The graph of R is shown in Figure 6-30. Notice that the point $O^*(0, 0) = P(-3, 3/2)$ in the xy-coordinate system.
(b) The domain of $R = \{x^* \mid x^* \geq 0\} = \{x \mid x \geq -3\}$, and the range of $R = \{y^* \mid y^* \in \{\text{reals}\}\} = \{y \mid y \in \{\text{reals}\}\}$.

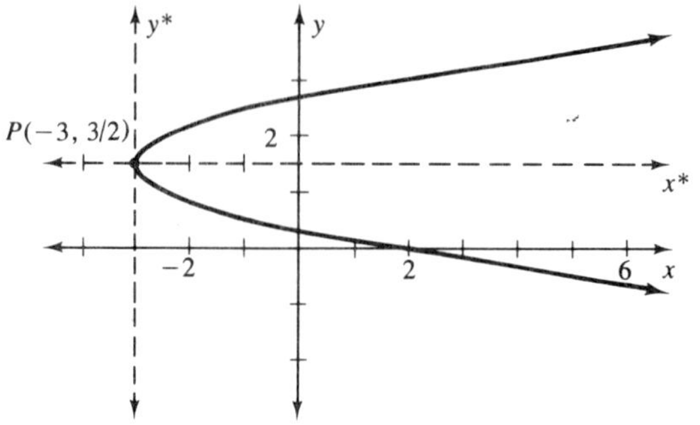

FIGURE 6-30 Graph of $R = \{(x, y) \mid 4y^2 - 2x - 12y + 3 = 0\}$

Exercise Set 6–6

For the following conic sections, (a) sketch the graphs, and (b) find the domain and range of each relation.

1. $3x^2 + 3y^2 - 6x + 12y - 33 = 0$
2. $6x^2 + 6y^2 + 12x = 42$
3. $x^2 + 8y - 2x - 7 = 0$
4. $y^2 + 6y + 2x = -5$
5. $x^2 + 4y^2 + 2x - 8y + 1 = 0$
6. $36x^2 + 96x + 100y^2 - 100y - 811 = 0$
7. $16x^2 - 16y^2 + 16y + 48x + 96 = 0$
8. $x^2 - 4y^2 - 2x + 8y - 2 = 0$
9. $3x - 2y^2 - 4y + 7 = 0$
10. $4x^2 + y^2 - 16x + 6y = -21$
11. $25x^2 + 9y^2 - 100x + 54y = 44$
12. $x^2 + 2x - 4y - 3 = 0$
13. $9y^2 - x^2 + 18y + 4x - 4 = 0$
14. $7x^2 + 14y^2 - x + 2y + 70 = 0$
15. $4x^2 + 4y^2 - 16x - 8y + 11 = 0$
16. $81x^2 - 4y^2 + 486x + 8y + 725 = 0$
17. $9x^2 + 4y^2 - 18x - 16y = 11$
18. $4y + 16 - 32x - y^2 - 16x^2 = 0$
19. $y^2 - 2y - 3x = 14$
20. $9x^2 - 90x + 25y^2 - 150y + 225 = 0$

SYSTEMS OF SECOND-DEGREE EQUATIONS IN TWO VARIABLES (optional) 6-7

In Section 4-4 we studied systems of linear equations, such as

$$L_1 = \{(x, y) \mid a_1x + b_1y = c_1\}$$
$$L_2 = \{(x, y) \mid a_2x + b_2y = c_2\}$$

We noted that a relationship exists between the solution of the system of equations and the graphs of L_1 and L_2. That is, *for each real solution of the system of equations there is a point of intersection of the graphs of L_1 and L_2.*

In this section we solve systems of equations that contain one or two quadratic equations in two variables. We solve such a system with the addition and substitution methods we used to solve systems of linear equations. A graphical sketch of the two equations is important in helping us to determine the *number of real solutions* to a given system. As we shall see, we can obtain algebraic solutions to a system of quadratic equations that are not real numbers. However, we will accept only the real solutions to a given system of equations. Therefore, for every point of intersection of the graphs of the equations in the system, there is an ordered pair of real numbers that is a solution of the system of equations.

Linear-Quadratic System

A **linear-quadratic system of equations** contains one linear equation and one quadratic equation. For example,

$$x - 2y = 5$$
$$x^2 + 4y^2 + 2x = 15$$

is a linear-quadratic system in which the quadratic equation is an ellipse.

The number of real solutions to a given linear-quadratic system of equations depends on the equations in the system. *But the number of real solutions to any linear-quadratic system is either 0, 1, or 2.* To illustrate, Figure 6-31 shows an ellipse E with three lines L_1, L_2, and L_3. If a system of equations is composed of E and L_1, E and L_2, or E and L_3, then the number of real solutions to each system

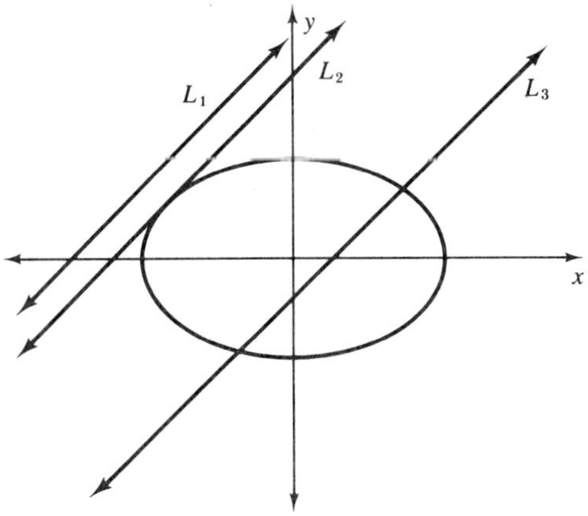

FIGURE 6-31 Graphs of E, L_1, L_2, and L_3

is 0, 1, and 2, respectively. That is, the number of real solutions to the system of equations is the same as the number of intersections of the graphs of the equations.

EXAMPLE 6–24 (a) Sketch the graphs and (b) find the real solutions of the system of equations

$$(1)\ x + 2y = 10$$
$$(2)\ x^2 - y + 2 = 0$$

Solutions:
(a) If $x + 2y = 10$, then

$$y = \frac{-1}{2}x + 5$$

and the graph is a line with y-intercept 5 and slope $-1/2$. If $x^2 - y + 2 = 0$, then

$$y = x^2 + 2$$

and the graph is a parabola with vertex $V(0, 2)$ and axis of symmetry $x = 0$. The graphs are shown in Figure 6-32. From the graph we can expect two real solutions for the given system of equations.

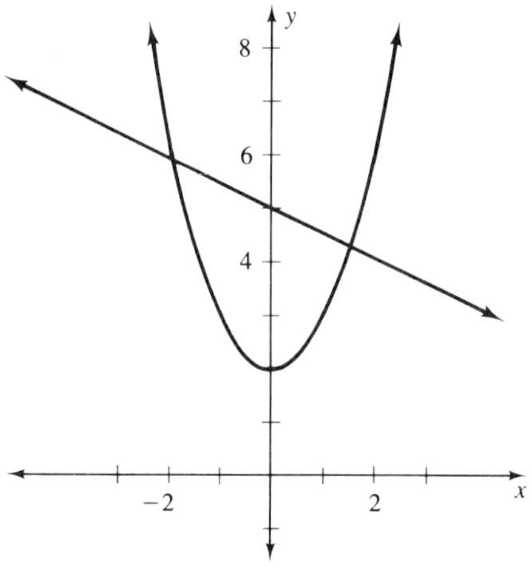

FIGURE 6-32 Graphs of $x + 2y = 10$ and $x^2 - y + 2 = 0$

(b) We solve the system of equations with the substitution method. In Equation (2), if $x^2 - y + 2 = 0$, then

$$y = x^2 + 2 \tag{3}$$

Substituting $x^2 + 2$ for y in Equation (1), we get

$$x + 2(x^2 + 2) = 10$$
$$2x^2 + x - 6 = 0$$
$$(2x - 3)(x + 2) = 0$$
$$x = 3/2 \quad \text{or} \quad x = -2$$

In Equation (3), if $x = 3/2$, then $y = (3/2)^2 + 2 = 17/4$; and if $x = -2$, then $y = (-2)^2 + 2 = 6$. Therefore, $\{(-2, 6), (3/2, 17/4)\}$ is the solution set for the given system of equations. We leave for you the check of these roots in the original equations.

Quadratic-Quadratic Systems

A **quadratic-quadratic system of equations** contains two quadratic equations. For example,

$$x^2 + 3y^2 - 4x - 9 = 0$$
$$x + y^2 - 6y + 10 = 0$$

is a quadratic-quadratic system in which the first equation is an ellipse ($AC > 0$) and the second equation is a parabola ($A = 0$ and $C = 1$).

The number of real solutions for a quadratic-quadratic system varies from 0 through 4. Again, *the number of real solutions to a given quadratic-quadratic system is reflected in the number of intersections in the graphs of the two equations in the system*. Exercise 23 in Exercise Set 6–7 asks for a sketch of two quadratic equations that demonstrates each of the five possible sets of real solutions to such a system of equations.

Are the number of real solutions to a quadratic-quadratic system the same as for a linear-quadratic system?

EXAMPLE 6–25 (a) Sketch the graphs and (b) find the real solutions for the system of equations

$$(1)\quad x^2 + y^2 = 16$$
$$(2)\quad 4x^2 - y^2 = 4$$

Solutions:

(a) Equation (1) defines a circle with center $O(0, 0)$ and radius 4. Equation (2) defines a standard hyperbola with foci on the x-axis. If $4x^2 - y^2 = 4$, then

$$\frac{x^2}{1} - \frac{y^2}{4} = 1$$

The graphs of the two equations are displayed in Figure 6-33.

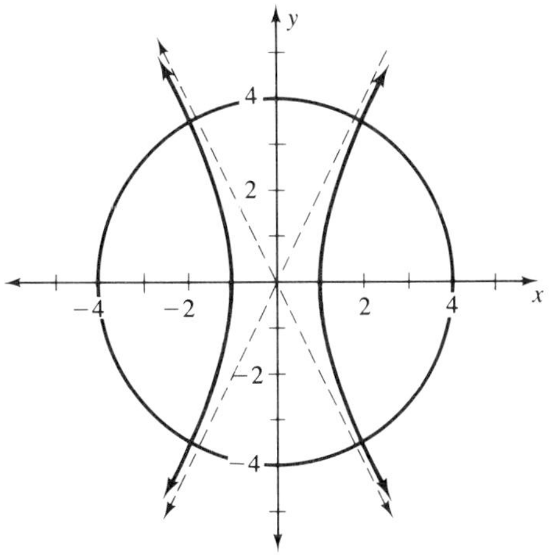

FIGURE 6-33 Graphs of $x^2 + y^2 = 16$ and $4x^2 - y^2 = 4$

(b) From the graphs it is apparent that the system of equations has four real solutions. We can solve the given system for x by adding the two equations:

$$\begin{array}{ll} (1)\ \ x^2 + y^2 &= 16 \\ (2)\ 4x^2 - y^2 &= 4 \\ \hline (3)\ 5x^2 &= 20 \\ \quad\ x &= \pm\sqrt{4} = \pm 2 \end{array}$$

If $x^2 = 4$ in Equation (1), then

$$4 + y^2 = 16$$
$$y = \pm\sqrt{12} = \pm 2\sqrt{3}$$

The set of real solutions to the given system of equations is $\{(-2, -2\sqrt{3}), (-2, 2\sqrt{3}), (2, -2\sqrt{3}), (2, 2\sqrt{3})\}$. Since both equations are symmetric with respect to the x and y axes, we need to check only one root to verify all four roots in the system. In Equation (1), if $x = 2$ and $y = 2\sqrt{3}$, then

$$(2)^2 + (2\sqrt{3})^2 = 16$$
$$4 + 12 = 16 \qquad \text{(True)}$$

In Equation (2), if $x = 2$ and $y = 2\sqrt{3}$, then

$$4(2)^2 - (2\sqrt{3})^2 = 4$$
$$16 - 12 = 4 \qquad \text{(True)}$$

Thus, the ordered pairs in the solution set are all roots of the given system of quadratic equations.

Exercise Set 6–7

For the following systems of equations, (a) identify the system as linear-quadratic or quadratic-quadratic; (b) identify the graphs of the equations in the system; (c) sketch the graphs of the equations to estimate the number of real solutions in the system; (d) solve the system of equations algebraically; and (e) check each solution in the equations in the system.

1. $2x - y - 2 = 0$
   $x^2 - y = 5$
2. $x = 1 - y$
   $y = x^2 - 2x - 5$
3. $x^2 + y^2 = 13$
   $y = 3x - 3$
4. $x^2 + y^2 - 4x + 10y + 4 = 0$
   $3x + y = 6$
5. $4x + 3y = 7$
   $16x^2 + 9y^2 = 25$
6. $x^2 + 4y^2 = 36$
   $x - 2y = 6$
7. $3x^2 + y^2 = 17$
   $2x^2 + y^2 = 13$
8. $x^2 + y^2 = 15$
   $x^2 - y^2 = 3$
9. $x^2 + y^2 = 36$
   $4x^2 + y^2 = 144$
10. $y^2 - 4x = 0$
    $y^2 - x^2 = 3$
11. $2x^2 + 7y^2 - 13 = 0$
    $3x^2 - 4y^2 = 34$
12. $3x^2 - y^2 = 2$
    $x^2 + 2y^2 = 3$

13. $x^2 - 4y = 0$
$y = x^2 - 3$

14. $2x^2 - y = 1$
$5x^2 + 2y = 2$

15. $3x^2 - 2y^2 = 6$
$5x^2 - 3y^2 = 11$

16. $4x^2 + 9y^2 = 36$
$x^2 - y^2 = 36$

17. $x^2 + y^2 = 10$
$3x + 4y = 15$

18. $4x^2 + 9y^2 = 36$
$12x^2 + 27y^2 = 108$

19. $x + y = 1$
$x^2 + y^2 + 4x - 2y - 5 = 0$

20. $y = x^2$
$2x - y + 3 = 0$

21. $\frac{x^2}{4} + \frac{y^2}{9} = 1$
$\frac{x^2}{4} - \frac{y^2}{9} = 1$

22. $x^2 + 4y^2 = 4$
$x + y = 2$

23. Sketch the graphs of a quadratic-quadratic system of equations, illustrating all possible solutions.

SUMMARY

Distance formula

Definition: If $P_1(x_1, y_1)$ and $P_2(x_2, y_2)$ are any two points in the coordinate plane, then the distance between P_1 and P_2, written $|P_1P_2|$, is given by the equation $|P_1P_2| = \sqrt{(x_2 - x_1)^2 + (y_2 - y_1)^2}$.

Theorem: The distance d from point $P(x_1, y_1)$ to line $L = \{(x, y) \mid Ax + By + C = 0\}$ is computed with the equation

$$d = \frac{|Ax_1 + By_1 + C|}{\sqrt{A^2 + B^2}}$$

Midpoint of a line segment

Definition: Point M is the midpoint of $\overline{P_1P_2}$ if and only if

(1) M is colinear with $P_1(x_1, y_1)$ and $P_2(x_2, y_2)$.
(2) $|P_1M| = |MP_2|$

Theorem: If $P_1(x_1, y_1)$ and $P_2(x_2, y_2)$ are any two points in the coordinate plane, then

$$M\left(\frac{x_1 + x_2}{2}, \frac{y_1 + y_2}{2}\right)$$

is the midpoint of $\overline{P_1P_2}$.

Circle

Definition: A circle is a set of points in a plane that are a fixed distance from a fixed point.

Theorem: The center-radius equation of the circle C with center $O(a, b)$ and radius r $(r > 0)$ is of the form $C = \{(x, y) \mid (x - a)^2 + (y - b)^2 = r^2\}$.

Ellipse

Definition: An ellipse is a set of points in a plane the sum of whose distances from two fixed points is a constant.

Standard ellipse

Definition: A standard ellipse has foci that are either $F_1(c, 0)$ and $F_2(-c, 0)$ or $F_1(0, c)$ and $F_2(0, -c)$, where $c \in \{\text{reals}\}$ and $c > 0$.

Theorem:
Case 1. If E is an ellipse with foci $F_1(c, 0)$ and $F_2(-c, 0)$, then

$$E = \left\{(x, y) \,\middle|\, \frac{x^2}{a^2} + \frac{y^2}{b^2} = 1\right\}, \quad \text{and } a > b > 0$$

Case 2. If E is an ellipse with foci $F_1(0, c)$ and $F_2(0, -c)$, then

$$E = \left\{(x, y) \,\middle|\, \frac{y^2}{a^2} + \frac{x^2}{b^2} = 1\right\}, \quad \text{and } a > b > 0$$

Parabola

Definition: A parabola is a set of points in a plane that are equally distant from a fixed point and a fixed line.

Vertex

Definition: The vertex is the point of the parabola closest to the directrix.

Axis of symmetry

Definition: The axis of symmetry of a parabola is the line containing the focus and perpendicular to the directrix.

Theorem: When the equation of the quadratic function $P = \{(x, y) \mid y = ax^2 + bx + c, a \neq 0\}$ is written in the form $P = \{(x, y) \mid y = a(x - h)^2 + k\}$, then the axis of symmetry is the line $x = h$, and the vertex is the point $V(h, k)$.

Hyperbola

Definition: A hyperbola is the set of points in a plane such that the difference in the distances between each point in the set and two fixed points is a constant.

Standard hyperbola

Definition: A standard hyperbola has foci that are either $F_1(c, 0)$ and $F_2(-c, 0)$ or $F_1(0, c)$ and $F_2(0, -c)$, where $c \in \{\text{reals}\}$ and $c > 0$.

Theorem:
Case 1. If H is a hyperbola with foci $F_1(c, 0)$ and $F_2(-c, 0)$, then

$$H = \left\{(x, y) \,\middle|\, \frac{x^2}{a^2} - \frac{y^2}{b^2} = 1\right\}$$

Case 2. If H is a hyperbola with foci $F_1(0, c)$ and $F_2(0, -c)$, then

$$H = \left\{(x, y) \,\middle|\, \frac{y^2}{a^2} - \frac{x^2}{b^2} = 1\right\}$$

Theorem: If the point $P(h, k)$ in an xy-coordinate system is the origin of an x^*y^*-coordinate system with similar units, then any point in the plane has two ordered pairs (x, y) and (x^*, y^*) such that

$$x^* = x - h \quad \text{and} \quad y^* = y - k \qquad \textbf{(1)}$$

$$x = x^* + h \quad \text{and} \quad y = y^* + k \qquad \textbf{(2)}$$

SELF-CHECKING EXERCISE

Find the length of $\overline{P_1P_2}$. **(6-1)**

1. $P_1(-1, 4)$, $P_2(3, -2)$
2. $P_1(-6, 0)$, $P_2(2, -2)$
3. $P_1(-3, 1)$, $P_2(1, -2)$
4. $P_1(7, 8)$, $P_2(-1, -7)$

Find the coordinates of the midpoint of $\overline{P_1P_2}$. **(6-1)**

5. $P_1(2, -1)$, $P_2(-4, 7)$
6. $P_1(10, 3)$, $P_2(-3, 4)$

Find the distance from each point to the line $3x - 4y + 8 = 0$. **(6-1)**

7. $P(5, 2)$
8. $P(-3, 1)$
9. $P(4, 5)$
10. $P(8, -2)$
11. Find the equation of the perpendicular bisector of the line segment joining (2, 3) and (5, 9). **(6-1)**

For the following circles, (a) state the coordinates of the center and the radius and (b) sketch a graph of the circles. **(6-2)**

12. $x^2 + y^2 - 2x + 4y - 11 = 0$
13. $3x^2 + 3y^2 + 9x - 15y = 1$

Sketch the graph of the following inequalities. **(6-2)**

14. $x^2 + y^2 < 9$
15. $x^2 + y^2 + 2x > 3$

Find the equation of the circles with the given properties. **(6-2)**

16. Center $(-4, 3)$; radius $\sqrt{5}$
17. Center (0, 2); radius 3
18. Center (4, 0) and containing $P(0, -3)$
19. Contains points $(-6, 2)$, $(1, 3)$ and $(2, -4)$

For the following ellipses, (a) state the coordinates of the vertices, (b) state the coordinates of the foci, and (c) sketch a graph. **(6-3)**

20. $x^2 + 25y^2 = 100$
21. $25x^2 + 2y^2 = 50$
22. Sketch the graph of $2 \leq x^2 + 2y^2 \leq 4$. **(6-3)**

For the following parabolas, (a) state the coordinates of the vertex, (b) state the equation of the axis of symmetry, and (c) sketch a graph. **(6-4)**

23. $y = 2x^2 - 4x + 5$
24. $x = 12y - 3y^2 - 8$

Find the equation of the following parabolas from the information given. **(6-4)**

25. Vertex (1, 3) and containing $P(3, 15)$
26. Vertex (3, 1) and directrix $y = -1$

For the following hyperbolas, (a) state the coordinates of the vertices, (b) state the equations of the asymptotes, and (c) sketch the graph. **(6-5)**

27. $x^2 - y^2 = 16$
28. $y^2 - 4x^2 = 100$

Determine an equation of a standard hyperbola from the given information. **(6-5)**

29. Focus (2, 0); x-intercept $= 1$
30. y-intercept $= 4$; $b = 5$

For the following equations, (a) determine an x^*y^*-coordinate system on which the graph of the relation is standard; (b) determine the domain and range of the relation in terms of x and y; and (c) sketch the graph of the relation and show both the xy and x^*y^* axes. **(6-6)**

31. $4x^2 - y^2 + 8x + 4y = 16$

32. $2x^2 + 3y^2 - 8x - 6y - 1 = 0$

Find all real solutions of the following systems of equations. **(6-7)**

33. $y = 2x + 3$
$x^2 + y^2 = 5$

34. $4x^2 + 3y^2 = 292$
$x^2 + y^2 = 89$

Answers to Self-Checking Exercise

1. $2\sqrt{13}$
2. $2\sqrt{17}$
3. 5
4. 17
5. $(-1, 3)$
6. $(7/2, 7/2)$
7. 3
8. 1
9. 0
10. 8
11. $2x + 4y = 31$

12. (a) $O(1, -2)$; $r = 4$
(b)

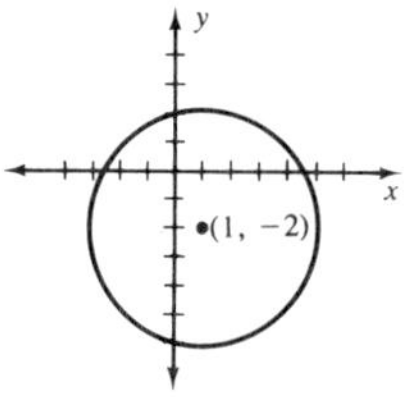

13. (a) $O(-3/2, 5/2)$; $r = \sqrt{318}/6 \approx 3$
(b)

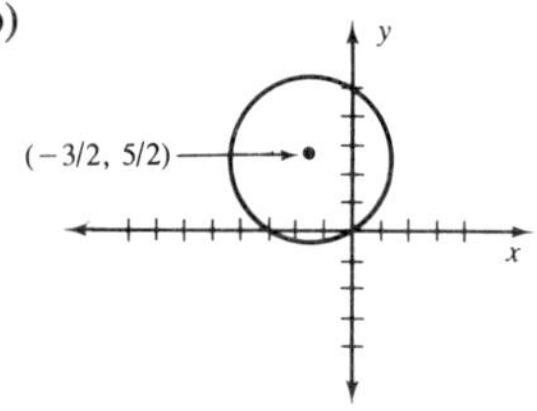

14.

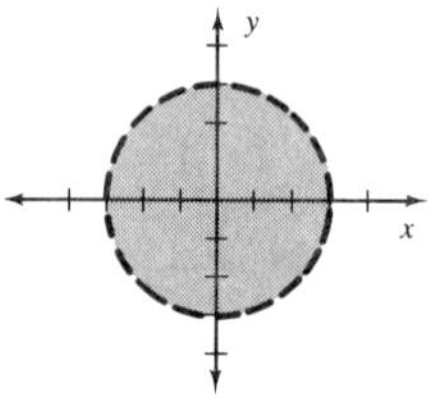

15.

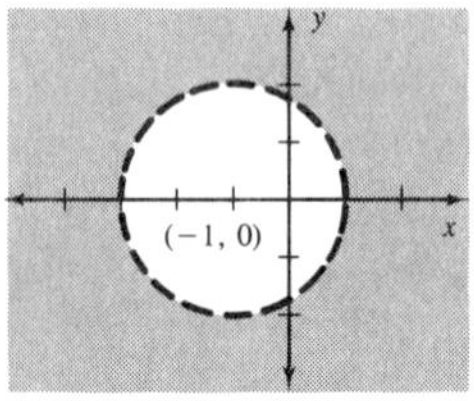

16. $(x + 4)^2 + (y - 3)^2 = 5$, or $x^2 + y^2 + 8x - 6y + 20 = 0$

17. $x^2 + (y - 2)^2 = 9$, or $x^2 + y^2 - 4y = 5$

18. $(x - 4)^2 + y^2 = 25$, or $x^2 + y^2 - 8x = 9$

19. $x^2 + y^2 + 4x + 2y - 20 = 0$

20. (a) $(10, 0)$, $(-10, 0)$
(b) $(4\sqrt{6}, 0)$, $(-4\sqrt{6}, 0)$
(c)

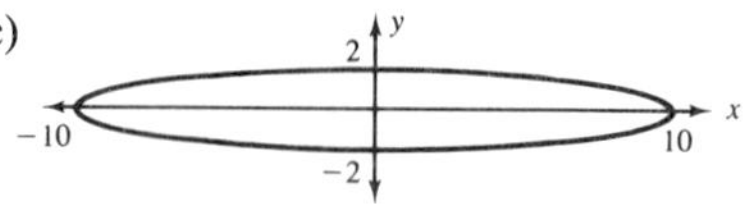

21. (a) $(0, 5)$, $(0, -5)$
(b) $(0, \sqrt{23})$, $(0, -\sqrt{23})$
(c)

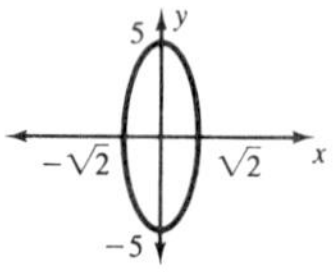

22.

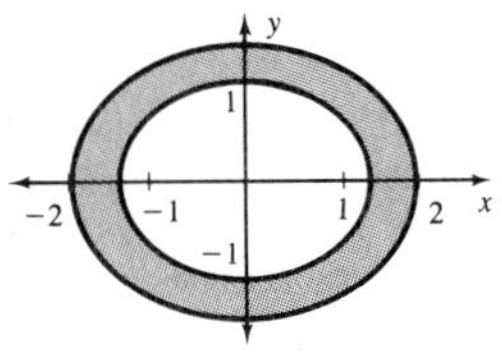

23. (a) (1, 3)
(b) $x = 1$
(c)

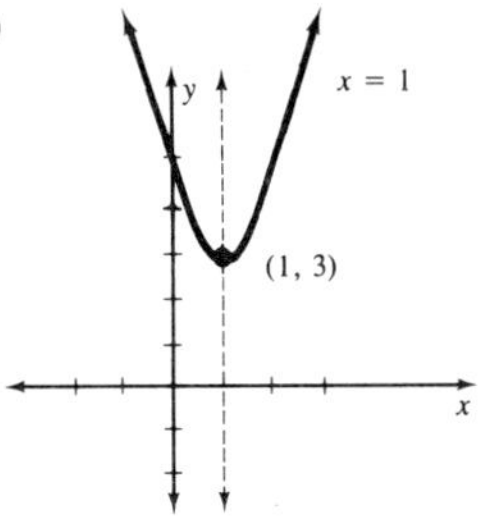

24. (a) (4, 2)
(b) $y = 2$
(c)

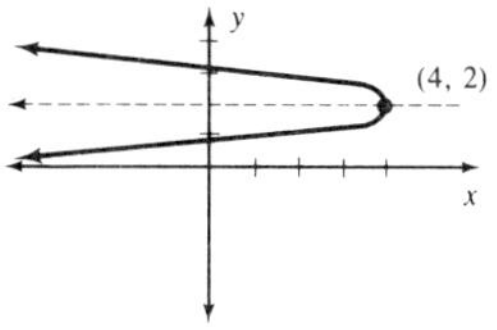

25. $y = 3x^2 - 6x + 6$

26. $8y = x^2 - 6x + 17$

27. (a) $(4, 0)$, $(-4, 0)$
(b) $y = x$ and $y = -x$
(c)

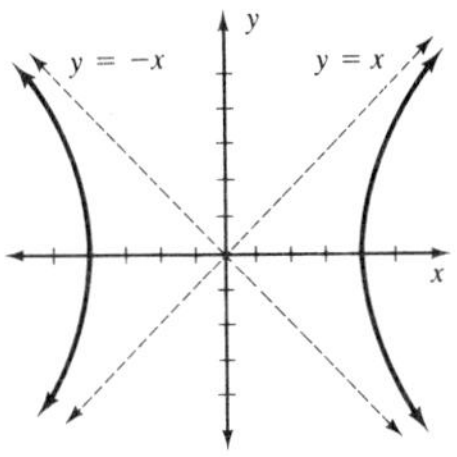

28. (a) $(0, 10)$, $(0, -10)$
(b) $y = 2x$ and $y = -2x$
(c)

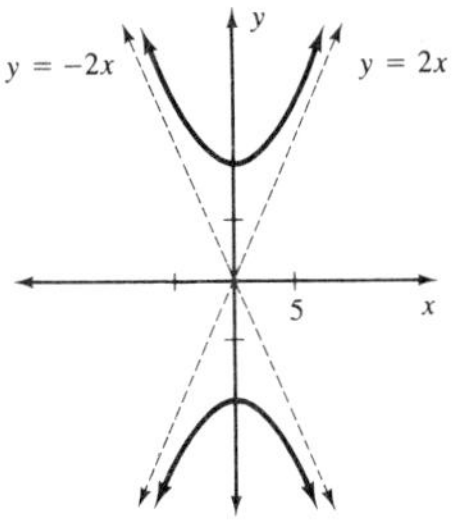

29. $3x^2 - y^2 = 3$

30. $25y^2 - 16x^2 = 400$

31. (a) $x^* = x + 1$
$y^* = y - 2$
(b) Domain = $\{x \mid x \leq -3 \text{ or } x \geq 1\}$
Range = $\{y \mid y \in \{\text{reals}\}\}$

(c)

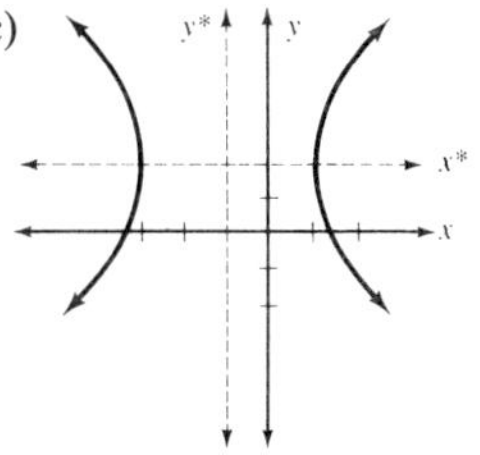

32. (a) $x^* = x - 2$
$y^* = y - 1$
(b) Domain = $\{x \mid 2 - \sqrt{6} \le x \le 2 + \sqrt{6}\}$
Range = $\{y \mid -1 \le y \le 3\}$
(c)

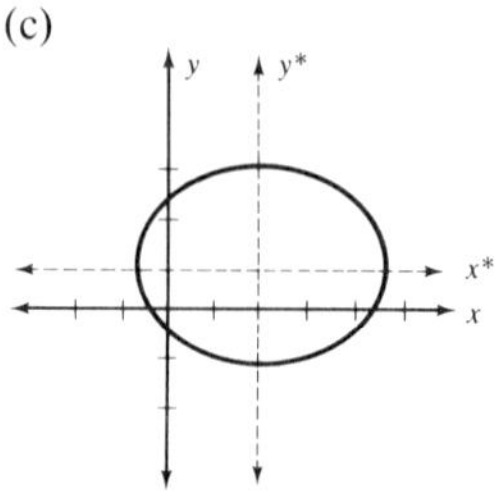

33. $\{(-2, -1), (-2/5, 11/5)\}$

34. $\{(5, 8), (5, -8), (-5, 8), (-5, -8)\}$

7 Trigonometric Functions

In past chapters we studied several relationships, such as linear, quadratic, polynomial of degree n, exponential, and logarithmic. In this chapter we study six relationships that are different in many ways from the ones studied thus far. These functions are known collectively as **trigonometric functions.** We first develop the basis for the relationship. We then discuss how these functions differ from other functions we have studied.

We should mention that the trigonometric functions can be defined in two different ways. We can establish the relationship between the set of real number coordinates of points on the number line and the coordinates of points on a circle in the plane. In this relationship the functions are frequently called **circular functions.**

A second approach to the trigonometric functions is to state the relationship between the set of angle measurements and the coordinates of points on a circle in the plane. We elect to use the circular function approach and then note the similarity between the two interpretations later in the chapter.

7-1 TRIGONOMETRIC FUNCTIONS OF REAL NUMBERS

In this section we have two principles to study, namely (1) the way in which arc lengths on a circle are measured with the set of real numbers and (2) the functional relationships that exist between the set of real numbers and the coordinates of points in the plane. To develop these principles we use a unit circle and a number line with unit length equal to the radius of the unit circle.

Unit circle

> **Definition:** A unit circle has a radius of one unit.

In Figure 7-1 a unit circle with center $O(0, 0)$ is drawn on a set of coordinate axes. A number line with positive numbers up and negative numbers down is also drawn with the zero point of the line on the unit circle at $(1, 0)$.

Imagine that the number line is a piece of string with infinite length. We now wrap the string around the circle so that the portions of the line with the positive

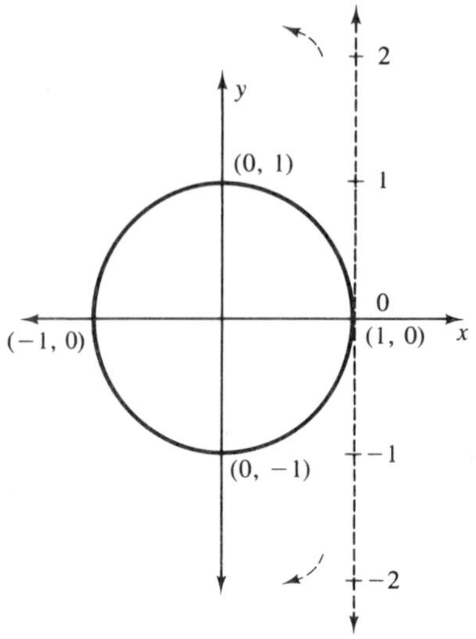

FIGURE 7-1 Unit circle and number line

numbers and the negative numbers are wrapped counterclockwise and clockwise, respectively. As the number line is wrapped around the circle, the points of the line pair themselves with points of the circle.

Each point of the number line is also paired with a real number that indicates the **distance** and **direction** from the zero point to the given point. Therefore, as each point of the circle is paired with a point of the number line, it is also paired with a real number that indicates the distance and direction along the arc of the circle from (1, 0). For example, there is a point P_1 on the circle that is paired with the point on the number line with coordinate 1. Thus, P_1 is 1 unit counterclockwise along the arc from (1, 0). In Figure 7-2, the approximate location of points P_1 through P_6 that are 1, 2, . . . , 6 units around the circle from (1, 0) are shown.

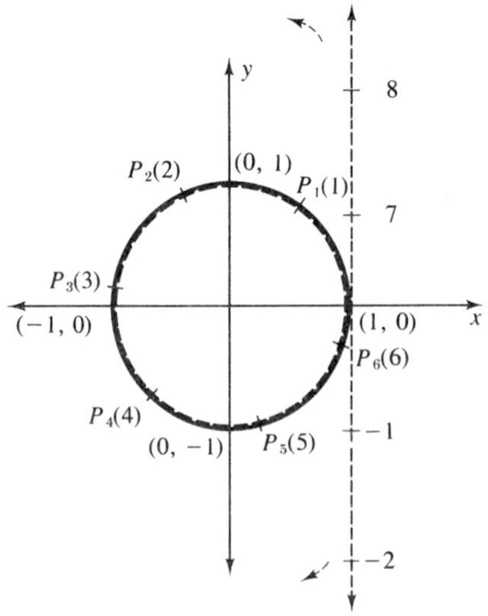

FIGURE 7-2 Unit circle and points P_1, P_2, P_3, P_4, P_5, and P_6

Remember that 0 is the first real number paired with (1, 0). Since the circumference of a circle is 2π times the radius of the circle, $C = 2\pi$ for the unit circle. Therefore, the real number 2π ($2\pi = 6.283185$ to six decimal places) is paired with (1, 0). Thus, 2π is the distance around the circle from (1, 0) back to the point (1, 0) on the circle.

What is the value of the real number that is paired with (1, 0)?

Furthermore, each time we wrap the line around the circle, each point of the circle is paired with another real number. *But each new number paired with a given point of the circle differs from the previous number by exactly* 2π. For example, in Figure 7-3 the points P_1 through P_6 are shown with the coordinates of the points on the number line with which they are paired when the line is wrapped the second time around the circle. On the third wrapping, P_1 is paired with a point of the number line whose coordinate is $1 + 2(2\pi)$ or $1 + 4\pi$. On the nth wrapping, P_1 is paired with a point on the number line whose coordinate is $1 + n(2\pi)$ or $1 + 2n\pi$.

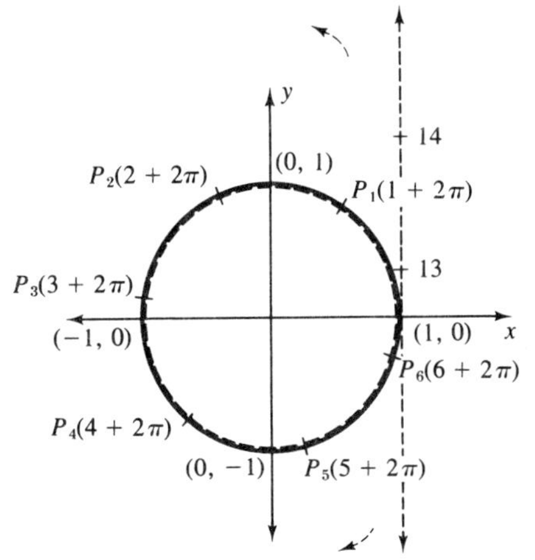

FIGURE 7-3 Unit circle and points P_1, P_2, P_3, P_4, P_5, and P_6

Every point on the circle is paired with an infinite number of real numbers. In general, *if P is a point on the unit circle, then P is paired with the real numbers* $s + 2n\pi$, *where s is a real number and* $0 \leq s < 2\pi$, *and n is an integer.*

Then every point on the circle is paired with an infinite number of real numbers?

Notice that we have established a functional relationship between the real numbers of the number line and the points of the circle. That is, for each number on the number line there is one and only one point of the circle with which the number is paired. Furthermore, the function is not one-to-one since any point of the circle is paired with infinitely many real numbers of the number line.

Let us restrict ourselves for the present to the real numbers s, where $0 \leq s < 2\pi$. *The real numbers s are paired with points of the circle for the first rotation of the number line around the circle.* The value of s that is paired with point P_1 of the circle depends on the fractional part of the rotation that P_1 is from (1, 0).

EXAMPLE 7–1 Find the real number s paired with the points P_1, P_2, P_3, and P_4 that are

(a) 1/3 (b) 1/2 (c) 5/8 (d) 5/6

of a rotation respectively from (1, 0). Graph the points on a unit circle.

Solutions:
(a) Since 1 rotation = 2π, then 1/3 rotation = $1/3(2\pi) = 2\pi/3$. Thus, P_1 is paired with $2\pi/3$ and P_1 is 1/3 of a rotation from (1, 0).
(b) 1/2 rotation = $1/2(2\pi) = \pi$, and the real number π is paired with P_2.
(c) 5/8 rotation = $5/8(2\pi) = 5\pi/4$, and P_3 is paired with $5\pi/4$.
(d) 5/6 rotation = $5/6(2\pi) = 5\pi/3$, and P_4 is paired with $5\pi/3$.
The graphs of the points are given in Figure 7-4.

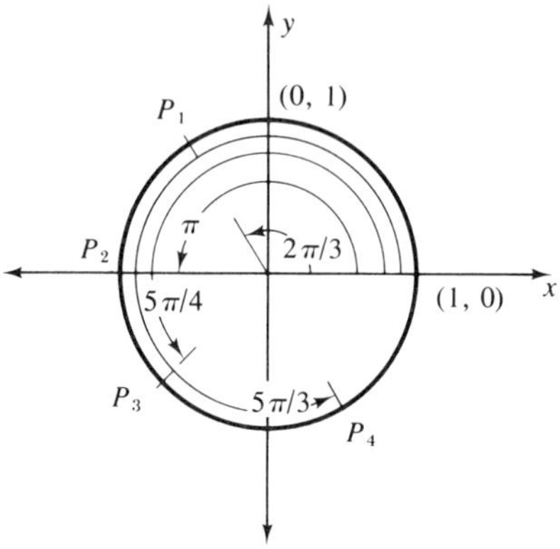

FIGURE 7-4 Unit circle and points P_1, P_2, P_3, and P_4

It is helpful to remember that for any real number s, $0 \leq s < 2\pi$, if s is paired

(1) with a point P in Quadrant I, then $0 < s < \pi/2$.
(2) with a point P in Quadrant II, then $\pi/2 < s < \pi$.
(3) with a point P in Quadrant III, then $\pi < s < 3\pi/2$.
(4) with a point P in Quadrant IV, then $3\pi/2 < s < 2\pi$.

EXAMPLE 7–2 Determine what fractional part of a rotation the following points are from (1, 0) if they are paired with the indicated real number s.
(a) P_1 is paired with $\pi/4$
(b) P_2 is paired with $5\pi/6$
(c) P_3 is paired with $4\pi/3$

Solutions:
(a) We use the following proportion:

$$\frac{\text{one rotation}}{2\pi} = \frac{x}{\pi/4}$$

and

$$x = \frac{(\pi/4)(\text{one rotation})}{2\pi} = \frac{1}{8} \text{ rotation}$$

Thus, P_1 is 1/8 of the rotation around the circle from (1, 0).

(b) $\dfrac{\text{one rotation}}{2\pi} = \dfrac{x}{5\pi/6}$, and $x = \dfrac{(5\pi/6)(\text{one rotation})}{2\pi} = \dfrac{5}{12}$ rotation

(c) $\dfrac{\text{one rotation}}{2\pi} = \dfrac{x}{4\pi/3}$, and $x = \dfrac{(4\pi/3)(\text{one rotation})}{2\pi} = \dfrac{2}{3}$ rotation

We now focus on the ordered pairs (x, y) that are the coordinates of the points on the unit circle. For example, in Figure 7-5, $P_1(1, 0)$, $P_2(0, 1)$, $P_3(-1, 0)$, and $P_4(0, -1)$ are shown. We have established that P_1, P_2, P_3, and P_4 are paired with the real numbers $2n\pi$, $\frac{\pi}{2} + 2n\pi$, $(2n + 1)\pi$, and $\frac{3\pi}{2} + 2n\pi$, respectively, with $n \in \{\text{integers}\}$, as the number line is wrapped around the circle. In general, for any point $P(x, y)$ on the unit circle, P is paired with a set of real numbers of the form

$$\{k \mid k = s + 2n\pi, \text{ and } 0 \leq s < 2\pi\}$$

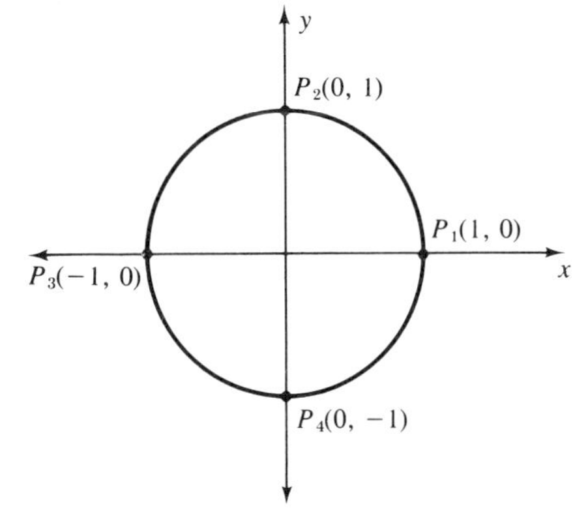

FIGURE 7-5 Points P_1, P_2, P_3, and P_4 on the unit circle

The coordinates x and y of point P are related to the real number k, and the relationship is a **function.** That is, for any real number k there is a point $P(x, y)$ on the unit circle paired with k. But a different point of the circle is paired with a number different from k, except numbers of the form $k + 2n\pi$. Thus, the coordinates of P, namely x and y, are functions of k.

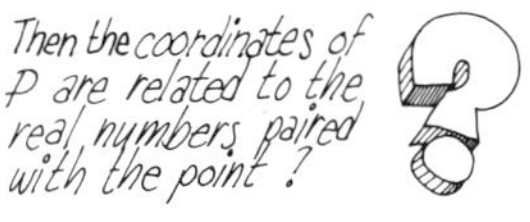

Cosine and sine functions

> **Definition:** If k is a real number paired with $P(x, y)$ on the unit circle, then
>
> $\{(k, x)\}$ is the cosine function and we write $x = \cos k$
>
> $\{(k, y)\}$ is the sine function and we write $y = \sin k$

The domain of both the cosine and the sine functions is the set of real numbers. Also, since the number line is wrapped on the unit circle, the maximum and minimum values of both x and y are 1 and -1. Therefore, the ranges of the cosine and sine functions are $\{x \mid -1 \leq x \leq 1\}$ and $\{y \mid -1 \leq y \leq 1\}$, respectively.

For any point $P(x, y)$ on the unit circle, P is paired with real numbers of the form $s + 2n\pi$, and $0 \le s < 2\pi$. Therefore,

$$x = \cos s = \cos(s + 2\pi) = \cos(s + 4\pi) = \cdots = \cos(s + 2n\pi)$$
$$y = \sin s = \sin(s + 2\pi) = \sin(s + 4\pi) = \cdots = \sin(s + 2n\pi)$$

Notice that the function values for cosine and sine repeat in regular cycles, or periods. A function with this property is called **periodic.** Furthermore, since

$$\cos s = \cos(s + 2\pi) \quad \text{and} \quad \sin s = \sin(s + 2\pi)$$

and 2π is the smallest positive number for which the equations are true, the period of these functions is 2π. As another example, the phases of the moon are periodic and the period is approximately 28 days.

Because of the periodic nature of the cosine and sine functions, we can concentrate on the real numbers s, with $0 \le s < 2\pi$, and yet develop the function values for all real numbers k. Let us therefore agree that *the symbol s will be used to represent a real number between* 0 *and* 2π ($0 \le s < 2\pi$), *and s is the length of an arc on a unit circle measured counterclockwise from* (1, 0).

The ratios of x and y and the reciprocals of x and y also define functions, which are identified below. These functions, together with the cosine and sine, are called the **trigonometric functions of real numbers.**

Cosine, sine, tangent, cotangent, secant, and cosecant functions

Definition: If s is a real number paired with $P(x, y)$ on the unit circle, then the six trigonometric functions are

(1) cosine $= \{(s, \cos s) \mid \cos s = x\}$

(2) sine $= \{(s, \sin s) \mid \sin s = y\}$

(3) tangent $= \{(s, \tan s) \mid \tan s = \dfrac{y}{x}, x \neq 0\}$

(4) cotangent $= \{(s, \cot s) \mid \cot s = \dfrac{x}{y}, y \neq 0\}$

(5) secant $= \{(s, \sec s) \mid \sec s = \dfrac{1}{x}, x \neq 0\}$

(6) cosecant $= \{(s, \csc s) \mid \csc s = \dfrac{1}{y}, y \neq 0\}$

These new functions must also be periodic, but are their periods also 2π?

The periods of $\sec s$ and $\csc s$ are 2π, but the periods of $\tan s$ and $\cot s$ are π. The domains and ranges of tangent, cotangent, secant, and cosecant are somewhat involved, and we will discuss them at a later time.

Example 7–3 If $P(3/5, 4/5)$ is paired with some real number s, find the values of the six trigonometric functions of s.

Solutions: With $x = 3/5$, $y = 4/5$, and the definitions above,

(a) $\cos s = 3/5$

(b) $\sin s = 4/5$

(c) $\tan s = \dfrac{4/5}{3/5} = \dfrac{4}{3}$

(d) $\cot s = \dfrac{3/5}{4/5} = \dfrac{3}{4}$

(e) $\sec s = \dfrac{1}{3/5} = \dfrac{5}{3}$ (f) $\csc s = \dfrac{1}{4/5} = \dfrac{5}{4}$

EXAMPLE 7–4 If $\cos s = -5/13$ and s is paired with a point in Quadrant II, find the other five trigonometric function values of s.

Solutions: For any $P(x, y)$ on the unit circle, $x^2 + y^2 = 1$. Since $x = -5/13$,

$$\left(\frac{-5}{13}\right)^2 + y^2 = 1$$

$$y^2 = \frac{144}{169}$$

$$y = \frac{\pm 12}{13}$$

Since $P(x, y)$ is in Quadrant II, $y = 12/13$. With $x = -5/13$ and $y = 12/13$,

(a) $\sin s = 12/13$ (b) $\tan s = -12/5$ (c) $\cot s = -5/12$

(d) $\sec s = -13/5$ (e) $\csc s = 13/12$

Exercise Set 7–1

Find the real number s that is paired with a point that is the given fractional part of one rotation on the unit circle from (1, 0).

1. 1/6 2. 1/4 3. 3/8 4. 5/12

5. 2/3 6. 3/4 7. 7/8 8. 11/12

Determine what fractional part of a rotation a point is from (1, 0) if it is paired with the given number s.

9. $\pi/3$ 10. $3\pi/4$ 11. π 12. $7\pi/6$

13. $5\pi/4$ 14. $3\pi/2$ 15. $11\pi/6$ 16. 2π

Approximate to two decimal places the fractional part of a rotation a point is from (1, 0) if it is paired with the given number s. Use $2\pi = 6.28$. (Use a hand calculator if possible.)

17. 1.00 18. 2.33 19. 3.00 20. 3.50

21. 4.00 22. 4.80 23. 5.00 24. 6.00

The following points are paired with some number s. Find the six trigonometric function values of s.

25. $(-3/5, 4/5)$ 26. $(-4/5, -3/5)$ 27. $(7/25, 24/25)$

28. $(\sqrt{2}/2, \sqrt{2}/2)$ 29. $(1/2, -\sqrt{3}/2)$ 30. $(-\sqrt{3}/2, -1/2)$

31. $(-8/17, 15/17)$ 32. $(-15/17, -8/17)$

Find the other five trigonometric function values of s, given that $P(x, y)$ paired with s is contained in the indicated quadrant.

33. $\sin s = 5/13$; I 34. $\cos s = \sqrt{3}/2$; I

35. $\tan s = -1$; II
36. $\sin s = 24/25$; II
37. $\cot s = \sqrt{3}$; III
38. $\sec s = -2$; III
39. $\cos s = 1/2$; IV
40. $\csc s = -5/4$; IV

Find the values of the real numbers k and l, with $2\pi \le k < 4\pi$ and $-2\pi \le l < 0$, that are paired with the same point on the unit circle as the given number.

41. $\pi/4$
42. $\pi/3$
43. $\pi/2$
44. $5\pi/6$
45. $5\pi/4$
46. $4\pi/3$
47. 0
48. $11\pi/6$

Find the value of $s (0 \le s < 2\pi)$ that is paired with the same point on the unit circle as the given number k.

49. $13\pi/6$
50. $13\pi/4$
51. $8\pi/3$
52. $29\pi/6$
53. $21\pi/4$
54. $11\pi/2$
55. $51\pi/12$
56. 33π

7-2 ANGLES, ANGLE MEASUREMENT, AND FUNCTIONS DEFINED ON ANGLES

In Section 7-1 a function relationship was established between the points on a unit circle and the coordinates of a number line. The relationship paired every real number k with a point $P(x, y)$ on the unit circle, with $k = 0$ paired with $(1, 0)$. Recall that k is also the length of the arc on the circle from $(1, 0)$ to $P(x, y)$. We know that *if* $k > 0$, *the arc is measured counterclockwise,* and *if* $k < 0$, *the arc is measured clockwise.*

In this section we use k to measure an angle with the unit circle. We define an angle in geometry as the union of two rays with a common starting point. Greek letters, such as θ (theta) and α (alpha), are used to symbolize angles. (A complete list of these symbols is on the endpapers of this book.) Since the values of the trigonometric functions are determined by k and the measure of an angle is measured by k, we will establish a relationship between the measure of an angle and the trigonometric functions of a real number.

In Figure 7-6, θ is drawn with its **vertex** at the center of a unit circle centered on a coordinate axis system. The sides of θ are $\overrightarrow{OQ}$ and $\overrightarrow{OP}$, called **rays.** The **initial side** and **terminal side** of θ are $\overrightarrow{OQ}$ and $\overrightarrow{OP}$, respectively.

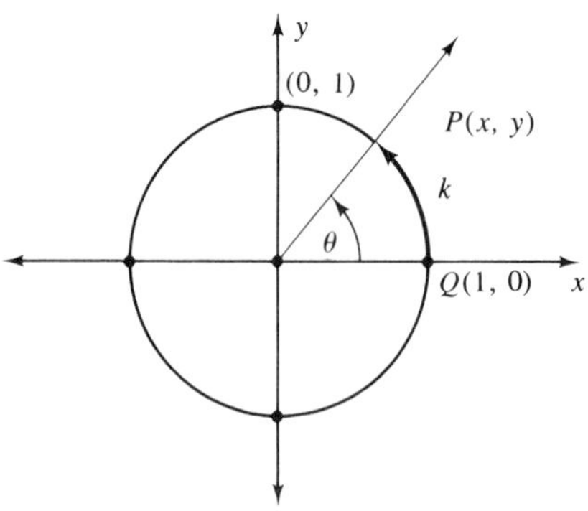

FIGURE 7-6 Standard position angle θ with measure k, $k > 0$

Standard position angle

> **Definition:** An angle is in standard position if the vertex is the origin of the coordinate system and the initial side is the positive x-axis.

In Figure 7-6, θ is in standard position. Let it be understood that all angles are in standard position, unless otherwise stated. Since θ **subtends** or cuts off an arc of the circle of length k, the measure of θ is equal to k.

Radian measure of an angle

> **Definition:** The measure of a standard position angle θ in radians is equal to the length of the arc of a unit circle subtended by the sides of the angle.

Since the length of the arc intercepted by θ in Figure 7-6 is k, then the measure of $\theta = k$ radians (also written k^r). In Figure 7-7, θ is drawn with a measure of one radian. That is, the length of the arc subtended by θ is equal to the radius of the circle. For simplicity we write

$$\theta = 1^r \quad \text{or} \quad \theta = 1$$

and the radian symbol is understood.

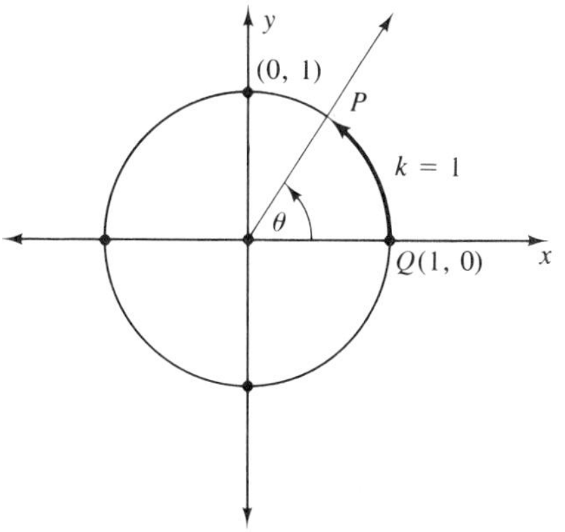

FIGURE 7-7 Measure of angle $\theta = 1$ radian

We use θ to represent the angle formed by the union of two rays. However, we are primarily interested in the *measure of* θ. Rather than create a new symbol or write "measure of θ," we will use the same symbol to represent both concepts. The dual use of the symbol will create no problem, since each use of the symbol will clearly indicate its meaning.

But doesn't θ represent the set of points that are the rays of the angle?

In Figure 7-6, the small arc in the interior of θ indicates that $\overrightarrow{OP}$ is rotated *counterclockwise* from $\overrightarrow{OQ}$ and θ is a *positive angle*. In Figure 7-8, θ is drawn with $\overrightarrow{OP}$ rotated *clockwise* from $\overrightarrow{OQ}$, and θ is a *negative angle*. It is significant to note that a positive angle has a positive measure and a negative angle has a negative measure.

Coterminal angles

> **Definition:** Two angles in standard position are coterminal if they have the same terminal side.

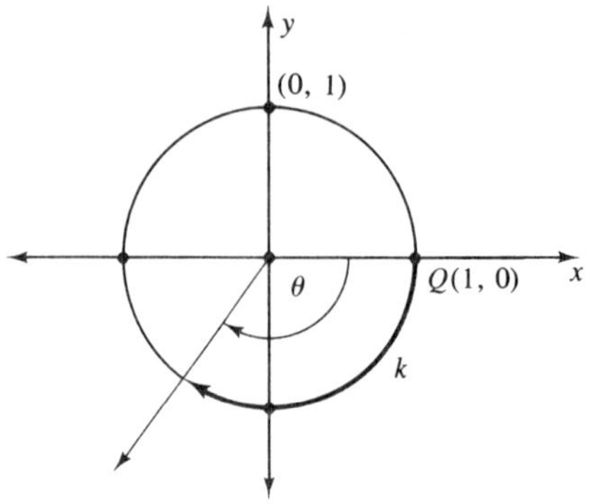

FIGURE 7-8 Negative angle θ with measure k, $k < 0$

In Figure 7-9 positive angle θ and negative angle α are shown as coterminal angles. Two angles are coterminal if their measures differ by an integral multiple of 2π. Thus, if $\theta = 3$ radians, then any angle with measure $3 + 2n\pi$ (with n an integer) is coterminal with θ.

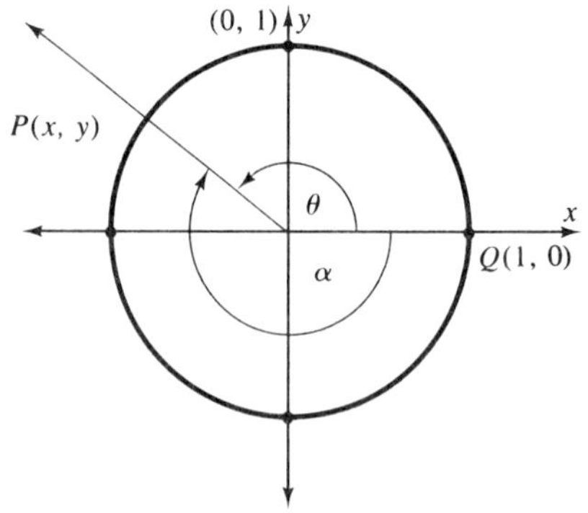

FIGURE 7-9 Coterminal angles θ and α

EXAMPLE 7–5 Find the measure of θ in radians if $\overrightarrow{OP}$ is rotated counterclockwise
(a) 1/8 (b) 7/12 (c) 7/6
of a rotation from $\overrightarrow{OQ}$. Graph the three angles.

Solutions:
(a) One rotation = 2π radians; therefore, 1/8 rotation = $1/8(2\pi$ radians$) = \pi/4$ radians, and $\theta_1 = \pi/4$.
(b) 7/12 rotation = 7/12 (2π radians) = $7\pi/6$ radians, and $\theta_2 = 7\pi/6$.
(c) 7/6 rotation = 7/6 (2π radians) = $7\pi/3$ radians, and $\theta_3 = 7\pi/3$.
The graphs of θ_1, θ_2, and θ_3 are shown in Figure 7-10. Notice that $\theta_3 > 2\pi$, since $7/6 > 1$.

Another method for measuring angles in standard position is with the degree.

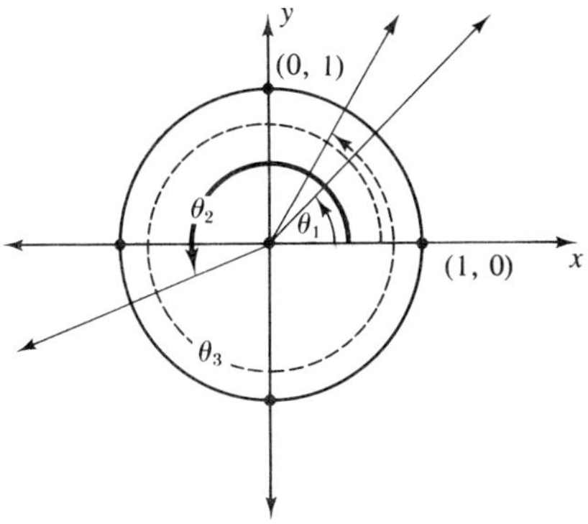

FIGURE 7-10 $\theta_1 = \pi/4$, $\theta_2 = 7\pi/6$, and $\theta_3 = 7\pi/3$

Degree

> **Definition:** One degree, written 1°, is equal to 1/360 of the circumference of a circle.

Minute

> **Definition:** One minute, written 1′, is equal to 1/60 of a degree.

Second

> **Definition:** One second, written 1″, is equal to 1/60 of a minute.

Recall that the number line divides the circumference of a circle into 2π radians. Therefore, 2π radians = 360° or π radians = 180°.

> If r and d are the measures of θ in radians and degrees, respectively, then
>
> $$\frac{r}{\pi} = \frac{d}{180}$$

EXAMPLE 7–6 Convert

(a) $2\pi/3$ (b) $-11\pi/2$ (c) 4.35 radians

to degrees.

Solutions:

(a) If $r = 2\pi/3$, then $\dfrac{2\pi/3}{\pi} = \dfrac{d}{180}$, and $d = \dfrac{2}{3}(180) = 120°$.

(b) If $r = -11\pi/2$, then $\dfrac{-11\pi/2}{\pi} = \dfrac{d}{180}$, and $d = \dfrac{-11}{2}(180) = -990°$.

(c) If $r = 4.35$ and we approximate π with 3.14, then $\dfrac{4.35}{3.14} = \dfrac{d}{180}$, and $d = 249.36°$ to the nearest hundredth of a degree.

EXAMPLE 7–7 Convert

(a) 225° (b) −630° (c) 140.6°

to radians.

Solutions:

(a) If $d = 225°$, then $\frac{r}{\pi} = \frac{225}{180}$, and $r = \left(\frac{225}{180}\right)\pi = \frac{5\pi}{4}$.

(b) If $d = -630°$, then $\frac{r}{\pi} = \frac{-630}{180}$, and $r = \left(\frac{-630}{180}\right)\pi = \frac{-7\pi}{2}$.

(c) $\frac{r}{\pi} = \frac{140.6}{180}$. Thus, $r = 0.78\pi$ or approximately 2.45 radians.

On many scientific hand-held calculators there are keys that convert radian measure to degrees, minutes, and seconds, and vice versa. The variations in the operations of such calculators prevent us from doing examples by this procedure. However, every calculator has a book of instructions that explains the details for a given machine, and conversion can be accomplished with minimal practice.

Cosine, sine, tangent, cotangent, secant, and cosecant functions

Definition: If $P(x, y)$ is the point of intersection of the terminal side of angle θ in standard position and the unit circle, then

(1) cosine $= \{(\theta, \cos\theta) \mid \cos\theta = x\}$

(2) sine $= \{(\theta, \sin\theta) \mid \sin\theta = y\}$

(3) tangent $= \left\{(\theta, \tan\theta) \mid \tan\theta = \frac{y}{x}, x \neq 0\right\}$

(4) cotangent $= \left\{(\theta, \cot\theta) \mid \cot\theta = \frac{x}{y}, y \neq 0\right\}$

(5) secant $= \left\{(\theta, \sec\theta) \mid \sec\theta = \frac{1}{x}, x \neq 0\right\}$

(6) cosecant $= \left\{(\theta, \csc\theta) \mid \csc\theta = \frac{1}{y}, y \neq 0\right\}$

EXAMPLE 7–8 If $P(-7/25, -24/25)$ is the point of intersection of the standard position angle θ and the unit circle, find the values of the six trigonometric functions.

Solutions: Using the definition above, we get

(a) $\cos\theta = -7/25$

(b) $\sin\theta = -24/25$

(c) $\tan\theta = \frac{-24/25}{-7/25} = \frac{24}{7}$

(d) $\cot\theta = \frac{-7/25}{-24/25} = \frac{7}{24}$

(e) $\sec\theta = \frac{1}{-7/25} = \frac{-25}{7}$

(f) $\csc\theta = \frac{1}{-24/25} = \frac{-25}{24}$

As we have seen, the values of x and y above determine the trigonometric function values of any angle θ, where $P(x, y)$ is the point of intersection of the terminal side of θ and the unit circle. The signs of x and y are a function of the quadrant in which $P(x, y)$ lies. For example if θ is a Quadrant III angle, then $x < 0$ and $y < 0$. Therefore, $\cos\theta < 0$, $\sin\theta < 0$, $\tan\theta > 0$, $\cot\theta > 0$, sec $\theta < 0$, and csc $\theta < 0$. In Figure 7-11 we identify the functions that are positive in each of the four quadrants.

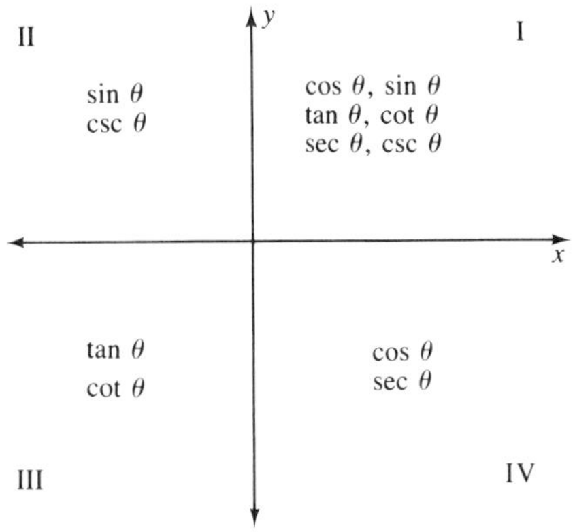

FIGURE 7-11 The functions that are positive in each of the four quadrants

An angle whose terminal side is one of the coordinate axes is called a **quadrantal angle.** The function values of a quadrantal angle are based on ordered pairs of (x, y) that are $(1, 0)$, $(0, 1)$, $(-1, 0)$, or $(0, -1)$. In the exercises for this section we complete a table of values for the quadrantal angles.

But what if the terminal side of θ lies on one of the coordinate axes?

EXAMPLE 7–9 If $\tan \theta = -8/15$ and θ is a Quadrant IV angle, find the other five function values of θ.

Solutions: In Figure 7-12, let $P(x, y)$ be the point of intersection of the terminal side of θ and the unit circle. Since $\tan \theta = y/x = -8/15$, then $x = 15/k$ and $y = -8/k$ such that

$$\frac{-8/k}{15/k} = \left(\frac{-8}{k}\right)\left(\frac{k}{15}\right) = \frac{-8}{15}$$

Since $x^2 + y^2 = 1$ on the unit circle, then

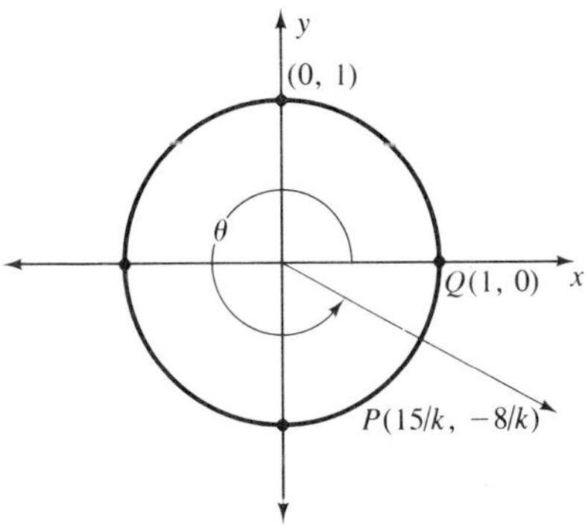

FIGURE 7-12 $\tan \theta = -8/15$

$$\left(\frac{15}{k}\right)^2 + \left(\frac{-8}{k}\right)^2 = 1$$

$$225 + 64 = k^2$$

$$k = 17$$

Therefore, $x = 15/17$, $y = -8/15$, and

(a) $\cos\theta = 15/17$
(b) $\sin\theta = -8/15$
(c) $\cot\theta = -15/8$
(d) $\sec\theta = 17/15$
(e) $\csc\theta = -15/8$

Exercise Set 7–2

Determine what fractional part of a complete rotation the terminal side of an angle is with the given measure.

1. $\pi/2$	2. $\pi/3$	3. $3\pi/4$	4. π
5. $5\pi/3$	6. $11\pi/6$	7. 30°	8. 120°
9. 150°	10. 225°	11. 270°	12. 300°

Find the measure of the angle θ in degrees and radians if the terminal side of θ is rotated the indicated number of rotations.

13. 1/4	14. 1/6	15. 1/3	16. 1/2
17. 2/9	18. 5/6	19. 5/12	20. 5/18
21. 7/10	22. 11/9	23. 13/3	24. 17/6

Express the following in degrees. If necessary, approximate answers to the nearest hundredth of a degree. Use $\pi = 3.14$.

25. $5\pi/8$	26. $19\pi/3$	27. $-2\pi/5$	28. $-16\pi/9$
29. 2.00	30. 2.75	31. 1.00	32. −0.25

Express the following in radians. Use $\pi = 3.14$ to approximate answers to the nearest hundredth of a radian.

33. 65°	34. 132°	35. 192°	36. 302°
37. 524°	38. 1320°	39. −298°	40. −390°

41. Complete the following table.

θ in degrees	θ in radians
0°	
	$\pi/6$
	$\pi/4$
60°	
90°	
120°	
	$3\pi/4$

θ in degrees	θ in radians
150°	
	π
	$7\pi/6$
	$5\pi/4$
240°	
270°	
	$5\pi/3$
	$7\pi/4$
330°	
	2π
390°	

Find the trigonometric function values of θ if the following points on the unit circle contain the terminal side of θ.

42. $(-3/5,\ 4/5)$ 43. $(-4/5,\ -3/5)$ 44. $(-8/17,\ 15/17)$

45. $(7/25,\ 24/25)$ 46. $(\sqrt{2}/2,\ -\sqrt{2}/2)$ 47. $(1/2,\ \sqrt{3}/2)$

48. $(-\sqrt{3}/2,\ -1/2)$ 49. $(-5/13,\ 12/13)$ 50. $(12/13,\ 5/13)$

Find the other five trigonometric function values of θ, given the quadrant that contains the terminal side of θ.

51. $\sin\theta = 6/10$; I 52. $\cos\theta = \sqrt{3}/2$; I

53. $\tan\theta = 1$; III 54. $\sin\theta = 12/13$; II

55. $\cos\theta = -15/17$; III 56. $\tan\theta = -1/\sqrt{3}$; IV

Find the measures of the two angles coterminal with the given angle whose terminal sides are (a) one positive rotation and (b) one negative rotation from the terminal side of the given angle.

57. 73° 58. 258° 59. −125° 60. −620°

61. $3\pi/5$ 62. $14\pi/3$ 63. $-6\pi/7$ 64. $-19\pi/8$

65. 3.77 66. −5.02

67. Complete the table of function values for the quadrantal angles 0°, 90°, 180°, 270°, and 360°. If a given function value is undefined, write "und."

θ	$\cos\theta$	$\sin\theta$	$\tan\theta$	$\cot\theta$	$\sec\theta$	$\csc\theta$
0°						
90°						
180°						
270°						
360°						

Suppose a circle is divided into 100 equal divisions called *centicircles*. How many centicircles are there in:

68. π radians 69. $\pi/4$ radians 70. 300° 71. 70°

7-3 TRIGONOMETRIC TABLES AND REFERENCE ANGLES

In Sections 7-1 and 7-2 we defined the six trigonometric functions of a real number k and an angle θ in terms of the coordinates of a point on the unit circle. We will not repeat the details of the definitions at this time. In this section we will find the function values of angles with measures $0° \leq \theta \leq 90°$ or $0 \leq \theta \leq \pi/2$ by using Table 4 in the Appendix or a hand-held calculator. The use of **reference angles** extends the values of Table 4 to any angle θ.

Trigonometric Tables

We accept the following theorem without proof, since it is easily verified with trigonometric identities developed later. The degree measure used in the theorem can be replaced with an equivalent radian measure.

Theorem: If $0° \leq \theta \leq 90°$, then

(1) $\sin \theta = \cos(90° - \theta)$ or $\cos \theta = \sin(90° - \theta)$

(2) $\tan \theta = \cot(90° - \theta)$ or $\cot \theta = \tan(90° - \theta)$

(3) $\sec \theta = \csc(90° - \theta)$ or $\csc \theta = \sec(90° - \theta)$

By definition, $90° - \theta$ is the complement of an acute angle θ. In Figure 7-13, θ and its complement are shown. Furthermore, we call sine and *co*-sine, tangent and *co*-tangent, and secant and *co*-secant **cofunctions.** The theorem states that for any acute angle θ, *the trigonometric function values of θ are equal to the cofunction values of its complement.*

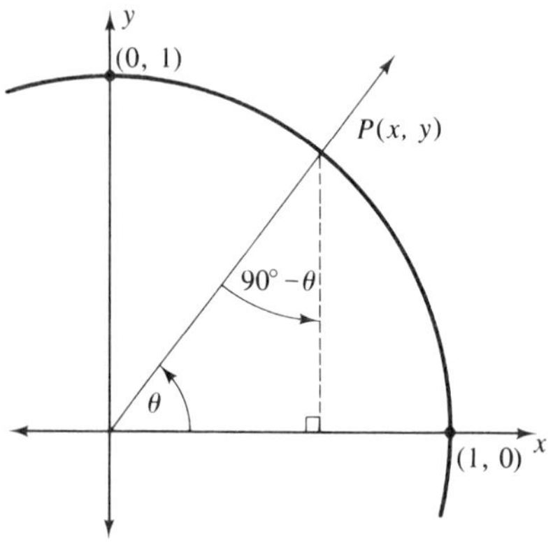

FIGURE 7-13 θ and $90° - \theta$, or $\frac{\pi}{2} - \theta$

EXAMPLE 7–10 Find
(a) $\cos 30°$, if $\sin 60° = 0.8660$ (b) $\tan 40°$, if $\cot 50° = 0.8391$

Solutions:
(a) Since $30° = 90° - 60°$, $30°$ and $60°$ angles are complementary. Therefore, $\cos 30° = \sin 60° = 0.8660$.

(b) Since $40° = 90° - 50°$, $40°$ and $50°$ angles are complementary. Thus, $\tan 40° = \cot 50° = 0.8391$.

The immediate consequence of the above theorem is to reduce the table of trigonometric function values of θ by 50 percent. Since $\cos \theta = \sin(90° - \theta)$, then for each value of $\cos \theta$ and $0° \leq \theta \leq 45°$, we automatically have the function values of $\sin(90° - \theta)$ and $45° \leq 90° - \theta \leq 90°$. Table 4 contains the function values of an acute angle θ. A portion of the table is given below.

Degrees	Radians	sin θ	cos θ	tan θ	cot θ	sec θ	csc θ		
12°00′	.2094	.2079	.9781	.2126	4.705	1.022	4.810	1.3614	78°00′
10′	123	108	775	156	638	023	745	584	50′
20′	153	136	769	186	574	024	682	555	40′
30′	.2182	.2164	.9763	.2217	4.511	1.024	4.620	1.3526	30′
40′	211	193	757	247	449	025	560	497	20′
50′	240	221	750	278	390	026	502	468	10′
13°00′	.2269	.2250	.9744	.2309	4.331	1.026	4.445	1.3439	77°00′
		cos θ	sin θ	cot θ	tan θ	csc θ	sec θ	Radians	Degrees

If $12°0' \leq \theta \leq 13°0'$, the entry in the table is the value of the function at the top of the column. For example,

$$\cos 12°20' = 0.9769$$
$$\tan 12°40' = 0.2247$$
$$\csc 12°10' = 4.745$$

If $77°0' \leq \theta \leq 78'0'$, then the entry in the table is the value of the function at the bottom of the column. For example,

$$\sin 77°20' = 0.9757$$
$$\cot 77°50' = 0.2156$$
$$\sec 77°10' = 4.502$$

Therefore, *any entry in the table is the value of two trigonometric functions, a function of* θ*, and the cofunction of the complement of* θ. We should further note that the values in Table 4 are *approximations* of the function values of θ correct to three or four decimal places. There are some angles whose function values can be expressed as exact, and we will study these angles in a later section.

EXAMPLE 7–11 Use a scientific hand calculator to obtain the following function values.

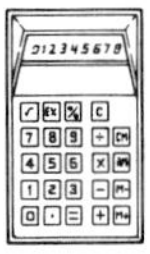

(a) $\sin 25°20'$ (b) $\tan 83°30'$

(c) $\cos \frac{\pi}{5}$ (d) $\sin 0.72$

Solutions:

(a) With the calculator set in the Degree register, enter 25.3333 and depress SIN key. Display reads 0.4279 to four decimal places. Note that $20' = 20/60 = 0.3333$ degrees to four decimal places.

(b) With the calculator set in the Degree register, enter 83.5 and depress TAN key. Display reads 8.7769 to four decimal places. Note that 30′ = 0.5 degrees.
(c) With the calculator set in the Radian register, enter π and divide by 5. Display reads 0.6283 to four decimal places. Depress COS key and display reads 0.8090.
(d) With the calculator set in the Radian register, enter 0.72 and depress SIN key. Display reads 0.6594.

How do you get values for sec θ, csc θ, and cot θ with a calculator? There are no keys for them!

Other relationships that exist between the trigonometric functions are the following:

(1) $\sec\theta = \dfrac{1}{\cos\theta}$, $\cos\theta \neq 0$

(2) $\csc\theta = \dfrac{1}{\sin\theta}$, $\sin\theta \neq 0$

(3) $\cot\theta = \dfrac{1}{\tan\theta}$, $\tan\theta \neq 0$

We can verify these **reciprocal relationships** with the definitions for the trigonometric functions. For example, $\cos\theta = x$, and for $x \neq 0$, $\dfrac{1}{\cos\theta} = \dfrac{1}{x}$. But $\dfrac{1}{x} = \sec\theta$; therefore, $\sec\theta = \dfrac{1}{\cos\theta}$, so long as $\cos\theta \neq 0$.

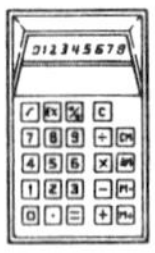

EXAMPLE 7–12 With a scientific hand calculator find
(a) csc 42°15′ (b) cot 1.38

Solutions:
(a) With the calculator set in the Degree register, enter 42.25 (15′ = 0.25°). Depress SIN key and the display reads 0.6724. Now depress $1/x$ key and the display reads 1.4873. Therefore, csc 42°15′ = 1.4873 to four decimal places.
(b) With the calculator set in the Radian register, enter 1.38. Depress TAN key and the display reads 5.1774. Now depress $1/x$ key and the display reads 0.1931. Thus, cot 1.38 = 0.1931 to four places.

Reference Angles

Recall that an angle in standard position can be identified by the quadrant that contains the terminal side of the angle. Table 4 provides the function values for Quadrant I angles. With a reference angle we can extend the values in Table 4 to angles in Quadrants II, III, and IV.

Reference angle

> **Definition:** Angle α is the reference angle of θ if $0° < \alpha < 90°$ and the trigonometric function values of α are the same as the trigonometric function values of θ in absolute value.

Therefore, if α is the reference angle of θ, then $\cos\alpha = |\cos\theta|$, $\sin\alpha = |\sin\theta|$, and so on. That is, *the same function value of any angle and its reference angle can differ by at most a sign.*

As a special set of angles, quadrantal angles have no reference angles. That is why we were able to compute the set of function values for the quadrantal angles in Exercise Set 7-2.

In Figure 7-14 a Quadrant II angle θ is drawn with reference angle α, and $\alpha = 180° - \theta$, or $\pi - \theta$. To verify that α is the reference angle of θ, angle α is also drawn in Quadrant I. Triangles OP_1A_1 and OP_2A_2 are congruent right triangles; therefore, the corresponding sides are equal in length, and

$$|x_1| = x_2 \quad \text{and} \quad y_1 = y_2$$

The x-coordinates of P_1 and P_2 are opposite in sign; thus, the absolute value symbols on x_1 are needed. However, the y-coordinates of P_1 and P_2 have the same sign, and $y_1 = y_2$. We apply the definitions of the trigonometric functions to θ and α to obtain the following equations:

(1) $\cos \theta = -\cos \alpha$ (2) $\sin \theta = \sin \alpha$

(3) $\tan \theta = -\tan \alpha$ (4) $\cot \theta = -\cot \alpha$

(5) $\sec \theta = -\sec \alpha$ (6) $\csc \theta = \csc \alpha$

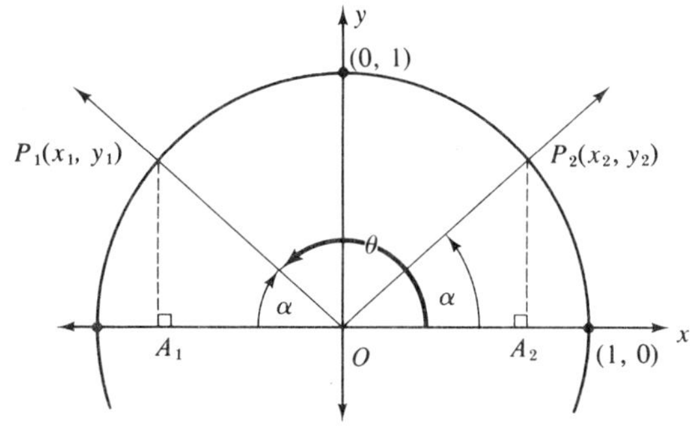

FIGURE 7-14 Quadrant II angle θ with reference angle $\alpha = 180° - \theta$, or $\alpha = \pi - \theta$

Notice that the signs of function values of α are always positive, since $0° < \alpha < 90°$. Thus, the functions in Quadrant II that have negative values have the opposite sign of the function values of α. *In Quadrant II, only the sine and cosecant function values are positive.*

EXAMPLE 7–13 Find the following with Table 4. Use $\pi = 3.1416$.
(a) $\tan 130°$ (b) $\sin 2.7285$

Solutions:
(a) With $\theta = 130°$, $\alpha = 180° - 130° = 50°$. Since $\tan 50° = 1.192$, then $\tan 130° = -\tan 50° = -1.192$.
(b) With $\theta = 2.7285$, $\alpha = 3.1416 - 2.7285 = 0.4131$. Since $\sin 0.4131 = 0.4014$, then $\sin 2.7285 = 0.4014$.

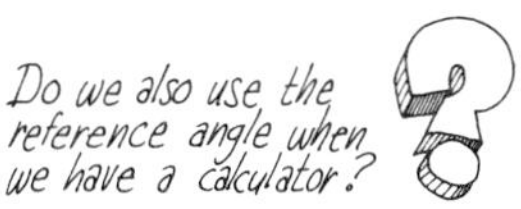

The scientific calculator gives the function value with the appropriate sign for Quadrant II angles without the use of a reference angle.

EXAMPLE 7–14 Find the following with a hand calculator.
(a) cos 128° (b) cot 2.75

Solutions:
(a) With the calculator set in the Degree register, enter 128. Depress COS key, and the display reads −0.6157 to four decimal places.
(b) With the calculator set in the Radian register, enter 2.75. Depress the TAN key, then the $1/x$ key, and the display reads −2.4218 to four decimal places.

In Figure 7-15 a Quadrant III angle θ is drawn with reference angle α, and $\alpha = \theta - 180°$, or $\theta - \pi$. Angle α is also shown in Quadrant I. Since $x_1 = -x_2$ and $y_1 = -y_2$, then for θ in Quadrant III,

(1) $\cos\theta = -\cos\alpha$ (2) $\sin\theta = -\sin\alpha$
(3) $\tan\theta = \tan\alpha$ (4) $\cot\theta = \cot\alpha$
(5) $\sec\theta = -\sec\alpha$ (6) $\csc\theta = -\csc\alpha$

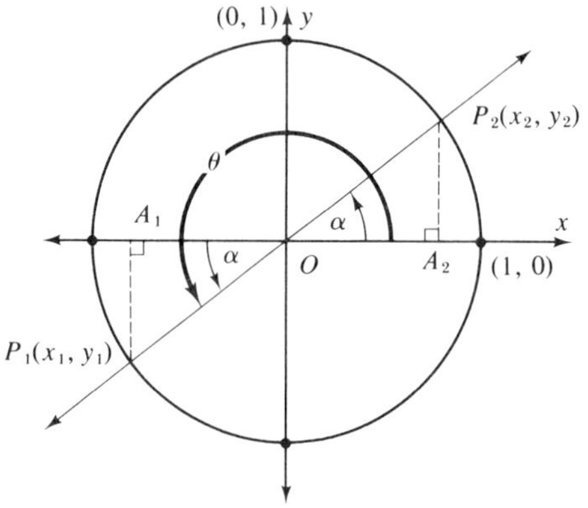

FIGURE 7-15 Quadrant III angle θ with reference angle $\alpha = \theta - 180°$, or $\alpha = \theta - \pi$

EXAMPLE 7–15 Find the following with Table 4.
(a) cot 253° (b) cos 3.8426

Solutions:
(a) If $\theta = 253°$, then $\alpha = 253° - 180° = 73°$. Since $\cot 73° = 0.3057$, then $\cot 253° = 0.3057$.
(b) If $\theta = 3.8426$, then $\alpha = 3.8426 - 3.1416 = 0.7010$. Since $\cos 0.7010 = 0.7642$, then $\cos 3.8426 = -0.7642$.

In Figure 7-16 a Quadrant IV angle θ is drawn with reference angle α, and $\alpha = 360° - \theta$, or $2\pi - \theta$. Angle α is also shown in Quadrant I. Since $x_1 = x_2$ and $y_1 = -y_2$, then for θ in Quadrant IV,

(1) $\cos\theta = \cos\alpha$ (2) $\sin\theta = -\sin\alpha$
(3) $\tan\theta = -\tan\alpha$ (4) $\cot\theta = -\cot\alpha$
(5) $\sec\theta = \sec\alpha$ (6) $\csc\theta = -\csc\alpha$

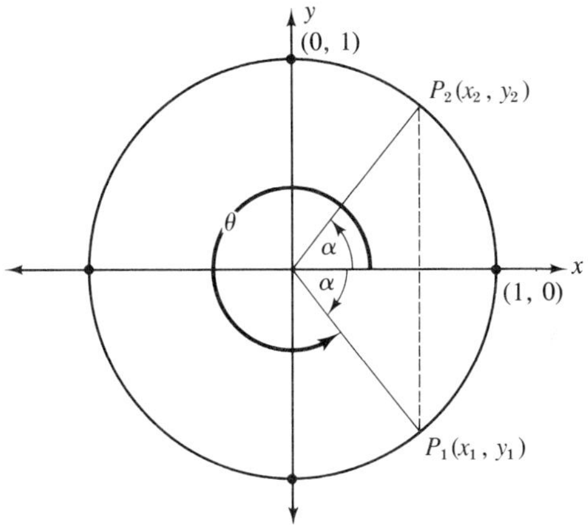

FIGURE 7-16 Quadrant IV angle θ with reference angle $\alpha = 360° - \theta$ or $\alpha = 2\pi - \theta$

EXAMPLE 7–16 Find the following with Table 4.
(a) tan 294° (b) sin 4.8608

Solutions:
(a) If $\theta = 294°$, then $\alpha = 360° - 294° = 66°$. Since tan 66° = 2.246, then tan 294° = −2.246.
(b) If $\theta = 4.8608$, *then* $\alpha = 6.2832 - 4.8608 = 1.4224$. Since sin 1.4224 = 0.9890, then sin 4.8608 = −0.9890.

To summarize, if θ is an angle in Quadrant II, III, or IV, we must first find the reference angle α. With Table 4 we find the function value of α. Depending on the quadrant and the function, either the function value of θ = the function value of α, or the function value of θ = the negative of the function value of α.

Exercise Set 7–3

Find the following function values, using Table 4 or a hand calculator.

1. sin 14°
2. cos 73°
3. tan 49°10′
4. cot 17°40′
5. cos 1.1432
6. cot 0.3869
7. sin 0.5701
8. sec 0.7912

Find angle θ in degrees, using Table 4 or a hand calculator, where $0° < \theta < 90°$.

9. $\sin \theta = 0.3145$
10. $\cos \theta = 0.6157$
11. $\tan \theta = 0.1944$
12. $\cot \theta = 0.6619$

Find angle θ in radians, using Table 4 or a hand calculator, where $0 < \theta < 1.5708$.

13. $\sec \theta = 1.504$
14. $\csc \theta = 1.052$
15. $\sin \theta = 0.3118$
16. $\tan \theta = 1.709$

Write each of the following function values as cofunctions of their complements.

17. $\sin 20^\circ$
18. $\tan 72^\circ$
19. $\sec 82^\circ$
20. $\cot 0.4363$
21. $\csc 0.2269$
22. $\tan \frac{5\pi}{12}$
23. $\sec \frac{4\pi}{9}$
24. $\cot \frac{\pi}{3}$

For the following angles in standard position, identify the quadrant in which the terminal side lies and state the reference angle.

25. 280°
26. 182°
27. 160°
28. 300°
29. $5\pi/4$
30. $11\pi/6$
31. $7\pi/12$
32. 4.2031
33. 2.0000

Find the following function values.

34. $\tan \frac{5\pi}{7}$
35. $\csc 117^\circ$
36. $\cot 175^\circ$
37. $\cos \frac{7\pi}{12}$
38. $\sin 2.7373$
39. $\cot 3.7525$
40. $\csc 2.266$
41. $\sin 119^\circ 10'$
42. $\sec \frac{23\pi}{18}$
43. $\sin 1.3497$
44. $\tan \frac{4\pi}{3}$
45. $\cot 120^\circ$

Express the following as functions of positive acute angles.

46. $\sin 315^\circ$
47. $\cos \frac{3\pi}{4}$
48. $\cot \frac{4\pi}{3}$
49. $\csc 100^\circ$
50. $\cos \frac{7\pi}{6}$
51. $\tan 330^\circ$

Identify the quadrant in which the terminal side of standard position angle θ lies if the following are known.

52. $\cos \theta < 0$, $\csc \theta > 0$
53. $\tan \theta < 0$, $\cos \theta > 0$
54. $\sin \theta < 0$, $\cos \theta < 0$
55. $\csc \theta > 0$, $\sec \theta > 0$

7-4 FUNCTION VALUES OF MULTIPLES OF 30° AND 45°

We noted that the values of the trigonometric functions given in Table 4 are approximated to three or four decimal places. However, *there is a set of values for θ whose function values can be stated exactly*. We frequently use these values of θ to develop new concepts in trigonometry, and they are also used extensively in applied mathematics courses. As a result, we will construct a table of values for these special angles in the exercise set for this section.

30° Angles

In Figure 7-17, θ and $-\theta$ are drawn on the unit circle, and each has a measure of 30°. When we draw $\overline{PP^*}$, an equilateral triangle (a triangle with three equal sides) is formed with sides of length one. The x-axis bisects $\overline{PP^*}$, since it bisects angle POP^*. Therefore, $|PA| = 1/2$, and the y-coordinate of point P is 1/2. On the unit circle, $x^2 + y^2 = 1$; therefore,

$$x^2 + \left(\frac{1}{2}\right)^2 = 1$$

$$x^2 = \frac{3}{4}$$

$$x = \frac{\sqrt{3}}{2}$$

since P is in Quadrant I. The coordinates of P are $(\sqrt{3}/2, 1/2)$, and the following function values are obtained:

(1) $\cos 30° = \sqrt{3}/2$ (2) $\sin 30° = 1/2$

(3) $\tan 30° = 1/\sqrt{3}$ or $\sqrt{3}/3$ (4) $\cot 30° = \sqrt{3}$

(5) $\sec 30° = 2/\sqrt{3}$ or $2\sqrt{3}/3$ (6) $\csc 30° = 2$

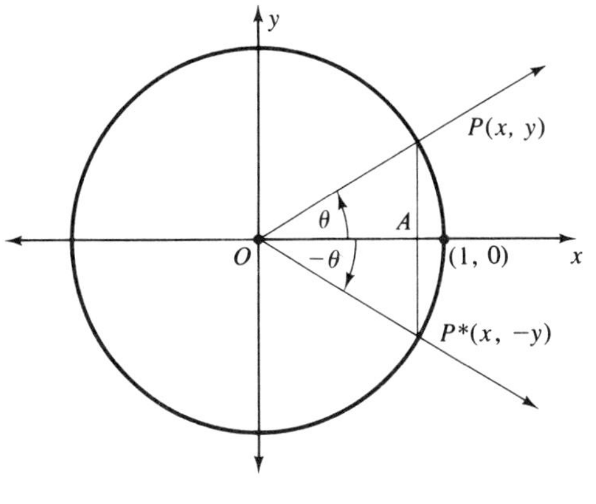

FIGURE 7-17 $\theta = 30°; -\theta = -30°$

EXAMPLE 7–17 Find the function values of $\theta = 7\pi/6$.

Solutions: Since θ is a Quadrant III angle, $\alpha = \frac{7\pi}{6} - \pi = \frac{\pi}{6}$. But $\frac{\pi}{6} = 30°$; therefore, the function values are the same as those obtained above with the appropriate signs for a Quadrant III angle.

(a) $\cos \frac{7\pi}{6} = \frac{-\sqrt{3}}{2}$ (b) $\sin \frac{7\pi}{6} = \frac{-1}{2}$

(c) $\tan \frac{7\pi}{6} = \frac{1}{\sqrt{3}}$ or $\frac{\sqrt{3}}{3}$ (d) $\cot \frac{7\pi}{6} = \sqrt{3}$

(e) $\sec \frac{7\pi}{6} = \frac{-2\sqrt{3}}{3}$ (f) $\csc \frac{7\pi}{6} = -2$

EXAMPLE 7–18 Find the function values of $\theta = 60°$.

Solutions: In Section 7-3 we stated the equations for the cofunctions of complementary angles. Since $60° = 90° - 30°$, then $\cos 60° = \sin 30°$, $\tan 60° = \cot 30°$, $\csc 60° = \sec 30°$, and so on. Therefore,

(a) $\cos 60° = 1/2$ (b) $\sin 60° = \sqrt{3}/2$

(c) $\tan 60° = \sqrt{3}$ (d) $\cot 60° = 1/\sqrt{3}$ or $\sqrt{3}/3$

(e) $\sec 60° = 2$ (f) $\csc 60° = 2/\sqrt{3}$ or $2\sqrt{3}/3$

In any 30°–60°–90° triangle, the ratio of the lengths of the three sides is $1:2:\sqrt{3}$, where 1, 2, and $\sqrt{3}$ are the lengths of the short leg (leg opposite the 30° angle), hypotenuse, and long leg (leg opposite the 60° angle), respectively. Since the actual length of the hypotenuse is the radius of the unit circle, the lengths of the three sides are 1/2, 1, and $\sqrt{3}/2$. Depending on whether the short leg is horizontal or vertical, the cosine or sine function is 1/2 or $-1/2$, and the remaining function values can be determined easily. See Figure 7-18 for two such examples.

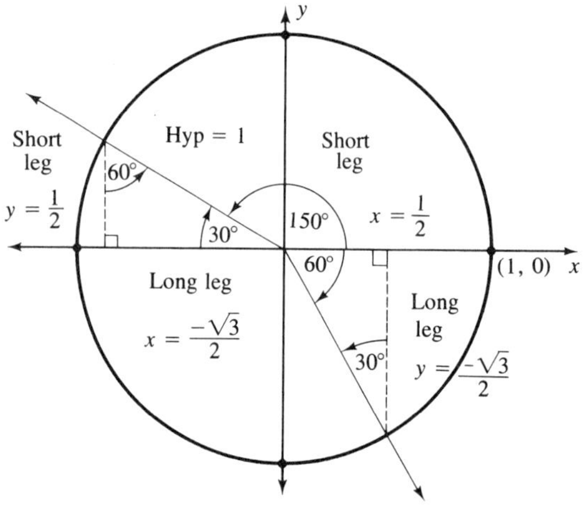

FIGURE 7-18 Two 30°–60°–90° triangles

The following equations state the function values for a 45° angle. The function values are readily obtained by constructing an isosceles right triangle in a unit circle in Quadrant I.

(1) $\cos 45° = 1/\sqrt{2}$ or $\sqrt{2}/2$ (2) $\sin 45° = 1/\sqrt{2}$ or $\sqrt{2}/2$

(3) $\tan 45° = 1$ (4) $\cot 45° = 1$

(5) $\sec 45° = \sqrt{2}$ (6) $\csc 45° = \sqrt{2}$

EXAMPLE 7–19 Find the function values of $\theta = 495°$.

Solutions: In Figure 7-19, θ is drawn as a Quadrant II angle with reference angle $\alpha = 45°$. In Quadrant II, $\sin\theta > 0$, $\csc\theta > 0$, and the other functions are negative. Therefore,

(a) $\cos 495° = -\sqrt{2}/2$ (b) $\sin 495° = \sqrt{2}/2$

(c) $\tan 495° = -1$ (d) $\cot 495° = -1$

(e) $\sec 495° = -\sqrt{2}$ (f) $\csc 495° = \sqrt{2}$

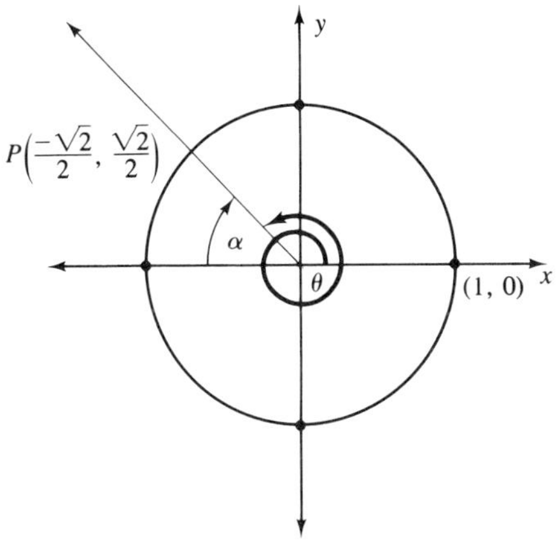

FIGURE 7-19 $\theta = 495°; \alpha = 45°$

If an angle θ is an integral multiple of $\pi/6$ or $\pi/4$ (30° or 45°) and a function value of θ is given, we can readily determine the value of θ. However, we must realize that even though we restrict θ to one positive rotation ($0 < \theta < 2\pi$), there are *two* angles that have the given function value. For example,

$$\cos\frac{\pi}{3} = \cos\frac{5\pi}{3} = \frac{1}{2} \quad \text{or} \quad \cos 60° = \cos 300° = \frac{1}{2}$$

$$\tan\frac{\pi}{4} = \tan\frac{5\pi}{4} = 1 \quad \text{or} \quad \tan 45° = \tan 225° = 1$$

Furthermore, if we consider all angles coterminal with θ and $0 \le \theta < 2\pi$ or $0° \le \theta < 360°$, then the number of angles is *infinite*.

EXAMPLE 7–20 Find all values of θ.

(a) $0° \le \theta < 360°$ if $\tan\theta = \sqrt{3}$ (b) θ is any angle

Solutions:
(a) If $\tan\theta = \sqrt{3}$ and $0° \le \theta < 360°$, then $\theta = 60°$ or $240°$. Notice that $\tan\theta > 0$ in Quadrants I and III.
(b) If $\tan\theta = \sqrt{3}$ and θ is any angle, then $\theta = 60° + 360°n$ or $\theta = 240° + 360°n$, and $n \in \{\text{integers}\}$. Notice, that since $240° = 60° + 180°$, the answer can also be written as $\theta = 60° + 180°n$, and $n \in \{\text{integers}\}$.

Exercise Set 7–4

1. Complete the following table of function values for θ, using only *exact* values. If a function value does not exist for some value of θ, write "und" for undefined.

θ in Degrees	θ in Radians	$\sin\theta$	$\cos\theta$	$\tan\theta$	$\cot\theta$	$\sec\theta$	$\csc\theta$
0	0						
30	$\pi/6$	1/2	$\sqrt{3}/2$	$1/\sqrt{3}$	$\sqrt{3}$	$2/\sqrt{3}$	2
45							
60							
90							
120							
135							
150							
180							
210							
225							
240							
270							
300							
315							
330							
360							

Find the value of the following expressions. Leave answers in simplified radical form.

2. $\dfrac{\tan 45^\circ + \tan 30^\circ}{1 - \tan 45^\circ \tan 30^\circ}$

3. $\cos 120^\circ \cos 45^\circ - \sin 120^\circ \sin 45^\circ$

4. $\dfrac{\cos 60^\circ + 1}{2}$

5. $\cos^2 225^\circ + \sin^2 225^\circ$

6. $\dfrac{\sin 4\pi/3}{1 + \cos 4\pi/3}$

7. $2\cos^2 \dfrac{5\pi}{6} - 1$

8. $\dfrac{1 - \cos 3\pi/4}{2}$

9. $\sin \dfrac{\pi}{6} \cos \dfrac{7\pi}{6} + \cos \dfrac{\pi}{6} \sin \dfrac{7\pi}{6}$

10. $\dfrac{\cot 5\pi/3 - \cot 3\pi/4}{1 + \cot 5\pi/3 \cot 3\pi/4}$

11. $2 \sin 90^\circ \cos 90^\circ$

Using *exact values,* evaluate the following.

12. $\sin 750^\circ$

13. $\tan \dfrac{13\pi}{4}$

14. $\sec \dfrac{-32\pi}{3}$

15. $\csc(-1200^\circ)$

16. $\sin \frac{17\pi}{3}$

17. $\cos 1125°$

18. $\tan(-540°)$

19. $\sec \frac{35\pi}{6}$

20. $\cos 480°$

21. $\cot \frac{-7\pi}{2}$

Find all values of θ, $0° \leq \theta \leq 90°$, such that:

22. $\sin \theta = 1/2$

23. $\cos \theta = 1$

24. $\tan \theta = \sqrt{3}$

25. $\sec \theta = 2\sqrt{3}/3$

26. $\csc \theta = 2$

27. $\sin \theta = \sqrt{2}/2$

Find all values of θ, (a) $0° \leq \theta < 360°$ and (b) θ is any angle, such that:

28. $\cos \theta = -\sqrt{3}/2$

29. $\sin \theta = -1$

30. $\tan \theta = -1$

31. $\cos \theta = -\sqrt{2}/2$

32. $\sec \theta = -2$

33. $\csc \theta = -\sqrt{2}$

34. $\cos \theta = \pm 1/2$

35. $\sin \theta = \pm\sqrt{2}/2$

GRAPHS OF $y = a \sin bx$ AND $y = a \cos bx$ 7–5

In this section we graph the equations $y = a \sin bx$ and $y = a \cos bx$. We first graph the equations for $a = b = 1$, that is, $y = \sin x$ and $y = \cos x$. We then consider the changes in these graphs for values of a and b other than one.

Notice that we used θ as the independent variable for the trigonometric functions and that x and y served as the dependent variables. For convenience we now use x as the independent variable and y as the only dependent variable in order to graph the trigonometric functions on the familiar x and y axes. Therefore, although by definition

$$x = \cos \theta \quad \text{and} \quad y = \sin \theta$$

we now write

$$y = \cos x \quad \text{and} \quad y = \sin x$$

and graph the functions on the (x, y) coordinate system.

$y = a \sin bx$

In Figure 7-20 the graph of $y = \sin x$ is shown for $0 \leq x \leq 2\pi$. The table of values used to draw the graph is given below.

x	0	$\pi/4$	$\pi/2$	$3\pi/4$	π	$5\pi/4$	$3\pi/2$	$7\pi/4$	2π
$\sin x$	0	0.7	1	0.7	0	−0.7	−1	−0.7	0

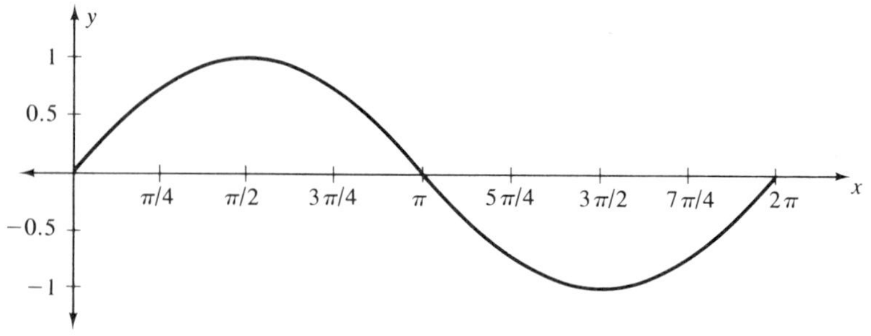

FIGURE 7-20 $y = \sin x,\ 0 \le x \le 2\pi$

The following observations are made from the graph:

1. For $0 < x < \pi/2$, the function *increases* from 0 to 1.
2. For $\pi/2 < x < 3\pi/2$, the function *decreases* from 1 to -1.
3. For $3\pi/2 < x < 2\pi$, the function *increases* from -1 to 0.
4. If the graph is extended for values of $x < 0$ or $x > 2\pi$, the given curve is repeated in intervals of 2π. The *periodic* nature of the sine function is the basis for this property of the graph. For $y = \sin x$, the period is 2π.
5. The *amplitude* of the function is the maximum distance above the x-axis. For $y = \sin x$, the amplitude $= 1$.
6. For $0 \le x \le 2\pi$, the graph of the function has
   (a) Three zero points, at $x = 0,\ \pi$, and 2π
   (b) One maximum point, at $x = \pi/2$
   (c) One minimum point, at $x = 3\pi/2$

   and these five key points are used to graph the sine function.

Why do you use radian measures for x and not degree measures?

Recall that the basis for the radian measure of an angle requires that the number line be wrapped around a unit circle. The correspondence between the points on the number line and the points of the unit circle enabled us to measure angles in terms of real numbers. In graphing the functions of θ, we utilize this correspondence to plot the real numbers that are the *function values* of θ with the real numbers that are the *measures* of θ.

EXAMPLE 7–21 Sketch a graph of $y = 3 \sin x,\ 0 \le x \le 2\pi$.

Solution: For the given equation, $a = 3$ and $b = 1$. The following table of values is used to sketch the graph in Figure 7-21.

x	0	$\pi/4$	$\pi/2$	$3\pi/4$	π	$5\pi/4$	$3\pi/2$	$7\pi/4$	2π
$3 \sin x$	0	2.1	3	2.1	0	−2.1	−3	−2.1	0

When we compare the graphs of $y = \sin x$ and $y = 3 \sin x$, we note the following:

1. The periods of both functions are 2π.

2. The amplitude of $y = 3 \sin x$ is three times the amplitude of $y = \sin x$.
3. The five key points used to graph both functions are obtained with the same values of x. That is,
   (a) The zero points at $x = 0, \pi$, and 2π
   (b) The maximum values at $x = \pi/2$
   (c) The minimum values at $x = 3\pi/2$

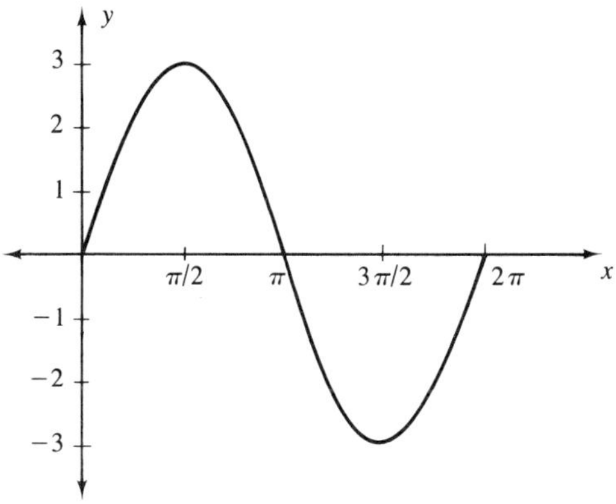

FIGURE 7-21 $y = 3 \sin x,\ 0 \le x \le 2\pi$

> **Theorem:** If $y = a \sin bx$, then the **amplitude** of the function is $|a|$.

EXAMPLE 7–22 Sketch a graph of $y = \sin 3x$ for $0 \le x \le 2\pi$.

Solution: For the given equation, $a = 1$ and $b = 3$. The following table of values is used to sketch the graph in Figure 7-22.

x	0	$\pi/12$	$\pi/6$	$\pi/4$	$\pi/3$	$5\pi/12$	$\pi/2$	$7\pi/12$	$2\pi/3$
$3x$	0	$\pi/4$	$\pi/2$	$3\pi/4$	π	$5\pi/4$	$3\pi/2$	$7\pi/4$	2π
$\sin 3x$	0	0.7	1	0.7	0	−0.7	−1	−0.7	0

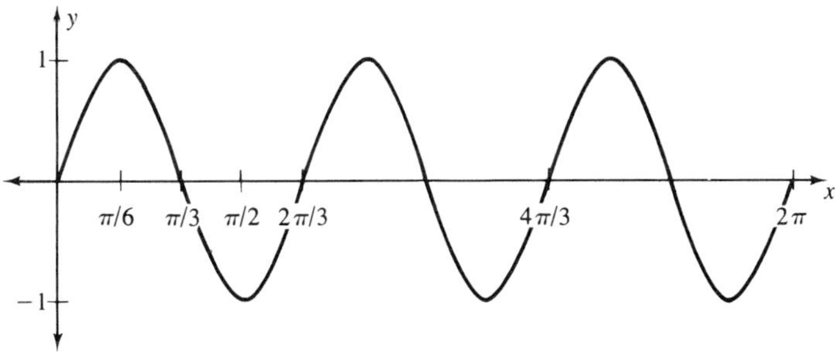

FIGURE 7-22 Graph of $y = \sin 3x,\ 0 \le x \le 2\pi$

When we compare the graphs of $y = \sin x$ and $y = \sin 3x$, we note the following:

1. The period of $y = \sin 3x$ is one-third the period of $y = \sin x$. Notice that in $y = \sin 3x$, $b = 3$ and $1/3$ is the reciprocal of 3.
2. The amplitude of both functions is one.
3. The five key points used to graph the function $y = \sin 3x$ are obtained with values of x that are each one-third the value of x used to obtain a similar point on the graph of $y = \sin x$. That is,
   (a) The zero points on $y = \sin x$ occur at $x = 0, \pi$, and 2π; and the zero points on $y = \sin 3x$ occur at $x = 0, \pi/3$, and $2\pi/3$.
   (b) A maximum value on $y = \sin x$ occurs at $x = \pi/2$, and a maximum value on $y = \sin 3x$ occurs at $x = \pi/6$.
   (c) A minimum value on $y = \sin x$ occurs at $x = 3\pi/2$, and a minimum value on $y = \sin 3x$ occurs at $x = \pi/2$.

Theorem: If $y = a \sin bx$, then the **period** of the function is

$$\frac{2\pi}{|b|}$$

In both theorems you put a and b in absolute value bars. Why?

In the two preceding theorems, a and b are expressed as absolute values because amplitude and period are defined in terms of positive values for a and b. However, a and b can be negative numbers. To analyze the behavior of the graphs of $y = a \sin bx$ for positive and negative values of a and b, we restrict ourselves to the interval $0 \le x \le \dfrac{2\pi}{|b|}$, the **first period** of the graph to the right of the y-axis.

If $a > 0$, the graph increases for

$$0 < x < \frac{1}{4}\left(\frac{2\pi}{|b|}\right)$$

It decreases for

$$\frac{1}{4}\left(\frac{2\pi}{|b|}\right) < x < \frac{3}{4}\left(\frac{2\pi}{|b|}\right)$$

and then increases again for

$$\frac{3}{4}\left(\frac{2\pi}{|b|}\right) < x < \frac{2\pi}{|b|}$$

If $a < 0$, then the graph obtained for $a > 0$ is **reflected** about the x-axis; and the graph first decreases, then increases, and then decreases again over the intervals obtained above.

As for $b < 0$, we shall verify that

$$\sin(-x) = -\sin x$$

Therefore, $a \sin bx = -a \sin(-bx)$. Thus, to graph $y = 2 \sin(-3x)$, we would graph $y = -2 \sin 3x$ instead. The following analysis of the equation $y = a \sin bx$ is an aid in graphing the function:

1. The amplitude of the function is $|a|$. Therefore, the maximum and minimum vertical displacements for the graph are $|a|$ units above and below the x-axis, respectively.
2. The period of the function is $\dfrac{2\pi}{|b|}$. Thus, one full period of the graph occurs on the interval $\left[0, \dfrac{2\pi}{|b|}\right]$.
3. Locate the five key points that determine the basic shape of the wave:
   (a) Zero points at $x = 0$, $x = \dfrac{\pi}{|b|}$, and $x = \dfrac{2\pi}{|b|}$. Notice that these points occur at the endpoints and the midpoint of the interval $\left[0, \dfrac{2\pi}{|b|}\right]$.
   (b) Maximum point at $x = \dfrac{1}{4}\left(\dfrac{2\pi}{|b|}\right)$ if $a > 0$, or minimum point at $x = \dfrac{1}{4}\left(\dfrac{2\pi}{|b|}\right)$ if $a < 0$.
   (c) Minimum point at $x = \dfrac{3}{4}\left(\dfrac{2\pi}{|b|}\right)$ if $a > 0$, or maximum point at $x = \dfrac{3}{4}\left(\dfrac{2\pi}{|b|}\right)$ if $a < 0$.

Notice that a maximum or minimum point occurs at a value of x that is 1/4 and 3/4 the length of the interval from $x = 0$ to $x = \dfrac{2\pi}{|b|}$.

EXAMPLE 7–23 Sketch the graph of $y = 2 \sin \dfrac{x}{2}$, $x \in [0, 4\pi]$.

Solution: For $y = 2 \sin \dfrac{x}{2}$, $a = 2$ and $b = 1/2$. Therefore, the amplitude is 2 and the period is $\dfrac{2\pi}{1/2} = 4\pi$.

(a) There are zero points at $x = 0$, $x = 2\pi$, and $x = 4\pi$.
(b) Since $a > 0$, there is a maximum of 2 at $x = 1/4(4\pi) = \pi$.
(c) There is a minimum of -2 at $x = 3/4(4\pi) = 3\pi$.

The curve is sketched in Figure 7-23.

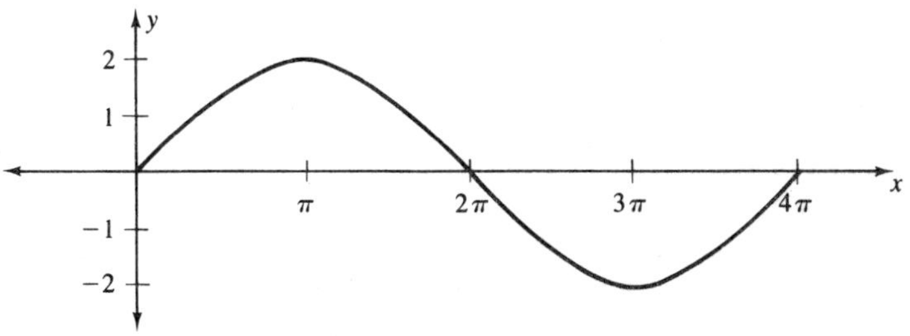

FIGURE 7-23 Graph of $y = 2 \sin \dfrac{x}{2}$, $0 \le x \le 4\pi$

$y = a \cos bx$

In Figure 7-24 the graph of $y = \cos x$ is shown for $0 \leq x \leq 2\pi$. The table of values used to draw the graph is given below.

x	0	$\pi/4$	$\pi/2$	$3\pi/4$	π	$5\pi/4$	$3\pi/2$	$7\pi/4$	2π
$\cos x$	1	0.7	0	-0.7	-1	-0.7	0	0.7	1

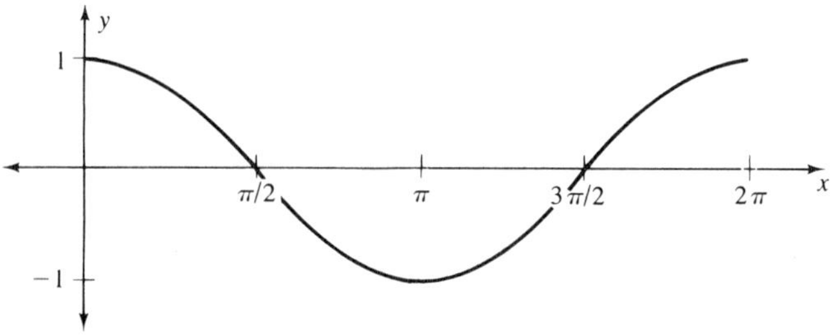

FIGURE 7-24 Graph of $y = \cos x$, $0 \leq x \leq 2\pi$

The following observations are made from the graph:

1. For $0 < x < \pi$, the function decreases from 1 to -1.
2. For $\pi < x < 2\pi$, the function increases from -1 to 1.
3. The periodic nature of the cosine function repeats the given curve for every interval of 2π.
4. The amplitude of the function is the maximum distance above the x-axis. For $y = \cos x$, the amplitude is 1.
5. For $0 \leq x \leq 2\pi$, the graph of the function has
   (a) Two zero points, at $x = \pi/2$ and $3\pi/2$
   (b) Two maximum points, at $x = 0$ and $x = 2\pi$
   (c) One minimum point, at $x = \pi$

and these five key points are used to graph the cosine function.

Do the constants a and b affect the graph of y = a cos bx the same as they affect the graph of y = a sin bx?

When a and b are not one, the effect on the graph of $y = \cos x$ is the same as the effect on the graph of $y = \sin x$.

> **Theorem:** If $y = a \cos bx$, the **amplitude** of the function is $|a|$.

> **Theorem:** If $y = a \cos bx$, the **period** of the function is $\dfrac{2\pi}{|b|}$.

As in the case of the sine function, the graph of $y = a \cos bx$ can be determined with five key points. The following analysis of the equation is an aid in graphing the function:

1. The amplitude of the function is $|a|$.
2. The period of the function is $\dfrac{2\pi}{|b|}$.
3. Locate the five key points that determine the basic shape of the wave:
   (a) Maximum points at $x = 0$ and $\dfrac{2\pi}{|b|}$ if $a > 0$, or minimum points at $x = 0$ and $\dfrac{2\pi}{|b|}$ if $a < 0$
   (b) Zero points at $x = \dfrac{1}{4}\left(\dfrac{2\pi}{|b|}\right)$ and $x = \dfrac{3}{4}\left(\dfrac{2\pi}{|b|}\right)$
   (c) Minimum point at $x = \dfrac{\pi}{|b|}$ if $a > 0$, or maximum point at $x = \dfrac{\pi}{|b|}$ if $a < 0$

EXAMPLE 7–24 Sketch the graph of $y = 2 \cos \dfrac{2x}{3}$ for $x \in [0, 6\pi]$.

Solution: For $y = 2 \cos \dfrac{2x}{3}$, $a = 2$ and $b = 2/3$. Therefore, the amplititude is 2 and the period is $\dfrac{2\pi}{2/3} = 3\pi$.

(a) There are maximum points at $x = 0$, 3π, and 6π.
(b) There are zero points at $x = 1/4(3\pi) = 3\pi/4$, $x = 3/4(3\pi) = 9\pi/4$, $x = 5/4(3\pi) = 15\pi/4$, and $x = 7/4(3\pi) = 21\pi/4$.
(c) There are minimum points at $x = \dfrac{\pi}{2/3} = \dfrac{3\pi}{2}$ and $x = \dfrac{3\pi}{2} + 3\pi = \dfrac{9\pi}{2}$.

The curve is sketched in Figure 7-25.

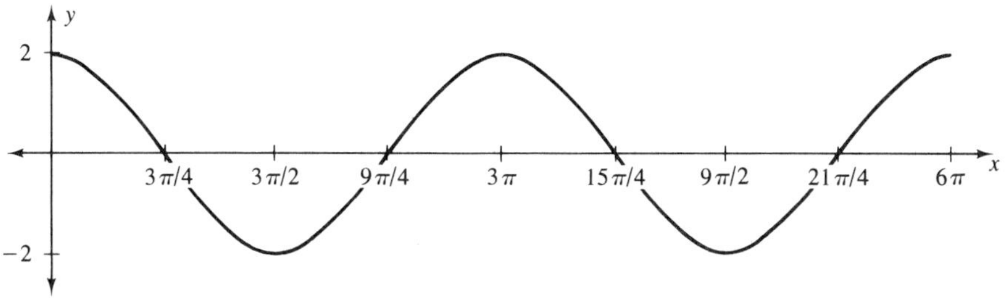

FIGURE 7-25 Graph of $y = 2 \cos \dfrac{2x}{3}$, $0 \le x \le 6\pi$

EXAMPLE 7–25 Determine the equation of the wave displayed in Figure 7-26.

Solution: The maximum point of the curve is 2.5; therefore, the amplitude is 2.5. One full period occurs on the interval from 0 to 6; therefore, $2\pi/b = 6$ and $b = 2\pi/6$, or $\pi/3$. Since the graph contains the origin, the equation is a sine function. With $a = 2.5$ and $b = \pi/3$, an equation of the curve is

$$y = 2.5 \sin \frac{\pi x}{3}$$

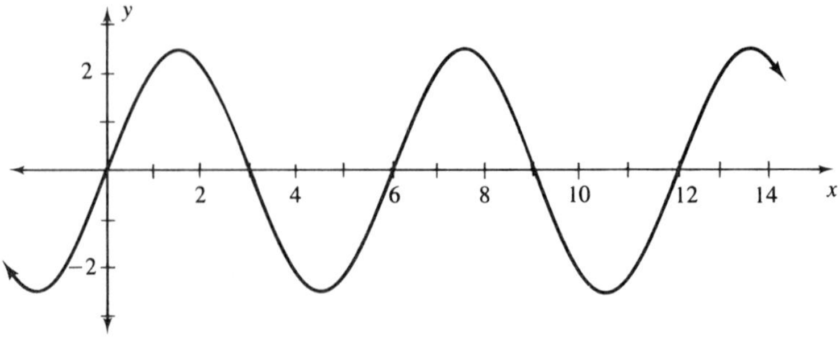

FIGURE 7-26 Graph of Example 7–25

Exercise Set 7–5

Identify the amplitude and period of each of the following functions. Do not graph.

1. $y = \cos 3x$
2. $y = 3 \cos 3x$
3. $y = \frac{1}{2} \sin x$
4. $y = \sin \frac{1}{2}x$
5. $y = 4 \cos 2x$
6. $y = -3 \sin \frac{1}{3}x$
7. $y = \frac{1}{9} \cos 5x$
8. $y = -2 \cos 7x$
9. $y = 3 \sin \pi x$
10. $y = \pi \sin 3x$
11. $y = \pi \cos 2x$
12. $y = \frac{3}{4} \cos 2\pi x$

Sketch the graph of the following functions for two periods of the function.

13. $y = 2 \cos 4x$
14. $y = \cos \frac{1}{2}x$
15. $y = 5 \cos 3x$
16. $y = \sin \frac{\pi x}{3}$
17. $y = 2 \sin \pi x$
18. $y = \frac{5}{2} \cos \frac{\pi x}{2}$
19. $y = \frac{2}{3} \cos 2\pi x$
20. $y = -3 \sin 4x$
21. $y = \frac{-4}{3} \sin 3x$
22. $y = \pi \cos x$

Graph the following pairs of equations on the same coordinate axes for $-2\pi \leq x \leq 2\pi$ and find the points of intersection.

23. $y = \sin 2x$; $y = 2 \sin x$

24. $y = \cos x$; $y = \sin \frac{1}{2}x$

Find a possible equation for the following sine and cosine graphs.

25.

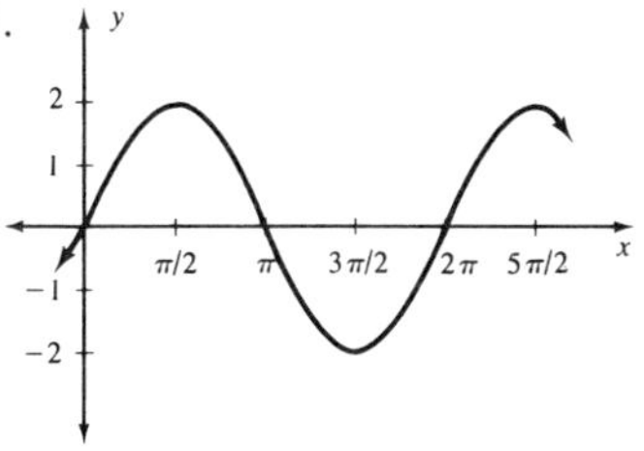

26.

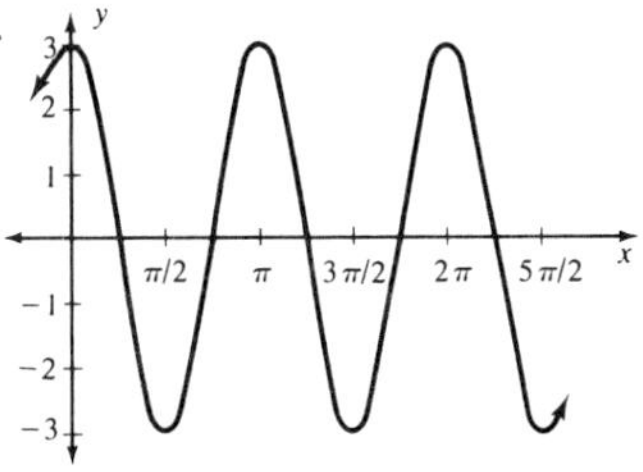

27.

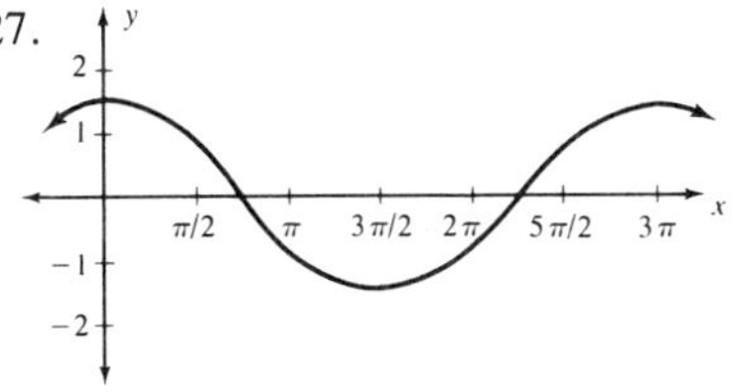

28.

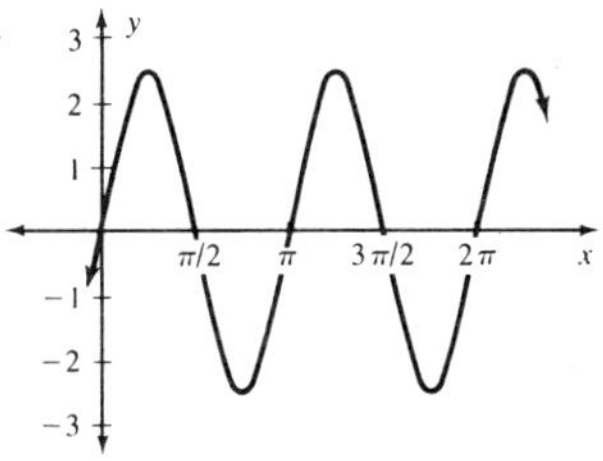

GRAPHS OF $y = \sin(x - c)$ AND $y = \cos(x - c)$ 7–6

When we compare the graphs of $y = \sin x$ and $y = \sin(x - c)$, we find that the two functions have the same amplitude and period. Furthermore, if the graph of $y = \sin x$ is shifted horizontally c units in the correct direction, then we can get the graphs of the two functions to coincide. The horizontal displacement between the two functions is called a *phase shift*.

Phase shift

> **Definition:** If $P(a, b)$ is a point of the graph of $y = \sin x$, then $P^*(a - c, b)$ is a point of the graph of $y = \sin(x - c)$, and a phase shift of c units has changed the graph of $y = \sin x$ to that of $y = \sin(x - c)$.

Does c represent only a positive value?

As we shall see, c can be negative as well as positive. The phase shift in the graph of $y = \sin(x - c)$ is to the right or left of the graph of $\sin x^*$ as c is positive or negative, respectively. Furthermore, if we replace sine everywhere with cosine in the above definition, we obtain the definition of phase shift for the cosine curve.

EXAMPLE 7–26 Sketch the graph of $y = \sin\left(x - \frac{\pi}{3}\right)$ over one period.

Solution: In the given equation, $c = \pi/3$. If we now let $x^* = x - \frac{\pi}{3}$, then $y = \sin x^*$. For the function $y = \sin x^*$, the amplitude is 1 and the period is 2π. Since $x^* = x - \frac{\pi}{3}$, then $x = x^* + \frac{\pi}{3}$. Therefore, the graph of $y = \sin x^*$ is shifted

$\pi/3$ units to the right to obtain the graph of $y = \sin\left(x - \frac{\pi}{3}\right)$. In particular,

$P_1^*(0, 0)$	is shifted to $P_1(\pi/3, 0)$
$P_2^*(\pi/2, 1)$	is shifted to $P_2(5\pi/6, 1)$
$P_3^*(\pi, 0)$	is shifted to $P_3(4\pi/3, 0)$
$P_4^*(3\pi/2, -1)$	is shifted to $P_4(11\pi/6, -1)$
$P_5^*(2\pi, 0)$	is shifted to $P_5(7\pi/3, 0)$

The graphs of $y = \sin x^*$ and $y = \sin\left(x - \frac{\pi}{3}\right)$ are shown in Figure 7-27 over the intervals $[0, 2\pi]$ and $[\pi/3, 7\pi/3]$, respectively.

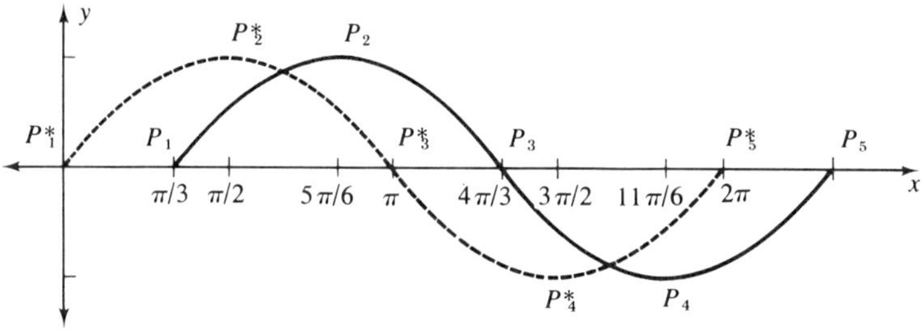

FIGURE 7-27 Graphs of $y = \sin x^*$ and $y = \sin\left(x - \frac{\pi}{3}\right)$

What effect does a different period have on a phase shift?

A different period has no effect on the phase shift. If $b \neq 1$, we simply graph $y = a \sin bx^*$. Then we shift the graph to the right or left to obtain the graph of the function that has been phase-shifted.

EXAMPLE 7–27 Sketch the graph of $y = 2\sin\left(\frac{\pi x}{2} - \pi\right)$ over three periods.

Solution: If $y = 2\sin\left(\frac{\pi x}{2} - \pi\right)$, then

$$y = 2\sin\frac{\pi}{2}(x - 2)$$

Let $x^* = x - 2$; then

$$y = 2\sin\frac{\pi x^*}{2}$$

The amplitude and period of $y = 2\sin\frac{\pi x^*}{2}$ are 2 and 4, respectively. Since $x^* = x - 2$, then $x = x^* + 2$; and each point of the graph of $y = 2\sin\frac{\pi x^*}{2}$ is shifted 2 units to the right to obtain the graph of

$$y = 2\sin\left(\frac{\pi x}{2} - \pi\right)$$

The graphs of both functions are shown in Figure 7-28 over the intervals [0, 12] and [2, 14], respectively.

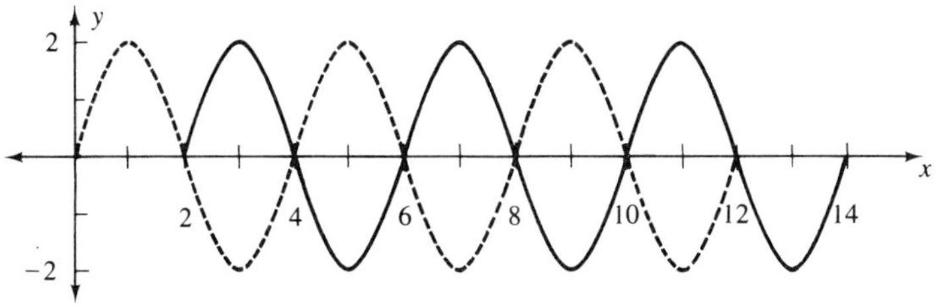

FIGURE 7-28 Graphs of $y = 2 \sin \frac{\pi x^*}{2}$ and $y = 2 \sin\left(\frac{\pi x}{2} - \pi\right)$

Now we can outline a procedure for graphing $y = a \sin(bx - c)$ or $y = a \cos(bx - c)$.

Step 1. Write $y = a \sin b\left(x - \frac{c}{b}\right)$ or $y = a \cos b\left(x - \frac{c}{b}\right)$.

Step 2. Let $x^* = x - \frac{c}{b}$ and $x = x^* + \frac{c}{b}$.

Step 3. Graph $y = a \sin bx^*$ or $y = a \cos bx^*$ with amplitude $|a|$ and period $\frac{2\pi}{|b|}$.

Step 4. To obtain the graph of the given function, shift horizontally the five key points of the graphs of $y = a \sin bx^*$ *or* $y = a \cos bx^*$

(a) $\frac{c}{b}$ units to the right if $\frac{c}{b} > 0$

(b) $\frac{c}{b}$ units to the left if $\frac{c}{b} < 0$

Exercise Set 7–6

Without graphing, find the (a) amplitude, (b) period, and (c) phase shift of the following functions.

1. $y = \sin\left(x - \frac{\pi}{2}\right)$
2. $y = \cos(x - 1)$
3. $y = \cos\left(x + \frac{\pi}{2}\right)$
4. $y = \sin\left(x + \frac{3\pi}{2}\right)$
5. $y = 3 \sin(2x - \pi)$
6. $y = \cos(3x + \pi)$
7. $y = \cos\left(\frac{5}{6}x + \frac{\pi}{2}\right)$
8. $y = 4 \sin\left(4x - \frac{\pi}{3}\right)$
9. $y = -2 \sin\left(2x + \frac{\pi}{2}\right)$
10. $y = 3 \cos 3\left(x - \frac{\pi}{4}\right)$
11. $y = \frac{3}{2} \cos(\pi x + \pi)$
12. $y = \frac{-4}{3} \sin 2\pi(x + 1)$
13. $y = 3 \sin 2(2x + \pi)$
14. $y = \cos\left(\frac{1}{2}x + \pi\right)$

Sketch one period of each of the following functions.

15. $y = \sin\left(x - \frac{\pi}{2}\right)$

16. $y = \cos\left(x + \frac{\pi}{4}\right)$

17. $y = \cos(x + 2)$

18. $y = \sin(x - 1)$

19. $y = 2\sin(2x + \pi)$

20. $y = \frac{3}{2}\cos(3x - \pi)$

21. $y = \cos\left(\frac{x}{2} - 1\right)$

22. $y = -\sin\left(\frac{1}{3}x + 1\right)$

23. $y = -2\sin\left(\frac{1}{2}x - \frac{\pi}{4}\right)$

24. $y = \frac{-2}{3}\cos\left(2x - \frac{\pi}{2}\right)$

25. $y = \cos\left(\frac{\pi}{2} - x\right)$

26. $y = \sin\left(x - \frac{\pi}{4}\right)$

Find an equation of a sine function with the following properties.

27. Period $= \pi$
Amplitude $= 3$
Phase shift $= \pi/2$ to the right

28. Period $= 2\pi/3$
Amplitude $= 1/2$
Phase shift $= 1$ to the left

29. Period $= 5\pi/3$
Amplitude $= 4$
Phase shift $= \pi/6$ to the right

30. Period $= 3$
Amplitude $= 5$
Phase shift $= 1$ to the left

Find an equation of a cosine function with the following properties.

31. Period $= 3\pi/2$
Amplitude $= 3/2$
Phase shift $= \pi/4$ to the right

32. Period $= \pi$
Amplitude $= 7$
Phase shift $= \pi/3$ to the left

33. Period $= 1.5$
Amplitude $= 3$
Phase shift $= \pi/5$ to the left

34. Period $= 3.2$
Amplitude $= 1.5$
Phase shift $= 0.5$ to the right

7-7 GRAPHS OF TANGENT, COTANGENT, SECANT, AND COSECANT FUNCTIONS

In this section we develop the graphs of the remaining trigonometric functions. The graphs of these functions differ from the graphs of the sine and cosine functions in that each graph has a *vertical asymptote*. As we shall see, determining the location of the asymptote is important in graphing these functions.

$y = a\tan(bx - c)$

To graph the function $y = \tan x$ for $0 \leq x \leq 2\pi$, we use the table of values below.

x	0	$\pi/4$	$\pi/2$	$3\pi/4$	π	$5\pi/4$	$3\pi/2$	$7\pi/4$	2π
$\tan x$	0	1	und	-1	0	1	und	-1	0

Since $\tan\frac{\pi}{2}$ and $\tan\frac{3\pi}{2}$ do not exist (the function is undefined), then the graph of $y = \tan x$ has vertical asymptotes at $x = \pi/2$ and $x = 3\pi/2$. The graph of $y = \tan x$ is shown in Figure 7-29. Notice that for the given equation, $a = b = 1$ and $c = 0$.

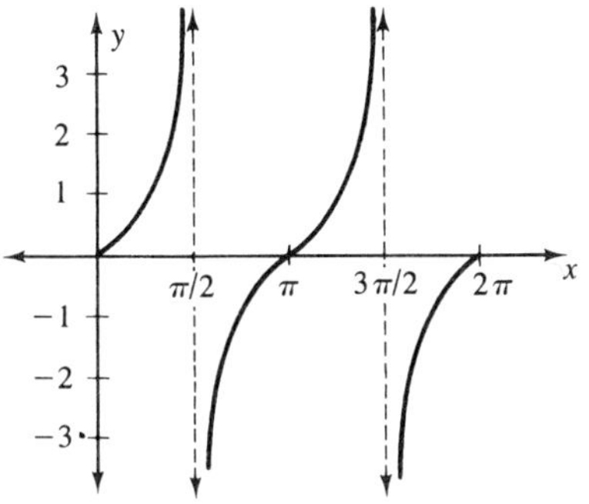

FIGURE 7-29 Graph of $y = \tan x$, $0 \le x \le 2\pi$

We can make the following observations about $y = \tan x$ and the graph of the function:

1. The *domain* of the function is the set of real numbers, except odd multiples of $\pi/2$. That is, $\left\{x \mid x \in \text{reals}, x \neq \frac{(2n+1)\pi}{2}, \text{ and } n \text{ is an integer}\right\}$.
2. The *range* of the function is the set of real numbers, that is, $\{y \mid y \in \text{reals}\}$.
3. The *period* of the function is π. Therefore, we obtain two periods of the function on the interval $[0, 2\pi]$.
4. The function *increases* throughout each period of the function. Thus, if $x_2 > x_1$ and both values of x are within the same period of the function, then $\tan x_2 > \tan x_1$.
5. The graph of any period of the function is separated from the graph of an adjoining period by an asymptote.
6. The graph of each period contains one zero point midway between the asymptotes.
7. With a period of π and no phase shift, an asymptote is drawn at each odd integral multiple of $\pi/2$.

What effect do values of a and b different from 1 have on the graph of the function?

The value of a alters the shape of the "kink" or bend in the curve. For $0 < a < 1$, the bend in the graph is *more pronounced* or visible. But for $a > 1$, an increase in the value of a *stretches out* or straightens the graph. If $a < 0$, the graph is reflected in the x-axis, and the function *decreases* throughout each period in the graph.

The value of b changes the period of the tangent function. If $y = \tan bx$, then the period is $\frac{\pi}{|b|}$. Furthermore, if $c = 0$ (the function does not have a phase shift), we draw vertical asymptotes at

$$x_1 = \pm\frac{1}{2}\left(\frac{\pi}{b}\right), \qquad x_2 = \pm\frac{3}{2}\left(\frac{\pi}{b}\right), \qquad x_3 = \pm\frac{5}{2}\left(\frac{\pi}{b}\right)$$

and so on. We use the preceding analysis in the following example.

EXAMPLE 7–28 Sketch the graph of $y = 3 \tan 2x$, $0 \leq x \leq 2\pi$.

Solution: With $b = 2$, the period of the function is $\pi/2$. Since $c = 0$, we construct asymptotes at

$$x_1 = \frac{1}{2}\left(\frac{\pi}{2}\right) = \frac{\pi}{4}$$

$$x_2 = \frac{3}{2}\left(\frac{\pi}{2}\right) = \frac{3\pi}{4}$$

$$x_3 = \frac{5}{2}\left(\frac{\pi}{2}\right) = \frac{5\pi}{4}$$

$$x_4 = \frac{7}{2}\left(\frac{\pi}{2}\right) = \frac{7\pi}{4}$$

Zero points in the graph occur at $x_5 = 0$ and midway between the asymptotes, namely at $x_6 = \pi/2$, $x_7 = \pi$, $x_8 = 3\pi/2$, and $x_9 = 2\pi$. Since $a = 3$, the curve is stretched and the graph is shown in Figure 7-30.

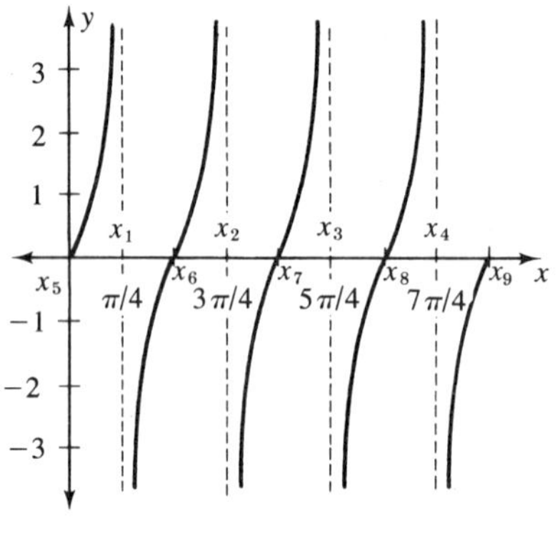

FIGURE 7-30 Graph of $y = 3 \tan 2x$, $0 \leq x \leq 2\pi$

EXAMPLE 7–29 Sketch $y = \frac{1}{2} \tan\left(x - \frac{\pi}{6}\right)$ over two periods.

Solution: With $b = 1$, the period of the function is π. Now let

$$x^* = x - \frac{\pi}{6} \quad \text{and} \quad x = x^* + \frac{\pi}{6}$$

For the equation $y = \frac{1}{2} \tan x^*$, asymptotes occur at

$$x_1^* = \frac{-1}{2}(\pi) = \frac{-\pi}{2}, \qquad x_2^* = \frac{\pi}{2}, \qquad \text{and} \quad x_3^* = \frac{3\pi}{2}$$

Since $x = x^* + \frac{\pi}{6}$, we draw asymptotes at

$$x_1 = \frac{-\pi}{2} + \frac{\pi}{6} = \frac{-\pi}{3}$$

$$x_2 = \frac{\pi}{2} + \frac{\pi}{6} = \frac{2\pi}{3}$$

$$x_3 = \frac{3\pi}{2} + \frac{\pi}{6} = \frac{5\pi}{3}$$

With zero points at $\pi/6$ and $7\pi/6$, and $a = 1/2$, the curve is sketched in Figure 7-31.

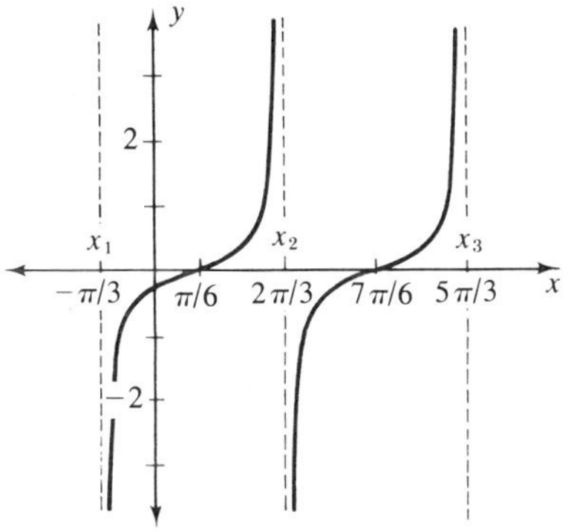

FIGURE 7-31 Graph of $y = \frac{1}{2}\tan\left(x - \frac{\pi}{6}\right)$, $-\pi/3 < x < 5\pi/3$

The asymptotes and zero points locate the curves of the graph. The value of a alters the basic shape of the curve, and the phase shift moves everything (asymptotes, zero points, and curve) c/b units to the left or right, depending on the sign of c/b.

All we must do for a phase shift in the tangent function is shift the asymptotes and zero points ?

EXAMPLE 7–30 Sketch $y = -2 \tan \frac{\pi x}{4}$ over two periods.

Solution: With $b = \frac{\pi}{4}$, the period of the function is $\frac{\pi}{\pi/4} = 4$. Therefore, for two periods, we graph the function over the interval [0, 8]. Since $c = 0$, there is no phase shift in the graph. We therefore sketch asymptotes at

$$x_1 = \frac{1}{2}(4) = 2 \quad \text{and} \quad x_2 = \frac{3}{2}(4) = 6$$

Zero points occur at $x_3 = 0$ and midway between the asymptotes, namely at $x_4 = 4$ and $x_5 = 8$. Since $a < 0$, the function decreases throughout each period of the graph. The graph is shown in Figure 7-32.

$y = a \cot(bx - c)$

To graph the function $y = \cot x$ for $0 \le x \le 2\pi$ in Figure 7-33, we use the following table of values.

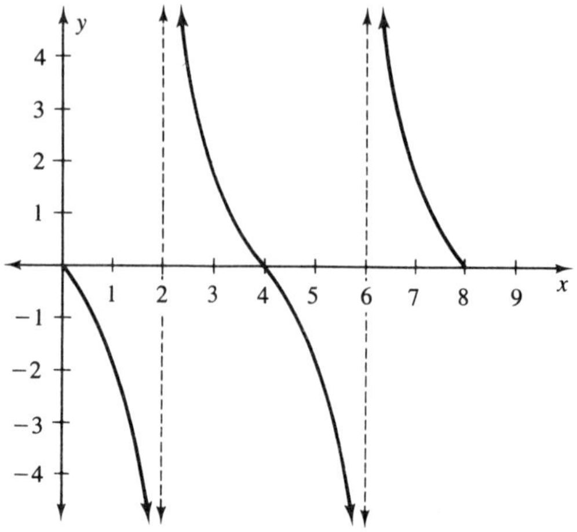

FIGURE 7-32 Graph of $y = -2 \tan \frac{\pi x}{4}$

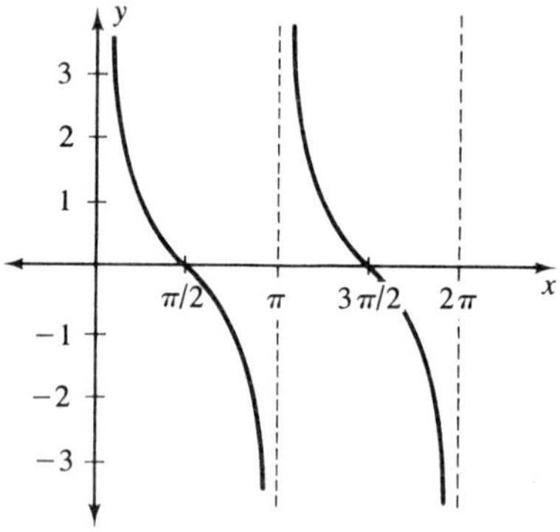

FIGURE 7-33 Graph of $y = \cot x,\ 0 \le x \le 2\pi$

x	0	$\pi/4$	$\pi/2$	$3\pi/4$	π	$5\pi/4$	$3\pi/2$	$7\pi/4$	2π
$\cot x$	und	1	0	-1	und	1	0	-1	und

We can make the following observations about $y = \cot x$ and the graph of the function:

1. The *domain* of the function is the set of real numbers, except integral multiples of π. That is, $\{x \mid x \in \text{reals}, x \neq n\pi, \text{ and } n \text{ is an integer}\}$.
2. The *range* of the function is the set of real numbers, that is, $\{y \mid y \in \text{reals}\}$.
3. The *period* of the function is π.
4. The function decreases throughout each period of the function. Therefore, if $x_2 > x_1$ and both values of x are within the same period of the function, then $\cot x_2 < \cot x_1$.
5. An asymptote separates the graphs of adjoining periods.

6. The graph of each period contains one zero point midway between the asymptotes.
7. With a period of π ($b = 1$) and no phase shift ($c = 0$), an asymptote is drawn at each integral multiple of π.

Our analysis of the cotangent function is similar to that for the tangent function, with two exceptions. First, the asymptotes for $y = \cot x$ are drawn at $x = n\pi$ and not $x = (2n + 1)(\pi/2)$. Second, the cotangent function *decreases* throughout each period, but the tangent function *increases* throughout each period for $a > 0$.

The analysis of the cotangent is like the one made for the tangent with only a couple of exceptions.

However, the similarity between the two functions yields the following analysis for the function $y = a \cot(bx + c)$:

1. The constant a alters the bend in the graph and determines whether the function increases or decreases throughout the period.
2. The period is $\dfrac{\pi}{|b|}$.
3. The phase shift is c/b units to the left if $c/b < 0$ and c/b units to the right if $c/b > 0$.
4. If $c = 0$, asymptotes are drawn at $x_1 = 0$, $x_2 = \dfrac{\pm\pi}{b}$, $x_3 = \dfrac{\pm 2\pi}{b}$, and so on.

EXAMPLE 7–31 Sketch the graph of $y = 2 \cot 3x$, $0 \le x \le 2\pi$.

Solution: With $b = 3$, the period of the function is $\pi/3$. Since $c = 0$, we draw asymptotes at $x_1 = 0, x_2 = \pi/3, x_3 = 2\pi/3, x_4 = \pi, x_5 = 4\pi/3, x_6 = 5\pi/3$, and $x_7 = 2\pi$. With $a = 2$, we sketch the curve in Figure 7-34.

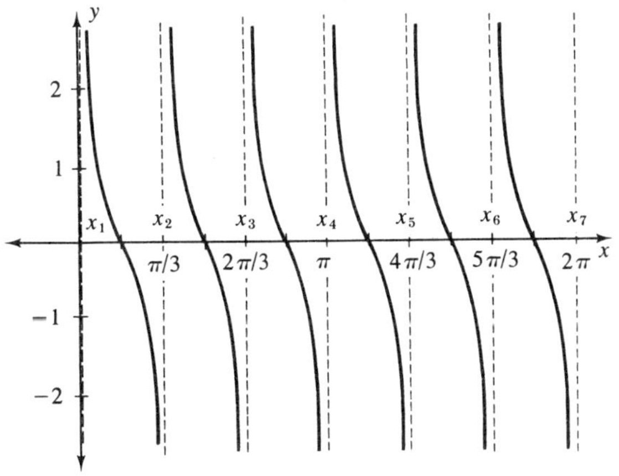

FIGURE 7-34 Graph of $y = 2 \cot 3x$, $0 \le x \le 2\pi$

We are not asked to graph the function in detail, but simply to sketch the graph. Therefore, once we locate the asymptotes and zero points and then note the value of a, a reliable picture of the function can be sketched.

Shouldn't some values for the function be used to construct these graphs?

$y = a \sec bx$

Previously we noted that the cosine and secant functions are reciprocal functions for all x's in the domain of the cosine function for which $\cos x \neq 0$. Therefore, $\sec x = \dfrac{1}{\cos x}$ when $\cos x \neq 0$; and $\sec x$ is undefined when $\cos x = 0$. In Figure 7-35 the graphs of $y = \cos x$ and $y = \sec x$ are shown for $0 \leq x \leq 2\pi$. We can make the following observations about the relationship between the two graphs:

1. The graph of $y = \sec x$ consists of *disjoint branches* separated by vertical asymptotes.
2. We draw an asymptote for each value of x such that $\cos x = 0$.
3. The *domain* of the secant function $= \{x \mid x \in \text{reals}, x \neq (2n + 1)(\pi/2)$, and n is an integer$\}$.
4. The *range* of the secant function $= \{y \mid |y| \geq 1\}$.
5. Since $a = 1$, *the minimum positive value of* $y = \sec x$ *is* 1 and *the maximum negative value is* -1. In general, for the function $y = a \sec x$ and $a > 0$, the minimum positive value and maximum negative value are a and $-a$, respectively.
6. For the function $y = a \sec bx$, the period is $\dfrac{2\pi}{|b|}$.

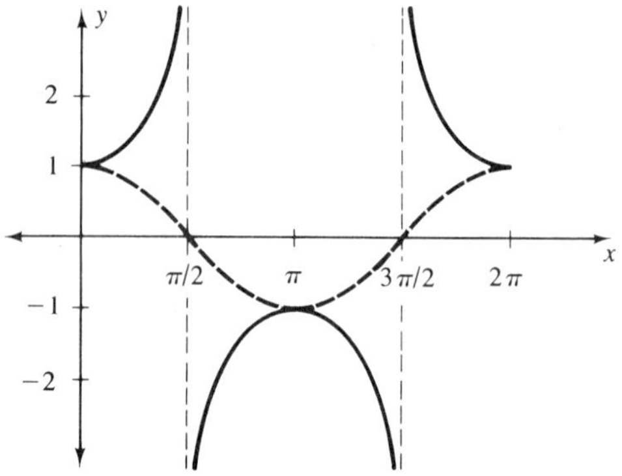

FIGURE 7-35 Graphs of $y = \cos x$ and $y = \sec x$, $0 \leq x \leq 2\pi$

EXAMPLE 7–32 Sketch the graph of $y = \dfrac{1}{2} \sec 2x$, $0 \leq x \leq 2\pi$.

Solution: We graph $y = \dfrac{1}{2} \cos 2x$ with amplitude $\dfrac{1}{2}$ and period π. We then draw asymptotes where $\dfrac{1}{2} \cos 2x$ crosses the x-axis and sketch the curves for $y = \dfrac{1}{2} \sec 2x$. The result is shown in Figure 7-36.

$y = a \csc bx$

Previously we noted that the sine and cosecant functions are reciprocal functions for all x's in the domain of the sine function for which $\sin x \neq 0$. Therefore,

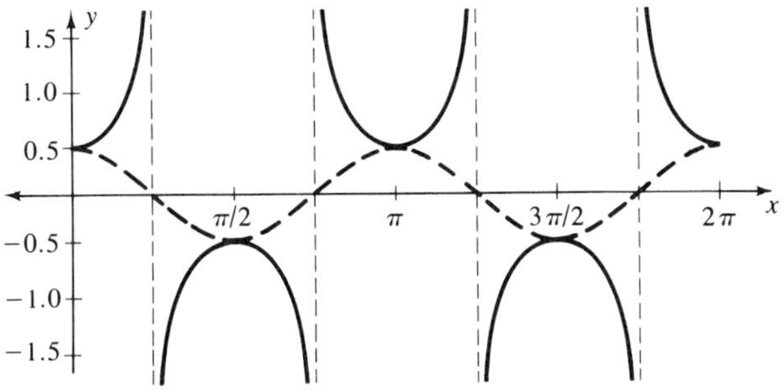

FIGURE 7-36 Graph of $y = \frac{1}{2}\sec 2x$

$\csc x = \dfrac{1}{\sin x}$ when $\sin x \neq 0$, and $\csc x$ is undefined when $\sin x = 0$. In Figure 7-37 the graphs of $y = \sin x$ and $y = \csc x$ are shown for $0 \leq x \leq 2\pi$. In the exercises for this section, we are required to make a summary of the relationship between the graphs of the two functions.

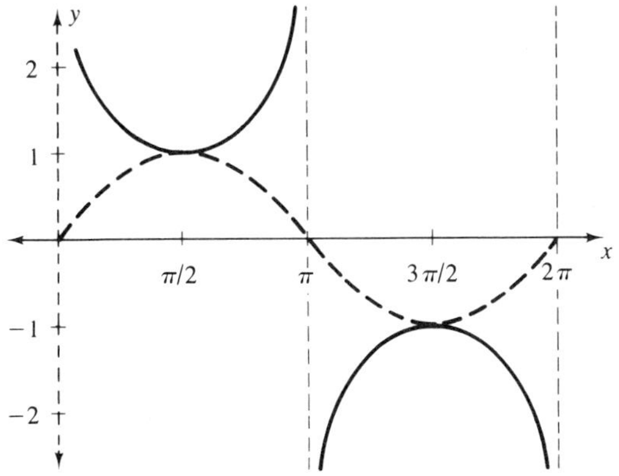

FIGURE 7-37 Graphs of $y = \sin x$ and $y = \csc x$, $0 \leq x \leq 2\pi$

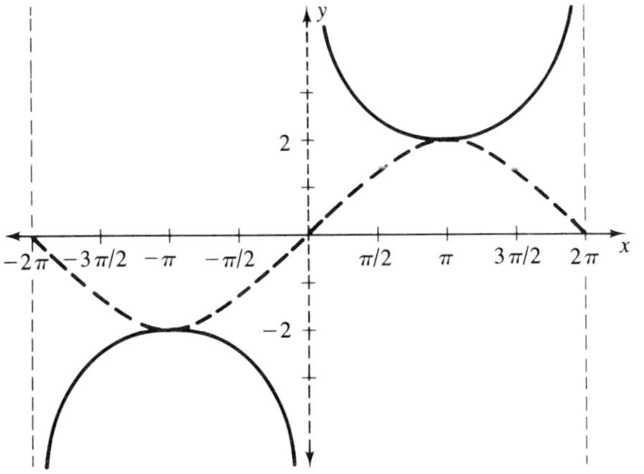

FIGURE 7-38 Graph of $y = 2\csc\frac{x}{2}$, $-2\pi \leq x \leq 2\pi$

EXAMPLE 7–33 Sketch the graph of $y = 2 \csc \frac{x}{2}$, $-2\pi \le x \le 2\pi$.

Solution: We graph $y = 2 \sin \frac{x}{2}$ with amplitude 2 and period 4π. We then draw asymptotes where $2 \sin \frac{x}{2}$ crosses the x-axis and sketch the curves for $y = 2 \csc \frac{x}{2}$. The result is shown in Figure 7-38.

Exercise Set 7–7

Without graphing, find (a) the period and (b) the horizontal shift for the following.

1. $y = \tan \frac{1}{3}x$
2. $y = 2 \cot \frac{1}{2}x$
3. $y = \tan(2x - \pi)$
4. $y = \tan\left(3x - \frac{\pi}{2}\right)$
5. $y = 3 \cot 4x$
6. $y = \frac{1}{2} \tan 7x$
7. $y = -\tan\left(2x + \frac{\pi}{3}\right)$
8. $y = -2 \cot\left(3x - \frac{\pi}{3}\right)$

Without graphing, locate (a) the zero points and (b) vertical asymptotes for $x \in [0, 2\pi]$.

9. $y = 3 \tan 2x$
10. $y = 8 \cot 4x$
11. $y = 8 \cot\left(2x - \frac{\pi}{4}\right)$
12. $y = \frac{1}{2} \tan\left(3x - \frac{\pi}{3}\right)$
13. $y = \cot\left(\frac{1}{2}x + \frac{\pi}{12}\right)$
14. $y = -\tan\left(\frac{1}{3}x - \frac{\pi}{2}\right)$
15. Summarize the procedures for graphing $y = a \csc bx$ as was done for the secant function. Be sure to include period, range, and location of asymptotes.

In each of the following, find (a) the period and (b) the location of the vertical asymptotes.

16. $y = \sec 2x$
17. $y = \csc 3x$
18. $y = 4 \csc \frac{1}{2}x$
19. $y = \sec \frac{2}{5}x$
20. $y = 3 \sec \frac{5\pi}{3}x$
21. $y = -2 \csc \frac{\pi}{3}x$

Graph the following for x such that $0 \le x \le 2\pi$.

22. $y = 2 \tan \frac{1}{2}x$
23. $y = 3 \sec x$
24. $y = -\tan\left(x - \frac{\pi}{3}\right)$
25. $y = \cot\left(\frac{1}{2}x + \frac{\pi}{3}\right)$
26. $y = \csc 4x$
27. $y = \frac{3}{2} \csc \frac{3}{2}x$

28. $y = 2 \sec 2x$

29. $y = -\cot 3x$

30. $y = 2 \cot 2\left(x - \frac{\pi}{4}\right)$

31. $y = \frac{1}{2} \tan \frac{1}{2}\left(x + \frac{\pi}{4}\right)$

Find the solution set of x, $0 \le x \le 2\pi$, such that the following inequalities are true. You can best obtain the solution by picturing the graph of the given trigonometric function and determining where the functional values satisfy the given criteria. For example, find x such that $\sin x \le 0$. Graph $y = \sin x$, $0 \le x \le 2\pi$.

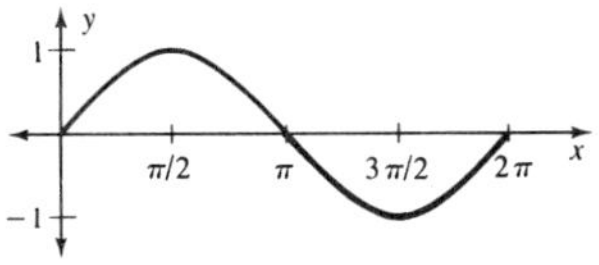

The solution set is x such that $\pi \le x \le 2\pi$.

32. $\cos x \le 0$

33. $\tan x \ge 1$

34. $\csc x \le -1$

35. $\sec x \ge 2$

36. $\cot x \ge 0$

37. $\sin 2x \le 0$

38. $\cos \frac{1}{2}x \ge 0$

39. $0 \le \tan x \le 1$

40. $\cos\left(x - \frac{\pi}{4}\right) \ge 0$

41. $\sin x \le \tan x$

SUMMARY

Unit circle

Definition: A unit circle has a radius of one unit.

Cosine and sine functions

Definition: If k is a real number paired with $P(x, y)$ on the unit circle, then (1) $\{(k, x)\}$ is the cosine function and we write $x = \cos k$, and (2) $\{(k, y)\}$ is the sine function and we write $y = \sin k$.

Cosine, sine, tangent, cotangent, secant, and cosecant functions

Definition: If s is a real number paired with $P(x, y)$ on the unit circle, then the six trigonometric functions are

(1) cosine $= \{(s, \cos s) \mid \cos s = x\}$

(2) sine $= \{(s, \sin s) \mid \sin s = y\}$

(3) tangent $= \left\{(s, \tan s) \mid \tan s = \frac{y}{x}, x \neq 0\right\}$

(4) cotangent $= \left\{(s, \cot s) \mid \cot s = \frac{x}{y}, y \neq 0\right\}$

(5) secant $= \left\{(s, \sec s) \mid \sec s = \frac{1}{x}, x \neq 0\right\}$

(6) cosecant $= \left\{(s, \csc s) \mid \csc s = \frac{1}{y}, y \neq 0\right\}$

Standard position angle

Definition: An angle is in standard position if the vertex is the origin of the coordinate system and the initial side is the positive x-axis.

Radian measure of an angle

Definition: The measure of a standard position angle θ in radians is equal to the length of the arc of a unit circle subtended by the sides of the angle.

Coterminal angles

Definition: Two angles in standard position are coterminal if they have the same terminal side.

Degree

Definition: One degree, written 1°, is equal to 1/360 of the circumference of a circle.

Minute

Definition: One minute, written $1'$, is equal to 1/60 of a degree.

Second

Definition: One second, written $1''$, is equal to 1/60 of a minute.

Cosine, sine, tangent, cotangent, secant, and cosecant functions

Definition: If $P(x, y)$ is the point of intersection of the terminal side of angle θ in standard position and the unit circle, then

(1) cosine $= \{(\theta, \cos\theta) \mid \cos\theta = x\}$

(2) sine $= \{(\theta, \sin\theta) \mid \sin\theta = y\}$

(3) tangent $= \left\{(\theta, \tan\theta) \mid \tan\theta = \frac{y}{x}, x \neq 0\right\}$

(4) cotangent $= \left\{(\theta, \cot\theta) \mid \cot\theta = \frac{x}{y}, y \neq 0\right\}$

(5) secant $= \left\{(\theta, \sec\theta) \mid \sec\theta = \frac{1}{x}, x \neq 0\right\}$

(6) cosecant $= \left\{(\theta, \csc\theta) \mid \csc\theta = \frac{1}{y}, y \neq 0\right\}$

Theorem: If $0^\circ < \theta < 90^\circ$, then

(1) $\sin\theta = \cos(90^\circ - \theta)$ or $\cos\theta = \sin(90^\circ - \theta)$

(2) $\tan\theta = \cot(90^\circ - \theta)$ or $\cot\theta = \tan(90^\circ - \theta)$

(3) $\sec\theta = \csc(90^\circ - \theta)$ or $\csc\theta = \sec(90^\circ - \theta)$

Reference angle

Definition: Angle α is the reference angle of θ if $0^\circ < \alpha < 90^\circ$ and the trigonometric function values of α are the same as the trigonometric function values of θ in absolute value.

Theorem: If $y = a \sin bx$, then the amplitude of the function is $|a|$.

Theorem: If $y = a \sin bx$, then the period of the function is $\frac{2\pi}{|b|}$.

Theorem: If $y = a \cos bx$, then the amplitude of the function is $|a|$.

Theorem: If $y = a \cos bx$, then the period of the function is $\frac{2\pi}{|b|}$.

Phase shift

Definition: If $P(a, b)$ is a point of the graph of $y = \sin x$, then $P^*(a - c, b)$ is a point of the graph of $y = \sin(x - c)$ and a phase shift of c units has changed the graph of $y = \sin x$ to that of $y = \sin(x - c)$.

SELF-CHECKING EXERCISE

Find the real number s ($0 \leq s < 2\pi$) that is paired with a point that is the given fractional part of one rotation on the unit circle from (1, 0). **(7-1)**

1. 1/8
2. 7/12

3. Find the six trigonometric function values of the real number s ($0 \leq s < 2\pi$) paired with the point $P(4/5, -3/5)$ on the unit circle. **(7-1)**

Find the other five trigonometric function values of s, given that $P(x, y)$ paired with s is contained in the indicated quadrant. **(7-1)**

4. $\tan s = -8/15$; II
5. $\cos s = 7/25$; IV
6. $\sec s = 3$; I
7. $\sin s = -1/5$; III

Find the values of the real numbers k and $-k$ ($2\pi \leq k < 4\pi$ and $-4\pi < -k \leq -2\pi$) that are paired with the same point on the unit circle as the given number. **(7-1)**

8. $7\pi/6$
9. $5\pi/9$
10. $-19\pi/12$
11. $-5\pi/7$

Sketch in standard position. **(7-2)**

12. 240°
13. $-5\pi/4$
14. −300°
15. $8\pi/3$

Express in radians. **(7-2)**

16. 300°
17. −100°

Express in degrees. **(7-2)**

18. $11\pi/18$
19. $-3\pi/5$

20. If $\cos\theta = -5/7$ for θ in Quadrant II, find the other five trigonometric functions of θ. **(7-3)**

Use Table 4 to find the following. **(7-3)**

21. cot 36°50′
22. cos 1.1897
23. sec 71°20′
24. tan 0.8523

Find θ (in radians) in Quadrant I for the following function values. **(7-3)**

25. $\cos\theta = 0.9194$
26. $\tan\theta = 2.605$
27. $\sin\theta = 0.8039$
28. $\sec\theta = 1.040$

Find the reference angle α for the following angles. **(7-3)**

29. 141°20′
30. −385°10′
31. $-17\pi/6$
32. 7.3382

Use Table 4 to find the following function values. **(7-3)**

33. tan 643°
34. cos(−900°)
35. sec 3.9997
36. $\sin(-12\pi/5)$

Find the values of θ (and $0 \leq \theta < 2\pi$) for the following. **(7-3)**

37. $\sin\theta = -0.3007$

38. $\cos\theta = 0.6361$

39. $\tan\theta = 0.1346$

40. $\cot\theta = -1.000$

Evaluate the following. **(7-4)**

41. $(\cos 60°)(\cos 30°) - (\sin 60°)(\sin 30°)$

42. $\dfrac{2\tan 45°}{1 - \tan 60°}$

43. $\left(\sec\dfrac{\pi}{3}\right)\left(\sin\dfrac{\pi}{6}\right) + \left(\tan\dfrac{3\pi}{4}\right)\left(\cot\dfrac{7\pi}{4}\right)$

44. $\dfrac{\cos\dfrac{7\pi}{6} + \sin\dfrac{5\pi}{3}}{\tan\dfrac{5\pi}{6}}$

45. In which quadrant is $\sin\theta < 0$ and $\sec\theta > 0$? **(7-4)**

State the amplitude of the following functions. **(7-5)**

46. $y = 2\sin(x + 3)$

47. $y = \dfrac{1}{2}\cos\dfrac{3}{2}x$

State the period of the following functions. **(7-5)**

48. $y = \cos 3x$

49. $y = \sin 5x$

Find the horizontal shift of the following functions. **(7-6)**

50. $y = \sin\left(x - \dfrac{\pi}{2}\right)$

51. $y = 3\cos(3x + \pi)$

Find the equations of the vertical asymptotes of each of the following over two periods. **(7-7)**

52. $y = 2\tan x$

53. $y = \cot 3x$

54. $y = 4\sec 2x$

State the period of the following functions. **(7-7)**

55. $y = \tan\dfrac{1}{2}x$

56. $y = \sec\dfrac{3}{4}x$

57. $y = \csc\dfrac{5}{3}x$

58. $y = \cot(2x - 3)$

Find the horizontal shift of the following functions. **(7-7)**

59. $y = \cot\left(3x + \dfrac{\pi}{3}\right)$

60. $y = -3\tan\dfrac{x}{2}$

Graph the following functions over one period. **(7-5, 7-6, 7-7)**

61. $y = 3\cos 2x$

62. $y = \sin\left(x - \dfrac{\pi}{4}\right)$

63. $y = \tan 2x$

64. $y = \dfrac{3}{2}\sec x$

65. $y = \csc\dfrac{x}{2}$

66. $y = \cot\left(x + \dfrac{\pi}{2}\right)$

Answers to Self-Checking Exercise

1. $\pi/4$

2. $7\pi/6$

3. $\cos s = 4/5$; $\sin s = -3/5$; $\tan s = -3/4$; $\sec s = 5/4$; $\csc s = -5/3$; $\cot s = -4/3$

4. $\cos s = -15/17$; $\sin s = 8/17$; $\cot s = -15/8$; $\sec s = -17/15$; $\csc s = 17/8$

5. $\sin s = -24/25$; $\tan s = -24/7$; $\cot s = -7/24$; $\sec s = 25/7$; $\csc s = -25/24$

6. $\cos s = 1/3$; $\sin s = 2\sqrt{2}/3$; $\tan s = 2\sqrt{2}$; $\cot s = \sqrt{2}/4$; $\csc s = 3\sqrt{2}/4$

7. $\cos s = -2\sqrt{6}/5$; $\tan s = \sqrt{6}/12$; $\cot s = 2\sqrt{6}$; $\sec s = -5\sqrt{6}/12$; $\csc s = -5$

8. $19\pi/6$ and $-17\pi/6$

9. $23\pi/9$ and $-31\pi/9$

10. $29\pi/12$ and $-43\pi/12$

11. $23\pi/7$ and $-19\pi/7$

12.

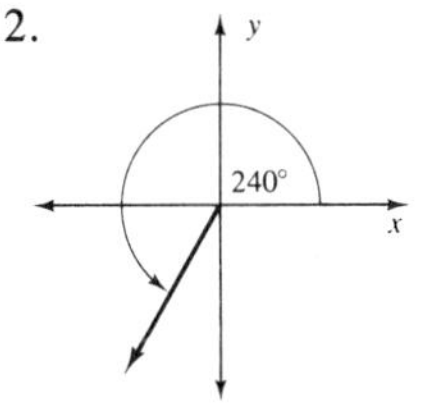

13.

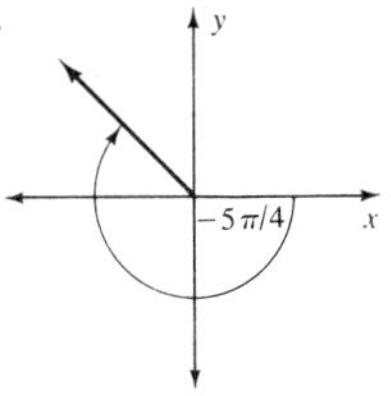

14.

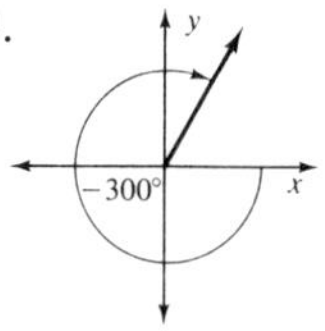

15.

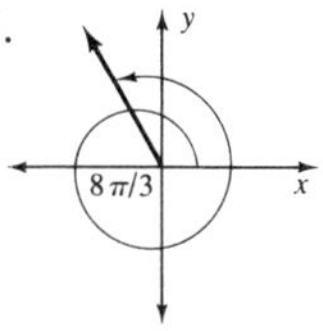

16. $5\pi/3$ 17. $-5\pi/9$ 18. $110°$ 19. $-108°$

20. $\sin\theta = 2\sqrt{6}/7$; $\tan\theta = -2\sqrt{6}/5$; $\cot\theta = -5\sqrt{6}/12$; $\sec\theta = -7/5$; $\csc\theta = 7\sqrt{6}/12$

21. 1.335 22. 0.3719 23. 3.124 24. 1.144

25. 0.4043 26. 1.2043 27. 0.9338 28. 0.2793

29. 38°40′ 30. 25°10′ 31. $\pi/6$ 32. 1.0550

33. −4.331 34. −1 35. −1.529 36. −0.9511

37. 3.4470; 5.9778 38. 0.8814; 5.4018 39. 0.1338; 3.2754

40. 2.3562; 5.4978 41. 0 42. $-1 - \sqrt{3}$ 43. 2

44. 3 45. IV 46. 2 47. 1/2

48. $2\pi/3$ 49. $2\pi/5$ 50. $\pi/2$ to the right 51. $\pi/3$ to the left

52. $x_1 = \pi/2$; $x_2 = 3\pi/2$

53. $x_1 = 0$; $x_2 = \pi/3$; $x_3 = 2\pi/3$

54. $x_1 = \pi/4$; $x_2 = 3\pi/4$; $x_3 = 5\pi/4$; $x_4 = 7\pi/4$

55. 2π 56. $8\pi/3$ 57. $6\pi/5$ 58. $\pi/2$

59. $\pi/9$ to the left

60. 0

61.

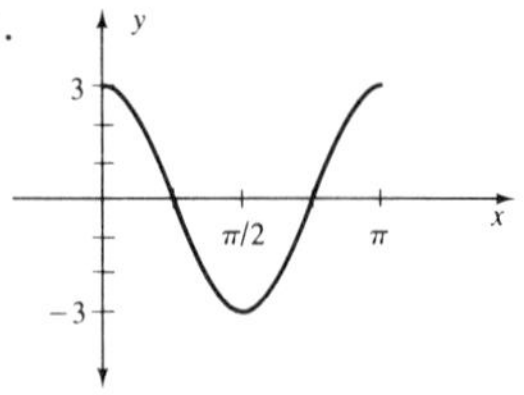

62.

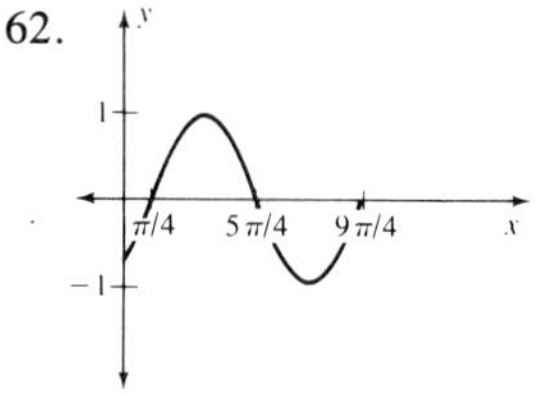

63.

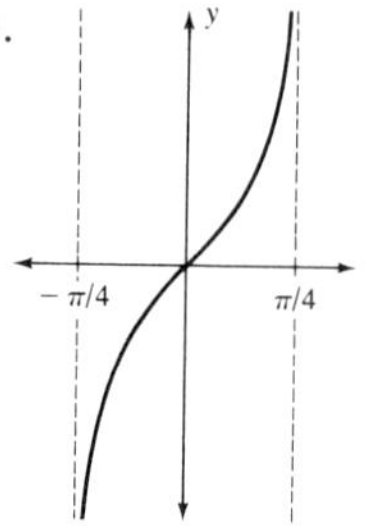

64.

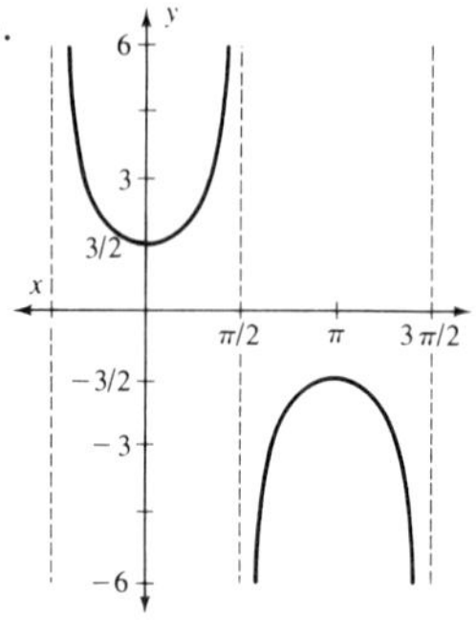

65.

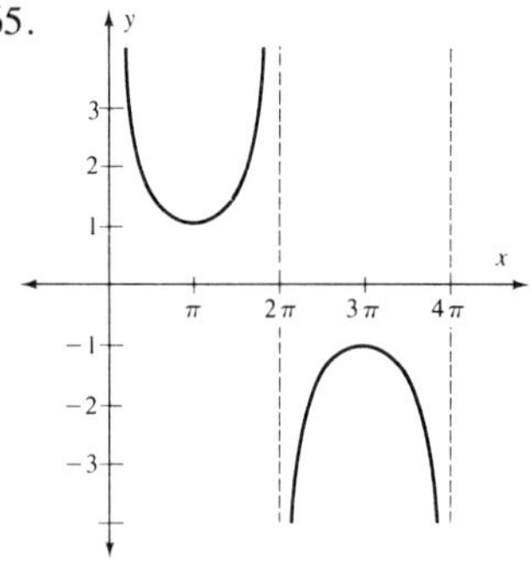

66.

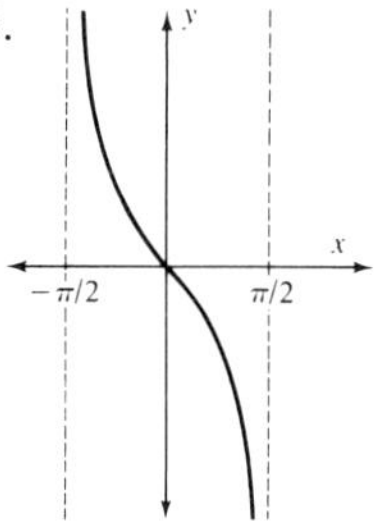

Trigonometric Identities and Conditional Equations

8

Throughout this text we have stressed the importance of solving equations and have noted that some topics were necessarily presented to improve our skills in solving equations. At this time we consider trigonometric equations. We divide this chapter into two parts based on the two types of trigonometric equations we will be dealing with: (1) identity equations and (2) conditional equations.

Identity

Definition: An identity is an equation that is true for any value of the variable for which its terms are defined.

Conditional equation

Definition: A conditional equation is one that is true for some but not all values of the variable.

Examples of algebraic identities are

$$x = x \quad \text{and} \quad 2(x + 1) = 2x + 2$$

Examples of conditional algebraic equations are

$$3(x + 1) = x + 9 \quad \text{and} \quad 2x^2 - x - 6 = 0$$

In this chapter we consider several important trigonometric identities, and we use some of the identities to solve conditional trigonometric equations. But trigonometric identities play a very important role in the study of calculus. To help us in learning these basic identities, we have prepared examples that require us to use the identities. Also, we develop new identities from established ones in order to improve our skill in manipulating them and in recognizing them in various forms.

We use the variables x and θ to represent the variable in the trigonometric equations. Also, let it be understood that *the variables x and θ represent a real number or the measure of an angle in standard position.*

8-1 RECIPROCAL, PYTHAGOREAN, AND NEGATIVE ANGLE IDENTITIES

The first five identities result from the definitions of the trigonometric functions. Since an identity must be true for all replacements of the variable, we must eliminate any values of the variable for which a given expression is not defined. We will note any restrictions on the variable in the statement of the identity, but the restrictions will be understood to exist when the identity is used in any subsequent equation.

Reciprocal Identities:

$$\sec\theta = \frac{1}{\cos\theta}, \quad \cos\theta \neq 0 \tag{1}$$

$$\csc\theta = \frac{1}{\sin\theta}, \quad \sin\theta \neq 0 \tag{2}$$

$$\cot\theta = \frac{1}{\tan\theta}, \quad \tan\theta \neq 0 \tag{3}$$

Tangent and Cotangent Identities:

$$\tan\theta = \frac{\sin\theta}{\cos\theta}, \quad \cos\theta \neq 0 \tag{4}$$

$$\cot\theta = \frac{\cos\theta}{\sin\theta}, \quad \sin\theta \neq 0 \tag{5}$$

EXAMPLE 8–1 Verify identity (1).

Solution: If $P(a, b)$ is the point of intersection of the terminal side of θ and the unit circle, and $a \neq 0$, then

$$\cos\theta = a \quad \text{and} \quad \sec\theta = \frac{1}{a}$$

Thus, on substitution, $\sec\theta = \dfrac{1}{\cos\theta}$.

Can't the restrictions on the identity also be stated in terms of θ?

If we choose to do so, we can state the restrictions on the identity in terms of θ. For example, in identity (1), if $a = 0$, then $\cos\theta = 0$ and $\theta = (2n + 1)(\pi/2)$, $n \in \{\text{integers}\}$. We can therefore write (1) as follows:

$$\sec\theta = \frac{1}{\cos\theta}, \quad \theta \neq (2n + 1)\frac{\pi}{2}, \ n \in \{\text{integers}\}$$

To determine the restricted values of θ, we set the denominator of the expression equal to 0 and solve for θ.

Pythagorean Identities:

$$\cos^2\theta + \sin^2\theta = 1 \quad \textbf{(6)}$$

$$1 + \tan^2\theta = \sec^2\theta \quad \textbf{(7)}$$

$$1 + \cot^2\theta = \csc^2\theta \quad \textbf{(8)}$$

In Figure 8-1, θ is drawn in standard position on a unit circle, with $P(x, y)$ the intersection of the terminal side of θ and the circle. By definition, $\cos\theta = x$ and $\sin\theta = y$. But on the unit circle, $x^2 + y^2 = 1$. Therefore,

$$(\cos\theta)^2 + (\sin\theta)^2 = 1$$

and we write

$$\cos^2\theta + \sin^2\theta = 1$$

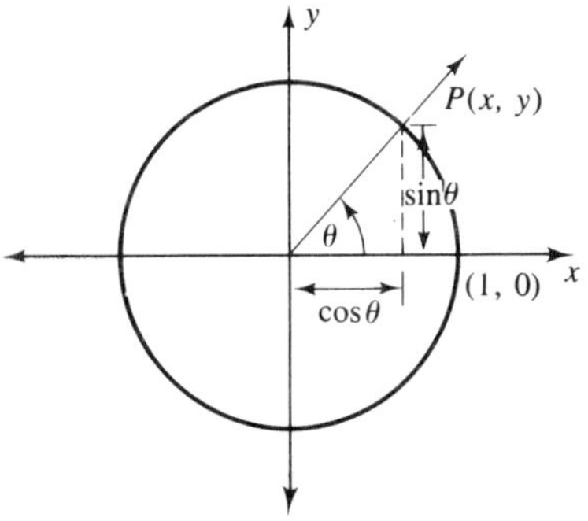

FIGURE 8-1 $\cos^2\theta + \sin^2\theta = 1$

EXAMPLE 8–2 Prove identity (7).

Solution: Since $\cos^2\theta + \sin^2\theta = 1$, then

$$\frac{1}{\cos^2\theta}(\cos^2\theta + \sin^2\theta) = 1\left(\frac{1}{\cos^2\theta}\right) \qquad (\text{for } \cos^2\theta \neq 0)$$

$$\frac{\cos^2\theta}{\cos^2\theta} + \frac{\sin^2\theta}{\cos^2\theta} = \frac{1}{\cos^2\theta} \qquad (\text{by the distributive axiom})$$

But $\dfrac{\sin\theta}{\cos\theta} = \tan\theta$ and $\dfrac{1}{\cos\theta} = \sec\theta$; therefore,

$$1 + \tan^2\theta = \sec^2\theta$$

We now provide some examples that require the identities to be manipulated. Although the examples do not appear to "do" anything, they are designed to improve our understanding of the identities.

EXAMPLE 8–3 Express the following in terms of $\cos\theta$:

(a) $\sin\theta$ (b) $\tan\theta$

Solutions:
(a) With identity (6), we get $\sin\theta = \pm\sqrt{1 - \cos^2\theta}$.

(b) With identity (4) and part (a) above, we get

$$\tan\theta = \frac{\pm\sqrt{1 - \cos^2\theta}}{\cos\theta}$$

In Figure 8-2, θ_1, $-\theta_1$, θ_2, and $-\theta_2$ are drawn on a unit circle. The following pairs of equations are obtained when we use the definition of $\cos\theta$ and the coordinates of P_1, P_1^*, P_2, and P_2^*:

$$\cos\theta_1 = a \quad \text{and} \quad \cos(-\theta_1) = a$$
$$\cos\theta_2 = c \quad \text{and} \quad \cos(-\theta_2) = c$$

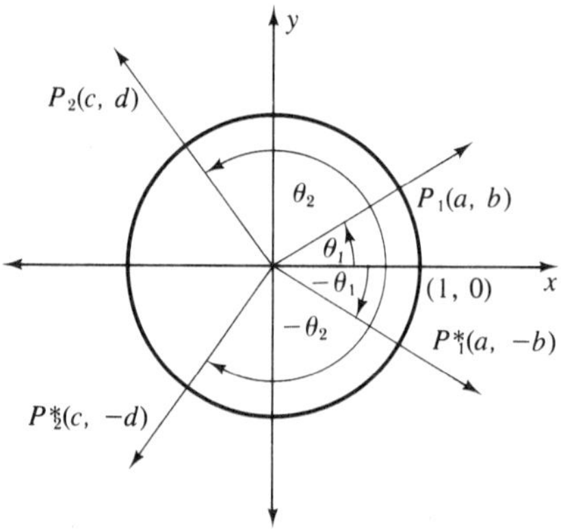

FIGURE 8-2 θ_1, $-\theta_1$, θ_2, and $-\theta_2$

When we generalize the above examples, we obtain identity (9).

$$\cos(-\theta) = \cos\theta \qquad \textbf{(9)}$$

The following pairs of equations are obtained when we use the definition of $\sin\theta$ and the coordinates of P_1, P_1^*, P_2, and P_2^*:

$$\sin\theta_1 = b \quad \text{and} \quad \sin(-\theta_1) = -b$$
$$\sin\theta_2 = d \quad \text{and} \quad \sin(-\theta_2) = -d$$

Therefore, in general, for any θ,

$$\sin(-\theta) = -\sin\theta \qquad \textbf{(10)}$$

Furthermore, since $\tan\theta = \dfrac{\sin\theta}{\cos\theta}$, then

$$\tan(-\theta) = -\tan\theta \qquad \textbf{(11)}$$

Identities (9), (10), and (11) are called the negative angle identities.

The two negative signs in each of the identities represent different things, don't they?

In $\sin(-\theta)$, the negative sign is used to show a *rotation that is opposite to the rotation of* θ. Thus, if θ is a positively rotated angle, then $-\theta$ is a negatively rotated angle, and vice versa. In $-\sin\theta$, the negative sign indicates the *opposite sign in the function value of* $\sin\theta$. That is, if $\sin\theta > 0$, then $-\sin\theta < 0$. But if $\sin\theta < 0$, then $-\sin\theta > 0$.

In the following examples we are instructed to verify that a given equation is an identity. To accomplish this task we use one or more of the basic identities to change the form of the given equation to one that is obviously an identity.

To illustrate what we mean, consider the equation in Example 8–4, $\tan^2 x = \sec^2 x(1 - \cos^2 x)$. We are asked to verify that the equation is an identity. It is not obvious that this equation is true for all acceptable values of x. But in the final step in the solution, we see the equation $\tan^2 x = \tan^2 x$. This equation is obviously true for any suitable value of x.

The following suggestions are offered as "helpful hints" for verifying an identity:

1. Convert one side of the equation to the form of the other side.
2. If possible, work on only one side of the equation.
3. Change the form of the more complicated expression to that of the simpler expression.
4. If several different functions are included in the equation, convert them to only sines and cosines.

EXAMPLE 8–4 Verify as an identity $\tan^2 x = \sec^2 x(1 - \cos^2 x)$.

Solution: The right side of the equation appears to be the more complicated expression. We therefore choose to try to change the right side of the equation to the form of the left side.

$$\begin{aligned}
\sec^2 x(1 - \cos^2 x) &= \frac{1}{\cos^2 x}(1 - \cos^2 x) && \left(\text{since } \sec x = \frac{1}{\cos x}\right)\\
&= \frac{\sin^2 x}{\cos^2 x} && (\text{since } (1 - \cos^2 x) = \sin^2 x)\\
&= \tan^2 x && \left(\text{since } \frac{\sin x}{\cos x} = \tan x\right)
\end{aligned}$$

Obviously, $\tan^2 x = \tan^2 x$ for all x in the domain of the tangent function; therefore, the given equation is an identity.

EXAMPLE 8–5 Verify as an identity $\cos^4 x - \sin^4 x = 1 - 2\sin^2 x$.

Solution: The left side of the equation is of the form $a^4 - b^4 = (a^2 + b^2)(a^2 - b^2)$. Therefore,

$$\begin{aligned}
\cos^4 x - \sin^4 x &= (\cos^2 x + \sin^2 x)(\cos^2 x - \sin^2 x)\\
&= 1(\cos^2 x - \sin^2 x) && (\text{since } \cos^2 x + \sin^2 x = 1)\\
&= (1 - \sin^2 x) - \sin^2 x && (\text{since } \cos^2 x = 1 - \sin^2 x)\\
&= 1 - 2\sin^2 x
\end{aligned}$$

Since $1 - 2\sin^2 x = 1 - 2\sin^2 x$ for all values of x, the given equation is an identity.

EXAMPLE 8–6 Verify as an identity $\dfrac{\tan\theta}{1+\sec\theta} = \dfrac{\sec\theta - 1}{\tan\theta}$.

Solution: The tangent and secant functions are related by the identity

$$1 + \tan^2\theta = \sec^2\theta$$

and

$$\tan^2\theta = \sec^2\theta - 1 = (\sec\theta - 1)(\sec\theta + 1)$$

We decide therefore to multiply the right side of the equation by $\left(\dfrac{\sec\theta + 1}{\sec\theta + 1}\right)$ to supply the missing factor.

$$\begin{aligned}\frac{\sec\theta - 1}{\tan\theta} &= \frac{\sec\theta - 1}{\tan\theta}\left(\frac{\sec\theta + 1}{\sec\theta + 1}\right)\\ &= \frac{\sec^2\theta - 1}{\tan\theta(\sec\theta + 1)}\\ &= \frac{\tan^2\theta}{\tan\theta(\sec\theta + 1)}\\ &= \frac{\tan\theta}{\sec\theta + 1}\end{aligned}$$

Since $\dfrac{\sec\theta - 1}{\tan\theta} = \dfrac{\tan\theta}{\sec\theta + 1}$, the given equation is an identity.

Exercise Set 8–1

For each of the following Pythagorean identities, write three other forms of the equations.

1. $\sin^2 x + \cos^2 x = 1$
2. $1 + \tan^2 x = \sec^2 x$
3. $1 + \cot^2 x = \csc^2 x$

Use the basic identities to replace the ? with the appropriate function.

4. $1 - \sin^2 x = ?$
5. $\sin^2 x + \cos^2 x = ?$
6. $\sec^2 x - 1 = ?$
7. $\dfrac{1}{\csc x} = ?$
8. $\cot^2 x + ? = \csc^2 x$
9. $\sec^2 x - \tan^2 x = ?$
10. $\dfrac{\cos x}{\sin x} = ?$
11. $\sec x \cos x = ?$
12. $\tan x(?) = 1$
13. $\sin x \csc x = ?$
14. $\dfrac{\sin x}{\cos x} = ?$
15. $\csc^2 x - \cot^2 x = ?$

Express each of the following in terms of $\sin x$.

16. $\cos x$
17. $\tan x$
18. $\cot x$
19. $\sec x$
20. $\csc x$

Express each of the following in terms of $\tan x$.

21. $\cot x$
22. $\sec x$
23. $\cos x$
24. $\csc x$
25. $\sin x$

Use the reciprocal and Pythagorean identities to express the following in terms of a single trigonometric function of a positive angle.

26. $\dfrac{\cos x}{\cot x}$
27. $\sin x(\cot^2 x + 1)$
28. $\dfrac{1 + \tan^2 x}{\tan^2 x}$
29. $\cos x \tan x \csc x \sec x$
30. $\cos^2 x - \cos x \sec x$
31. $(\csc^2 x - 1)\tan x$
32. $\csc x \cot x \sin x$
33. $\sin(-x)\cos(-x) + \tan(-x)\cos^2 x$
34. $\dfrac{\cos^2 x}{1 + \sin x}$
35. $\dfrac{\sin x(\csc^2 x - \cot^2 x)}{\cos x \sec x}$
36. $\dfrac{\csc(-x)}{\cot(-x) + \tan(-x)}$
37. $(\cot x - \csc x)(1 + \cos x)$
38. If $y = \sqrt{1 - z^2}$, $z = \sin\theta$, and $\theta \in [0, \pi/2]$, express y in terms of θ.
39. If $y = \sqrt{z^2 + 4}$, $z = 2\tan\theta$, and $\theta \in [0, \pi/2]$, express y in terms of θ.
40. If $y = \sqrt{z^2 - 9}$, $z = 3\sec\theta$, and $\theta \in [0, \pi/2]$, express y in terms of θ.

Verify the following identities.

41. $\sec x = \cos x + \sin x \tan x$
42. $\tan x - \sec x = \dfrac{\sin x - 1}{\cos x}$
43. $\sin^2 x - \cos^2 x = 2\sin^2 x - 1$
44. $\sec^2 x + \csc^2 x = (\tan x + \cot x)^2$
45. $\sin x \tan x = \dfrac{1 - \cos^2 x}{\cos x}$
46. $\dfrac{\sin^2 x}{1 - \cos x} = \dfrac{1 + \sec x}{\sec x}$
47. $\dfrac{\tan x - 1}{\tan x + 1} = \dfrac{1 - \cot x}{1 + \cot x}$
48. $\dfrac{1 + \cot x}{\csc x} = \dfrac{1 + \tan x}{\sec x}$
49. $\dfrac{\sin x}{\cot x} - \dfrac{\cos x}{\sin x} = \dfrac{\tan x - \csc x \cot x}{\csc x}$
50. $\cos^4 x - \sin^4 x = 2\cos^2 x - 1$
51. $\dfrac{\sin x}{1 + \cos x} = \dfrac{1 - \cos x}{\sin x}$
52. $\csc^2 x - \cot x \csc x = \dfrac{1}{1 + \cos x}$
53. $\dfrac{\sec x + \csc x}{1 + \tan x} = \csc x$
54. $\dfrac{\cot x}{\csc x - 1} = \dfrac{\csc x + 1}{\cot x}$

55. $\cos x = (1 + \sin x)(\sec x - \tan x)$

56. $3 \sin x = \dfrac{2 \sin^2 x - \cos^2 x + 1}{\sin x}$

57. $\sec^4 x - \sec^2 x = \tan^4 x + \tan^2 x$

58. $\dfrac{2 \cos x + 2}{\sec x + \tan x + 1} = 1 - \sin x + \cos x$

59. $\dfrac{1}{\cos x - \sin x} = \dfrac{2 \sec x \tan x + \sec x}{\tan x - 2 \sec^2 x + 3}$

60. $\dfrac{1 - \sin x}{\cos x} = \dfrac{1}{\sec x + \tan x}$

8-2 COSINE, SINE, AND TANGENT OF THE SUM AND DIFFERENCE OF ANGLES

Cosine of the Sum and Difference of Angles

In Figure 8-3 angles α, β, and $\alpha - \beta$ are shown with the unit circle. In Figure 8-4, $\alpha - \beta$ is again drawn as an angle in standard position. By definition, $P_1(\cos \alpha, \sin \alpha)$, $P_2(\cos \beta, \sin \beta)$, and $P_3(\cos(\alpha - \beta), \sin(\alpha - \beta))$ are the intersections of the terminal sides of α, β, and $\alpha - \beta$, respectively.

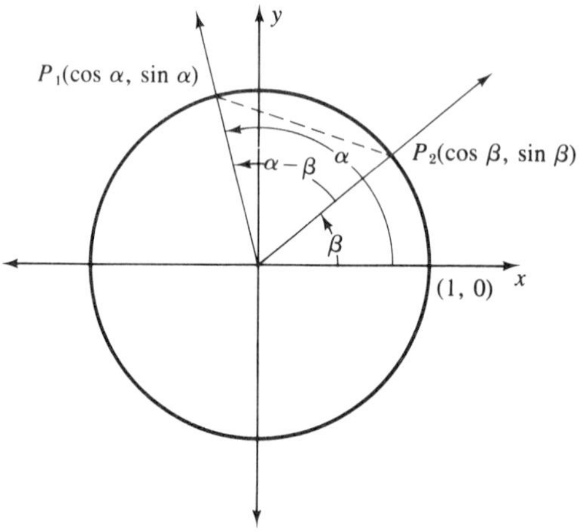

FIGURE 8-3 Angles α, β, and $\alpha - \beta$

The two angles shown as $\alpha - \beta$ subtend arcs of equal lengths on the unit circle. Since equal arcs have chords of equal lengths, then $|P_1P_2| = |P_3P_4|$. Recall that the distance between points $A(x_1, y_1)$ and $B(x_2, y_2)$ is given with the equation

$$|AB| = \sqrt{(x_2 - x_1)^2 + (y_2 - y_1)^2}$$

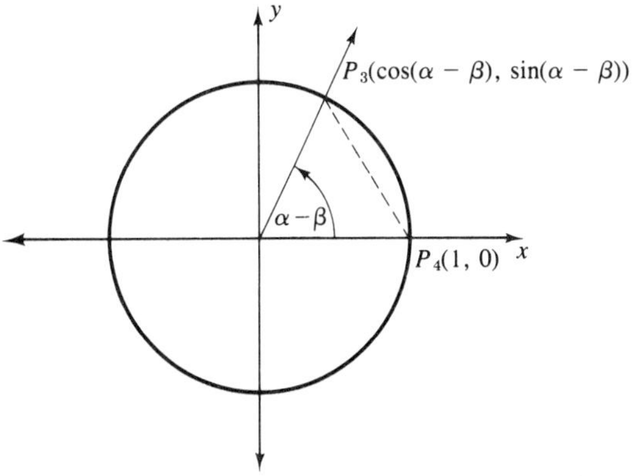

FIGURE 8-4 Angle $\alpha - \beta$ in standard position

When we substitute the coordinates of P_1, P_2, P_3, and P_4 into the above equation, we get the following pair of equations:

$$|P_1P_2| = \sqrt{(\cos\alpha - \cos\beta)^2 + (\sin\alpha - \sin\beta)^2}$$
$$|P_3P_4| = \sqrt{(\cos(\alpha - \beta) - 1)^2 + (\sin(\alpha - \beta) - 0)^2}$$

Since $|P_1P_2| = |P_3P_4|$, then $|P_1P_2|^2 = |P_3P_4|^2$, and

$$(\cos\alpha - \cos\beta)^2 + (\sin\alpha - \sin\beta)^2 = (\cos(\alpha - \beta) - 1)^2 + \sin^2(\alpha - \beta)$$

Expanding the indicated binomials, we get

$$\cos^2\alpha - 2\cos\alpha\cos\beta + \cos^2\beta + \sin^2\alpha - 2\sin\alpha\sin\beta + \sin^2\beta$$
$$= \cos^2(\alpha - \beta) - 2\cos(\alpha - \beta) + 1 + \sin^2(\alpha - \beta)$$

But

$$\cos^2\alpha + \sin^2\alpha = \cos^2\beta + \sin^2\beta = \cos^2(\alpha - \beta) + \sin^2(\alpha - \beta) = 1$$

Therefore,

$$2 - 2\cos\alpha\cos\beta - 2\sin\alpha\sin\beta = 2 - 2\cos(\alpha - \beta)$$
$$\cos\alpha\cos\beta + \sin\alpha\sin\beta = \cos(\alpha - \beta)$$

The result of the above development is identity (12).

$$\cos(\alpha - \beta) = \cos\alpha\cos\beta + \sin\alpha\sin\beta \qquad \textbf{(12)}$$

The identity relates the cosine of the difference of two angles to the cosines and sines of the given angles. For example, if $\alpha = 60°$ and $\beta = 45°$, then $\cos(60° - 45°) = \cos 15°$ can be computed with identity (12). Notice that the equation also implies that the inequality $\cos(\alpha - \beta) \neq \cos\alpha - \cos\beta$ is true. That is, *cosine is not distributive over the quantity* $\alpha - \beta$.

EXAMPLE 8–7 Find the exact value of cos 15°.

Solution: Since $15° = 45° - 30°$ and the exact function values of 45° and 30°

are known, we use the identity for $\cos(\alpha - \beta)$ with $\alpha = 45°$ and $\beta = 30°$. We get

$$\begin{aligned}\cos(45° - 30°) &= \cos 45° \cos 30° + \sin 45° \sin 30° \\ &= \left(\frac{\sqrt{2}}{2}\right)\left(\frac{\sqrt{3}}{2}\right) + \left(\frac{\sqrt{2}}{2}\right)\left(\frac{1}{2}\right) \\ &= \frac{\sqrt{6} + \sqrt{2}}{4}\end{aligned}$$

and the result is the exact value for $\cos 15°$.

In Chapter 7 we stated three equations that were needed to read Table 4. We can now verify one of these equations with identity (12).

EXAMPLE 8–8 Verify that $\cos(90° - x) = \sin x$ is an identity.

Solution: With $\alpha = 90°$ and $\beta = x$ and the identity for $\cos(\alpha - \beta)$, we have

$$\begin{aligned}\cos(90° - x) &= \cos 90° \cos x + \sin 90° \sin x \\ &= 0(\cos x) + 1(\sin x) \\ &= \sin x\end{aligned}$$

Since $\alpha + \beta = \alpha - (-\beta)$, then $\cos(\alpha + \beta) = \cos(\alpha - (-\beta))$. With identity (12),

$$\cos(\alpha - (-\beta)) = \cos \alpha \cos(-\beta) + \sin \alpha \sin(-\beta)$$

But $\cos(-\beta) = \cos \beta$ with identity (9) and $\sin(-\beta) = -\sin \beta$ with identity (10); therefore,

$$\cos(\alpha + \beta) = \cos \alpha \cos \beta - \sin \alpha \sin \beta \qquad \textbf{(13)}$$

EXAMPLE 8–9 Find the exact value of $\cos 105°$.

Solution: Since $105° = 60° + 45°$ and the exact function values of $60°$ and $45°$ are known, then with $\alpha = 60°$ and $\beta = 45°$ and the identity for $\cos(\alpha + \beta)$, we have

$$\begin{aligned}\cos 105° &= \cos 60° \cos 45° - \sin 60° \sin 45° \\ &= \left(\frac{1}{2}\right)\left(\frac{\sqrt{2}}{2}\right) - \left(\frac{\sqrt{3}}{2}\right)\left(\frac{\sqrt{2}}{2}\right) \\ &= \frac{\sqrt{2} - \sqrt{6}}{4} \qquad \text{(the exact value for } \cos 105°\text{)}\end{aligned}$$

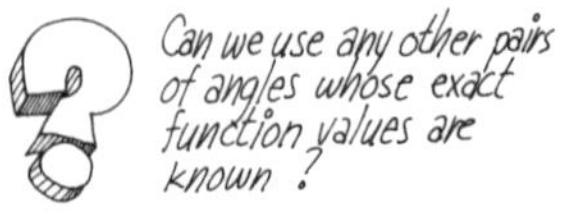

Any pair of angles whose exact function values are known can be used, so long as their sum or difference is $105°$. For example, $105° = 135° - 30°$ or $225° - 120°$, and so on.

Sine and Tangent of the Sum and Difference of Angles

The four identities below state the sine and tangent of the sum and difference of angles in terms of functions of the angles being added or subtracted.

$$\sin(\alpha - \beta) = \sin \alpha \cos \beta - \cos \alpha \sin \beta \qquad \textbf{(14)}$$

$$\sin(\alpha + \beta) = \sin \alpha \cos \beta + \cos \alpha \sin \beta \qquad \textbf{(15)}$$

$$\tan(\alpha - \beta) = \frac{\tan \alpha - \tan \beta}{1 + \tan \alpha \tan \beta} \qquad \textbf{(16)}$$

$$\tan(\alpha + \beta) = \frac{\tan \alpha + \tan \beta}{1 - \tan \alpha \tan \beta} \qquad \textbf{(17)}$$

EXAMPLE 8–10 Verify the identity $\sin(\alpha - \beta) = \sin \alpha \cos \beta - \cos \alpha \sin \beta$.

Solution: Since $\cos(90° - \theta) = \sin \theta$, then $\sin(\alpha - \beta) = \cos(90° - (\alpha - \beta))$ and $\sin(\alpha - \beta) = \cos((90° - \alpha) + \beta)$. In the identity for $\cos(\alpha + \beta)$, replace α with $(90° - \alpha)$. Then,

$$\cos((90° - \alpha) + \beta) = \cos(90° - \alpha)\cos \beta - \sin(90° - \alpha)\sin \beta$$

But $\cos(90° - \alpha) = \sin \alpha$ and $\sin(90° - \alpha) = \cos \alpha$; therefore,

$$\sin(\alpha - \beta) = \sin \alpha \cos \beta - \cos \alpha \sin \beta$$

EXAMPLE 8–11 Find the exact value of sin 345°.

Solution: Since $345° = 210° + 135°$, then with the identity for $\sin(\alpha + \beta)$ and $\alpha = 210°$ and $\beta = 135°$, we have

$$\begin{aligned}\sin 345° = \sin(210° + 135°) &= \sin 210° \cos 135° + \cos 210° \sin 135° \\ &= \left(\frac{-1}{2}\right)\left(\frac{-\sqrt{2}}{2}\right) + \left(\frac{-\sqrt{3}}{2}\right)\left(\frac{\sqrt{2}}{2}\right) \\ &= \frac{\sqrt{2} - \sqrt{6}}{4} \quad \text{(the exact value for sin 345°)}\end{aligned}$$

The identity automatically gives the correct sign to the function value. This is true regardless of the pair of angles used to compute the desired value. In Example 8–11, 345° is a Quadrant IV angle; thus, $\sin 345° < 0$. Since $\sqrt{6} > \sqrt{2}$, then $\dfrac{\sqrt{2} - \sqrt{6}}{4} < 0$ as required.

Do we need to consider putting a negative or positive with the answer to obtain the correct sign of the function?

EXAMPLE 8–12 Simplify the following expressions.

(a) $\sin(\pi + x)$ (b) $\tan(2\pi - x)$

Solutions:

(a) By identity (15),

$$\begin{aligned}\sin(\pi + x) &= \sin \pi \cos x + \cos \pi \sin x\\ &= 0(\cos x) + (-1)\sin x\\ &= -\sin x\end{aligned}$$

(b) By identity (16),

$$\begin{aligned}\tan(2\pi - x) &= \frac{\tan 2\pi - \tan x}{1 + \tan 2\pi \tan x}\\ &= \frac{0 - \tan x}{1 + 0(\tan x)}\\ &= -\tan x\end{aligned}$$

Exercise Set 8–2

Express the following as functions of x.

1. $\cos(90° - x)$
2. $\sin\left(\frac{\pi}{2} + x\right)$
3. $\tan(180° - x)$
4. $\cos(180° + x)$
5. $\sin\left(\frac{3\pi}{2} + x\right)$
6. $\cos(270° - x)$
7. $\sin\left(\frac{\pi}{6} - x\right)$
8. $\tan\left(x + \frac{\pi}{3}\right)$
9. $\tan\left(x - \frac{\pi}{4}\right)$
10. $\sin\left(x + \frac{3\pi}{4}\right)$
11. $\cos\left(x + \frac{\pi}{3}\right) + \cos\left(x - \frac{\pi}{3}\right)$
12. $\cos\left(\frac{\pi}{6} - x\right) - \cos\left(\frac{\pi}{6} + x\right)$

Find the exact values of the following functions.

13. $\cos \frac{\pi}{12}$
14. $\sin \frac{7\pi}{12}$
15. $\tan \frac{11\pi}{12}$
16. $\cos \frac{5\pi}{12}$
17. $\sin 285°$
18. $\tan 75°$
19. $\cos 195°$
20. $\cos 255°$
21. Verify identity (15).
22. Verify identity (16).
23. Verify identity (17).

Verify the following identities.

24. $\sin(x + y)\sin(x - y) = \cos^2 y - \cos^2 x$
25. $\dfrac{\cos(x + y)}{\cos(x - y)} = \dfrac{1 - \tan x \tan y}{1 + \tan x \tan y}$
26. $\tan x - \tan y = \dfrac{\sin(x - y)}{\cos x \cos y}$

27. $\sin(x + y + z) = \sin x \cos y \cos z + \cos x \sin y \cos z + \cos x \cos y \sin z - \sin x \sin y \sin z$

28. $\cot(x - y) = \dfrac{\cot x \cot y + 1}{\cot y - \cot x}$

29. $\sin(x + y) + \sin(x - y) = 2 \sin x \cos y$

MORE IDENTITIES 8–3

In this section we derive identities that fall into distinct categories, such as double-angle identities, half-angle identities, and so on. In general, these identities are a consequence of the basic identities presented in Section 8-2.

Double-Angle Identities

The following identities identify the relationships that exist between the function of twice an angle and the functions of the angle.

Double-Angle Identities:

$$\cos 2x = \cos^2 x - \sin^2 x \quad \textbf{(18a)}$$
$$= 2 \cos^2 x - 1 \quad \textbf{(18b)}$$
$$= 1 - 2 \sin^2 x \quad \textbf{(18c)}$$
$$\sin 2x = 2 \sin x \cos x \quad \textbf{(19)}$$
$$\tan 2x = \frac{2 \tan x}{1 - \tan^2 x} \quad \textbf{(20)}$$

EXAMPLE 8–13 Verify identity (18a).

Solution: Since $2x = x + x$, then $\cos 2x = \cos(x + x)$. By identity (13),

$$\cos(x + x) = \cos x \cos x - \sin x \sin x$$
$$= \cos^2 x - \sin^2 x$$

Identities (18b) and (18c) are obtained with a substitution from the identity $\cos^2 x + \sin^2 x = 1$.

EXAMPLE 8–14 If $\cos x = -5/13$ and $\pi/2 < x < \pi$, find
(a) $\sin 2x$ (b) $\tan 2x$

Solutions: If $\cos x = -5/13$, then

$$\sin^2 x = 1 - \left(\frac{-5}{13}\right)^2$$
$$= \frac{169 - 25}{169} = \frac{144}{169}$$
$$\sin x = \frac{12}{13}$$

Since x is in Quadrant II, $\sin x = 12/13$.

(a) By the identity $\sin 2x = 2 \sin x \cos x$, we have

$$\sin 2x = 2\left(\frac{12}{13}\right)\left(\frac{-5}{13}\right) = \frac{-120}{169}$$

(b) If $\sin x = 12/13$ and $\cos x = -5/13$, then $\tan x = -12/5$. Therefore, by the identity $\tan 2x = \dfrac{2 \tan x}{1 - \tan^2 x}$, we have

$$\tan 2x = \frac{2(-12/5)}{1 - (-12/5)^2} = \frac{-24/5}{-119/25} = \frac{120}{119}$$

Do these identities also give the correct sign to the value of the function?

These identities automatically give the correct sign to the function value. In Example 8–14, x is given as a Quadrant II angle. Thus, if $\pi/2 < x < \pi$, then $\pi < 2x < 2\pi$, and $2x$ can be a Quadrant III or a Quadrant IV angle. Since $\tan 2x > 0$, we may correctly assert that $2x$ is in Quadrant III.

EXAMPLE 8–15 Express $\cos 3x$ in terms of functions of $\cos x$ only.

Solution: Let $3x = 2x + x$; then $\cos 3x = \cos(2x + x)$. By the identity for $\cos(\alpha + \beta)$ with $\alpha = 2x$ and $\beta = x$, then

$$\begin{aligned}
\cos(2x + x) &= \cos 2x \cos x - \sin 2x \sin x \\
&= (2 \cos^2 x - 1)\cos x - (2 \sin x \cos x)\sin x \\
&= 2 \cos^3 x - \cos x - 2 \sin^2 x \cos x \\
&= 2 \cos^3 x - \cos x - 2 \cos x(1 - \cos^2 x) \\
\cos 3x &= 4 \cos^3 x - 3 \cos x
\end{aligned}$$

Half-Angle Identities

The following identities express the relationships that exist between the functions of half an angle and the functions of the angle.

Half-Angle Identities:

$$\cos^2\left(\frac{x}{2}\right) = \frac{1 + \cos x}{2} \qquad \textbf{(21)}$$

$$\sin^2\left(\frac{x}{2}\right) = \frac{1 - \cos x}{2} \qquad \textbf{(22)}$$

$$\tan \frac{x}{2} = \frac{1 - \cos x}{\sin x} \qquad \textbf{(23a)}$$

$$= \frac{\sin x}{1 + \cos x} \qquad \textbf{(23b)}$$

EXAMPLE 8–16 Verify identity (21).

Solution: By the identity $\cos 2\theta = 2\cos^2\theta - 1$, we have

$$\cos^2\theta = \frac{1 + \cos 2\theta}{2}$$

Now let $\theta = x/2$ and $2\theta = x$; therefore,

$$\cos^2\left(\frac{x}{2}\right) = \frac{1 + \cos x}{2}$$

EXAMPLE 8–17 Find the exact value of $\cos \dfrac{3\pi}{8}$.

Solution: Since $\dfrac{3\pi}{8} = \dfrac{1}{2}\left(\dfrac{3\pi}{4}\right)$, we can use the half-angle identity for cosine. Therefore,

$$\cos^2\left(\frac{1}{2}\right)\left(\frac{3\pi}{4}\right) = \frac{1 + \cos \dfrac{3\pi}{4}}{2}$$

$$= \frac{1 + \left(\dfrac{-\sqrt{2}}{2}\right)}{2}$$

$$= \frac{2 - \sqrt{2}}{4}$$

$$\cos \frac{3\pi}{8} = \sqrt{\frac{2 - \sqrt{2}}{4}} = \frac{\sqrt{2 - \sqrt{2}}}{2}$$

and the result is the exact value for $\cos \dfrac{3\pi}{8}$.

We should note that both a positive and a negative square root are possible for the solution of Example 8–17. Since

How did you know to use the positive, and not the negative, square root in Example 8-17 ?

$$\cos^2\left(\frac{x}{2}\right) = \frac{1 + \cos x}{2}$$

$$\cos \frac{x}{2} = \pm\sqrt{\frac{1 + \cos x}{2}}$$

We take the positive square root if $x/2$ is a Quadrant I or IV angle and use the negative square root if $x/2$ is a Quadrant II or III angle. Since $3\pi/8$ is in Quadrant I, we take the positive square root of the radical. A similar analysis will give us the correct sign for $\sin \dfrac{x}{2}$ in identity (22).

EXAMPLE 8–18 Express $\cos^4 x$ in terms of linear powers of $\cos x$.

Solution: In the double-angle identity for cosine, $\cos 2x = 2\cos^2 x - 1$. Thus,

$$\cos^2 x = \frac{1 + \cos 2x}{2}$$

Since $\cos^4 x = (\cos^2 x)^2$, then

$$\begin{aligned}\cos^4 x &= \left(\frac{1 + \cos 2x}{2}\right)^2 \\ &= \frac{1}{4}(1 + 2\cos 2x + \cos^2 2x)\end{aligned}$$

To eliminate the $\cos^2 2x$ term, we write

$$\cos^2 2x = \frac{1 + \cos 2(2x)}{2}$$

and

$$\begin{aligned}\cos^4 x &= \frac{1}{4}\left(1 + 2\cos 2x + \frac{1 + \cos 4x}{2}\right) \\ &= \frac{1}{8}(3 + 4\cos 2x + \cos 4x)\end{aligned}$$

In the integral calculus we are sometimes required to *integrate* (an operation studied in integral calculus) higher powers of trigonometric functions. In order to do this operation on even powers of the functions, and yet use only elementary methods of integration, the above conversion is necessary. Thus, Example 8–18 demonstrates a process that is very useful in calculus.

Product, Sum, and Difference Identities

The following set of identities follow directly from identities (12), (13), (14), and (15). With these identities we can express the product of sines and cosines as sums or differences of sines and cosines. This group of identities assume their greatest value in the study of calculus.

Product Identities:

$$\cos\alpha \cos\beta = \frac{1}{2}(\cos(\alpha + \beta) + \cos(\alpha - \beta)) \qquad \textbf{(24)}$$

$$\sin\alpha \sin\beta = \frac{-1}{2}(\cos(\alpha + \beta) - \cos(\alpha - \beta)) \qquad \textbf{(25)}$$

$$\sin\alpha \cos\beta = \frac{1}{2}(\sin(\alpha + \beta) + \sin(\alpha - \beta)) \qquad \textbf{(26)}$$

$$\cos\alpha \sin\beta = \frac{1}{2}(\sin(\alpha + \beta) - \sin(\alpha - \beta)) \qquad \textbf{(27)}$$

EXAMPLE 8–19 Verify identity (24).

Solution: We add the following system of equations obtained with the identities for $\cos(\alpha + \beta)$ and $\cos(\alpha - \beta)$:

$$\begin{array}{rl} \cos(\alpha + \beta) = \cos\alpha\cos\beta - \sin\alpha\sin\beta & \textbf{(13)} \\ \underline{\cos(\alpha - \beta) = \cos\alpha\cos\beta + \sin\alpha\sin\beta} & \textbf{(12)} \\ \cos(\alpha + \beta) + \cos(\alpha - \beta) = 2\cos\alpha\cos\beta & \end{array}$$

and

$$\cos\alpha\cos\beta = \frac{1}{2}(\cos(\alpha + \beta) + \cos(\alpha - \beta)) \qquad \textbf{(24)}$$

EXAMPLE 8–20 Express $\sin 5x \cos 3x$ as a sum or difference of functions.

Solution: The given product $\sin 5x \cos 3x$ has the form of identity (26) with $\alpha = 5x$ and $\beta = 3x$. Therefore,

$$\begin{aligned} \sin 5x \cos 3x &= \frac{1}{2}(\sin(5x + 3x) + \sin(5x - 3x)) \\ &= \frac{1}{2}(\sin 8x + \sin 2x) \end{aligned}$$

The final four identities developed in this text are derived by making two substitutions in identities (24), (25), (26), and (27). These identities express the sum and difference of sines and cosines as products of sines and cosines.

Sum and Difference Identities:

$$\cos A + \cos B = 2\cos\left(\frac{A + B}{2}\right)\cos\left(\frac{A - B}{2}\right) \qquad \textbf{(28)}$$

$$\cos A - \cos B = -2\sin\left(\frac{A + B}{2}\right)\sin\left(\frac{A - B}{2}\right) \qquad \textbf{(29)}$$

$$\sin A + \sin B = 2\sin\left(\frac{A + B}{2}\right)\cos\left(\frac{A - B}{2}\right) \qquad \textbf{(30)}$$

$$\sin A - \sin B = 2\cos\left(\frac{A + B}{2}\right)\sin\left(\frac{A - B}{2}\right) \qquad \textbf{(31)}$$

EXAMPLE 8–21 Verify identity (28).

Solution: In the identity $\cos\alpha\cos\beta = \frac{1}{2}(\cos(\alpha + \beta) + \cos(\alpha - \beta))$, let

$$A = \alpha + \beta \quad \text{and} \quad B = \alpha - \beta$$

Then

$$A + B = 2\alpha \qquad \text{and} \qquad A - B = 2\beta$$

$$\alpha = \frac{A + B}{2} \qquad \text{and} \qquad \beta = \frac{A - B}{2}$$

We now substitute in identity (24) for α, β, $\alpha + \beta$, and $\alpha - \beta$ the expressions obtained in terms of A and B, and

$$\cos\left(\frac{A+B}{2}\right)\cos\left(\frac{A-B}{2}\right) = \frac{1}{2}(\cos A + \cos B)$$

Multiplying both sides of the equation by 2, we get identity (28):

$$\cos A + \cos B = 2\cos\left(\frac{A+B}{2}\right)\cos\left(\frac{A-B}{2}\right)$$

EXAMPLE 8–22 Verify as an identity $\cot 4x = \dfrac{\cos 5x + \cos 3x}{\sin 5x + \sin 3x}$.

Solution: In the identities for $\cos A + \cos B$ and $\sin A + \sin B$, let $A = 5x$ and $B = 3x$. Then

$$\frac{A+B}{2} = \frac{5x+3x}{2} = 4x \quad \text{and} \quad \frac{A-B}{2} = \frac{5x-3x}{2} = x$$

Then

$$\begin{aligned} \frac{\cos 5x + \cos 3x}{\sin 5x + \sin 3x} &= \frac{2\cos 4x \cos x}{2\sin 4x \cos x} \\ &= \frac{\cos 4x}{\sin 4x} \\ &= \cot 4x \end{aligned}$$

Since $\cot 4x = \cot 4x$ is an obvious identity, the given equation is verified as an identity.

Exercise Set 8–3

1. Verify identity for $\sin 2x$.
2. Verify identity for $\tan 2x$.
3. Verify identity for $\sin^2\left(\dfrac{x}{2}\right)$.
4. Verify both identities for $\tan \dfrac{x}{2}$.
5. Verify identity for $\sin\alpha \sin\beta$.
6. Verify identity for $\sin\alpha \cos\beta$.
7. Verify identity for $\cos\alpha \sin\beta$.
8. Verify identity for $\cos A - \cos B$.
9. Verify identity for $\sin A + \sin B$.
10. Verify identity for $\sin A - \sin B$.

Find the exact values of the following.

11. $1 - 2\sin^2 165°$
12. $\dfrac{2\tan 5\pi/12}{1 - \tan^2 5\pi/12}$
13. $2\sin 22.5° \cos 22.5°$
14. $2\cos^2 67.5° - 1$
15. $\sin 112.5°$
16. $\tan 15°$
17. $\cos \dfrac{5\pi}{12}$
18. $\sin \dfrac{5\pi}{8}$
19. $\cos^2\left(\dfrac{7\pi}{12}\right) - \sin^2\left(\dfrac{7\pi}{12}\right)$
20. $\tan \dfrac{3\pi}{8}$

If $\cos x = -3/5$, $\pi < x < 3\pi/2$, find the following.

21. $\cos 2x$
22. $\tan 2x$
23. $\sin^2\left(\frac{x}{2}\right)$
24. $\cos^2\left(\frac{x}{2}\right)$
25. $\tan \frac{x}{2}$
26. $\csc 2x$

If $\tan x = -24/7$, $3\pi/2 < x < 2\pi$, find the following.

27. $\cos \frac{x}{2}$
28. $\sin \frac{x}{2}$
29. $\sin 2x$
30. $\cot 2x$

Verify the following identities.

31. $\frac{\sin 2x}{1 + \cos 2x} = \tan x$
32. $\frac{1}{\cos^2 x - \sin^2 x} = \sec 2x$
33. $\cot \frac{\theta}{2} = \frac{\sin \theta}{1 - \cos \theta}$
34. $\cos 4x = 1 - 8 \sin^2 x \cos^2 x$
35. $\frac{2}{1 - \cos \theta} = \csc^2\left(\frac{\theta}{2}\right)$
36. $\sin^4 x = \frac{3}{8} - \frac{1}{2} \cos 2x + \frac{1}{8} \cos 4x$
37. $\sin 3x = 3 \sin x - 4 \sin^3 x$
38. $1 - \cos 2\theta = 8 \sin^2\left(\frac{\theta}{2}\right) \cos^2\left(\frac{\theta}{2}\right)$

Express as a sum or difference.

39. $\sin 42° \cos 32°$
40. $\sin 139° \sin 36°$
41. $\cos 16° \cos 56°$
42. $\cos 67° \sin 69°$
43. $\sin 3x \sin 2x$
44. $\cos 5x \cos 3x$

Express as a product.

45. $\sin 13° + \sin 9°$
46. $\cos 32° + \cos 18°$
47. $\cos 119° - \cos 35°$
48. $\sin 214° - \sin 274°$
49. $\cos 4x + \cos 2x$
50. $\sin \frac{5\theta}{2} - \sin \frac{3\theta}{2}$

Find the exact values of the following.

51. $\cos 75° + \cos 15°$
52. $\sin 45° \sin 15°$
53. $\cos 45° \sin 105°$
54. $\sin 165° - \sin 105°$

CONDITIONAL TRIGONOMETRIC EQUATIONS 8–4

We have noted that much emphasis in the study of mathematics is given to obtaining solution sets of open sentences. We therefore turn our attention to trigonometric equations that are conditionally true. In this section we will consider only basic equations of one trigonometric function.

EXAMPLE 8–23 Solve $2 \cos x - 1 = 0$ for x, and $0 \leq x < 2\pi$.

Solution: In Figure 8-5 the graph of the function $y = 2 \cos x - 1$ is shown for $x \in [0, 2\pi]$. The roots of the equation $2 \cos x - 1 = 0$ are the x-coordinates of the points of intersection of $y = 2 \cos x - 1$ and the x-axis. From the graph it is apparent that two such points exist. If $2 \cos x - 1 = 0$, then $\cos x = 1/2$ and x is a number whose cosine is 1/2. From the table of exact function values,

$$\cos \frac{\pi}{3} = \cos \frac{5\pi}{3} = \frac{1}{2}$$

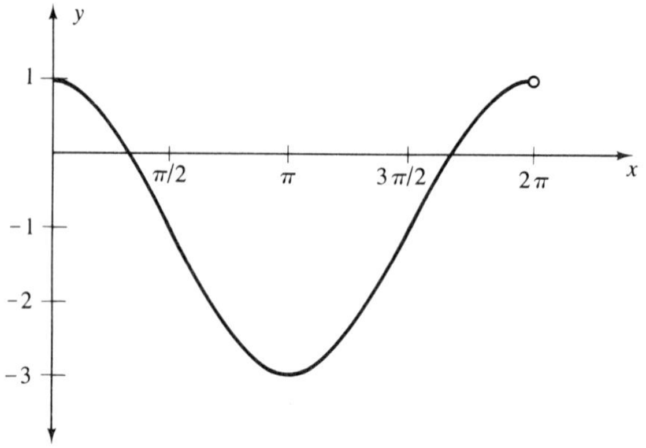

FIGURE 8-5 Graph of $y = 2 \cos x - 1$, $0 \leq x < 2\pi$

Check. If $x = \pi/3$ or $5\pi/3$,

$$2\left(\frac{1}{2}\right) - 1 = 0 \qquad \text{(True)}$$

Thus, the solution set is $\{\pi/3, 5\pi/3\}$.

Notice that to graph the function $y = a \cos bx + d$, we vertically shift by d units the graph of $y = a \cos bx$. If $d > 0$, then the shift is up. But if $d < 0$, then the shift is down. In Example 8–23, $d = -1$; therefore, the graph of $y = 2 \cos x$ is shifted down one unit as shown.

EXAMPLE 8–24 Solve $\tan x - \sqrt{3} = 0$ for $x \in \{\text{reals}\}$.

Solution: If $\tan x - \sqrt{3} = 0$, then $\tan x = \sqrt{3}$, and x is a number whose tangent is $\sqrt{3}$. From the table of exact function values, $\tan \dfrac{\pi}{3} = \sqrt{3}$. Since the period of the function $y = \tan x - \sqrt{3}$ is π, then any integral multiple of π added to $\pi/3$ is also a root of the equation. Therefore, the solution set is $\{\pi/3 + n\pi\}$, and n is an integer.

EXAMPLE 8–25 Solve $\sin 2x - 1 = 0$, and $0 \le x < 2\pi$.

Solution: If $0 \le x < 2\pi$, then $0 \le 2x < 4\pi$. If

$$\sin 2x - 1 = 0$$

then

$$\sin 2x = 1$$

and $2x$ is an angle whose sine is 1, and $0 \le 2x < 4\pi$. Since $\sin \dfrac{\pi}{2} =$ $\sin \dfrac{5\pi}{2} = 1$, and both $\pi/2$ and $5\pi/2$ are on the desired interval, then

$$2x = \pi/2 \quad \text{or} \quad 2x = 5\pi/2$$
$$x = \pi/4 \quad \text{or} \quad x = 5\pi/4$$

Check. If $x = \pi/4$ or $5\pi/4$, then

$$\sin 2\left(\frac{\pi}{4}\right) - 1 = \sin 2\left(\frac{5\pi}{4}\right) - 1 = 0 \qquad \text{(True)}$$

The solution set is $\{\pi/4, 5\pi/4\}$.

We expanded the interval initially because the double angle $2x$ is used in the equation. However, the final pair of equations in the example converted the roots for $2x$ on the interval 0 to 4π to roots for x on the interval 0 to 2π.

The roots of the equation are on the interval 0 to 2π, even though we used the interval 0 to 4π. How did that happen?

Also notice in Example 8–25 that the period of the function $y = \sin 2x - 1$ is $2\pi/2$ or π. Furthermore, the two roots $\pi/4$ and $5\pi/4$ differ by the period π. In general, *if k is the root of a trigonometric equation for which the period of the corresponding function is p, then k + np is also a root of the given equation, with* $n \in \{\text{integers}\}$.

Thus, if in Example 8–25 we let x be any real number, then . . . , $-7\pi/4$, $-3\pi/4, \pi/4, 5\pi/4, 9\pi/4, 13\pi/4, \ldots, \pi/4 + n\pi, \ldots$ (n an integer) are all roots of the equation $\sin 2x - 1 = 0$. But for $x \in [0, 2\pi)$, we have only the two roots $\pi/4$ and $5\pi/4$.

The following examples illustrate the procedure for solving quadratic equations involving trigonometric functions.

EXAMPLE 8–26 Solve $2\cos^2 x + \cos x - 1 = 0$ for $0 \le x < 2\pi$.

Solution: As an aid for factoring the given equation, we let $a = \cos x$. Then we have

$$\begin{aligned} 2\cos^2 x + \cos x - 1 &= 2a^2 + a - 1 \\ &= (2a - 1)(a + 1) \\ &= (2\cos x - 1)(\cos x + 1) \end{aligned}$$

With the zero-product factor theorem,

$$2\cos x - 1 = 0 \quad \text{or} \quad \cos x + 1 = 0$$
$$\cos x = 1/2 \quad \text{or} \quad \cos x = -1$$

Thus, x is an angle whose cosine is 1/2 or -1; and for $0 \le x < 2\pi$, $x = \pi/3$, π, or $5\pi/3$.

Check. If $x = \pi/3$ or $5\pi/3$,

$$2\cos^2\left(\frac{\pi}{3}\right) + \cos\frac{\pi}{3} - 1 = 0$$

$$2\left(\frac{1}{2}\right)^2 + \frac{1}{2} - 1 = 0 \qquad \text{(True)}$$

If $x = \pi$, then

$$2\cos^2\pi + \cos\pi - 1 = 0$$

$$2(-1)^2 + (-1) - 1 = 0 \qquad \text{(True)}$$

Therefore, the solution set is $\{\pi/3, \pi, 5\pi/3\}$.

EXAMPLE 8–27 Solve for θ to the nearest hundredth of a degree.

$$\tan^2 2\theta - 4\tan 2\theta + 3 = 0, \quad \text{and } 0° \le \theta < 360°$$

Solution: If $0° \le \theta < 360°$, then $0° \le 2\theta < 720°$. Now let $\tan 2\theta = a$. Then we have

$$\begin{aligned}\tan^2 2\theta - 4\tan 2\theta + 3 &= a^2 - 4a + 3\\ &= (a - 1)(a - 3)\\ &= (\tan 2\theta - 1)(\tan 2\theta - 3)\end{aligned}$$

and $(\tan 2\theta - 1)(\tan 2\theta - 3) = 0$. Thus,

$$\tan 2\theta = 1 \qquad \textbf{(1)}$$

or

$$\tan 2\theta = 3 \qquad \textbf{(2)}$$

In Equation (1), $\tan 2\theta = 1$ and $0° \le 2\theta < 720°$, and

$$2\theta = 45°, 225°, 405°, 585°$$

$$\theta = 22.50°, 112.50°, 202.50°, 292.50°$$

In Equation (2), $\tan 2\theta = 3$ and $0° \le 2\theta < 720°$. In Table 4 the closest tangent value to 3.000 is 2.989, and the reference angle of 2θ is $71°30'$ or $71.50°$. Therefore,

$$2\theta = 71.50°, 251.50°, 431.50°, 611.50°$$

$$\theta = 35.75°, 125.75°, 215.75°, 305.75°$$

We omit the check and the solution set is {22.50°, 35.75°, 112.50°, 125.75°, 202.50°, 215.75°, 292.50°, 305.75°}.

Can we use a hand calculator to find the value of 2θ rather than using Table 4?

We can use a hand calculator instead of Table 4 to find the value of 2θ; however, the theory involved in such a process is not covered until Chapter 9. For those of us who have scientific calculators, there is a key marked TAN^{-1}. When we enter 3.000 and depress this key, the display reads 71.57° for 2θ, and then $\theta = 35.78°$. The value of 35.75° approximated with Table 4 compares favorably to the calculator value. If we use the calculator and the COS^{-1}, SIN^{-1}, and TAN^{-1} keys to approximate values of x or θ, the values will differ slightly from the approximated values we get from Table 4.

EXAMPLE 8–28 The path of a particle moving in a plane is a function of time. The path is described with the equation

$$y = \cos^2 t - 2\cos t - 1$$

where t is time measured in seconds and y is the vertical displacement from the t-axis. For what values of t to the nearest hundredth of a second does the particle cross the t-axis and $t \in [0, 7]$?

Solution: When the particle is on the t-axis, then $y = 0$, and

$$\cos^2 t - 2\cos t - 1 = 0$$

To solve for t, we use the quadratic formula with $a = 1$, $b = -2$, and $c = -1$:

$$\cos t = \frac{-(-2) \pm \sqrt{(-2)^2 - 4(1)(-1)}}{2(1)}$$

$$= 1 \pm \sqrt{2}$$

With Table 1 or a hand calculator, $\sqrt{2} \approx 1.4142$, and

$$\cos t = 2.4142 \qquad \textbf{(1)}$$

or

$$\cos t = -0.4142 \qquad \textbf{(2)}$$

Since $\cos t \neq 2.4142$, the solution set for Equation (1) is $\varnothing$. If $\cos t = -0.4142$ in Equation (2), then from Table 4 the reference value for t is approximately 1.1432. Since $\cos t < 0$, then t is in Quadrants II and III. Therefore,

$$t = 3.1416 + 1.1432 \quad \text{or} \quad t = 3.1416 - 1.1432$$

$$t = 4.2848 \quad \text{or} \quad t = 1.9984$$

Thus, at time $t = 4.28$ seconds and $t = 2.00$ seconds, the particle is on the t-axis.

Exercise Set 8–4

Solve the following equations for x, $0 \leq x < 2\pi$.

1. $\cos x = 0$
2. $\sin x = -1$
3. $\tan x - 1 = 0$
4. $2\sin x - \sqrt{2} = 0$
5. $2\tan x + 2\sqrt{3} = 0$
6. $\dfrac{\sqrt{3}\cos x}{7} = \dfrac{3}{14}$

Solve the following equations for $x \in \{\text{reals}\}$.

7. $\sin x = 0$
8. $\sec x + 1 = 0$
9. $\tan x = -1$
10. $2\cos x + 1 = 0$
11. $2 - \csc x = 1$
12. $\sqrt{3}\cot x + 7 = 8$

Solve for x, where $x \in [0, 2\pi)$.

13. $2 \cos \frac{x}{2} = 1$

14. $\cos 3x = 1$

15. $\sin 3x = \frac{\sqrt{3}}{2}$

16. $2 \sin \frac{x}{2} = \sqrt{2}$

17. $\tan 2x = 1$

18. $\sqrt{3} \sec 2x + 2 = 0$

19. $7 - \csc 4x = 8$

20. $\sqrt{3} \cot \frac{x}{2} = -1$

Solve for x, where $x \in \{\text{reals}\}$.

21. $\sin \frac{x}{2} = -1$

22. $\cot 2x = -1$

23. $2 \cos \frac{3}{2}x + \sqrt{3} = 0$

24. $4 \tan 2x = 3 \tan 2x$

Solve the following equations for $0 \leq x < 2\pi$. When necessary, use Table 4 or a scientific calculator to approximate a value of x to the nearest hundredth of a radian.

25. $\sin^2 x - 1 = 0$

26. $2 \cos^2 x - \cos x = 0$

27. $\cot^2 x - \cot x = 0$

28. $\tan^2 x - 3 = 0$

29. $2 \sin^2 x - \sin x - 1 = 0$

30. $5 \cos^2 x + 2 \cos x = 3$

31. $\tan^2 x - 3 = -2 \tan x$

32. $2 \sin^2 3x = 1$

Solve the following equations for $x \in [0°, 360°)$.

33. $2 \sin^3 x - \sin x = 0$

34. $4 \sin^2 2x = 3$

35. $\cos^2 x - 2 \cos x = 0$

36. $\cos^2 x - \cos x - 6 = 0$

37. $3 \tan^3 x - \tan x = 0$

38. $2 \cos^2 3x + \sqrt{2} \cos 3x = 0$

39. $\tan x - \tan^2 x = 0$

40. $\cos^2 2x - \cos 2x = 0$

41. The path of a particle moving in a t, y plane is described by the equation $y = \cos^2 t - 3 \cos t + 2$, where t is time in seconds. For what value(s) of t to the nearest hundredth of a second is the particle on the t-axis, and $t \in [0, 7]$?

42. The path of another particle moving in the t, y plane is described by the equation $y = 2 \sin t \cos t - 2 \cos t + \sin t - 1$, where t is time in seconds. For what value(s) of t to the nearest hundredth of a second is the particle on the t-axis, and $t \in [0, 7]$?

43. For the particle in Exercise 41, for what value(s) of t to the nearest hundredth of a second is the vertical displacement of the particle two units above the t-axis?

44. For the particle in Exercise 42, for what value(s) of t to the nearest hundredth of a second is the vertical displacement of the particle equal to $\sin t - 1$?

MULTIPLE FUNCTION EQUATIONS 8-5

In Section 8-4 we studied equations with only one trigonometric function. There are times, however, when we must solve trigonometric equations with two or more different functions. We call such an equation a multiple function equation. The following are examples of multiple function equations:

$$\sin x + \cos x = 1$$
$$\cos 3x + 2 \cos x = 0$$

These two equations, in their present form, cannot be solved by the methods developed in Section 8-4.

If a multiple function equation can be expressed as the product of factors equal to zero, and each factor contains but one trigonometric function, then the zero-product factor property can be used to solve the equation. Example 8–29 illustrates this method of solving a multiple function equation.

EXAMPLE 8–29 Solve $\sin x \tan^2 x - \sin x = 0$, and $0 \le x < 2\pi$.

Solution: Let $\sin x = a$ and $\tan x = b$. Then we have

$$\begin{aligned} \sin x \tan^2 x - \sin x &= ab^2 - a \\ &= a(b^2 - 1) \\ &= a(b + 1)(b - 1) \\ &= \sin x(\tan x + 1)(\tan x - 1) \end{aligned}$$

Therefore, $\sin x(\tan x + 1)(\tan x - 1) = 0$. Thus, $\sin x = 0$ or $\tan x = -1$ or $\tan x = 1$. If $0 \le x < 2\pi$ and $\sin x = 0$, then $x = 0$ or π. If $0 \le x < 2\pi$ and $\tan x = -1$, then $x = 3\pi/4$ or $7\pi/4$. If $0 \le x < 2\pi$ and $\tan x = 1$, then $x = \pi/4$ or $5\pi/4$.

Check. If $x = 0$,

$$\sin 0 \tan^2 0 - \sin 0 = 0 \quad \text{and} \quad 0 - 0 = 0 \qquad \text{(True)}$$

If $x = \pi/4$,

$$\sin \frac{\pi}{4} \tan^2\left(\frac{\pi}{4}\right) - \sin \frac{\pi}{4} = 0 \quad \text{and} \quad \frac{\sqrt{2}}{2}(1) - \frac{\sqrt{2}}{2} = 0 \qquad \text{(True)}$$

The remaining roots are similarly checked, and the solution set is $\{0, \pi/4, 3\pi/4, \pi, 5\pi/4, 7\pi/4\}$.

If the equation cannot be readily factored, then we use one or more of the identities to try to convert the given equation to an equivalent equation in one function that can be solved. Examples 8–30 through 8–36 demonstrate this process for solving a multiple function equation.

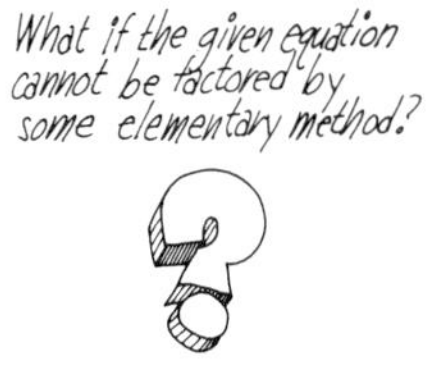

EXAMPLE 8–30 Solve $1 + \sin x - 2\cos^2 x = 0$, and $x \in [0, 2\pi)$.

Solution: The given equation cannot be factored in its present form. However, the identity $\cos^2 x + \sin^2 x = 1$ can be used to express $\cos^2 x$ in terms of $\sin^2 x$. That is,

$$\begin{aligned} 1 + \sin x - 2\cos^2 x &= 1 + \sin x - 2(1 - \sin^2 x) \\ &= 1 + \sin x - 2 + 2\sin^2 x \end{aligned}$$

and

$$2\sin^2 x + \sin x - 1 = 0$$
$$(2\sin x - 1)(\sin x + 1) = 0$$

Thus,

$$\sin x = 1/2 \quad \text{or} \quad \sin x = -1$$

If $0 \le x < 2\pi$ and $\sin x = 1/2$, then $x = \pi/6$ or $5\pi/6$. If $0 \le x < 2\pi$ and $\sin x = -1$, then $x = 3\pi/2$.

Check. If $x = \pi/6$, then

$$1 + \sin\frac{\pi}{6} - 2\cos^2\left(\frac{\pi}{6}\right) = 0$$
$$1 + \frac{1}{2} - 2\left(\frac{\sqrt{3}}{2}\right)^2 = 0$$
$$\frac{3}{2} - 2\left(\frac{3}{4}\right) = 0 \qquad \text{(True)}$$

The remaining roots are similarly checked, and the solution set is $\{\pi/6, 5\pi/6, 3\pi/2\}$.

EXAMPLE 8–31 Solve for x to four decimal places:

$$2\sec^2 x + \tan x - 3 = 0, \quad \text{and } x \in [0, 2\pi)$$

Solution: The identity $1 + \tan^2 x = \sec^2 x$ can be used to express $2\sec^2 x$ in terms of $\tan^2 x$. That is,

$$\begin{aligned} 2\sec^2 x + \tan x - 3 &= 2(1 + \tan^2 x) + \tan x - 3 \\ &= 2\tan^2 x + \tan x - 1 \\ &= (2\tan x - 1)(\tan x + 1) \end{aligned}$$

Thus, $(2\tan x - 1)(\tan x + 1) = 0$, and

$$\tan x = 0.5000 \qquad \textbf{(1)}$$

or

$$\tan x = -1 \qquad \textbf{(2)}$$

In Equation (1), x is an angle whose tangent is 0.5000. In Table 4, $\tan 0.4625 = 0.4986$; thus, the reference angle for x is approximately 0.4625. Since $\tan x > 0$, x is in Quadrants I and III, and

$$x = 0.4625 \quad \text{or} \quad x = 3.1416 + 0.4625 = 3.6041$$

In Equation (2), $\tan x = -1$, and

$$x = 3\pi/4 \quad \text{or} \quad x = 7\pi/4$$

Check. If $x = 0.4625$, then

$$2\sec^2 0.4625 + \tan 0.4625 - 3 = 0$$
$$2(1.117)^2 + 0.4986 - 3 = 0$$
$$-0.0060 = 0 \qquad \text{(True)}$$

The remaining roots are similarly checked, and the solution set is $\{0.4625, 3.6041, 3\pi/4, 7\pi/4\}$.

The function values we get from Table 4 or a hand calculator are *approximate values* correct to a given number of decimal places. Therefore, when we use such an approximation in checking the equation, we can obtain a value close to, but not exactly, the desired value. Thus, the value -0.0060 is probably the result of an approximation of $\sec x$ or $\tan x$, or both, rather than an error in the solving of the equation.

Why did you say true to the equation -0.0060=0?

Another type of multiple function equation requires introducing a radical term. We solve such an equation as we solve algebraic equations that contain radical terms.

Example 8–32 Solve for θ to the nearest tenth of a degree:

$$2\cos\theta - \sin\theta = 1, \quad \text{and } 0° \le \theta < 360°$$

Solution: Since $\cos\theta = \pm\sqrt{1 - \sin^2\theta}$, we substitute the radical expression for $\cos\theta$ in the given equation:

$$2\left(\pm\sqrt{1 - \sin^2\theta}\right) - \sin\theta = 1$$
$$\pm 2\sqrt{1 - \sin^2\theta} = 1 + \sin\theta$$

Now square both sides. Then we have

$$4(1 - \sin^2\theta) = 1 + 2\sin\theta + \sin^2\theta$$

or

$$5\sin^2\theta + 2\sin\theta - 3 = 0$$
$$(5\sin\theta - 3)(\sin\theta + 1) = 0$$

Therefore,

$$\sin\theta = 0.6000 \qquad \textbf{(1)}$$

or

$$\sin\theta = -1 \qquad \textbf{(2)}$$

In Equation (1), if $\sin\theta = 0.6000$, then from Table 4 the reference angle for θ is $36°50'$ or $36.8°$ to the nearest tenth of a degree. With θ in Quadrants I and II,

$$\theta = 36.8° \quad \text{or} \quad 180° - 36.8° = 143.2°$$

In Equation (2), if $\sin\theta = -1$, then $\theta = 270°$.

Check. If $\theta = 36.8°$, then

$$2\cos 36.8° - \sin 36.8° = 1$$
$$2(0.8004) - 0.5995 = 1$$
$$1.6008 - 0.5995 = 1 \qquad \text{(True)}$$

If $\theta = 143.2°$, then

$$2 \cos 143.2° - \sin 143.2° = 1$$
$$-1.6008 - 0.5995 = 1 \qquad \text{(False)}$$

If $\theta = 270°$, then

$$2 \cos 270° - \sin 270° = 1$$
$$2(0) - (-1) = 1 \qquad \text{(True)}$$

The solution set is $\{36.8°, 270°\}$. Notice that we obtained an extraneous root since we squared both sides of the equation in solving the equation.

The following example illustrates the process of solving a multiple function equation that involves fractions.

EXAMPLE 8–33 Solve $\sec x - 2 \cos x = \tan x$, and $x \in [0, 2\pi)$.

Solution: From the fundamental identities,

$$\sec x = \frac{1}{\cos x} \quad \text{and} \quad \tan x = \frac{\sin x}{\cos x}$$

Thus, $\sec x - 2 \cos x = \tan x$ is equivalent to

$$\frac{1}{\cos x} - 2 \cos x = \frac{\sin x}{\cos x}, \quad \text{and } \cos x \neq 0$$

If $\cos x \neq 0$, we can multiply both sides of the equation by $\cos x$, and

$$1 - 2 \cos^2 x = \sin x$$

But $\cos^2 x = 1 - \sin^2 x$; therefore,

$$1 - 2(1 - \sin^2 x) = \sin x$$
$$2 \sin^2 x - \sin x - 1 = 0$$
$$(2 \sin x + 1)(\sin x - 1) = 0$$

Thus,

$$\sin x = -1/2 \quad \text{or} \quad \sin x = 1$$

If $\sin x = -1/2$, then $x = 7\pi/6$ or $11\pi/6$. If $\sin x = 1$, then $x = \pi/2$. But $\sec \frac{\pi}{2}$ and $\tan \frac{\pi}{2}$ are undefined, and we reject $\pi/2$ as a root. Thus, the solution set is $\{7\pi/6, 11\pi/6\}$.

Exercise Set 8–5

Solve the following equations for x, $x \in [0°, 360°)$. Use Table 4 where necessary and approximate x to the nearest tenth of a degree.

1. $\sin x \cos x - \cos x = 0$
2. $2 \tan x \sin x - \tan x = 0$
3. $\sin x \sec^2 x - 2 \sin x = 0$
4. $4 \sin x \cos x = 0$

5. $2\cos^2 x + \sin x + 1 = 0$
6. $\tan^2 x + 3\sec x + 3 = 0$
7. $\csc^2 x - \cot x - 1 = 0$
8. $3 + 3\cos x - 2\sin^2 x = 0$
9. $\cot^2 x + \csc^2 x = 7$
10. $\cos^2 x - 4\sin^2 x = 0$

Solve the following equations for x, $x \in [0, 2\pi)$. Use Table 4 where necessary and approximate x to the nearest hundredth of a radian.

11. $\tan x = \sin x \cos x$
12. $\tan x = \cot x$
13. $2\cot x = 1 - \csc x$
14. $2\cos x \cot x - 3 = 0$
15. $\sin x + 4\csc x + 5 = 0$
16. $2\sin x + \cot x = \csc x$
17. $\sin x - \cos x = 1$
18. $\sec x + \tan x = 1$
19. $4\sin x \cos x - 2\cos x - 2\sin x + 1 = 0$
20. $\sin x \tan x - \cos x \cot x = 0$

Solve the following equations for x, $x \in \{\text{reals}\}$. When necessary, use Table 4 to approximate the roots to the nearest hundredth of a radian.

21. $4\sin^2 x - 3\tan^2 x + 1 = 0$
22. $\sin x - \sqrt{3}\cos x - 1 = 0$
23. $2\cos^4 x - \sin^2 x = 0$
24. $\sin x = 4\cos^2 x - 1$
25. $4\sin^2 x + 1 = 5\sin x$
26. $5\tan^2 x + 2\tan x = 3$
27. $10\cos^2 x = 3\cos x + 4$
28. $\sec x + \tan x = \sqrt{3}$
29. $\cot x - \csc x = 2$
30. $\sin 2x + \cos 2x = 1$

MULTIPLE ANGLE EQUATIONS 8-6

In Section 8-5 we solved multiple function equations that contained terms with different trigonometric functions. The equation $\cos 3x + \cos x = 1$ is also multiple function because $\cos 3x$ and $\cos x$ are functions of different angles. We must first use an identity to express $\cos 3x$ in terms of $\cos x$ before the equation can be solved by methods used in previous sections. The following examples illustrate the procedure we use to solve equations of this type.

EXAMPLE 8–34 Solve $\cos 2\theta + 3\cos\theta + 2 = 0$, and $0° \le \theta < 360°$.

Solution: We use the double-angle identity $\cos 2\theta = 2\cos^2\theta - 1$ to write the given equation in terms of only $\cos\theta$:

$$(2\cos^2\theta - 1) + 3\cos\theta + 2 = 0$$
$$2\cos^2\theta + 3\cos\theta + 1 = 0$$
$$(2\cos\theta + 1)(\cos\theta + 1) = 0$$

Therefore,

$$\cos\theta = -1/2 \quad \text{or} \quad \cos\theta = -1$$

If $0° \le \theta < 360°$ and $\cos\theta = -1/2$, then $\theta = 120°$ or $240°$. If $0° \le \theta < 360°$ and $\cos\theta = -1$, then $\theta = 180°$.

Check. If $\theta = 120°$, then

$$\cos 2(120°) + 3\cos 120° + 2 = 0$$

$$\frac{-1}{2} + 3\left(\frac{-1}{2}\right) + 2 = 0 \qquad \text{(True)}$$

The remaining roots are similarly checked, and the solution set is $\{120°, 180°, 240°\}$.

EXAMPLE 8–35 Solve $\sin 3x + \sin x = 0$, and $x \in \{\text{reals}\}$.

Solution: Since the coefficients of both $\sin 3x$ and $\sin x$ are 1, we can solve the given equation by using the identity

$$\sin A + \sin B = 2\sin\left(\frac{A+B}{2}\right)\cos\left(\frac{A-B}{2}\right)$$

with $A = 3x$ and $B = x$. If $\sin 3x + \sin x = 0$, then

$$2\sin\left(\frac{3x+x}{2}\right)\cos\left(\frac{3x-x}{2}\right) = 0$$

$$2\sin 2x\cos x = 0$$

Thus,

$$\sin 2x = 0 \quad \text{or} \quad \cos x = 0$$

If $\sin 2x = 0$, then

$$2x = 0 + 2n\pi \quad \text{or} \quad 2x = \pi + 2n\pi$$

$$x = n\pi \quad \text{or} \quad x = \pi/2 + n\pi$$

If $\cos x = 0$, then $x = \pi/2 + n\pi$. But these roots are part of the solution obtained for $\sin 2x = 0$. Thus, the solution set is $\{n\pi, \pi/2 + n\pi\}$ and $n \in \{\text{integers}\}$.

EXAMPLE 8–36 Solve for x:

$$\tan 2x - \frac{2}{1 - \tan^2 x} + 2 = 0, \quad \text{and } x \in \{\text{reals}\}$$

Solution: In the given equation, $1 - \tan^2 x \ne 0$; thus, $\tan x \ne \pm 1$. With the identity

$$\tan 2x = \frac{2\tan x}{1 - \tan^2 x}$$

we substitute for $\tan 2x$ in the given equation:

$$\frac{2\tan x}{1 - \tan^2 x} - \frac{2}{1 - \tan^2 x} + 2 = 0$$

If $\tan x \neq \pm 1$, we can multiply both sides of the equation by $1 - \tan^2 x$:

$$2 \tan x - 2 + 2(1 - \tan^2 x) = 0$$
$$2 \tan x - 2 + 2 - 2 \tan^2 x = 0$$
$$\tan x(1 - \tan x) = 0$$

If $\tan x = 0$, then $x = 0 + n\pi$, or simply $n\pi$. If $1 - \tan x = 0$, then $\tan x = 1$. But the original restriction on the given equation states that $\tan x \neq \pm 1$. Thus, we reject the roots of $\tan x = 1$.

Check. If $x = 0$, then

$$\tan 2(0) - \frac{2}{1 - (\tan 0)^2} + 2 = 0$$
$$0 - \frac{2}{1} + 2 = 0 \qquad \text{(True)}$$

Thus, the solution set is $\{n\pi\}$.

Exercise Set 8–6

Solve for x, $0 \leq x < 2\pi$.

1. $\cos 2x = 0$
2. $\sin 2x = -\sin x$
3. $\tan 2x + \tan x = 0$
4. $\cos 2x - \sin x = 0$
5. $\tan \frac{x}{2} = \sin x$
6. $\sin 2x + 2 \sin^2\left(\frac{x}{2}\right) = 1$
7. $\sin 4x + \sin 2x = 0$
8. $\tan 2x = \tan 4x$
9. $\sin 3x \cos x + \cos 3x \sin x = 0$
10. $\cos 3x = \cos x$
11. $\cos 4x - \cos 2x = -\sin x$
12. $\sin 3x = \cos 3x$
13. $\sin x = 2 \cos \frac{x}{2}$
14. $\tan \frac{x}{2} - \csc x + 2 \sin 2x = 0$

SUMMARY

Identity

Definition: An identity is an equation that is true for any value of the variable for which each term of the equation is defined.

Conditional equation

Definition: A conditional equation is one that is true for some but not all values of the variable.

Reciprocal Identities:

$$\sec \theta = \frac{1}{\cos \theta}, \quad \cos \theta \neq 0$$
$$\csc \theta = \frac{1}{\sin \theta}, \quad \sin \theta \neq 0$$
$$\cot \theta = \frac{1}{\tan \theta}, \quad \tan \theta \neq 0$$

Tangent and Cotangent Identities:

$$\tan\theta = \frac{\sin\theta}{\cos\theta}, \quad \cos\theta \neq 0$$

$$\cot\theta = \frac{\cos\theta}{\sin\theta}, \quad \sin\theta \neq 0$$

Pythagorean Identities:

$$\cos^2\theta + \sin^2\theta = 1$$
$$1 + \tan^2\theta = \sec^2\theta$$
$$1 + \cot^2\theta = \csc^2\theta$$

Negative Angle Identities:

$$\cos(-\theta) = \cos\theta$$
$$\sin(-\theta) = -\sin\theta$$
$$\tan(-\theta) = -\tan\theta$$

cos $(\alpha - \beta)$ Identity: $\cos(\alpha - \beta) = \cos\alpha\cos\beta + \sin\alpha\sin\beta$

cos $(\alpha + \beta)$ Identity: $\cos(\alpha + \beta) = \cos\alpha\cos\beta - \sin\alpha\sin\beta$

Sine and Tangent of the Sum and Difference of Angles:

$$\sin(\alpha - \beta) = \sin\alpha\cos\beta - \cos\alpha\sin\beta$$
$$\sin(\alpha + \beta) = \sin\alpha\cos\beta + \cos\alpha\sin\beta$$
$$\tan(\alpha - \beta) = \frac{\tan\alpha - \tan\beta}{1 + \tan\alpha\tan\beta}$$
$$\tan(\alpha + \beta) = \frac{\tan\alpha + \tan\beta}{1 - \tan\alpha\tan\beta}$$

Double-Angle Identities:

$$\begin{aligned}\cos 2x &= \cos^2x - \sin^2x \\ &= 2\cos^2x - 1 \\ &= 1 - 2\sin^2x\end{aligned}$$
$$\sin 2x = 2\sin x\cos x$$
$$\tan 2x = \frac{2\tan x}{1 - \tan^2x}$$

Half-Angle Identities:

$$\cos^2\left(\frac{x}{2}\right) = \frac{1 + \cos x}{2}$$
$$\sin^2\left(\frac{x}{2}\right) = \frac{1 - \cos x}{2}$$
$$\begin{aligned}\tan\frac{x}{2} &= \frac{1 - \cos x}{\sin x} \\ &= \frac{\sin x}{1 + \cos x}\end{aligned}$$

Product Identities:

$$\cos \alpha \cos \beta = \frac{1}{2}(\cos(\alpha + \beta) + \cos(\alpha - \beta))$$

$$\sin \alpha \sin \beta = \frac{-1}{2}(\cos(\alpha + \beta) - \cos(\alpha - \beta))$$

$$\sin \alpha \cos \beta = \frac{1}{2}(\sin (\alpha + \beta) + \sin(\alpha - \beta))$$

$$\cos \alpha \sin \beta = \frac{1}{2}(\sin(\alpha + \beta) - \sin(\alpha - \beta))$$

Sum and Difference Identities:

$$\cos A + \cos B = 2 \cos\left(\frac{A + B}{2}\right)\cos\left(\frac{A - B}{2}\right)$$

$$\cos A - \cos B = -2 \sin\left(\frac{A + B}{2}\right)\sin\left(\frac{A - B}{2}\right)$$

$$\sin A + \sin B = 2 \sin\left(\frac{A + B}{2}\right)\cos\left(\frac{A - B}{2}\right)$$

$$\sin A - \sin B = 2 \cos\left(\frac{A + B}{2}\right)\sin\left(\frac{A - B}{2}\right)$$

SELF-CHECKING EXERCISE

Express each of the following in terms of only $\sec x$. **(8-1)**

1. $\cos x$
2. $\sin x$
3. $\tan x$
4. $\cot x$
5. $\csc x$

Express the following as a single trigonometric function of positive angles only. **(8-1)**

6. $\tan(90° - x)$
7. $\cos x(\sec x) + \cot^2 x$
8. $\cos^2 x(\sec^2 x - 1)$
9. $\dfrac{\tan(-x)}{\sin(-x)}$
10. $1 + \dfrac{\sin x}{\csc(-x)}$
11. $\cos(-x)\cos x - 1$

Verify the following identities. **(8-1)**

12. $\cos x \csc x \tan x = 1$
13. $\csc^2 x(\sin^2 x - 1) = -\cot^2 x$
14. $\sec^2 x - \csc^2 x = \dfrac{\csc^2 x - \cot^2 x \csc^2 x}{\cot^2 x}$
15. $\tan x \csc x + \tan x = \sqrt{\dfrac{1 + \sin x}{1 - \sin x}}$

Express the following as functions of θ. **(8-2)**

16. $\cos(\pi + \theta)$
17. $\sin(2\pi - \theta)$
18. $\tan(\pi - \theta)$
19. $\tan(2\pi + \theta)$
20. $\sec\left(\dfrac{3\pi}{6} + \theta\right)$
21. $\csc\left(\dfrac{\pi}{2} + \theta\right)$

Find the exact values of the following functions. **(8-2)**

22. $\tan\left(\frac{\pi}{6} + \frac{\pi}{4}\right)$

23. $\sin\left(\frac{2\pi}{3} - \frac{\pi}{4}\right)$

24. $\cos(210° - 135°)$

25. $\sec 15°$

26. $\sin \frac{\pi}{12}$

27. $\tan \frac{11\pi}{12}$

If $\sin \theta = -8/17$ and $270° < \theta < 360°$, find the following function values. **(8-3)**

28. $\cos 2\theta$

29. $\tan 2\theta$

30. $\csc 2\theta$

31. $\sin^2\left(\frac{\theta}{2}\right)$

32. $\tan \frac{\theta}{2}$

33. $\sec^2\left(\frac{\theta}{2}\right)$

Find the exact values of the following functions. **(8-3)**

34. $1 - 2 \sin^2 67.5°$

35. $\cot \frac{\pi}{8}$

36. $\cos^2\left(\frac{5\pi}{8}\right) - \sin^2\left(\frac{5\pi}{8}\right)$

37. $\sin 165° \cos 165°$

38. $\sin 75° + \sin 15°$

39. $\cos \frac{\pi}{4} \cos \frac{\pi}{12}$

Verify the following identities. **(8-3)**

40. $\dfrac{\sin x - \cos 2x}{\sin x + 1} = 2 \sin x - 1$

41. $\dfrac{\sin 2x - \sin x}{\cos 2x + \cos x} = \tan \dfrac{x}{2}$

42. $\tan \frac{x}{2} + \tan x = \csc x - 2 \cot 2x$

43. $\dfrac{1 + \tan(\theta/2)}{1 - \tan(\theta/2)} = \dfrac{\sin \theta + 1}{\cos \theta}$

44. $\dfrac{\sin \theta}{2} = \sin \dfrac{\theta}{2} \cos \dfrac{\theta}{2}$

45. $\dfrac{\sin 5\theta - \sin 3\theta}{\sin 5\theta + \sin 3\theta} = \dfrac{\cot 4\theta}{\cot \theta}$

46. $\dfrac{\sin^3 x - \cos^3 x}{\sin x - \cos x} = 1 + \sin x \cos x$

47. $\dfrac{\tan^2(x/2)}{\cos^2 x} = \dfrac{\tan^2 x}{4 \cos^4(x/2)}$

Solve the following equations for $x \in [0, 2\pi)$. Where necessary, use Table 4 to approximate x to the nearest hundredth of a radian. **(8-4)**

48. $4 \cos^2 x = 3$

49. $\sin^2 x \tan x - \tan x + \sin^2 x = 1$

50. $5 \cos^2 x + 2 \cos x = 3$

51. $2 \sec^2 x + \sec x - 6 = 0$

Solve the following equations for $\theta \in [0°, 360°)$. Where necessary, use Table 4 to approximate θ to the nearest ten minutes. **(8-5)**

52. $\cot \theta = \sin \theta \cos \theta$

53. $2 \sec^2 \theta = 5 \tan \theta$

54. $\sin \theta + \cos \theta = 1$

55. $\sin 2\theta - \cos \theta - 2 \sin \theta + 1 = 0$

Solve the following equations for $x \in \{\text{reals}\}$. **(8-6)**

56. $3 \cos x - 1 = 2 \sin^2\left(\frac{x}{2}\right)$

57. $\cos 4x - 5 \sin 2x + 2 = 0$

Answers to Self-Checking Exercise

1. $1/\sec x$

2. $\dfrac{\pm\sqrt{\sec^2 x - 1}}{\sec x}$

3. $\pm\sqrt{\sec^2 x - 1}$

4. $\dfrac{\pm 1}{\sqrt{\sec^2 x - 1}}$

5. $\dfrac{\pm \sec x}{\sqrt{\sec^2 x - 1}}$

6. $\cot x$

7. $\csc^2 x$

8. $\sin^2 x$

9. $\sec x$

10. $\cos^2 x$

11. $-\sin^2 x$

12. $\cos x \csc x \tan x = \cos x\left(\dfrac{1}{\sin x}\right)\left(\dfrac{\sin x}{\cos x}\right) = \left(\dfrac{\sin x}{\sin x}\right)\left(\dfrac{\cos x}{\cos x}\right) = 1$

13. $\csc^2 x(\sin^2 x - 1) = \left(\dfrac{1}{\sin^2 x}\right)(-\cos^2 x) = -\cot^2 x$

14.
$$\begin{aligned}\frac{\csc^2 x - \cot^2 x \csc^2 x}{\cot^2 x} &= \frac{\csc^2 x}{\cot^2 x} - \frac{\cot^2 x \csc^2 x}{\cot^2 x}\\ &= \frac{1/\sin^2 x}{\cos^2 x/\sin^2 x} - \csc^2 x\\ &= \frac{1}{\cos^2 x} - \csc^2 x\\ &= \sec^2 x - \csc^2 x\end{aligned}$$

15.
$$\begin{aligned}\sqrt{\frac{1 + \sin x}{1 - \sin x}} &= \frac{\sqrt{1 + \sin x}}{\sqrt{1 - \sin x}}\left(\frac{\sqrt{1 + \sin x}}{\sqrt{1 + \sin x}}\right)\\ &= \frac{1 + \sin x}{\cos x}\\ &= \frac{1}{\cos x} + \frac{\sin x}{\cos x}\\ &= \frac{1}{\cos x}\cdot\left(\frac{\sin x}{\sin x}\right) + \tan x\\ &= \tan x \csc x + \tan x\end{aligned}$$

16. $-\cos \theta$

17. $-\sin \theta$

18. $-\tan \theta$

19. $\tan \theta$

20. $-\csc \theta$

21. $\sec \theta$

22. $2 + \sqrt{3}$

23. $\dfrac{\sqrt{6} + \sqrt{2}}{4}$

24. $\dfrac{\sqrt{6} - \sqrt{2}}{4}$

25. $\sqrt{6} - \sqrt{2}$

26. $\dfrac{\sqrt{6} - \sqrt{2}}{4}$

27. $\sqrt{3} - 2$

28. $161/289$

29. $-240/161$

30. $-289/240$

31. $1/17$

32. $-1/4$

33. $17/16$

34. $-\sqrt{2}/2$

35. $1 + \sqrt{2}$

36. $-\sqrt{2}/2$

37. $-1/4$

38. $\sqrt{6}/2$

39. $\dfrac{1 + \sqrt{3}}{4}$

40.
$$\begin{aligned}\frac{\sin x - \cos 2x}{\sin x + 1} &= \frac{\sin x - (1 - 2\sin^2 x)}{\sin x + 1}\\ &= \frac{(2\sin x - 1)(\sin x + 1)}{\sin x + 1}\\ &= 2\sin x - 1\end{aligned}$$

41. $$\frac{\sin 2x - \sin x}{\cos 2x + \cos x} = \frac{2 \sin x \cos x - \sin x}{(2 \cos^2 x - 1) + \cos x}$$
$$= \frac{\sin x(2 \cos x - 1)}{(\cos x + 1)(2 \cos x - 1)}$$
$$= \tan \frac{x}{2}$$

42. $$\csc x - 2 \cot 2x = \frac{1}{\sin x} - 2\left(\frac{1 - \tan^2 x}{2 \tan x}\right)$$
$$= \frac{1 - \cos x(1 - \tan^2 x)}{\sin x}$$
$$= \frac{1 - \cos x}{\sin x} + \frac{\sin^2 x}{\sin x \cos x}$$
$$= \tan \frac{x}{2} + \tan x$$

43. $$\frac{1 + \tan(\theta/2)}{1 - \tan(\theta/2)} = \frac{1 + \dfrac{\sin(\theta/2)}{\cos(\theta/2)}}{1 - \dfrac{\sin(\theta/2)}{\cos(\theta/2)}}$$
$$= \frac{\cos(\theta/2) + \sin(\theta/2)}{\cos(\theta/2) - \sin(\theta/2)}\left(\frac{\cos(\theta/2) + \sin(\theta/2)}{\cos(\theta/2) + \sin(\theta/2)}\right)$$
$$= \frac{\cos^2(\theta/2) + 2 \cos(\theta/2)\sin(\theta/2) + \sin^2(\theta/2)}{\cos^2(\theta/2) - \sin^2(\theta/2)}$$
$$= \frac{1 + \sin \theta}{\cos \theta}$$

44. $$\sin(\theta/2)\cos(\theta/2) = \frac{2 \sin(\theta/2)\cos(\theta/2)}{2} = \frac{\sin \theta}{2}$$

45. $$\frac{\sin 5\theta - \sin 3\theta}{\sin 5\theta + \sin 3\theta} = \frac{2 \cos 4\theta \sin \theta}{2 \sin 4\theta \cos \theta} = \cot 4\theta \tan \theta = \frac{\cot 4\theta}{\cot \theta}$$

46. $$\frac{\sin^3 x - \cos^3 x}{\sin x - \cos x} = \frac{(\sin x - \cos x)(\sin^2 x + \sin x \cos x + \cos^2 x)}{\sin x - \cos x}$$
$$= 1 + \sin x \cos x$$

47. $$\frac{\tan^2 x}{4 \cos^4(x/2)} = \frac{\sin^2 x/\cos^2 x}{4\left(\dfrac{1 + \cos x}{2}\right)^2}$$
$$= \left(\frac{\sin x}{1 + \cos x}\right)^2\left(\frac{1}{\cos^2 x}\right)$$
$$= \frac{\tan^2(x/2)}{\cos^2 x}$$

48. $\{\pi/6, 5\pi/6, 7\pi/6, 11\pi/6\}$

49. $\{\pi/2, 3\pi/4, 3\pi/2, 7\pi/4\}$

50. $\{0.93, \pi, 5.36\}$

51. $\{0.84, 2\pi/3, 4\pi/3, 5.44\}$

52. $\{90°, 270°\}$

53. $\{26°30', 63°30', 206°30', 243°30'\}$

54. $\{0°, 90°\}$

55. $\{0°, 30°, 150°\}$

56. $\{\pi/3 + 2n\pi, 5\pi/3 + 2n\pi\}$

57. $\{\pi/12 + n\pi, 5\pi/12 + n\pi\}$

9 Inverse Trigonometric Functions

In Chapter 4 we developed the theory of inverse functions. In this chapter we define the inverse trigonometric functions that are very important in the study of calculus. The elements of the theory on inverse functions that are important for our development here are reviewed as follows:

1. If $f = \{(x, y) \mid y = f(x)\}$, then f is a one-to-one function if for every y in the range of f there is only one x in the domain.
2. If $f = \{(x, y) \mid y = f(x)\}$ is a one-to-one function, then there exists a function $f^{-1} = \{(x, y) \mid x = f(y)\}$. Furthermore, we write $y = f^{-1}(x)$ if and only if $x = f(y)$.
3. If f and f^{-1} are inverse functions, then the domain of f is the range of f^{-1}, and the range of f is the domain of f^{-1}.
4. If f and f^{-1} are inverse functions, then
$$f \circ f^{-1} = f^{-1} \circ f = I \quad \text{and} \quad I = \{(x, y) \mid y = x\}$$
5. If f and f^{-1} are inverse functions, then the graphs of f and f^{-1} are symmetric with respect to the line $y = x$.

The periodic nature of the trigonometric functions results in a *many-to-one relationship*. That is, there are many values of x paired with a given function value for each of the trigonometric functions. However, we define **primary functions** for each of the trigonometric functions that are one-to-one over a subset of the domains of each function. Since the primary functions are one-to-one, then each of the functions has an inverse that is also a function.

Are the trigonometric functions one-to-one so that their inverses are also functions?

INVERSE SINE AND COSECANT FUNCTIONS 9-1

In Figure 9-1 the graphs of $y = \sin x$ and $y = \csc x$ are shown. Notice that any horizontal line $y = k,\ -1 \le k \le 1$, will intersect the sine curve in infinitely many points. In particular, the line $y = 0.5$ is shown intersecting the curve at $x_1 = \pi/6,\ x_2 = 5\pi/6,\ x_3 = 13\pi/6, \ldots$. Furthermore, any horizontal line $y = 1/k, k \neq 0$, will intersect the cosecant curve for the same values of x that the line $y = k$ intersects the sine curve. For example, if $k = 0.5$, then $1/k = 2$, and the line $y = 2$ intersects the cosecant curve at $x_1 = \pi/6$, $x_2 = 5\pi/6$, $x_3 = 13\pi/6, \ldots$.

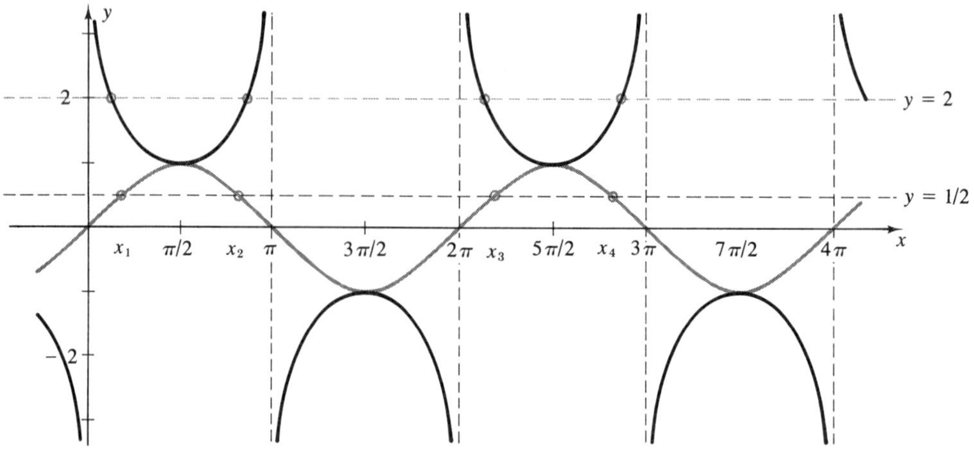

FIGURE 9-1 Graphs of $y = \sin x$, $y = \csc x$, $y = 1/2$, and $y = 2$

To convert the many-to-one functions to one-to-one, we select a subset of the domains of the functions over which the funcions are one-to-one. The following definitions identify the intervals selected for this purpose.

Primary sine function

Definition: The primary sine function is

$$\text{Sine} = \{(x, y) \mid y = \text{Sin}\, x\}$$

with domain $= \{x \mid -\pi/2 \le x \le \pi/2\}$ and range $= \{y \mid -1 \le y \le 1\}$.

Primary cosecant function

Definition: The primary cosecant function is

$$\text{Cosecant} = \{(x, y) \mid y = \text{Csc}\, x\}$$

with domain $= \{x \mid -\pi/2 \le x \le \pi/2,\ x \ne 0\}$ and range $= \{y \mid |y| \ge 1\}$.

The capital letter is the symbol used to indicate the primary function ?

The most common method used to indicate the primary trigonometric functions is simply to capitalize the function. The significant distinction between the functions $y = \sin x$ and $y = \text{Sin}\, x$ is the difference in the domains of the two functions. For $y = \sin x$, $x \in \{\text{reals}\}$; but for $y = \text{Sin}\, x$, $x \in [-\pi/2, \pi/2]$.

EXAMPLE 9–1 Solve for x.

(a) $\text{Sin}\, x = \sqrt{3}/2$ (b) $\text{Csc}\, x = \sqrt{2}$

Solutions:

(a) If $\text{Sin}\, x = \sqrt{3}/2$ and $-\pi/2 \le x \le \pi/2$, then $x = \pi/3$.
(b) If $\text{Csc}\, x = \sqrt{2}$ and $-\pi/2 \le x \le \pi/2$, $x \ne 0$, then $x = \pi/4$.

Since Sine and Cosecant are one-to-one functions, there exist inverse functions such that the composition of each function and its inverse is the identity function.

Sine $^{-1}$

> **Definition:** The inverse function of Sine is
> $$\text{Sine}^{-1} = \{(x, y) \mid x = \text{Sin } y\}$$
> with domain $= \{x \mid -1 \le x \le 1\}$ and range $=$ $\{y \mid -\pi/2 \le y \le \pi/2\}$.

Notation. $y = \text{Sin}^{-1}x$ if and only if $x = \text{Sin } y$.

Cosecant $^{-1}$

> **Definition:** The inverse function of Cosecant is
> $$\text{Cosecant}^{-1} = \{(x, y) \mid x = \text{Csc } y\}$$
> with domain $= \{x \mid |x| \ge 1\}$ and range $=$ $\{y \mid -\pi/2 \le y \le \pi/2, y \ne 0\}$.

Notation. $y = \text{Csc}^{-1}x$ if and only if $x = \text{Csc } y$.

In Figure 9-2 the graphs of $y = \text{Sin } x$ and $y = \text{Sin}^{-1}x$ are shown, and in Figure 9-3 the graphs of $y = \text{Csc } x$ and $y = \text{Csc}^{-1}x$ are drawn. In both graphs the identity function is shown as a dashed line.

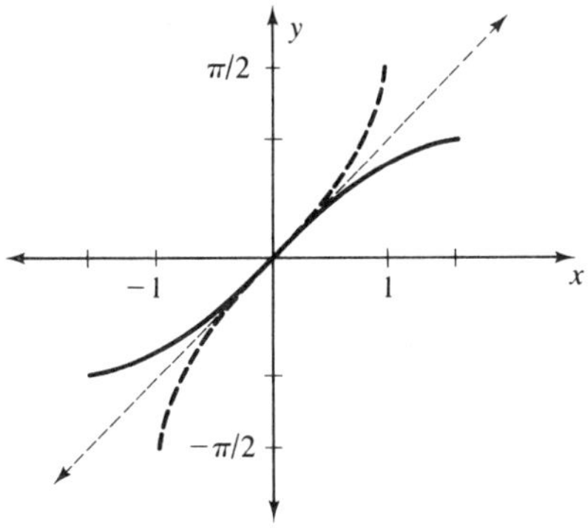

FIGURE 9-2 Graphs of $y = \text{Sin } x$ and $y = \text{Sin}^{-1}x$

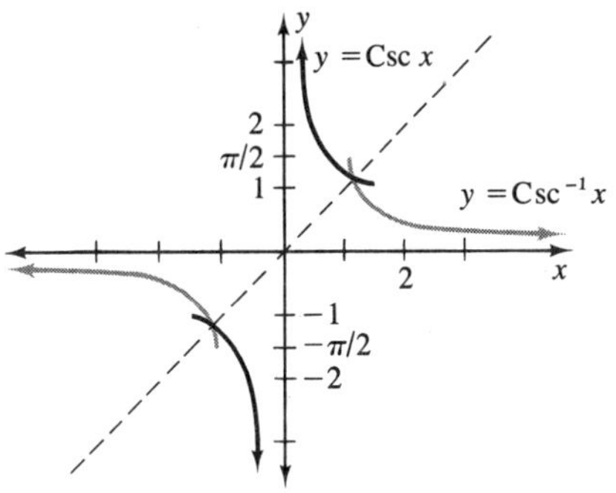

FIGURE 9-3 Graphs of $y = \text{Csc } x$ and $y = \text{Csc}^{-1}x$

It is important to distinguish between the roles of x and y in the equations $y = \text{Sin}\, x$ and $y = \text{Sin}^{-1}x$:

1. If $y = \text{Sin}\, x$, then x is a real number paired with point P on a unit circle centered at (0, 0) and y is the ordinate of P. Furthermore, $x \in \{-\pi/2, \pi/2\}$ and $y \in \{-1, 1\}$.
2. If $y = \text{Sin}^{-1}x$, then x is the ordinate of P and y is the real number paired with P. Now $x \in \{-1, 1\}$ and $y \in \{-\pi/2, \pi/2\}$. An alternate notation for $\text{Sin}^{-1}x$ is Arcsin x. Thus, if $y = \text{Arcsin}\, x$, then *y is the arc whose sine is x.*

In the equation $y = \text{Sin}^{-1}x$, when y is interpreted as an angle measured in radians, there are corresponding intervals for the angle measured in degrees. That is, since $1° = \dfrac{180°}{\pi}$, then $-\pi/2 \le y \le \pi/2$ corresponds to $-90° \le y \le 90°$.

EXAMPLE 9–2 Solve for y:
(a) $y = \text{Sin}^{-1}1$ (b) $y = \text{Csc}^{-1}(-2)$

Solutions:

(a) Since $\text{Sin}\, \dfrac{\pi}{2} = 1$ and $\pi/2$ is in the range of the Sine^{-1} function, then $y = \pi/2$.

(b) Since $\text{Csc}\, \dfrac{-\pi}{6} = -2$ and $-\pi/6$ is in the range of the Cosecant^{-1} function, then $y = -\pi/6$.

If x is not an exact function value, then we can use Table 4 or a scientific hand calculator to approximate the value of y. But we must realize that in either case — table or calculator — *the value of y is approximated to a specified number of decimal places.*

EXAMPLE 9–3 Solve for y, using Table 4:
(a) $y = \text{Sin}^{-1}0.3987$ (b) $y = \text{Csc}^{-1}1.853$

Solutions:
(a) If $y = \text{Sin}^{-1}0.3987$, then $\text{Sin}\, y = 0.3987$ and $y = 0.4102$ or $y = 23°30'$.
(b) If $y = \text{Csc}^{-1}1.853$, then $\text{Csc}\, y = 1.853$ and $y = 0.5701$ or $y = 32°40'$.

Notice that y can be expressed in terms of either a real number or a degree measure. That is, the range of $\text{Sine}^{-1} = \{y \mid -\pi/2 \le y \le \pi/2\}$ or $\{y \mid -90° \le y \le 90°\}$. Similarly, the range of Cosecant^{-1} can be stated in terms of degrees.

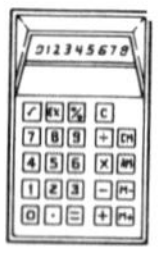

EXAMPLE 9–4 Solve for y to the nearest hundredth of a degree, using a hand calculator.
(a) $y = \text{Sin}^{-1}(-0.4592)$ (b) $y = \text{Csc}^{-1}2.490$

Solutions:
(a) With calculator set in Degree register, enter -0.4592 and depress SIN^{-1} key. Display reads -27.34. Thus, $y = -27.34°$ to the nearest hundredth of a degree.

(b) With calculator set in Degree register, enter 2.490. Depress $1/x$ key and display reads 0.4016. Now depress SIN^{-1} key, and $y = 23.68°$ to the nearest hundredth of a degree.

Sine and Sine^{-1} are inverse functions; therefore,

$$\text{Sin}(\text{Sin}^{-1}x) = x, \quad \text{for } -1 \le x \le 1$$
$$\text{Sin}^{-1}(\text{Sin } x) = x, \quad \text{for } -\pi/2 \le x \le \pi/2, \text{ or } -90° \le x \le 90°$$

EXAMPLE 9–5 Evaluate
(a) $\text{Sin}(\text{Sin}^{-1}0.2)$ (b) $\text{Sin}^{-1}(\text{Sin } 37°)$

Solutions:
(a) Since 0.2 is in the domain of Sine^{-1}, then $\text{Sin}(\text{Sin}^{-1}0.2) = 0.2$. That is, the sine of the angle whose sine is 0.2 is 0.2.
(b) Since 37° is in the domain of Sine, then $\text{Sin}^{-1}(\text{Sin } 37°) = 37°$. That is, the angle whose sine is the sine of 37° is 37°.

EXAMPLE 9–6 Evaluate
(a) $\text{Csc}(\text{Sin}^{-1}0.5)$ (b) $\text{Sin}(\text{Csc}^{-1}(-\sqrt{2}))$

Solutions:
(a) $\text{Csc}(\text{Sin}^{-1}0.5)$ can be thought of as $\text{Csc } \theta$, where $\text{Sin } \theta = 0.5$. Since $\text{Sin } 30° = 0.5$, then $\text{Csc } 30° = 2$, and $\text{Csc}(\text{Sin}^{-1}0.5) = 2$.
(b) $\text{Sin}(\text{Csc}^{-1}(-\sqrt{2}))$ can be interpreted as $\text{Sin } \theta$ where $\text{Csc } \theta = -\sqrt{2}$. Since $\text{Csc } \frac{-\pi}{4} = -\sqrt{2}$, then $\theta = -\pi/4$, and $\text{Sin } \frac{-\pi}{4} = -1/\sqrt{2}$. Thus,

$$\text{Sin}(\text{Csc}^{-1}(-\sqrt{2})) = -1/\sqrt{2}$$

We can use the inverse function to state the solution set of a trigonometric equation as an exact value. For example, if $3 \text{ Sin } x - 1 = 0$, then $\text{Sin } x = 1/3$. In Chapter 8 we approximated the value of x with Table 4. However, the exact value of x is stated as follows:

$$x = \text{Sin}^{-1}\left(\frac{1}{3}\right)$$

Can we apply the inverse function to anything we have studied already in trigonometry?

We will also use the inverse functions to review the basic identities studied in Chapter 8.

Exercise Set 9–1

Solve for x in each of the following primary functions.

1. $\text{Sin } x = 1$
2. $\text{Sin } x = -1/2$
3. $\text{Csc } x = -2$
4. $\text{Csc } x = -1$
5. $\text{Csc } x = 2$
6. $\text{Sin } x = 1/\sqrt{2}$

Find the value of y in degrees to the nearest ten minutes. Use Table 4 or a calculator.

7. $y = \text{Sin}^{-1}0.1908$
8. $y = \text{Csc}^{-1}3.420$

9. $y = \text{Csc}^{-1}(-57.30)$
10. $y = \text{Sin}^{-1}(-0.6202)$
11. $y = \text{Sin}^{-1}0.4950$
12. $y = \text{Csc}^{-1}(-1.163)$

Find the value of y in radians to the nearest hundredth. Use Table 4 or a calculator.

13. $y = \text{Csc}^{-1}1.052$
14. $y = \text{Sin}^{-1}0.9925$
15. $y = \text{Sin}^{-1}(-0.7451)$
16. $y = \text{Csc}^{-1}(-1.188)$
17. $y = \text{Csc}^{-1}(-2.107)$
18. $y = \text{Sin}^{-1}0.0204$

Evaluate the following.

19. $\text{Csc}^{-1}\left(\text{Csc}\,\frac{\pi}{5}\right)$
20. $\text{Sin}^{-1}\left(\text{Sin}\,\frac{2\pi}{7}\right)$
21. $\text{Sin}\left(\text{Sin}^{-1}\left(\frac{3}{4}\right)\right)$
22. $\text{Csc}(\text{Csc}^{-1}10)$
23. $\text{Sin}(\text{Csc}^{-1}2)$
24. $\text{Csc}\left(\text{Sin}^{-1}\frac{1}{\sqrt{2}}\right)$
25. $\text{Csc}(\text{Sin}^{-1}(-1))$
26. $\text{Sin}(\text{Csc}^{-1}(-2\sqrt{3}/3))$

Solve the following equations for exact values of the variables, using inverse function notation when necessary. Do not approximate the roots.

27. $4\ \text{Sin}\,x + 1 = 0$
28. $\frac{\text{Csc}\,x}{4} - 1 = 0$
29. $3\ \text{Sin}\,x - 2 = 0$
30. $2\ \text{Csc}\,x + \sqrt{5} = 0$
31. $6\ \text{Sin}^2x - 5\ \text{Sin}\,x + 1 = 0$
32. $4\ \text{Sin}^2x - 3\ \text{Sin}\,x - 1 = 0$
33. $5\ \text{Sin}^3y + \text{Sin}^2y - 5\ \text{Sin}\,y - 1 = 0$
34. $2\ \text{Sin}^3y = \text{Sin}\,y$
35. $2\ \text{Csc}^2\theta - 5\ \text{Csc}\,\theta = 12$
36. $3\ \text{Csc}^2x + 8\ \text{Csc}\,x = 3$
37. $4\ \text{Sin}^2 3\theta + 3\ \text{Sin}\,3\theta = 1$
38. $10\ \text{Sin}^2 5x + \text{Sin}\,5x = 2$
39. $\text{Csc}^2 4y = \text{Csc}\,4y + 6$
40. $2\ \text{Csc}^2 6\theta + 12 = 11\ \text{Csc}\,6\theta$

9-2 INVERSE COSINE AND SECANT FUNCTIONS

The cosine and secant functions — like the sine and cosecant functions — are many-to-one. In Figure 9-4 the graph of $y = \cos x$ and $y = \sec x$ are shown. The horizontal lines $y = -1/2$ and $y = -2$ are also drawn, with several points of intersection of the lines and curves indicated. Thus, for a given value of y, and $y \in \{\text{reals}\}$, there are infinitely many values of x such that either $\cos x = y$ or $\sec x = y$.

Therefore, to form the inverse functions of cosine and secant, we must first convert the functions to one-to-one functions. The following definitions identify the domains over which the two functions are one-to-one.

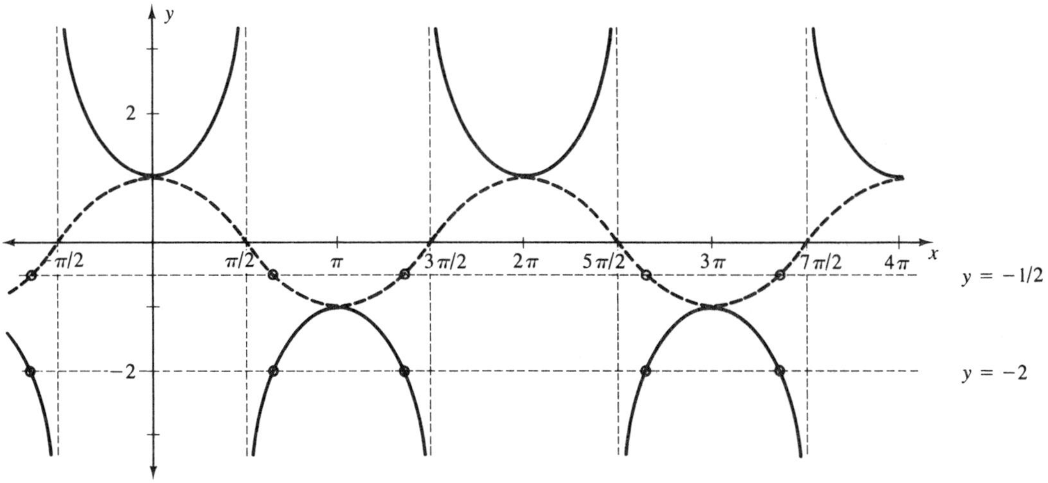

FIGURE 9-4 Graphs of $y = \cos x$, $y = \sec x$, $y = -1/2$, and $y = -2$

Primary cosine function

> **Definition:** The primary cosine function is
>
> $$\text{Cosine} = \{(x, y) \mid y = \text{Cos } x\}$$
>
> with domain $= \{x \mid 0 \le x \le \pi\}$ and range $=$ $\{y \mid -1 \le y \le 1\}$.

Primary secant function

> **Definition:** The primary secant function is
>
> $$\text{Secant} = \{(x, y) \mid y = \text{Sec } x\}$$
>
> with domain $= \{x \mid 0 \le x \le \pi, x \ne \pi/2\}$ and range $=$ $\{y \mid |y| \ge 1\}$.

There are infinitely many intervals that can be used for the domains of any of the primary trigonometric functions. However, the domains are part of the definitions; therefore, we accept the stated intervals.

Aren't there many intervals over which these functions are one-to-one?

Notice that the interval used for the primary cosine and secant functions is different from the interval used for the primary sine and cosecant functions. In terms of quadrants, for $\text{Sin } x$ and $\text{Csc } x$, x is in Quadrants I or IV. But for $\text{Cos } x$ and $\text{Sec } x$, x is in Quadrants I or II.

Over the indicated intervals, Cosine and Secant are one-to-one functions. Therefore, Cosine and Secant have inverse functions identified in the following definitions.

Cosine^{-1}

> **Definition:** The inverse function of Cosine is
>
> $$\text{Cosine}^{-1} = \{(x, y) \mid x = \text{Cos } y\}$$
>
> with domain $= \{x \mid -1 \le x \le 1\}$ and range $=$ $\{y \mid 0 \le y \le \pi\}$.

Notation. $y = \text{Cos}^{-1}x$ if and only if $x = \text{Cos } y$.

Secant^{-1}

> **Definition:** The inverse function of Secant is
> $$\text{Secant}^{-1} = \{(x, y) \mid x = \text{Sec } y\}$$
> with domain $= \{x \mid |x| \geq 1\}$ and range $=$ $\left\{y \mid 0 \leq y \leq \pi, y \neq \frac{\pi}{2}\right\}$.

Notation. $y = \text{Sec}^{-1}x$ if and only if $x = \text{Sec } y$.

In Figure 9-5 the graphs of $y = \text{Cos } x$ and $y = \text{Cos}^{-1}x$ are shown; and in Figure 9-6 the graphs of $y = \text{Sec } x$ and $y = \text{Sec}^{-1}x$ are drawn. In both graphs the identity function is displayed as a dashed line.

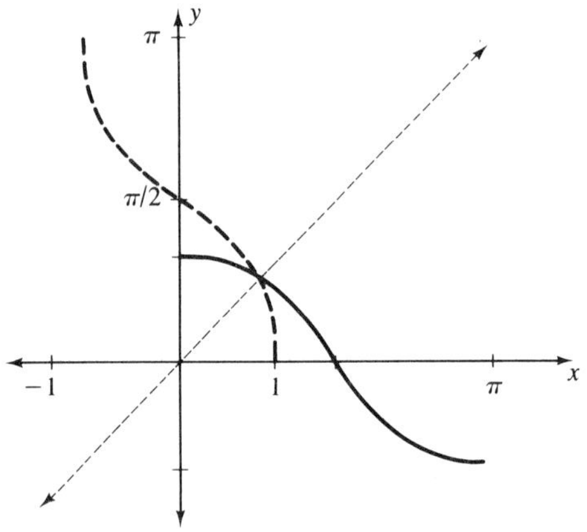

FIGURE 9-5 Graphs of $y = \text{Cos } x$ and $y = \text{Cos}^{-1}x$

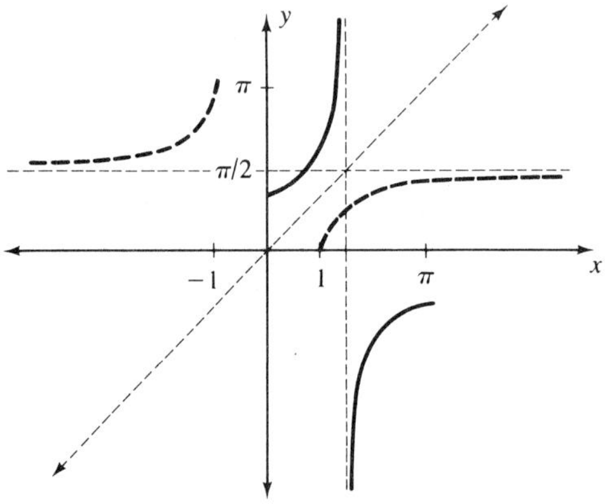

FIGURE 9-6 Graphs of $y = \text{Sec } x$ and $y = \text{Sec}^{-1}x$

EXAMPLE 9–7 Solve for y:

(a) $y = \text{Cos}^{-1}0.5$ (b) $y = \text{Sec}^{-1}(-\sqrt{2})$

Solutions:

(a) Since $\text{Cos}\,\dfrac{\pi}{3} = 0.5$, then $y = \pi/3$, and $\pi/3 = \text{Cos}^{-1}0.5$.

(b) Since $\text{Sec}\,\dfrac{3\pi}{4} = -\sqrt{2}$, then $y = 3\pi/4$, and $3\pi/4 = \text{Sec}^{-1}(-\sqrt{2})$.

EXAMPLE 9–8 Approximate the value of y to the nearest ten minutes, using Table 4.

(a) $y = \text{Sec}^{-1}1.171$ (b) $y = \text{Cos}^{-1}(-0.3007)$

Solutions:

(a) Since $x > 0$, y is in Quadrant I and $\text{Sec}\ y = 1.171$. From Table 4, $\text{Sec}\ 31°20' = 1.171$; therefore, $y = 31°20'$ to the nearest ten minutes.

(b) Since $x < 0$, y is in Quadrant II and the reference angle of y has a Cosine value of 0.3007. From Table 4, $\text{Cos}\ 72°30' = 0.3007$; therefore, $y = 180° - 72°30' = 107°30'$.

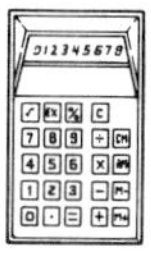

EXAMPLE 9–9 Approximate the value of y to the nearest hundredth of a degree, using a hand calculator.

(a) $y = \text{Cos}^{-1}0.8515$ (b) $y = \text{Sec}^{-1}(-2.385)$

Solutions:

(a) With calculator set in Degree register, enter 0.8515 and depress COS^{-1} key. Display reads 31.62°, the value of y to the nearest hundredth of a degree.

(b) With calculator set in Degree register, enter -2.385 and depress $1/x$ key. Display reads -0.419. Now depress COS^{-1} key and display reads 114.79°. Thus, $y \approx 114.79°$. Notice that the calculator automatically yields a Quadrant II angle when $x < 0$ for Cos^{-1}.

Since Cosine and Cosine^{-1} are inverse functions, then

$$\text{Cos}(\text{Cos}^{-1}x) = x, \quad \text{for } -1 \le x \le 1$$
$$\text{Cos}^{-1}(\text{Cos}\,x) = x, \quad \text{for } 0 \le x \le \pi$$

EXAMPLE 9–10 Simplify

(a) $\text{Cos}(\text{Cos}^{-1}(-0.73))$ (b) $\text{Cos}^{-1}\left(\text{Cos}\,\dfrac{3\pi}{5}\right)$

Solutions:

(a) Since $x = -0.73$ and -0.73 is in the domain of Cosine^{-1}, then $\text{Cos}(\text{Cos}^{-1}(-0.73)) = -0.73$.

(b) With $x = 3\pi/5$, the given value of x is in the domain of Cosine; thus, $\text{Cos}^{-1}\left(\text{Cos}\,\dfrac{3\pi}{5}\right) = 3\pi/5$.

EXAMPLE 9–11 Simplify

(a) $\text{Sec}\left(\text{Cos}^{-1}\left(\frac{\sqrt{3}}{2}\right)\right)$ (b) $\text{Cos}(\text{Sec}^{-1}(-\sqrt{2}))$

Solutions:

(a) $\text{Sec}\left(\text{Cos}^{-1}\left(\frac{\sqrt{3}}{2}\right)\right)$ can be interpreted as Sec θ where Cos $\theta = \sqrt{3}/2$. Since $\theta = \pi/6$ and $\text{Sec}\,\frac{\pi}{6} = 2/\sqrt{3}$, then $\text{Sec}\left(\text{Cos}^{-1}\left(\frac{\sqrt{3}}{2}\right)\right) = 2/\sqrt{3}$.

(b) $\text{Cos}(\text{Sec}^{-1}(-\sqrt{2}))$ can be interpreted as Cos θ where Sec $\theta = -\sqrt{2}$. Since $\theta = 3\pi/4$ and $\text{Cos}\,\frac{3\pi}{4} = -1/\sqrt{2}$, then $\text{Cos}(\text{Sec}^{-1}(-\sqrt{2})) = -1/\sqrt{2}$.

EXAMPLE 9–12 Find the exact value of x when $2\,\text{Cos}\,x + 0.4 = 0$.

Solution: If $2\,\text{Cos}\,x + 0.4 = 0$, then

$$\text{Cos}\,x = -0.2$$
$$x = \text{Cos}^{-1}(-0.2)$$

Exercise Set 9–2

Solve for x in each of the following primary functions.

1. $\text{Cos}\,x = 1/2$
2. $\text{Sec}\,x = -2\sqrt{3}/3$
3. $\text{Sec}\,x = -2$
4. $\text{Cos}\,x = 0$
5. $\text{Cos}\,x = -\sqrt{3}/2$
6. $\text{Sec}\,x = \sqrt{2}$

Use Table 4 or a calculator to approximate the following in degrees to the nearest ten minutes.

7. $y = \text{Sec}^{-1}1.681$
8. $y = \text{Cos}^{-1}0.9474$
9. $y = \text{Cos}^{-1}(-0.5299)$
10. $y = \text{Sec}^{-1}(-1.029)$
11. $y = \text{Sec}^{-1}4.876$
12. $y = \text{Cos}^{-1}0.5250$

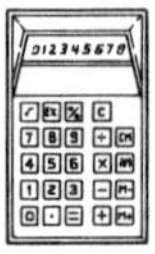

Use Table 4 or a calculator to approximate the following in radians to the nearest hundredth.

13. $y = \text{Cos}^{-1}0.9976$
14. $y = \text{Sec}^{-1}2.381$
15. $y = \text{Sec}^{-1}1.127$
16. $y = \text{Cos}^{-1}0.6383$
17. $y = \text{Cos}^{-1}(-0.3360)$
18. $y = \text{Sec}^{-1}(-1.040)$

Evaluate the following functions.

19. $\text{Sec}\left(\text{Sec}^{-1}\left(\frac{3}{2}\right)\right)$
20. $\text{Cos}^{-1}\left(\text{Cos}\,\frac{3\pi}{5}\right)$
21. $\text{Cos}\left(\text{Cos}^{-1}\left(\frac{-7}{8}\right)\right)$
22. $\text{Sec}^{-1}\left(\text{Sec}\,\frac{5\pi}{8}\right)$
23. $\text{Cos}(\text{Sec}^{-1}\sqrt{3})$
24. $\text{Sec}(\text{Cos}^{-1}0.2)$

25. $\text{Sec}\left(\text{Cos}^{-1}\left(\frac{-1}{3}\right)\right)$

26. $\text{Cos}\left(\text{Sec}^{-1}\left(\frac{-13}{7}\right)\right)$

Solve the following equations for exact values of the variables, using inverse function notation when necessary. Do not approximate the roots.

27. $3 \text{ Cos } x + 1 = 0$

28. $-2 \text{ Sec } x - 3 = 2$

29. $\frac{3}{4} \text{ Sec } x + \frac{2}{3} = 7$

30. $5 \text{ Cos } \theta - 2 = 0$

31. $6 \text{ Cos}^2 y + \text{Cos } y = 2$

32. $2 \text{ Cos}^3 y = \text{Cos } y$

33. $2 \text{ Sec}^2\theta + \text{Sec } \theta = 15$

34. $2 \text{ Sec}^2\theta + 7 \text{ Sec } \theta + 5 = 0$

35. $10 \text{ Cos}^2 3x - 2 = \text{Cos } 3x$

36. $3 \text{ Cos}^2 2x + \text{Cos } 2x = 2$

37. $2 \text{ Sec}^2 4x + \text{Sec } 4x = 10$

38. $\text{Sec}^3\theta - 2 \text{ Sec } \theta = 0$

39. Sketch a graph of $y = 2 \text{ Cos}^{-1} x$.

40. Sketch a graph of $y = \text{Cos}^{-1} 2x$.

INVERSE TANGENT AND COTANGENT FUNCTIONS 9–3

To form the inverse tangent and cotangent functions, we first define the primary tangent and cotangent functions. As in the case of the other primary trigonometric functions, the primary tangent and cotangent functions are one-to-one over the intervals selected for the functions. Notice, however, that the intervals are not the same for both functions.

Primary tangent function

Definition: The primary tangent function is

$$\text{Tangent} = \{(x, y) \mid y = \text{Tan } x\}$$

with domain $= \{x \mid -\pi/2 < x < \pi/2\}$ and range $= \{y \mid y \in \text{reals}\}$.

Primary cotangent function

Definition: The primary cotangent function is

$$\text{Cotangent} = \{(x, y) \mid y = \text{Cot } x\}$$

with domain $= \{x \mid 0 < x < \pi\}$ and range $= \{y \mid y \in \text{reals}\}$.

So the domain of the Tangent function is the same as the Sine and Cosecant, but the Cotangent function is the same as the Cosine and Secant?

We will list for comparison the domains of all six functions.

Group I

Function	Domain	Special Feature
Cosine	$0 \le x \le \pi$	Continuous interval, endpoints included
Secant	$0 \le x \le \pi$, $x \ne \pi/2$	Excluded value in the middle of the interval
Cotangent	$0 < x < \pi$	Continuous interval, endpoints excluded

Group II

Function	Domain	Special Feature
Sine	$-\pi/2 \le x \le \pi/2$	Continuous interval, endpoints included
Cosecant	$-\pi/2 \le x \le \pi/2$, $x \ne 0$	Excluded value in the middle of the interval
Tangent	$-\pi/2 < x < \pi/2$	Continuous interval, endpoints excluded

The functions Tangent and Cotangent are one-to-one as defined; therefore, they have inverse functions. We identify these functions in the following definitions.

Tangent^{-1}

Definition: The inverse function of Tangent is

$$\text{Tangent}^{-1} = \{(x, y) \mid x = \text{Tan } y\}$$

with domain $= \{x \mid x \in \text{reals}\}$ and range $= \{y \mid -\pi/2 < y < \pi/2\}$.

Notation. $y = \text{Tan}^{-1}x$ if and only if $x = \text{Tan } y$.

Cotangent^{-1}

Definition: The inverse function of Cotangent is

$$\text{Cotangent}^{-1} = \{(x, y) \mid x = \text{Cot } y\}$$

with domain $= \{x \mid x \in \text{reals}\}$ and range $= \{y \mid 0 < y < \pi\}$.

Notation. $y = \text{Cot}^{-1}x$ if and only if $x = \text{Cot } y$.

The graphs of Tangent and Tangent^{-1} are given in Figure 9-7. The graphs of Cotangent and Cotangent^{-1} are shown in Figure 9-8.

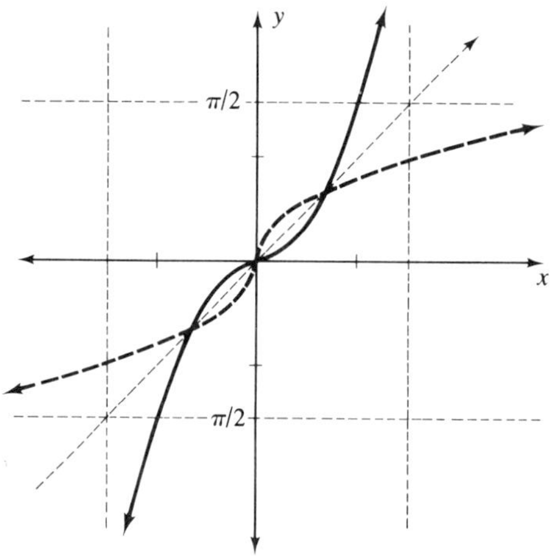

FIGURE 9-7 Graphs of $y = \text{Tan } x$ and $y = \text{Tan}^{-1}x$

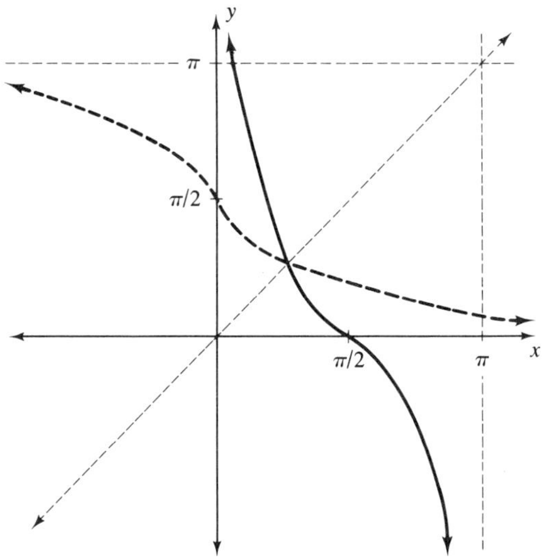

FIGURE 9-8 Graphs of $y = \text{Cot}\, x$ and $y = \text{Cot}^{-1}x$

We must remember that the stated range for each inverse function is necessary to guarantee an inverse function. This principle is of primary importance, even though it may occasionally cause some difficulty. For example,

$$\text{Tan}\,\frac{\pi}{6} = \frac{1}{\sqrt{3}} \quad \text{and} \quad \text{Cot}\,\frac{\pi}{6} = \sqrt{3}$$

Therefore,

$$\text{Tan}^{-1}\left(\frac{1}{\sqrt{3}}\right) = \text{Cot}^{-1}\sqrt{3} = \frac{\pi}{6}$$

However,

$$\text{Tan}\,\frac{-\pi}{6} = \frac{-1}{\sqrt{3}} \quad \text{and} \quad \text{Cot}\,\frac{5\pi}{6} = -\sqrt{3}$$

Thus, $\text{Tan}^{-1}\left(\frac{-1}{\sqrt{3}}\right) \neq \text{Cot}^{-1}(-\sqrt{3})$.

The reason for the equality in the first case and the nonequality in the next is found in the comparison of the ranges of the two inverse functions. The two functions have a common range for $x > 0$, but they have different ranges for $x < 0$. It is therefore very important that we keep in mind the ranges of each of the six inverse functions.

EXAMPLE 9–13 Use Table 4 to approximate y to the nearest hundredth of a radian.

(a) $y = \text{Tan}^{-1}0.7536$ (b) $y = \text{Cot}^{-1}(-0.5095)$

Solutions:

(a) If $y = \text{Tan}^{-1}0.7536$, then $\text{Tan}\, y = 0.7536$ and $0 < y < \pi/2$. With Table 4, $y = 0.65$ to the nearest hundredth of a radian.

(b) If $y = \text{Cot}^{-1}(-0.5095)$, then $\text{Cot}\, y = -0.5095$ and $\pi/2 < y < \pi$. From Table 4, the reference angle for y is 1.0996; therefore, $y = \pi - 1.0996 = 2.04$ to two decimal places.

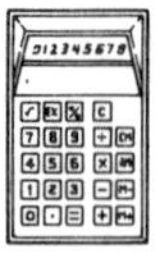

EXAMPLE 9–14 Use a hand calculator to approximate y to the nearest hundredth of a degree.

(a) $y = \text{Tan}^{-1}(-3.7055)$ (b) $y = \text{Cot}^{-1}0.4488$

Solutions:

(a) With calculator set in the Degree register, enter -3.7055. Depress TAN^{-1} key and display reads -74.90. Thus, $y = -74.90°$ to the nearest hundredth of a degree. Notice that for $x < 0$, the calculator automatically gives us a negative value for y and places y in Quadrant IV.

(b) Enter 0.4488, depress $1/x$ key, and display reads 2.2282. Now depress TAN^{-1} key and we read 65.83. Therefore, $\text{Cot}^{-1}0.4488 = 65.83°$ to the nearest hundredth of a degree.

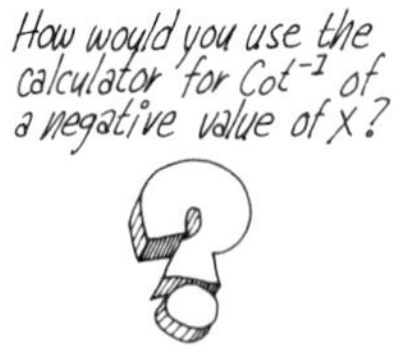

If $y = \text{Cot}^{-1}x$ and $x < 0$, then we know that $90° < y < 180°$. But we use the TAN^{-1} key to obtain $\text{Cot}^{-1}x$, and for $1/x < 0$ and $\text{Tan}^{-1}\left(\frac{1}{x}\right)$, then $-90° < y < 0°$. We therefore add $\text{Tan}^{-1}\left(\frac{1}{x}\right)$ to 180° or π to obtain $\text{Cot}^{-1}x$. For example,

$$\begin{aligned}\text{Cot}^{-1}(-1.5780) &= 180° + \text{Tan}^{-1}(-0.6337)\\ &= 180° + (-32.36°)\\ &= 147.64° \qquad \text{(to two decimal places)}\end{aligned}$$

EXAMPLE 9–15 Find the exact value of x.

(a) $3\,\text{Tan}\,x - 4 = 0$ (b) $\frac{2}{3}\,\text{Cot}\,x + 8 = 0$

Solutions:

(a) If $3\,\text{Tan}\,x - 4 = 0$, then

$$\begin{aligned}\text{Tan}\,x &= \frac{4}{3}\\ x &= \text{Tan}^{-1}\left(\frac{4}{3}\right)\end{aligned}$$

(b) If $\frac{2}{3}\,\text{Cot}\,x + 8 = 0$, then

$$\begin{aligned}\text{Cot}\,x &= -12\\ x &= \text{Cot}^{-1}(-12)\end{aligned}$$

Exercise Set 9–3

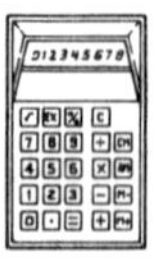

Find the approximate value of y to two decimal places. Use Table 4 or a hand calculator.

1. $y = \text{Tan}\ 0.4245$
2. $y = \text{Tan}^{-1}0.4245$
3. $y = \text{Cot}^{-1}1.540$
4. $y = \text{Cot}\ 1.540$

5. $y = \text{Tan}^{-1}(-1.111)$
6. $y = \text{Cot}^{-1}(-0.1051)$
7. $y = \text{Cot}^{-1}57.29$
8. $y = \text{Tan}^{-1}22.90$
9. $y = \text{Tan}^{-1}(-0.9325)$
10. $y = \text{Cot}^{-1}(-0.7813)$

Evaluate without tables or a calculator. State only exact values.

11. $\text{Cot}^{-1}1.000$
12. $\text{Tan}^{-1}(-1.000)$
13. $\text{Tan}^{-1}(-\sqrt{3})$
14. $\text{Cot}^{-1}\left(\frac{-1}{\sqrt{3}}\right)$
15. $\text{Cot}^{-1}(-1.000)$
16. $\text{Tan}^{-1}0$
17. $\text{Tan}\left(\text{Tan}^{-1}\left(\frac{1}{3}\right)\right)$
18. $\text{Cot}^{-1}\left(\text{Cot}\,\frac{\pi}{3}\right)$
19. $\text{Cot}\left(\text{Cot}^{-1}\left(\frac{-2}{5}\right)\right)$
20. $\text{Tan}^{-1}\left(\text{Tan}\left(\frac{-3\pi}{8}\right)\right)$
21. $\text{Tan}\left(\text{Cot}^{-1}\left(\frac{2}{3}\right)\right)$
22. $\text{Cot}^{-1}\left(\text{Tan}\left(\frac{-\pi}{4}\right)\right)$
23. $\text{Cot}(\text{Tan}^{-1}1)$
24. $\text{Tan}^{-1}\left(\text{Cot}\,\frac{3\pi}{4}\right)$
25. $\text{Tan}(\text{Cot}^{-1}\sqrt{3})$
26. $\text{Cot}^{-1}\left(\text{Tan}\left(\frac{-\pi}{3}\right)\right)$

Solve the following equations for exact values of the variables, using inverse function notation when necessary. Do not approximate the roots.

27. $2\,\text{Tan}\,x + 3 = 0$
28. $-4\,\text{Cot}\,x + \frac{3}{4} = 7$
29. $\text{Tan}^2x - 1 = 0$
30. $\text{Cot}^2x - 3 = 0$
31. $3\,\text{Cot}^2y + 4\,\text{Cot}\,y = 0$
32. $\text{Tan}^2y - 5\,\text{Tan}\,y + 6 = 0$
33. $\text{Tan}^2 2\theta - 5\,\text{Tan}\,2\theta = 0$
34. $6\,\text{Cot}^2 3\theta + \text{Cot}\,3\theta = 1$
35. $5\,\text{Cot}^2 4x + 3\,\text{Cot}\,4x = 2$
36. $20\,\text{Tan}^2 3x - 9 = 24\,\text{Tan}\,3x$

TRIANGLE INTERPRETATIONS OF INVERSE FUNCTIONS 9–4

In the first three sections of this chapter we defined the primary trigonometric functions and their inverse functions. We also showed some examples of uses for the inverse functions. Specifically, we used the inverse function to state the solution of a trigonometric equation as an exact value. Now, in Sections 9-4 and 9-5, we will study in greater depth the inverse trigonometric functions and their relationships to other aspects of trigonometry. To help us in this task, we offer the following alternate definition of the trigonometric functions.

Trigonometric functions in terms of *a*, *b*, and *r*

> **Definition:** If θ is a standard position angle with $P(a, b)$ on the terminal side of θ, then
>
> (1) $\cos\theta = \dfrac{a}{r}$ (2) $\sin\theta = \dfrac{b}{r}$
>
> (3) $\tan\theta = \dfrac{b}{a}$, $a \neq 0$ (4) $\cot\theta = \dfrac{a}{b}$, $b \neq 0$
>
> (5) $\sec\theta = \dfrac{r}{a}$, $a \neq 0$ (6) $\csc\theta = \dfrac{r}{b}$, $b \neq 0$
>
> where $r = \sqrt{a^2 + b^2}$.

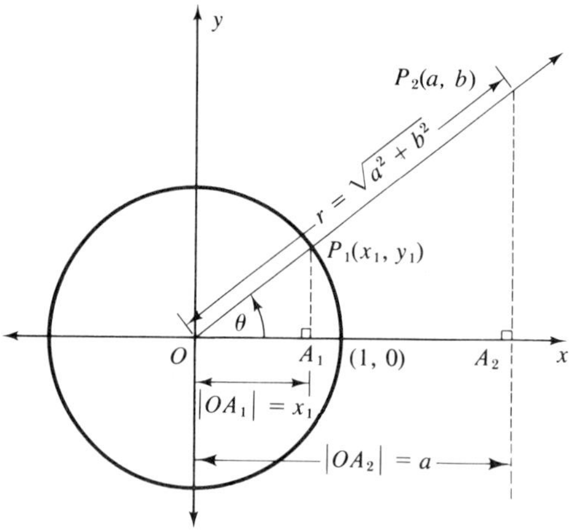

FIGURE 9-9 Angle θ in standard position with $P_2(a, b)$ on the terminal side

How does this definition of the trigonometry functions tie in with the previous definition using the coordinates of points on the unit circle?

To demonstrate the relationship between the two definitions, we show angle θ in standard position on a unit circle in Figure 9-9. As indicated, $P_1(x_1, y_1)$ is the intersection of the terminal side of θ and the unit circle. Therefore, by definition, $\cos\theta = x_1$. We now select a point $P_2(a, b)$ on the terminal side of θ, and

$$|OP_2| = r = \sqrt{a^2 + b^2}$$

Since θ is a common acute angle of right triangles OP_1A_1 and OP_2A_2, the two triangles are similar. Thus, *the ratios of the lengths of the corresponding sides of the two triangles are equal.* In particular,

$$\frac{|OA_1|}{|OP_1|} = \frac{|OA_2|}{|OP_2|}$$

But $|OA_1| = x_1$, $|OP_1| = 1$ (the radius of the unit circle is one), $|OA_2| = a$, and $|OP_2| = r$. Substituting these values in the above equation, we get

$$\frac{x_1}{1} = \frac{a}{r} \quad \text{and} \quad \cos\theta = x_1 = \frac{a}{r}$$

Therefore, $\cos\theta$ can be determined with the coordinates of any point on the

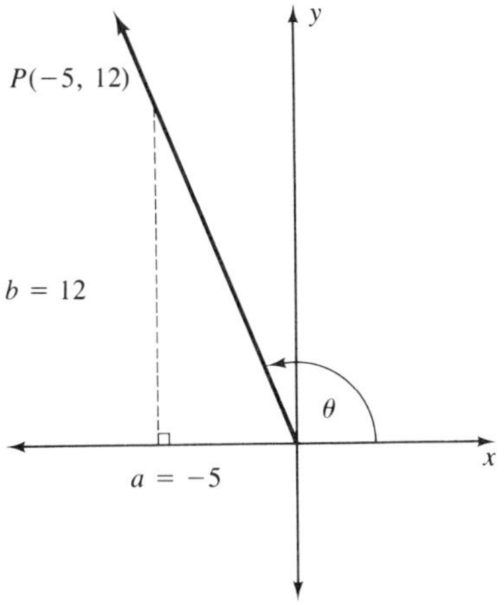

FIGURE 9-10 Standard position angle θ

terminal side of θ in standard position. Similarly, we can show that a and b can be used to compute the values of the other five trigonometric functions.

EXAMPLE 9–16 If $P(-5, 12)$ is a point on the terminal side of a standard position angle θ, find the six trigonometric function values of θ.

Solution: Angle θ is sketched in Figure 9-10. If $a = -5$ and $b = 12$, then $r = \sqrt{(-5)^2 + 12^2} = \sqrt{169} = 13$. The ratios of the definition are used to obtain the following function values:

(1) $\cos\theta = \dfrac{a}{r} = \dfrac{-5}{13}$ (2) $\sin\theta = \dfrac{b}{r} = \dfrac{12}{13}$

(3) $\tan\theta = \dfrac{b}{a} = \dfrac{-12}{5}$ (4) $\cot\theta = \dfrac{a}{b} = \dfrac{-5}{12}$

(5) $\sec\theta = \dfrac{r}{a} = \dfrac{-13}{5}$ (6) $\csc\theta = \dfrac{r}{b} = \dfrac{13}{12}$

The absolute values of a and b are always the legs of a right triangle, so long as point $P(a, b)$ is not on one of the coordinate axes (that is, $a \neq 0$ and $b \neq 0$). Otherwise, $|a|$ is the length of the *horizontal leg* and $|b|$ is the length of the *vertical leg* of the right triangle whose hypotenuse is contained in the terminal side of θ. Notice that the signs of a and b determine the quadrant in which the triangle is contained. For example, if $a < 0$ and $b > 0$, then we sketch the triangle in Quadrant II. But if $a > 0$ and $b < 0$, then the triangle lies in Quadrant IV.

Are the values of a and b always the lengths of the legs of a right triangle?

We now turn our attention to the primary trigonometric functions defined in the last three sections of this chapter. Recall that the domains of the primary functions are restricted. Therefore, for each primary function, the triangle determined by a function value must be in one of only two quadrants. For example, for $\text{Cos}\,x = a/r$, if $a/r > 0$, the triangle is sketched in Quadrant I. If $a/r < 0$, the triangle is sketched in Quadrant II, since the domain of $y = \text{Cos}\,x$ is $0 \leq x \leq \pi$. Similarly, the domain of $y = \text{Tan}\,x$ is $-\pi/2 < x < \pi/2$. Thus, for $\text{Tan}\,x > 0$, $0 < x < \pi/2$, and we sketch the triangle in Quadrant I. But for $\text{Tan}\,x < 0$, $-\pi/2 < x < 0$, and the triangle is contained in Quadrant IV.

EXAMPLE 9–17 If Sin $x = -8/17$, find the other five trigonometric function values of x.

Solution: If Sin $x = -8/17$, then $P(a, -8)$ is a point on the terminal side of x in standard position, and $r = 17$. Therefore,

$$a^2 + (-8)^2 = 17^2$$
$$a^2 = 225$$
$$a = \pm 15$$

Since Sin $x < 0$, we sketch the triangle (Figure 9-11) with acute angle x in Quadrant IV. Thus, $a = 15$. With $a = 15$, $b = -8$, $r = 17$, and the definition above, we have

(1) $\cos x = 15/17$ (2) $\tan x = -8/15$ (3) $\cot x = -15/8$
(4) $\sec x = 17/15$ (5) $\csc x = -17/8$

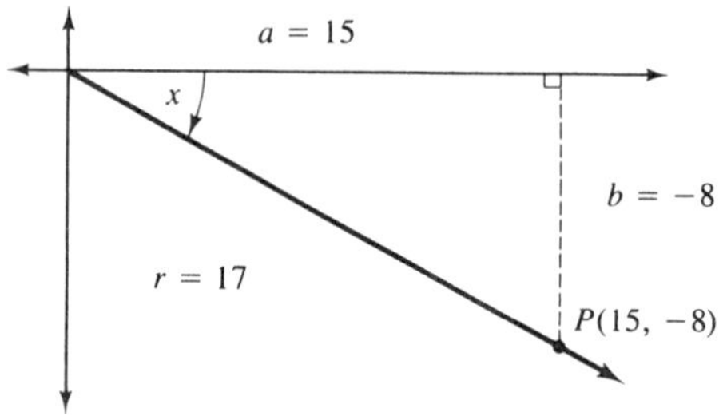

FIGURE 9-11 Sin $x = -8/17$

The inverse trigonometric functions can also be defined in terms of the coordinates of point $P(a, b)$ on the terminal side of an acute angle θ.

Inverse trigonometric functions in terms of *a*, *b*, and *r*

Definition: If $0 \le \theta \le \pi/2$ or $0° \le \theta \le 90°$ with $P(a, b)$ on the terminal side of θ, and $r = \sqrt{a^2 + b^2}$, then

(1) $\theta = \text{Cos}^{-1}\left(\frac{a}{r}\right)$ (2) $\theta = \text{Sin}^{-1}\left(\frac{b}{r}\right)$

(3) $\theta = \text{Tan}^{-1}\left(\frac{b}{a}\right)$, $a \neq 0$ (4) $\theta = \text{Cot}^{-1}\left(\frac{a}{b}\right)$, $b \neq 0$

(5) $\theta = \text{Sec}^{-1}\left(\frac{r}{a}\right)$, $a \neq 0$ (6) $\theta = \text{Csc}^{-1}\left(\frac{r}{b}\right)$, $b \neq 0$

EXAMPLE 9–18 Find $\sin\left(\text{Cos}^{-1}\left(\frac{4}{5}\right)\right)$.

Solution: We can interpret $\sin\left(\text{Cos}^{-1}\left(\frac{4}{5}\right)\right)$ as $\sin\theta$, where $\theta = \text{Cos}^{-1}\left(\frac{4}{5}\right)$ and Cos $\theta = 4/5$. That is, we want the function value of the sine of θ when it is known that the Cosine of θ is 4/5.

Since Cos $\theta > 0$, we sketch θ in Quadrant I, and $P(4, b)$ is on the terminal side of θ in Figure 9-12. Notice that $b > 0$. Then we have

$$4^2 + b^2 = 5^2$$
$$b = 3$$

Thus,

$$\sin \theta = \frac{b}{r} = \frac{3}{5}$$
$$\sin\left(\text{Cos}^{-1}\left(\frac{4}{5}\right)\right) = \frac{3}{5}$$

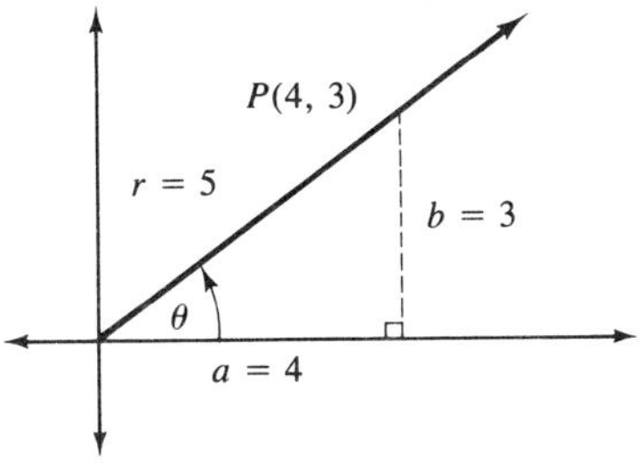

FIGURE 9-12 $\theta = \text{Cos}^{-1}\left(\frac{4}{5}\right)$

The sine function in Example 9–18 is not the primary function. However, since for Cos $\theta > 0$ and Sin $\theta > 0$ we know that $0 < \theta < \pi/2$, then

$$\text{Sin}\left(\text{Cos}^{-1}\left(\frac{4}{5}\right)\right) = \frac{3}{5}$$

Does the lowercase letter s in Example 9-18 mean that the function is not the primary sine function?

But the primary functions do not always share the same domains. For example, if Cos $\theta < 0$, then $\pi/2 < \theta < \pi$; and for Sin $\theta < 0$, then $-\pi/2 < \theta < 0$. Thus,

$$\sin\left(\text{Cos}^{-1}\left(\frac{-1}{2}\right)\right) = \sin\frac{2\pi}{3} = \frac{\sqrt{3}}{2}$$

But $\text{Sin}\left(\text{Cos}^{-1}\left(\frac{-1}{2}\right)\right)$ is undefined, since $2\pi/3$ is not in the domain of the Sine function.

Examples 9–19 through 9–21 illustrate applications of the inverse functions. Where needed, a triangle is sketched to assist us in obtaining a solution to a given problem.

EXAMPLE 9–19 Approximate to two decimal places the value of tan y if $y = 3 \sec x - 5 \sin x$ and $x = \text{Cos}^{-1}\left(\frac{24}{25}\right)$.

Solution: If $x = \text{Cos}^{-1}\left(\frac{24}{25}\right)$, then $\text{Cos}\, x = \frac{24}{25}$. Thus, $P(24, b)$ is on the terminal side of x with $r = 25$. Then we have

$$24^2 + b^2 = 25^2$$
$$b = 7$$

The triangle is sketched in Figure 9-13. Now we have

$$\sec x = \frac{r}{a} = \frac{25}{24} \quad \text{and} \quad \sin x = \frac{b}{r} = \frac{7}{25}$$

Substituting these values in the given equation, we get

$$y = 3\left(\frac{25}{24}\right) - 5\left(\frac{7}{25}\right)$$
$$= \frac{69}{40} = 1.7250$$

Thus, $\tan y = \tan 1.725 = -6.43$ to two decimal places.

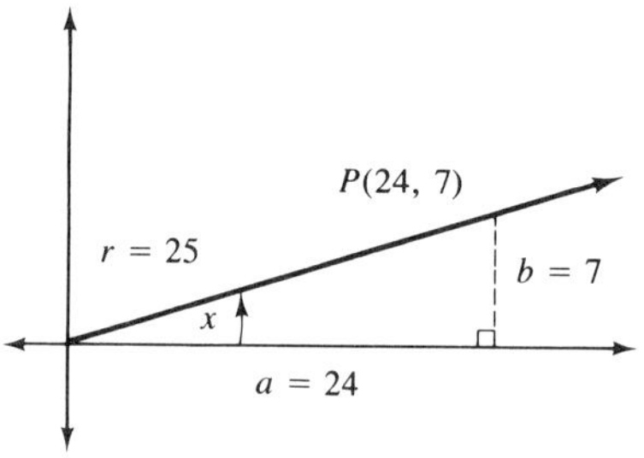

FIGURE 9-13 $x = \text{Cos}^{-1}\left(\frac{24}{25}\right)$

EXAMPLE 9–20 If a point $P(x, y)$ moves in a plane according to the equations $x = 2\cos t$ and $y = 1 + \sin t$, find the coordinates of P when $t = \text{Tan}^{-1}\left(\frac{12}{5}\right)$.

Solution: If $t = \text{Tan}^{-1}\left(\frac{12}{5}\right)$, then $a = 5$, $b = 12$, and $r = \sqrt{5^2 + 12^2} = 13$. Therefore, $\cos t = 5/13$ and $\sin t = 12/13$. Since

$$x = 2\cos t \quad \text{and} \quad y = 1 + \sin t$$

then

$$x = 2(5/13) \quad \text{and} \quad y = 1 + 12/13$$
$$x = 10/13 \quad \text{and} \quad y = 25/13$$

Thus, for $t = \text{Tan}^{-1}\left(\frac{12}{5}\right)$, $P(x, y) = P(10/13, 25/13)$.

EXAMPLE 9–21 Evaluate $\text{Cos}(\text{Sin}^{-1} z)$, and $0 < z < 1$.

Solution: If $\theta = \text{Sin}^{-1} z$, and $0 < z < 1$, then θ is a Quadrant I angle. The triangle is sketched in Figure 9-14. With $b = z$ and $r = 1$, then

$$a^2 + z^2 = 1$$
$$a = \sqrt{1 - z^2}$$

Therefore,

$$\text{Cos}\,\theta = \frac{a}{r} = \frac{\sqrt{1 - z^2}}{1} = \sqrt{1 - z^2}$$

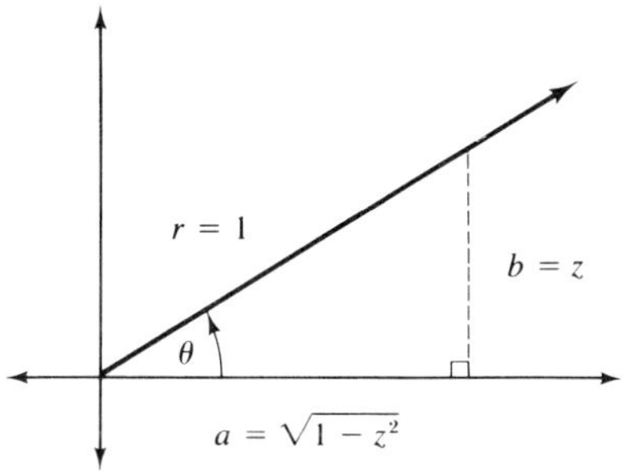

FIGURE 9-14 $\theta = \text{Sin}^{-1} z$

Exercise Set 9–4

Find the six trigonometric function values of a standard position angle θ with the given point on the terminal side of the angle.

1. $P(4, 3)$
2. $P(-8, 15)$
3. $P(7, -24)$
4. $P(-3, -4)$
5. $P(5, 12)$
6. $P(-15, 8)$
7. $P(-1, 3)$
8. $P(2, -3)$
9. $P(-4, -1)$
10. $P(3, -5)$

For the given function value of x, find the other five trigonometric function values of x. Notice that the other functions are not primary functions.

11. $\text{Cos}\,x = -12/13$
12. $\text{Sin}\,x = 5/13$
13. $\text{Tan}\,x = -15/8$
14. $\text{Sec}\,x = 25/7$
15. $\text{Cos}\,x = 0.3$
16. $\text{Tan}\,x = 2$
17. $\text{Sin}\,x = -0.1$
18. $\text{Csc}\,x = -3$
19. $\text{Cot}\,x = 3/2$
20. $\text{Cos}\,x = 2/3$

Evaluate without a table or calculator. When necessary, sketch a triangle to assist in determining a given function.

21. $\text{Sin}^{-1}\left(\cos \frac{3\pi}{4}\right)$
22. $\sec\left(\text{Cot}^{-1}\left(\frac{5}{12}\right)\right)$
23. $\text{Sin}^{-1}\left(\cos \frac{\pi}{3}\right)$
24. $\cos\left(\text{Sin}^{-1}\left(\frac{2}{3}\right)\right)$

25. $\cot\left(\text{Cos}^{-1}\left(\frac{15}{17}\right)\right)$

26. $\tan\left(\text{Sin}^{-1}\left(\frac{-1}{2}\right)\right)$

27. $\tan\left(\text{Sec}^{-1}\left(\frac{5}{4}\right)\right)$

28. $\text{Cos}^{-1}\left(\sin\frac{2\pi}{3}\right)$

29. $\text{Cos}^{-1}\left(\tan\left(\frac{-\pi}{4}\right)\right)$

30. $\tan\left(\text{Sin}^{-1}\left(\frac{8}{17}\right)\right)$

Find the value of y. When possible, state the value of y as an exact value. When necessary, approximate the value of y to two decimal places.

31. $y = \cos x - 2\sin x$, where $x = \text{Cot}^{-1}\left(\frac{4}{3}\right)$

32. $y = \sec x + \tan x$, where $x = \text{Sin}^{-1}\left(\frac{7}{25}\right)$

33. $y = 3\sin x - \tan x$, where $x = \text{Csc}^{-1}\left(\frac{-17}{8}\right)$

34. $y = \cos z$, where $z = \dfrac{\sin x}{\cos x}$ and $x = \text{Tan}^{-1}\left(\frac{1}{2}\right)$

35. $y = \sin(2\cos x + 3\tan x)$, where $x = \text{Cot}^{-1}\left(\frac{3}{4}\right)$

36. $y = \tan\left(\text{Sin}^{-1}\left(\frac{7}{25}\right)\right) + \cos\left(\text{Sin}^{-1}\left(\frac{-7}{25}\right)\right)$

37. If a point $P(x, y)$ moves in a plane according to the equations $x = 2\sin t - 4$ and $y = 2\cos t - 3$, find the coordinates when $t = \text{Sec}^{-1}\left(\frac{13}{5}\right)$.

38. A point $Q(x, y)$ moves in a plane according to the equations $x = \frac{3}{2}\sec t$ and $y = \frac{-4}{5}\cos t + 1$. Find the coordinates of Q when $t = \text{Sin}^{-1}\left(\frac{1}{2}\right)$.

Evaluate in terms of z.

39. $\sin(\text{Cos}^{-1} z)$, $0 < z < 1$

40. $\tan(\text{Sec}^{-1} 2z)$, $z > \frac{1}{2}$

41. $\csc(\text{Cot}^{-1} z)$, $z < 0$

42. $\cos\left(\text{Tan}^{-1}\left(\frac{z}{3}\right)\right)$, $z > 0$

43. $\sin(\text{Sin}^{-1} z)$, $0 < z < 1$

44. $\cos(\text{Sec}^{-1} z)$, $z > 1$

9-5 TRIGONOMETRIC IDENTITIES AND INVERSE FUNCTIONS

The inverse trigonometric functions are used extensively in the study of calculus. It is essential, therefore, that you have a good understanding of the basic concepts of these functions and are able to manipulate expressions containing them.

In this section we will study expressions and equations that include the inverse trigonometric functions. We will also use the triangle interpretation of the value of an inverse function to solve several examples in this section.

EXAMPLE 9–22 Evaluate $\cos\left(\text{Sin}^{-1}\left(\frac{3}{5}\right) - \text{Tan}^{-1}\left(\frac{5}{12}\right)\right)$.

Solution: The given expression can be interpreted as $\cos(\alpha - \beta)$, where $\alpha = \text{Sin}^{-1}\left(\frac{3}{5}\right)$ and $\beta = \text{Tan}^{-1}\left(\frac{5}{12}\right)$ and α and β are standard position angles. Then,

$$\cos\left(\text{Sin}^{-1}\left(\frac{3}{5}\right) - \text{Tan}^{-1}\left(\frac{5}{12}\right)\right) = \cos(\alpha - \beta)$$

and

$$\cos(\alpha - \beta) = \cos\alpha\cos\beta + \sin\alpha\sin\beta$$

That is,

$$\cos(\alpha - \beta) = \cos\left(\text{Sin}^{-1}\left(\frac{3}{5}\right)\right)\cos\left(\text{Tan}^{-1}\left(\frac{5}{12}\right)\right) + \sin\left(\text{Sin}^{-1}\left(\frac{3}{5}\right)\right)\sin\left(\text{Tan}^{-1}\left(\frac{5}{12}\right)\right)$$

In Figure 9-15(a) and (b), we sketch α and β as acute angles of triangles in Quadrant I. Since $\alpha = \text{Sin}^{-1}\left(\frac{3}{5}\right)$, then in triangle (a), $b = 3$, $r = 5$, and $a = \sqrt{5^2 - 3^2} = 4$. Furthermore, if $\beta = \text{Tan}^{-1}\left(\frac{5}{12}\right)$, then in triangle (b), $b = 5$, $a = 12$, and $r = \sqrt{5^2 + 12^2} = 13$.

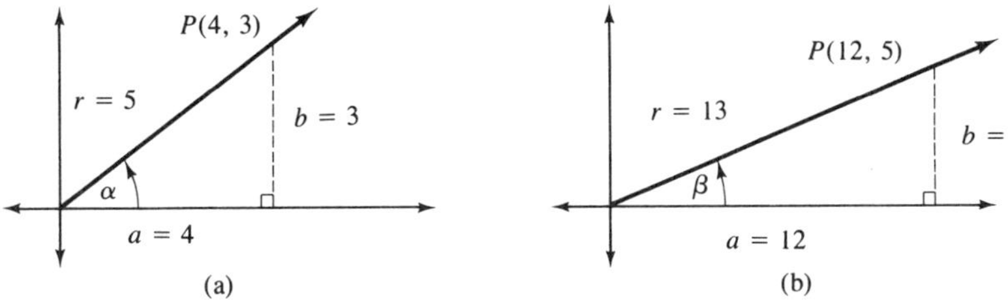

FIGURE 9-15 (a) $\alpha = \text{Sin}^{-1}\left(\frac{3}{5}\right)$; (b) $\beta = \text{Tan}^{-1}\left(\frac{5}{12}\right)$

Therefore, $\cos\alpha = 4/5$, $\sin\alpha = 3/5$, $\cos\beta = 12/13$, and $\sin\beta = 5/13$. Substituting these values in the equation for $\cos(\alpha - \beta)$, we get

$$\cos\left(\text{Sin}^{-1}\left(\frac{3}{5}\right) - \text{Tan}^{-1}\left(\frac{5}{12}\right)\right) = \left(\frac{4}{5}\right)\left(\frac{12}{13}\right) + \left(\frac{3}{5}\right)\left(\frac{5}{13}\right)$$

$$= \frac{48}{65} + \frac{15}{65} = \frac{63}{65}$$

We must remind ourselves of the *circular function* interpretation of trigonometry. That is, $\text{Sin}^{-1}\left(\frac{3}{5}\right)$ might be the sine of a real number x and $\text{Tan}^{-1}\left(\frac{5}{12}\right)$ the tangent of a real number y. The expression in Example 9–22 is then the cosine of the difference of two real numbers and not the difference in the measures of two angles in standard position. The angle interpretation provides a geometric model that helps us to determine the value of an expression without using tables or a calculator.

Why did you say that the expression in Example 9-22 can be interpreted as the cosine of the difference of two angles in standard position?

EXAMPLE 9–23 If $x = \text{Tan}^{-1}2$, find $\cos 2x$.

Solution: If $x = \text{Tan}^{-1}2$, then $\text{Tan}\, x = 2$ and $0 < x < \pi/2$. In Figure 9-16, x is shown as an acute angle of a triangle in Quadrant I. With $a = 1$ and $b = 2$, then $r = \sqrt{5}$. Now

$$\cos x = \frac{a}{r} = \frac{1}{\sqrt{5}}$$

and by the identity $\cos 2x = 2\cos^2 x - 1$, we have

$$\cos 2x = 2\left(\frac{1}{\sqrt{5}}\right)^2 - 1$$
$$= \frac{-3}{5}$$

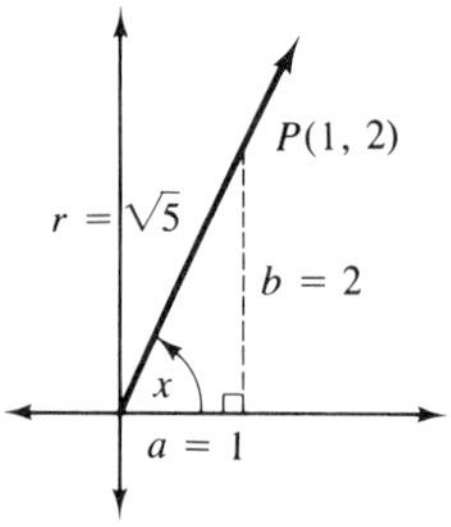

FIGURE 9-16 $x = \text{Tan}^{-1}2$

How can cos 2x be a negative number when x is between 0 and π/2?

If x is an acute angle in Quadrant I, then $2x$ can be an obtuse angle in Quadrant II. That is, if $\pi/4 < x < \pi/2$, then $\pi/2 < 2x < \pi$. We can check the value obtained in Example 9–23 very easily with a calculator set in the Radian register. Enter 2 in the calculator and depress TAN^{-1} key. Display reads 1.1071 to four decimal places. Multiply by 2 (since we want the cosine of $2x$) and the display reads 2.2143. Now depress COS key and display reads -0.6000, that is, $-3/5$.

EXAMPLE 9–24 Verify that the following equation is true:

$$\text{Cos}^{-1}\left(\frac{-4}{5}\right) - \text{Sin}^{-1}\left(\frac{5}{13}\right) = \text{Tan}^{-1}\left(\frac{-56}{33}\right)$$

Solution: With the angle measure interpretation of an inverse function, the given equation is of the form

$$\alpha - \beta = \gamma$$

where $\alpha = \text{Cos}^{-1}\left(\frac{-4}{5}\right)$, $\beta = \text{Sin}^{-1}\left(\frac{5}{13}\right)$, and $\gamma = \text{Tan}^{-1}\left(\frac{-56}{33}\right)$. To establish that $\alpha - \beta = \gamma$, we will show that a *trigonometric function value of* $\alpha - \beta$ *is equal to the same function value of* γ. Any trigonometric function would work, but we use the tangent function, since $\tan(\text{Tan}^{-1}x) = x$ for any x in the domain of the inverse tangent function. Thus, we will attempt to show that

$$\tan(\alpha - \beta) = \tan \gamma$$

or

$$\tan\left(\text{Cos}^{-1}\left(\frac{-4}{5}\right) - \text{Sin}^{-1}\left(\frac{5}{13}\right)\right) = \tan\left(\text{Tan}^{-1}\left(\frac{-56}{33}\right)\right)$$

In Figure 9-17(a) and (b) the triangle representations of α and β are shown. From the given figures we see that $\tan\alpha = -3/4$ and $\tan\beta = 5/12$. In the identity

$$\tan(\alpha - \beta) = \frac{\tan\alpha - \tan\beta}{1 + \tan\alpha\tan\beta}$$

we substitute the values for $\tan\alpha$ and $\tan\beta$, which yields

$$\tan(\alpha - \beta) = \frac{\frac{-3}{4} - \frac{5}{12}}{1 + \left(\frac{-3}{4}\right)\left(\frac{5}{12}\right)}$$

Multiplying numerator and denominator by 48/48, we get

$$\tan(\alpha - \beta) = \frac{-36 - 20}{48 - 15} = \frac{-56}{33}$$

Since $\tan\left(\text{Tan}^{-1}\left(\frac{-56}{33}\right)\right) = \frac{-56}{33}$, the given equation is verified as true.

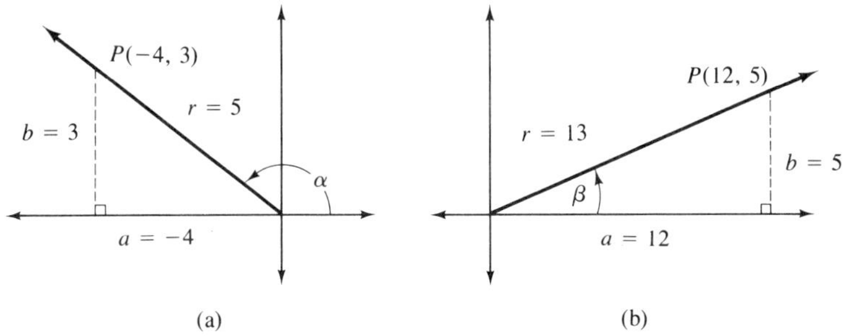

FIGURE 9-17 (a) $\alpha = \text{Cos}^{-1}\left(\frac{-4}{5}\right)$; (b) $\beta = \text{Sin}^{-1}\left(\frac{5}{13}\right)$

Sometimes we encounter conditional equations in which at least one term of the equation is an inverse trigonometric function. The following example illustrates a method of solving such an equation.

EXAMPLE 9–25 Solve $\text{Cos}^{-1}x - \frac{\pi}{2} = \text{Sin}^{-1}(x - 1)$ for x.

Solution: The given equation is of the form $\alpha - \beta = \gamma$. With this interpretation, $\alpha = \text{Cos}^{-1}x$, $\beta = \pi/2$, and $\gamma = \text{Sin}^{-1}(x - 1)$. If $x - 1$ is in the domain of the Sine^{-1} function, then

$$-1 \le x - 1 \le 1$$
$$0 \le x \le 2$$

If x is in the domain of the Cosine^{-1} function, then $-1 \leq x \leq 1$. To simultaneously satisfy both conditions, the acceptable range of values for x is between 0 and 1, that is, $0 \leq x \leq 1$. With $0 \leq x \leq 1$ and the equation $\alpha - \beta = \gamma$, then

$$\sin(\alpha - \beta) = \sin \gamma$$

$$\sin\left(\text{Cos}^{-1}x - \frac{\pi}{2}\right) = \sin(\text{Sin}^{-1}(x - 1))$$

$$\sin(\text{Cos}^{-1}x)\cos\frac{\pi}{2} - \cos(\text{Cos}^{-1}x)\sin\frac{\pi}{2} = \sin(\text{Sin}^{-1}(x - 1)) \qquad \textbf{(1)}$$

But $\cos\frac{\pi}{2} = 0$; thus, $\sin(\text{Cos}^{-1}x)\cos\frac{\pi}{2} = 0$. Furthermore, $\cos(\text{Cos}^{-1}x) = x$ for $0 \leq x \leq 1$ and $\sin\frac{\pi}{2} = 1$; thus, $\cos(\text{Cos}^{-1}x)\sin\frac{\pi}{2} = x$. Finally, for $0 \leq x \leq 1$, $\sin(\text{Sin}^{-1}(x - 1)) = x - 1$. When we substitute each of these values in Equation (1), we get

$$0 - x = x - 1$$

$$2x = 1$$

$$x = \frac{1}{2}$$

Check. If $x = 1/2$, then

$$\text{Cos}^{-1}\left(\frac{1}{2}\right) - \frac{\pi}{2} = \text{Sin}^{-1}\left(\frac{1}{2} - 1\right)$$

$$\frac{\pi}{3} - \frac{\pi}{2} = \frac{-\pi}{6}$$

$$\frac{-\pi}{6} = \frac{-\pi}{6} \qquad \text{(True)}$$

Thus, the solution set is $\{1/2\}$. Notice that $x = 1/2$ and $x - 1 = -1/2$ are both values in the domains of the Cosine^{-1} and Sine^{-1} functions. If the computed value of x did not occur in both domains, the solution set would be the null set.

Exercise Set 9–5

Evaluate the following expressions. Use exact values only; do not approximate answers with tables or a calculator.

1. $\sin 2\left(\text{Tan}^{-1}\left(\frac{8}{15}\right)\right)$
2. $\cos 2\left(\text{Cos}^{-1}\left(\frac{7}{25}\right)\right)$
3. $\tan 2\left(\text{Cos}^{-1}\left(\frac{-1}{2}\right)\right)$
4. $\sin 2\left(\text{Sin}^{-1}\left(\frac{-1}{3}\right)\right)$
5. $\cos 2(\text{Sec}^{-1}(-3))$
6. $\tan 2(\text{Cot}^{-1}4)$
7. $\cos\frac{1}{2}\left(\text{Sin}^{-1}\left(\frac{3}{5}\right)\right)$
8. $\sin\frac{1}{2}\left(\text{Tan}^{-1}\left(\frac{-1}{4}\right)\right)$
9. $\tan\frac{1}{2}\left(\text{Cos}^{-1}\left(\frac{3}{4}\right)\right)$
10. $\cos\frac{1}{2}\left(\text{Cos}^{-1}\left(\frac{2}{3}\right)\right)$
11. $\sin\frac{1}{2}(\text{Cos}^{-1}x)$, $x > 0$
12. $\tan\frac{1}{2}(\text{Sin}^{-1}x)$, $x > 0$

13. $\cos\left(\text{Sin}^{-1}\left(\frac{3}{5}\right) + \text{Cos}^{-1}\left(\frac{12}{13}\right)\right)$

14. $\sin(\text{Tan}^{-1}(\sqrt{3}) + \text{Cot}^{-1}(-1))$

15. $\tan\left(\text{Cos}^{-1}\left(\frac{1}{2}\right) - \text{Csc}^{-1}2\right)$

16. $\cos(\text{Tan}^{-1}x - \text{Cot}^{-1}x),\ x > 0$

Verify that the following equations are true.

17. $\text{Sin}^{-1}\left(\frac{1}{2}\right) + \text{Cos}^{-1}\left(\frac{1}{2}\right) = \frac{\pi}{2}$

18. $\text{Cos}^{-1}\left(\frac{3}{\sqrt{10}}\right) - \text{Sin}^{-1}\left(\frac{-1}{\sqrt{5}}\right) = \frac{\pi}{4}$

19. $\text{Tan}^{-1}1 + \text{Cot}^{-1}(-1) = \pi$

20. $\text{Sec}^{-1}3 + \text{Csc}^{-1}3 = \frac{\pi}{2}$

21. $\text{Sin}^{-1}\left(\frac{3}{\sqrt{13}}\right) + \text{Tan}^{-1}\left(\frac{3}{2}\right) = \text{Cos}^{-1}\left(\frac{-5}{13}\right)$

22. $\text{Sin}^{-1}\left(\frac{4}{5}\right) - \text{Sin}^{-1}\left(\frac{5}{13}\right) = \text{Sin}^{-1}\left(\frac{33}{65}\right)$

23. $2\ \text{Tan}^{-1}2 - \frac{3\pi}{4} = \text{Tan}^{-1}\left(\frac{-1}{7}\right)$

24. $2\ \text{Cos}^{-1}\left(\frac{3}{5}\right) + \frac{\pi}{4} = \text{Cos}^{-1}\left(\frac{-31\sqrt{2}}{50}\right)$

Solve for x.

25. $\text{Sin}^{-1}x - \text{Cos}^{-1}x = 0$

26. $\text{Sin}^{-1}2x + \text{Sin}^{-1}x = 0$

27. $\text{Sec}^{-1}x = \pi - \text{Sec}^{-1}(-x)$

28. $\text{Tan}^{-1}x = \text{Tan}^{-1}2 - \text{Tan}^{-1}3x$

29. $\text{Sin}^{-1}x + \text{Cos}^{-1}(1 - x) = 0$

30. $\text{Sin}^{-1}\left(\frac{1}{x}\right) - \text{Sin}^{-1}\left(\frac{4}{5}\right) = 2\ \text{Tan}^{-1}\left(\frac{1}{2}\right)$

SUMMARY

Definition: $y = \text{Sin}\ x,\ -\pi/2 \leq x \leq \pi/2,\ -1 \leq y \leq 1$ — **Primary sine function**

Definition: $y = \text{Csc}\ x,\ -\pi/2 \leq x \leq \pi/2,\ x \neq 0,\ |y| \geq 1$ — **Primary cosecant function**

Definition: $y = \text{Cos}\ x,\ 0 \leq x \leq \pi,\ -1 \leq y \leq 1$ — **Primary cosine function**

Definition: $y = \text{Sec}\ x,\ 0 \leq x \leq \pi,\ x \neq \pi/2,\ |y| \geq 1$ — **Primary secant function**

Definition: $y = \text{Tan}\ x,\ -\pi/2 < x < \pi/2,\ y \in \{\text{reals}\}$ — **Primary tangent function**

Definition: $y = \text{Cot}\ x,\ 0 < x < \pi,\ y \in \{\text{reals}\}$ — **Primary cotangent function**

Definition: The inverse function of Sine is — **Sine^{-1}**

$$y = \text{Sin}^{-1}x,\ -1 \leq x \leq 1,\ -\pi/2 \leq y \leq \pi/2$$

Cosecant^{-1}

Definition: The inverse function of Cosecant is

$$y = \text{Csc}^{-1}x,\ |x| \geq 1,\ -\pi/2 \leq y \leq \pi/2,\ y \neq 0$$

Cosine^{-1}

Definition: The inverse function of Cosine is $y = \text{Cos}^{-1}x,\ -1 \leq x \leq 1,\ 0 \leq y \leq \pi$.

Secant^{-1}

Definition: The inverse function of Secant is

$$y = \text{Sec}^{-1}x,\ |x| \geq 1,\ 0 \leq y \leq \pi,\ y \neq \pi/2$$

Tangent^{-1}

Definition: The inverse function of Tangent is

$$y = \text{Tan}^{-1}x,\ x \in \{\text{reals}\},\ -\pi/2 < y < \pi/2$$

Cotangent^{-1}

Definition: The inverse function of Cotangent is $y = \text{Cot}^{-1}x,\ x \in \{\text{reals}\},\ 0 < y < \pi$.

Trigonometric functions in terms of *a*, *b*, and *r*

Definition: If θ is a standard position angle with $P(a, b)$ on the terminal side of θ, then

(1) $\cos\theta = a/r$ (2) $\sin\theta = b/r$

(3) $\tan\theta = b/a,\ a \neq 0$ (4) $\cot\theta = a/b,\ b \neq 0$

(5) $\sec\theta = r/a,\ a \neq 0$ (6) $\csc\theta = r/b,\ b \neq 0$

where $r = \sqrt{a^2 + b^2}$.

Inverse trigonometric functions in terms of *a*, *b*, and *r*

Definition: If $0 \leq \theta \leq \pi/2$ or $0° \leq \theta \leq 90°$ with $P(a, b)$ on the terminal side of θ, and $r = \sqrt{a^2 + b^2}$, then

(1) $\theta = \text{Cos}^{-1}\left(\frac{a}{r}\right)$ (2) $\theta = \text{Sin}^{-1}\left(\frac{b}{r}\right)$

(3) $\theta = \text{Tan}^{-1}\left(\frac{b}{a}\right),\ a \neq 0$ (4) $\theta = \text{Cot}^{-1}\left(\frac{a}{b}\right),\ b \neq 0$

(5) $\theta = \text{Sec}^{-1}\left(\frac{r}{a}\right),\ a \neq 0$ (6) $\theta = \text{Csc}^{-1}\left(\frac{r}{b}\right),\ b \neq 0$

SELF-CHECKING EXERCISE

Solve for x. **(9-1)**

1. $\text{Csc}\,x = 1$
2. $\text{Sin}\,x = 0.5$
3. $\text{Sin}\,x = -1$
4. $\sqrt{3}\,\text{Csc}\,x = 2$
5. $x = \text{Sin}^{-1}0$
6. $x = \text{Csc}^{-1}(-2)$

Using Table 4, find the value of y in degrees. **(9-1)**

7. $y = \text{Sin}^{-1}0.6157$
8. $y = \text{Csc}^{-1}1.099$
9. $y = \text{Sin}^{-1}(-0.9799)$
10. $y = \text{Csc}^{-1}(-1.630)$

Solve for x. **(9-2)**

11. $\text{Cos}\,x = -0.5$
12. $\text{Sec}\,x = 1$
13. $\text{Sec}\,x = -\sqrt{2}$
14. $2\,\text{Cos}\,x = -\sqrt{3}$
15. $x = \text{Cos}^{-1}0$
16. $x = \text{Sec}^{-1}2$

Using Table 4, find the value of y in degrees. **(9-2)**

17. $y = \text{Cos}^{-1}(-0.6428)$
18. $y = \text{Sec}^{-1}1.082$
19. $y = \text{Sec}^{-1}(-5.575)$
20. $y = \text{Cos}^{-1}0.5783$

Solve for x. **(9-3)**

21. $\text{Cot}\, x = -1$
22. $\text{Tan}\, x = -1$
23. $\text{Tan}\, x = \sqrt{3}$
24. $\text{Cot}\, x = \sqrt{3}$
25. $x = \text{Cot}^{-1}0$
26. $x = \text{Tan}^{-1}0$

Using Table 4, find the value of y in degrees. **(9-3)**

27. $y = \text{Cot}^{-1}(-1.327)$
28. $y = \text{Tan}^{-1}2.006$
29. $y = \text{Tan}^{-1}(-0.2035)$
30. $y = \text{Cot}^{-1}0.4699$

Find the six trigonometric function values of θ if the given point is on θ and θ is in standard position. **(9-4)**

31. $P(-7, -24)$
32. $P(5, -1)$

For the given function value of x, find the other five function values. **(9-4)**

33. $\text{Tan}\, x = 4/3$
34. $\text{Sec}\, x = -5$

Evaluate. **(9-4)**

35. $\csc\left(\text{Sin}^{-1}\left(\frac{3}{7}\right)\right)$
36. $\text{Cos}^{-1}\left(\text{Cos}\frac{\pi}{5}\right)$
37. $\cot(\text{Tan}^{-1}4)$
38. $\tan(\text{Sec}^{-1}(-\sqrt{3}))$
39. $\sin\left(\text{Cot}^{-1}\left(\frac{5}{12}\right)\right)$
40. $\cos\left(\text{Sin}^{-1}\left(\frac{\sqrt{2}}{5}\right)\right)$

Evaluate the following expressions. **(9-5)**

41. $\cos 2\left(\text{Tan}^{-1}\left(\frac{4}{3}\right)\right)$
42. $\sin 2(\text{Sec}^{-1}(-3))$
43. $\tan\left(\text{Cos}^{-1}\left(\frac{8}{17}\right) + \text{Sin}^{-1}\left(\frac{3}{5}\right)\right)$
44. $\cos(\text{Sec}^{-1}(-2) - \text{Csc}^{-1}2)$
45. $\tan\frac{1}{2}\left(\text{Sin}^{-1}\left(\frac{5}{13}\right)\right)$
46. $\sin\frac{1}{2}\left(\text{Tan}^{-1}\left(\frac{-12}{5}\right)\right)$

Solve the following equations for x. **(9-5)**

47. $\text{Tan}\, x + \sqrt{1 - 2\,\text{Tan}^2 x} = 0$
48. $\text{Sin}^{-1}2x + \text{Sin}^{-1}x = \pi/3$
49. Verify $\text{Tan}^{-1}\left(\frac{1}{3}\right) + \text{Tan}^{-1}\left(\frac{1}{7}\right) = \text{Tan}^{-1}\left(\frac{1}{2}\right)$ **(9-5)**

Answers to Self-Checking Exercise

1. $\pi/2$	2. $\pi/6$	3. $-\pi/2$	4. $\pi/3$
5. 0	6. $-\pi/6$	7. 38°	8. 65°30′
9. −78°30′	10. −37°50′	11. $2\pi/3$	12. 0
13. $3\pi/4$	14. $5\pi/6$	15. $\pi/2$	16. $\pi/3$
17. 130°	18. 22°30′	19. 100°20′	20. 54°40′
21. $3\pi/4$	22. $\pi/4$	23. $\pi/3$	24. $\pi/6$
25. $\pi/2$	26. 0	27. 143°	28. 63°30′
29. −11°30′	30. 64°50′		

31. $\cos\theta = -7/25$; $\sin\theta = -24/25$; $\tan\theta = 24/7$; $\cot\theta = 7/24$; $\sec\theta = -25/7$; $\csc\theta = -25/24$

32. $\cos\theta = 5/\sqrt{26}$; $\sin\theta = -1/\sqrt{26}$; $\tan\theta = -1/5$; $\cot\theta = -5$; $\sec\theta = \sqrt{26}/5$; $\csc\theta = -\sqrt{26}$

33. $\cos x = 3/5$; $\sin x = 4/5$; $\cot x = 3/4$; $\sec x = 5/3$; $\csc x = 5/4$

34. $\cos x = -1/5$; $\sin x = (2\sqrt{6})/5$; $\tan x = -2\sqrt{6}$; $\cot x = -1/(2\sqrt{6})$; $\csc x = 5/(2\sqrt{6})$

35. 7/3	36. $\pi/5$	37. 1/4	38. $-\sqrt{2}$
39. 12/13	40. $\sqrt{23}/5$	41. −7/25	42. $(-4\sqrt{2})/9$
43. −84/13	44. 0	45. 1/5	46. $(2\sqrt{13})/13$

47. $\{\pi/6, -\pi/6\}$

10 Applications of Trigonometry

Trigonometric concepts have applications in many areas of mathematics that are quite different from the narrow view of trigonometry we have seen thus far. Therefore, in this chapter we will study several topics that require some application of trigonometry. In this way we intend to increase our appreciation of trigonometry as a discipline. We will also consider some of the applications of trigonometry in the physical world.

10-1 SOLUTION OF RIGHT TRIANGLES AND THE LAW OF SINES

In Sections 10-1 and 10-2 we study the process known as **solving a triangle.** We first learn to solve right triangles. We then study two principles known as the **Law of Sines** and the **Law of Cosines,** which enable us to solve most triangles for which certain information is known.

We know that any triangle contains six parts — three angles and three sides. To solve a triangle we must determine, if possible, the measures of the three angles and the lengths of the three sides.

What is meant by solving a triangle?

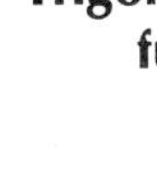

Solution of Right Triangles

The definition of the trigonometric functions stated in Chapter 9 is restated below. We use the ratios in the equations to solve a right triangle.

Trigonometric functions

> **Definition:** If $P(a, b)$ is a point on the terminal side of a standard position angle θ and $r = \sqrt{a^2 + b^2}$, then
>
> (1) $\cos\theta = \dfrac{a}{r}$ (2) $\sin\theta = \dfrac{b}{r}$
>
> (3) $\tan\theta = \dfrac{b}{a},\ a \neq 0$ (4) $\cot\theta = \dfrac{a}{b},\ b \neq 0$
>
> (5) $\sec\theta = \dfrac{r}{a},\ a \neq 0$ (6) $\csc\theta = \dfrac{r}{b},\ b \neq 0$

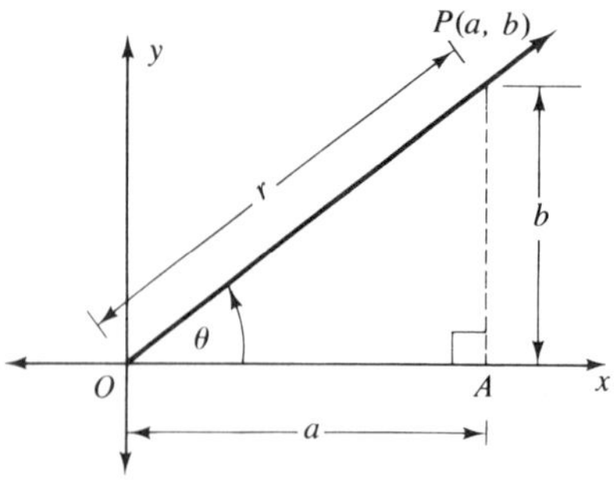

FIGURE 10-1 Acute angle θ. $\overline{OA}$ is leg adjacent θ; $\overline{AP}$ is leg opposite θ; $\overline{OP}$ is hypotenuse of triangle OAP

In Figure 10-1 an acute angle θ is shown with $P(a, b)$ on the terminal side. When line segment $\overline{AP}$ is drawn, right triangle OAP is formed. Line segment $\overline{OA}$ is called the **leg adjacent** angle θ. Line segment $\overline{AP}$ is called the **leg opposite** angle θ, and $\overline{OP}$ is the **hypotenuse** of the triangle. From the figure we note that

1. $a = |OA|$ = length of the leg adjacent θ
2. $b = |AP|$ = length of the leg opposite θ
3. $r = |OP|$ = length of the hypotenuse

When the word interpretations for a, b, and r are substituted in the equations in the definition, we get the following equations:

(1) $\cos\theta = \dfrac{\text{length of leg adj}}{\text{length of hyp}}$ (2) $\sin\theta = \dfrac{\text{length of leg opp}}{\text{length of hyp}}$

(3) $\tan\theta = \dfrac{\text{length of leg opp}}{\text{length of leg adj}}$ (4) $\cot\theta = \dfrac{\text{length of leg adj}}{\text{length of leg opp}}$

(5) $\sec\theta = \dfrac{\text{length of hyp}}{\text{length of leg adj}}$ (6) $\csc\theta = \dfrac{\text{length of hyp}}{\text{length of leg opp}}$

The principal advantage to these equations over the ones in the definition is that the right triangle is not tied to a coordinate system. That is, in a right triangle, *the function values of an acute angle θ are related to the lengths of the sides of the triangle*. Therefore, we can solve a right triangle without first placing the triangle on a coordinate axis.

By definition, a right triangle has a 90° angle. The unknown values of three parts of the triangle can be determined if the values of two parts are known, and the parts are either (a) the lengths of any two sides of the triangle, or (b) the length of any side and the measure of one of the acute angles.

The following examples illustrate a procedure for solving a right triangle. The flowchart that parallels the solution to Example 10–1 outlines a recommended format. We use the familiar convention that in right triangle ABC, angle C is the right angle. Furthermore, lowercase letters a, b, and c are the lengths of the sides opposite A, B, and C, respectively. Finally, *we will compute the values of the unknown parts of the triangle to the same degree of accuracy as the values of the given parts.*

EXAMPLE 10–1 Solve right triangle ABC if $A = 35°$ and $c = 10.0$.

Solution:

(1) Construct a right triangle and indicate the values of the given parts on the figure. Also, label the unknown parts.

(1) In Figure 10-2 triangle ABC is drawn with the known and unknown values labeled.

(2) Where possible, select an equation that uses the values of the two known parts and each of the unknown parts. In addition, choose the function in which the unknown value is the numerator of the ratio.

(2) To find a, we use the ratio

$$\sin A = \frac{\text{length of leg opp}}{\text{length of hyp}} = \frac{a}{c}$$

Thus, $a = (10.0)(\sin 35°) = (10.0)(0.5736) = 5.7$.
To find b, we use the ratio

$$\cos A = \frac{\text{length of leg adj}}{\text{length of hyp}} = \frac{b}{c}$$

Thus, $b = (10.0)(\cos 35°) = (10.0)(0.8192) = 8.2$.

(3) There are two exceptions to Step (2):
(a) When the measure of an acute angle is given, the unknown angle is the complement of the given angle; thus, we subtract the given measure from 90°.
(b) When the lengths of two sides are given, the Pythagorean Theorem is used to find the length of the third side.

(3) Since $A + B = 90°$, then $35° + B = 90°$ and $B = 55°$.

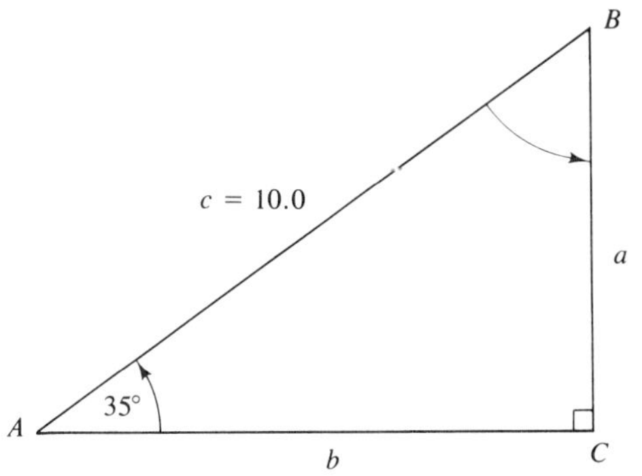

FIGURE 10-2 Right triangle ABC with $A = 35°$ and $c = 10.0$

Were the values of a and b rounded to the nearest tenth because the value of c is given to the nearest tenth?

Since $c = 10.0$, we round off the values of a and b to the nearest tenth. If the value of c had been given as 10, then we would have rounded the values of a and b to the nearest whole number.

EXAMPLE 10–2 Solve right triangle ABC if $a = 15.0$ and $c = 22.0$. Find the measures of A and B to the nearest 10 minutes.

Solution: Triangle ABC is shown in Figure 10-3. To find A, we use $\sin A = a/c$ with $a = 15.0$ and $c = 22.0$. Therefore,

$$\sin A = \frac{15.0}{22.0} = 0.6818$$

With Table 4, $A = 43°00'$ to the nearest 10 minutes. With a hand calculator, $A = \text{Sin}^{-1}0.6818 = 42.98°$, or approximately 43°.

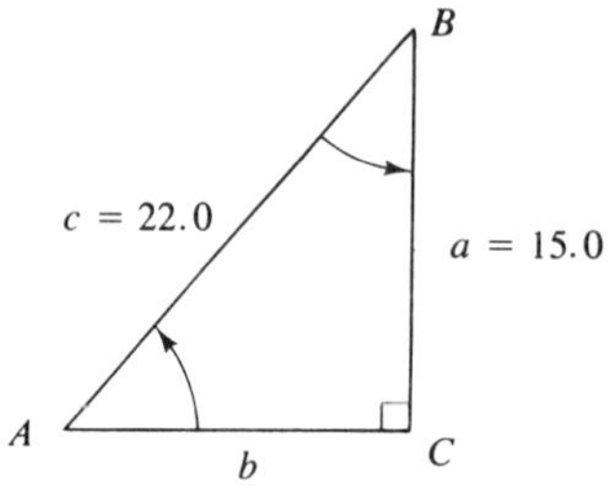

FIGURE 10-3 Right triangle ABC with $a = 15.0$ and $c = 22.0$

To find B, we use $\cos B = a/c$. Thus,

$$\cos B = 0.6818$$
$$B = 47°00' \text{ to the nearest 10 minutes}$$

Notice that we did not compute the value of B by using the measure of A. As a rule, we avoid — if possible — using a computed value to calculate the value of an unknown quantity.

To find b, we use the Pythagorean Theorem. Thus,

$$15^2 + b^2 = 22^2$$
$$b = 16.1$$

A physical problem may involve an **angle of elevation** or an **angle of depression.** In Figure 10-4 the two angles are indicated as **alternate interior angles** between two horizontal lines. The transversal to the horizontal lines is sometimes called the **line of sight.**

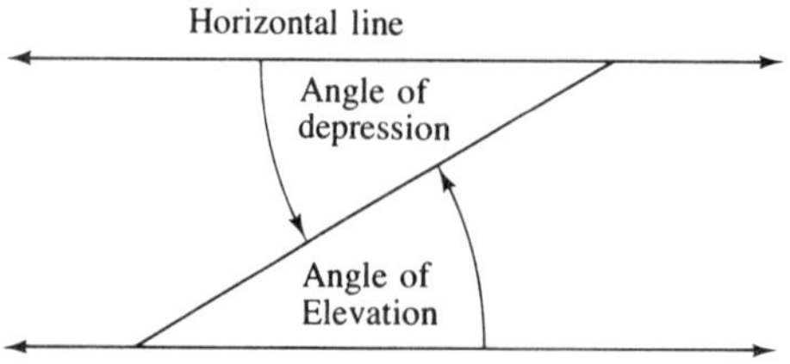

FIGURE 10-4 Diagram indicating the angles of elevation and depression

EXAMPLE 10–3 The shadow of a redwood tree measures 40.0 feet when the angle of elevation of the sun is 72°. Find the height of the tree to the nearest foot.

Solution: In Figure 10-5 a sketch of the given information is shown. From the diagram we observe that

$$\tan 72° = \frac{\text{height of tree}}{\text{length of shadow}} = \frac{b}{a}$$

Thus,

$$b = (40.0 \text{ feet})(\tan 72°) = 123 \text{ feet}$$

and the height of the tree is 123 feet to the nearest foot.

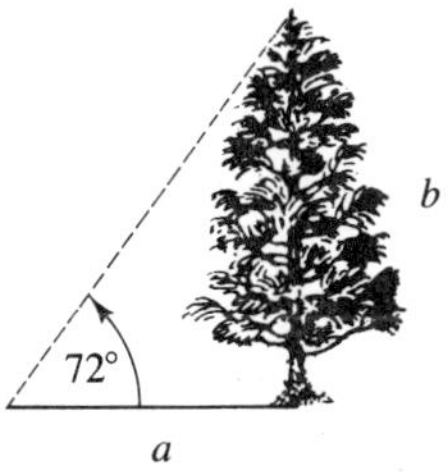

FIGURE 10-5 a = length of shadow, b = height of tree, and angle of elevation = 72°

Law of Sines

In Figure 10-6 triangle ABC is drawn, with $\overline{CD}$ an altitude to $\overline{AB}$. The lengths of the sides of the triangle are labeled with lowercase letters that correspond to the capital letters used to label the angles opposite the sides. Notice that the length of $\overline{CD}$ is h.

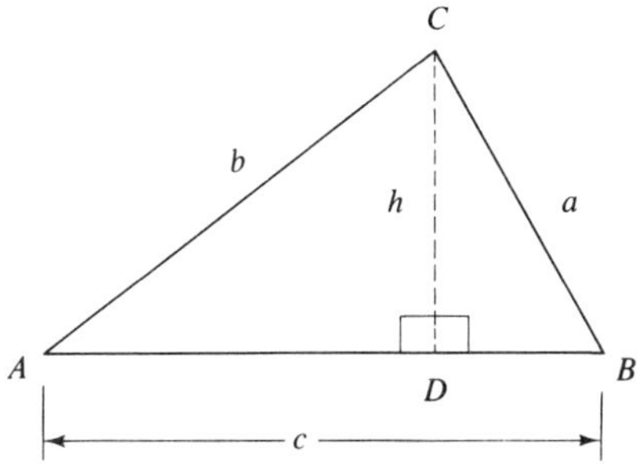

FIGURE 10-6 Triangle ABC with altitude $\overline{CD}$, and $|CD| = h$

The area of any triangle is obtained with the equation

$$A = \frac{1}{2}bh$$

where b is the length of a side (base) of the triangle and h is the length of the altitude drawn to the side. Thus, for triangle ABC, we have

$$A = \frac{1}{2}ch$$

But

$$\sin A = \frac{h}{b} \quad \text{and} \quad \sin B = \frac{h}{a}$$

since $\overline{CD}$ forms two right triangles ACD and BCD. Therefore,

$$h = b \sin A \quad \text{and} \quad h = a \sin B$$

$$b \sin A = a \sin B$$

$$\frac{\sin A}{a} = \frac{\sin B}{b}$$

When the altitude is drawn from point A to side $\overline{BC}$, we can establish the following equation:

$$\frac{\sin c}{c} = \frac{\sin B}{b}$$

When we combine the two equations, we obtain the following theorem.

Law of Sines: In any triangle, if A, B, and C are angles opposite the sides of lengths a, b, and c, respectively, then

$$\frac{\sin A}{a} = \frac{\sin B}{b} = \frac{\sin C}{c}$$

EXAMPLE 10–4 Solve triangle ABC if $a = 25$, $A = 72°$, and $B = 50°$.

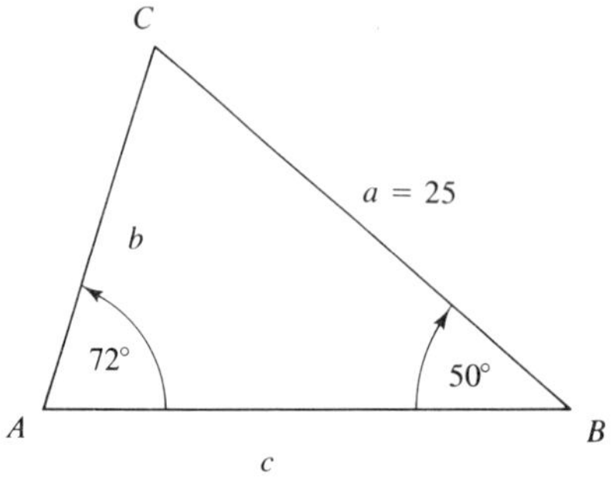

FIGURE 10-7 $A = 72°$, $B = 50°$, and $a = 25$

Solution: Triangle ABC is sketched in Figure 10-7. To find C, we use the following equation:

$$A + B + C = 180°$$

$$72° + 50° + C = 180°$$

$$C = 58°$$

To find b, we use the equation

$$\frac{\sin A}{a} = \frac{\sin B}{b}$$

and

$$b = \frac{25 \sin 50°}{\sin 72°} = 20$$

To find c, we use the equation

$$\frac{\sin A}{a} = \frac{\sin C}{c}$$

and

$$c = \frac{25 \sin 58°}{\sin 72°} = 22$$

Unfortunately, the Law of Sines does not solve any triangle for which the values of two sides and an angle or two angles and one side are given. If the values given for the parts of a triangle consist of the lengths of two sides and a **nonincluded angle** (that is, the angle is not between the sides whose lengths are given), we have what is called **the ambiguous case.** The given data may determine two triangles, one triangle, or no triangles.

Does the Law of Sines solve any triangle for which the values of two sides and an angle, or two angles and one side, are given?

To decide whether the given information determines two, one, or no triangles, we apply the following tests. Let the values of a, b, and A be the given information on triangle ABC.

Case 1. If $a < b \sin A$, then triangle ABC does not exist. See Figure 10-8.

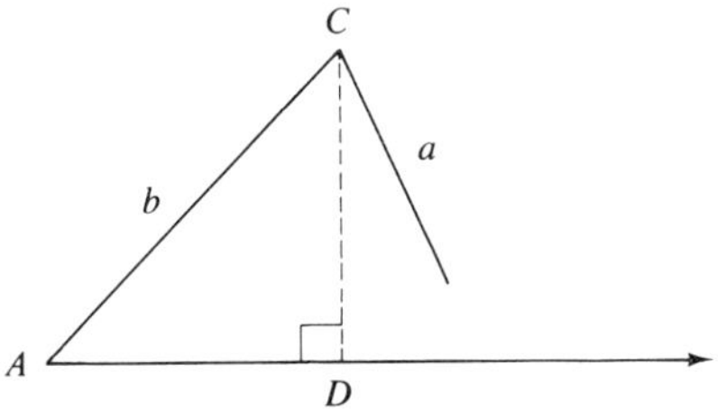

FIGURE 10-8 $|CD| = b \sin A$ and $a < b \sin A$

Case 2. If $a = b \sin A$, then triangle ABC is a right triangle. See Figure 10-9.

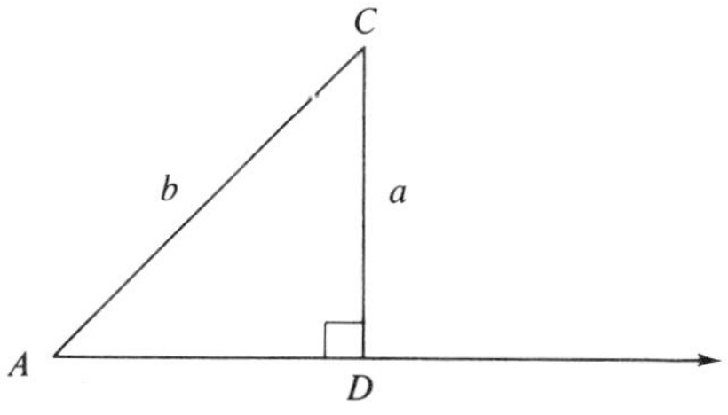

FIGURE 10-9 $|CD| = b \sin A = a$

Case 3. If $b \sin A < a < b$, then there are two different triangles. See Figure 10-10.

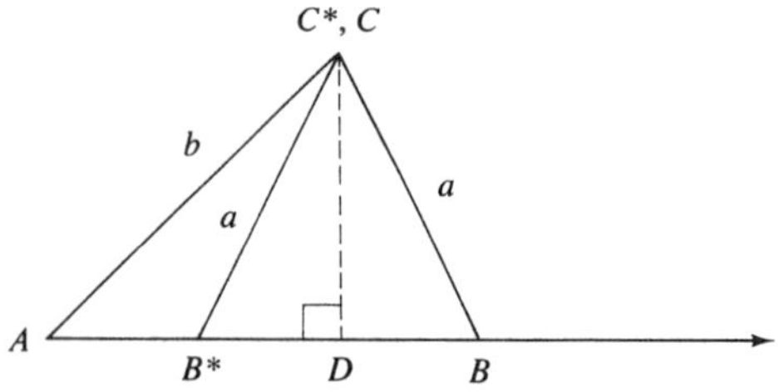

FIGURE 10-10 $|CD| = b \sin A$ and $b \sin A < a < b$; therefore, triangles ABC and AB^*C^* are formed

Case 4. If $a \geq b$, then there is one triangle ABC. See Figure 10-11.

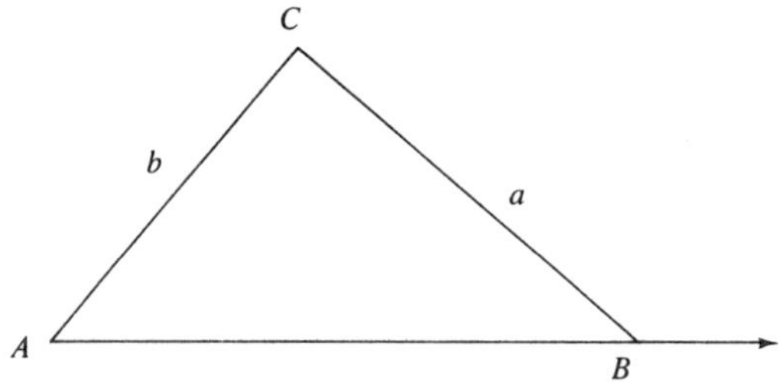

FIGURE 10-11 $a \geq b$

EXAMPLE 10–5 Determine the number of triangles for the given data.

(a) $a = 15; b = 20; A = 40°$ (b) $a = 10; b = 25; A = 30°$

Solutions:

(a) $\sin 40° = 0.6428$ and $b \sin A = 20(0.6428) = 12.856$. Since $a = 15$ and $12.856 < 15 < 20$, the given data determine two distinct triangles ABC and AB^*C^*.

(b) $\sin 30° = 0.5000$ and $b \sin A = 25(0.5000) = 12.5$. Since $a = 10$ and $10 < 12.5$, the given data do not determine a triangle ABC.

EXAMPLE 10–6 Solve triangle ABC if $a = 18.0$, $b = 21.0$, and $A = 37°30'$.

Solution: Since A is a nonincluded angle, we need to determine whether the given data yield two, one, or no triangles. Since

$$b \sin A = 21.0 \sin 37°30' = 21(0.6088) = 12.78$$

and $12.78 < 18 < 21$, we conclude that two triangles, ABC and AB^*C^*, are formed. The triangles are shown in Figure 10-12. For both triangles,

$$\frac{0.6088}{18} = \frac{\sin B}{21} \quad \text{or} \quad \frac{0.6088}{18} = \frac{\sin B^*}{21}$$

and $\sin B = \sin B^* = 0.7103$. To the nearest ten minutes,

$$B = 45°20' \quad \text{and} \quad B^* = 179°60' - 45°20' = 134°40'$$

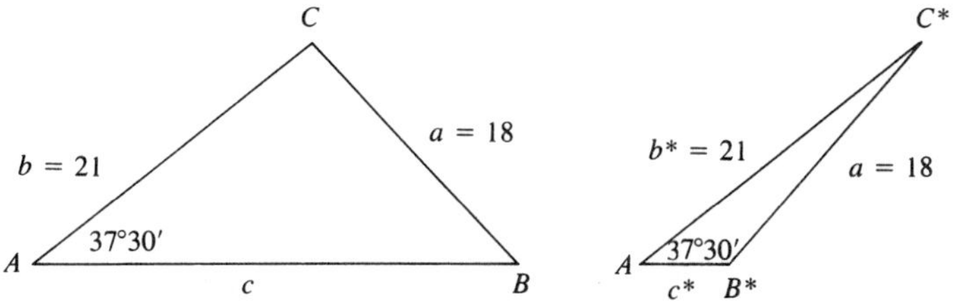

FIGURE 10-12 Triangles ABC and AB^*C^*, with $a = 18$, $b = b^* = 21$, and $A = 37°30'$

In triangle ABC,

$$37°30' + 45°20' + C = 180°$$
$$C = 97°10'$$

Now we have

$$\frac{0.6088}{18} = \frac{\sin 97°10'}{c}$$

$$c = \frac{18(0.9922)}{0.6088} = 29.3$$

In triangle AB^*C^*,

$$37°30' + 134°40' + C^* = 180°$$
$$C^* = 7°50'$$

Now we have

$$\frac{0.6088}{18} = \frac{\sin 7°50'}{c^*}$$

$$c^* = \frac{18(0.1363)}{0.6088} = 4.0$$

Notice that since both c^* and C^* are unknown, we are forced to use the computed value of B^* to determine the value of C^*. Thus, in triangle ABC, $B = 45°20'$, $C = 97°10'$, and $c = 29.3$. In triangle AB^*C^*, $B^* = 134°40'$, $C^* = 7°50'$, and $c^* = 4.0$.

Exercise Set 10–1

Solve the following right triangles. Round answers according to the accuracy of the given data.

1. $b = 45$; $A = 51°$
2. $a = 5.2$; $A = 34°$
3. $A = 37°50'$; $c = 114$
4. $b = 48$; $c = 62$
5. $B = 39°00'$; $b = 33.9$
6. $a = 253$; $b = 397$
7. $a = 13.1$; $B = 49°40'$
8. $b = 89.6$; $c = 93.1$
9. Find the altitude of an equilateral triangle whose side is 100 feet.
10. Find the height of a flagpole whose shadow is 40.0 feet long when the angle of elevation of the sun is 36°40′.

11. From the top of a lighthouse 86.0 feet above the water, the angle of depression of a ship is $6°20'$. How far is the ship from the lighthouse?

12. A jet plane flying at an elevation of 8000 feet flies directly over an observer. If the plane travels 1820 feet from the point overhead, at what angle of elevation will the observer see the plane?

13. In order to measure the width of a lake at its widest point, a surveyor takes a point on one shore and sights across the lake to the widest point. Then walking 200 meters along the perpendicular to the original sighting, she takes another sighting of 82°. Find the width of the lake.

14. From a point on level ground the top of a mountain is sighted at an angle of elevation of 6°. Walking 2000 meters directly away from the mountain, an observer takes a new sighting of 5°. How high is the mountain?

15. The mast of a television antenna is 8 meters. What is the length of a support cable that makes an angle of $63°20'$ to the flat roof on which the antenna is mounted?

For the following ambiguous case problems, find the number of triangles determined. Do not solve the triangles.

16. $a = 3$; $b = 4$; $A = 37°$

17. $a = 11$; $b = 13$; $A = 80°$

18. $a = 36$; $b = 24$; $A = 31°$

19. $b = 19.8$; $c = 21.6$; $B = 50°$

20. $b = 4.23$; $c = 2.57$; $C = 50°$

Solve the following triangles, using the Law of Sines. If no triangle is formed, state this as the answer. If two triangles are formed, solve both triangles.

21. $C = 80°$; $A = 20°$; $b = 6.3$

22. $B = 37°$; $A = 113°$; $b = 3.6$

23. $C = 57°30'$; $B = 24°30'$; $c = 359$

24. $B = 64°20'$; $A = 65°20'$; $a = 32.2$

25. $a = 2.4$; $b = 3.6$; $A = 28°$

26. $a = 5$; $b = 3$; $A = 30°$

27. $a = 109$; $c = 233$; $A = 27°30'$

28. $a = 54.0$; $c = 75.0$; $A = 29°20'$

29. The diagonal of a parallelogram is 36 inches long and meets the sides at angles of 65° and 70°. Find the lengths of the sides.

30. A ship at point A sights a lighthouse at $47°20'$ east of North. After traveling 2.5 miles north to point B, the ship sights the lighthouse at $71°50'$ east of North. How far is it from point B to the lighthouse?

31. A 16-foot ladder can reach a window of a building if it has a 65° angle with the ground. At what angle will a 22-foot ladder meet the ground?

32. The mast of a television antenna makes an angle of 120° with the roof. A guidewire 15 feet away from the base of the antenna makes an angle of 35° with the roof. If all measurements are given in the same vertical plane, how tall is the mast?

LAW OF COSINES 10-2

When the data given for a triangle that is not a right triangle consist of the lengths of the three sides, or the lengths of two sides and the measure of the included angle, then the Law of Sines cannot be used to solve the triangle. With such data, we must use the Law of Cosines to solve the triangle.

In Figure 10-13 triangle ABC is placed on the coordinate system so that A is in standard position with point C on the initial side of A. The coordinates of the vertices of triangle ABC are $A(0, 0)$, $B(c \cos A, c \sin A)$, and $C(b, 0)$. We know that the length of $\overline{BC}$ is a. We now compute $|BC|$ with the distance formula and the coordinates of points B and C:

$$\begin{aligned} a^2 &= (x_2 - x_1)^2 + (y_2 - y_1)^2 \\ &= (c \cos A - b)^2 + (c \sin A - 0)^2 \\ &= c^2\cos^2 A - 2bc \cos A + b^2 + c^2\sin^2 A \\ &= b^2 + c^2(\cos^2 A + \sin^2 A) - 2bc \cos A \end{aligned}$$

But $\cos^2 A + \sin^2 A = 1$; therefore,

$$a^2 = b^2 + c^2 - 2bc \cos A$$

When angles B and C are each placed in standard position and the distance formula is used to compute $|AC|$ and $|AB|$, then the other equations in the Law of Cosines are obtained.

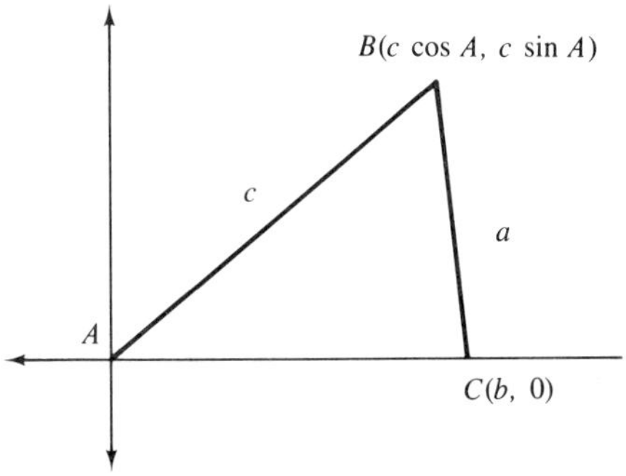

FIGURE 10-13 Triangle ABC with A in standard position

Law of Cosines: If a, b, and c are the lengths of the sides of any triangle ABC, then

$$a^2 = b^2 + c^2 - 2bc \cos A \quad (1)$$

$$b^2 = a^2 + c^2 - 2ac \cos B \quad (2)$$

$$c^2 = a^2 + b^2 - 2ab \cos C \quad (3)$$

In Equation (1), how does the term $2bc \cos A$ change the value of a?

If we keep the values of b and c constant, notice what happens to the value of a as A increases from 0° to 180°.

Case 1. For $A < 90°$, $\cos A > 0$ and $2bc \cos A > 0$. Thus, a positive number is subtracted from $b^2 + c^2$. As A increases in measure, the length of $\overline{BC}$ increases. Similarly, the value of a increases, since $\cos A$ decreases and a smaller value is subtracted from $b^2 + c^2$.

Case 2. For $A = 90°$, $\cos A = 0$ and $a^2 = b^2 + c^2$. This equation is the Pythagorean Theorem, since a is the length of the hypotenuse of right triangle ABC.

Case 3. For $90° < A < 180°$, $\cos A < 0$ and $2bc \cos A < 0$. Thus, the term $2bc \cos A$ is added to $b^2 + c^2$. As A increases in measure, the length of $\overline{BC}$ increases. Similarly, the value of a increases, since $\cos A$ decreases through negative values and a larger positive number is added to $b^2 + c^2$.

EXAMPLE 10–7 Solve triangle ABC if $a = 10.0$, $c = 8.0$, and $B = 60°00'$.

Solution: Triangle ABC is sketched in Figure 10-14. To find b, we use the Law of Cosines with $a = 10.0$, $c = 8.0$, and $\cos 60°00' = 0.5$:

$$b^2 = 10.0^2 + 8.0^2 - 2(10.0)(8.0)(0.5) = 84$$
$$b = 9.2$$

To find A, we can use either the Law of Sines or Law of Cosines. To illustrate, we use both methods in this example.

(a) If $a^2 = b^2 + c^2 - 2bc \cos A$, then

$$\cos A = \frac{b^2 + c^2 - a^2}{2bc} = \frac{84 + 64 - 100}{2(\sqrt{84})(8)} = 0.3273$$

and $A = 70°50'$.

(b) If $\dfrac{\sin A}{a} = \dfrac{\sin B}{b}$, then

$$\sin A = \frac{10(0.8660)}{\sqrt{84}} = 0.9449$$

and $A = 70°50'$.

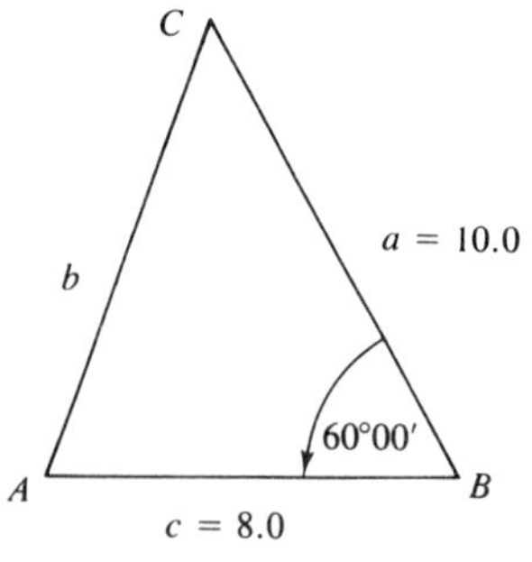

FIGURE 10-14 Triangle ABC with $a = 10.0$, $c = 8.0$, and $B = 60°00'$

To find C we use the Law of Sines:

$$C = \frac{8.0 \sin 60^\circ}{\sqrt{84}} = 49^\circ 10'$$

There is an advantage in using the Law of Cosines that we should note. In the above example, for $\cos A = 0.3273$, we know that A is the acute angle $70^\circ 50'$. However, for $\sin A = 0.9449$, A can be the acute angle $70^\circ 50'$ or the obtuse angle $180^\circ - 70^\circ 50' = 109^\circ 10'$. This conclusion follows from the fact that the sine of an acute angle and obtuse angle are both positive. However, *the cosine of an acute angle is positive, but the cosine of an obtuse angle is negative. The Law of Cosines yields the correct sign for any given angle of a triangle.*

Is there an advantage in using the Law of Cosines over the Law of Sines when we have a choice of using either one?

Notice that for the data in Example 10–7, since a and b are approximately equal, we would conclude that the measures of A and B must also be approximately equal. Therefore, A must be the acute angle $70^\circ 50'$.

EXAMPLE 10–8 Find the measures of the angles of triangle ABC to the nearest ten minutes if $a = 7.00$, $b = 4.00$, and $c = 5.00$.

Solution: Triangle ABC is sketched in Figure 10-15. To find A, we use the following equation:

$$\cos A = \frac{b^2 + c^2 - a^2}{2bc} = \frac{16 + 25 - 49}{2(4)(5)} = -0.2000$$

and $A = 180^\circ - 78^\circ 30' = 101^\circ 30'$. To find B,

$$\cos B = \frac{a^2 + c^2 - b^2}{2ac} = \frac{49 + 25 - 16}{2(7)(5)} = 0.8286$$

and $B = 34^\circ 00'$. To find C,

$$\cos C = \frac{a^2 + b^2 - c^2}{2ab} = \frac{49 + 16 - 25}{2(7)(4)} = 0.7143$$

and $C = 44^\circ 20'$.

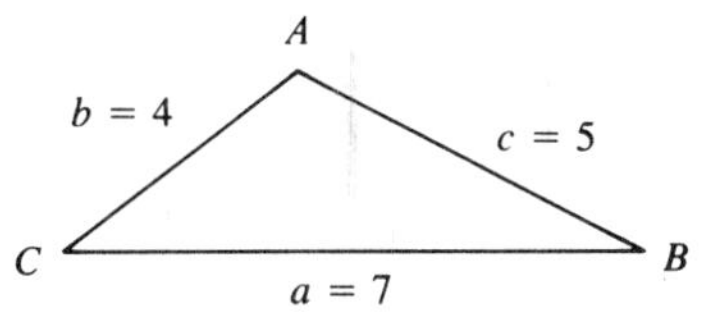

FIGURE 10-15 Triangle ABC with lengths of three sides given

Notice that $A + B + C = 101°30' + 34°00' + 44°20' = 179°50'$ and not 180°, as expected. The reason for the apparent error is in the approximations for the measures of A, B, and C to the nearest ten minutes.

EXAMPLE 10–9 Fresno, California is 135 miles south of Sacramento. An aircraft left Fresno, heading 40° east of the line from Fresno to Sacramento at a ground speed of 280 miles per hour. How far is the aircraft from Sacramento 1.25 hours after departure from Fresno?

Solution: A sketch of the given information is shown in Figure 10-16. Notice that at a speed of 280 miles per hour and a time of 1.25 hours, the aircraft is 280(1.25), or 350, miles from Fresno. If x is the distance the plane is from Sacramento, then by the Law of Cosines,

$$\begin{aligned} x^2 &= 135^2 + 350^2 - 2(135)(350)\cos 40° \\ &= 18{,}225 + 122{,}500 - 72{,}387 = 68{,}338 \end{aligned}$$

and $x = 260$ miles.

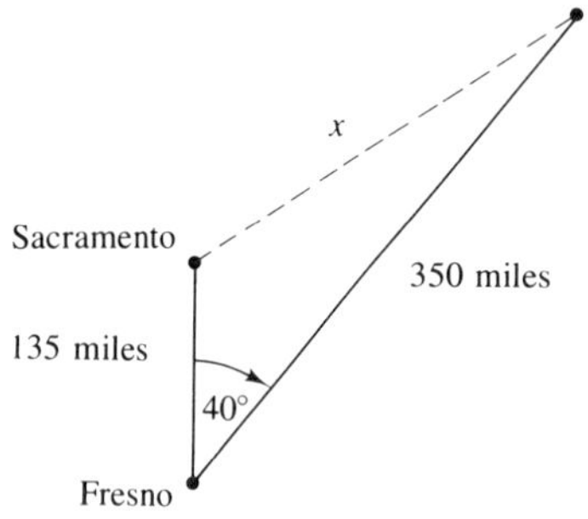

FIGURE 10-16 Diagram for data of Example 10–9

Exercise Set 10–2

Solve the following triangles.

1. $a = 28$; $b = 35$; $C = 49°00'$
2. $b = 15.0$; $c = 23.0$; $A = 28°00'$
3. $a = 4.0$; $b = 5.0$; $c = 6.0$ (approximate angle measures to the nearest ten minutes)
4. $c = 23.0$; $a = 18.0$; $b = 30.0$ (approximate angle measures to the nearest ten minutes)
5. $c = 10.0$; $a = 5.0$; $B = 60°00'$
6. $b = 150$; $c = 170$; $a = 80$ (approximate angle measures to the nearest ten minutes)

Find the required part of the following triangles.

7. Find a to the nearest tenth if $c = 10.0$, $b = 6.0$, and $A = 155°30'$.

8. Find A to the nearest ten minutes if $b = 14.80$, $a = 16.80$, and $c = 9.92$.

9. Find C to the nearest whole degree if $a = 20$, $c = 24$, and $b = 10$.

10. Find B to the nearest ten minutes if $a = 11.8$, $b = 15.6$, and $A = 34°20'$.

11. Find a to the nearest whole number if $c = 119$, $A = 30°20'$, and $B = 120°30'$.

12. Find b to the nearest whole number if $A = 70°$, $a = 20$, and $C = 80°$.

13. Find b to the nearest tenth if $B = 92°00'$, $a = 14.0$, and $c = 12.0$.

14. Find B to the nearest ten minutes if $b = 15.6$, $a = 11.8$, and $A = 34°20'$.

15. In a circle with a diameter of 40 inches, find to the nearest ten minutes the central angle which intercepts a chord 28 inches long.

16. Find the lengths of the sides of a parallelogram with diagonals of lengths 18 and 32 inches that intersect at an angle of 34°.

17. How far is it from the pitcher's mound to first base on a baseball diamond (square) where home plate is 90 feet from first base and 60 feet from the pitcher's mound?

18. Two planes leave the same airport at the same time. One plane flies due north at 130 miles per hour while the other plane flies 63° east of North at 108 miles per hour. How far apart are the planes at the end of two hours?

19. How long (to the nearest meter) is a "lunge line" (a rope used to work a young horse in a circle) if the distance between two points on the circle is 10 meters when the central angle is 67°30'?

20. To measure the distance d between two points A and B, a surveyor is forced to measure a distance of 115 meters from A to point C. She then measures the distance from C to B as 220 meters, and C is 112°40'. Find d to the nearest whole meter.

21. A tower 15 meters high supports high voltage wires. The tower makes an angle of 75°30' with the side of the hill on which it is standing. A cable attached to the top of the tower is also anchored to the hill 8 meters below the tower. How long is the cable to the nearest tenth of a meter?

22. A billiard ball X is struck by a cue stick at point A on the table. It travels 2.4 feet and strikes another ball at point B. Ball X is deflected through an angle θ and moves 4.3 feet to point C on the cushion. If $\overline{AC}$ is 5.4 feet, find θ to the nearest whole degree.

23. A highway traffic sign is in the shape of a regular hexagon (all six sides are of equal length). If the length of an edge of the sign is 30 centimeters, determine the height of the sign to the nearest centimeter.

24. The sides of a rhombus (an equilateral parallelogram) are 18 centimeters long. To the nearest degree, find the measure of the obtuse angle of the rhombus if the longer diagonal is 33 centimeters long.

25. An airplane is traveling due north at 480 miles per hour. It travels from city A for 1.25 hours to city B, then changes direction toward the east for 2.0 hours to city C. If the straight-line distance from city A to city C is 1300 miles, determine to the nearest ten minutes what azimuth the plane flew in going from city B to city C. (*Note:* An azimuth is measured clockwise from the direction due north.)

10-3 POLAR COORDINATE SYSTEM

In Chapter 4 we developed the cartesian coordinate system, also called the rectangular coordinate system. In this section we consider the polar coordinate system. The basis of this coordinate system is a fixed point O, called the **origin** or **pole,** and a ray OP, written $\overrightarrow{OP}$, called the **polar axis.** The positive horizontal axis of a rectangular system is frequently used as the polar axis of a polar coordinate system.

In Figure 10-17 a point R is shown in a plane with a polar axis $\overrightarrow{OP}$. Line segment $\overline{OR}$ is also drawn. We can locate R relative to O with (1) the polar axis as the initial side of an angle θ with terminal side $\overline{OR}$, and (2) the distance r from O to R, that is, $r = |OR|$. In Figure 10-17 the positively rotated angle θ and $|OR|$ locate R relative to O. We designate the polar coordinates of R as an ordered pair of the form (r, θ), where $r = |OR|$ and θ is an angle with initial side $\overrightarrow{OP}$ and terminal side $\overline{OR}$.

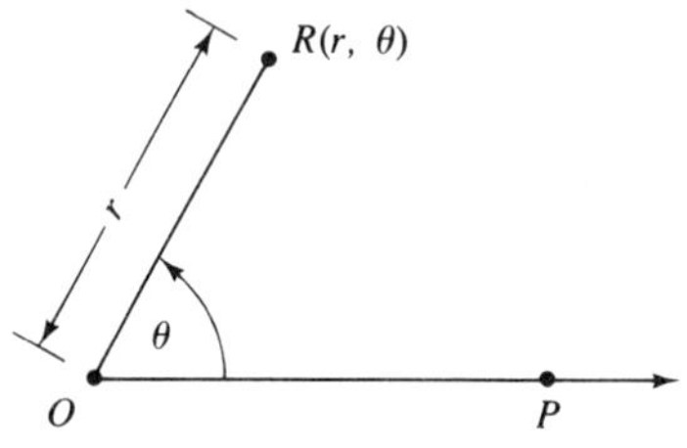

FIGURE 10-17 Polar coordinates (r, θ) of point R, where $r = |OR|$

Is an ordered pair (r, θ) unique for a given point R as it is in the rectangular system?

In the polar coordinate system, the ordered pair (r, θ) is not unique for R. In fact, there are infinitely many ordered pairs of the form (r, θ) that locate R relative to O and $\overrightarrow{OP}$. First, *any angle coterminal with* θ $(0 \leq \theta < 2\pi)$ *can also be paired with r as a polar coordinate of R.* For example, in Figure 10-18, $(2, \pi/4)$, $(2, 9\pi/4)$, and $(2, -7\pi/4)$ are all polar coordinates of R.

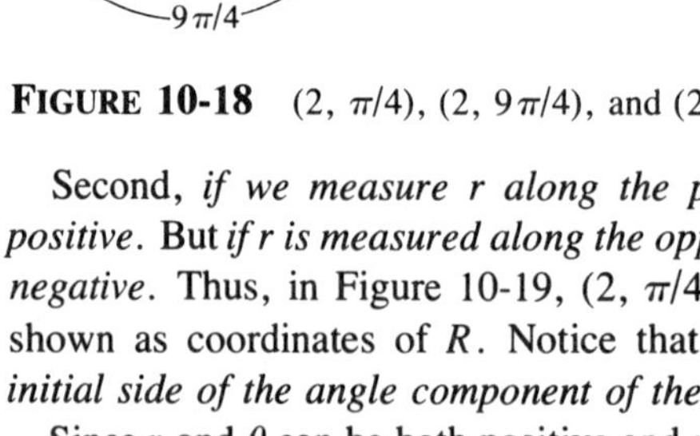

FIGURE 10-18 $(2, \pi/4)$, $(2, 9\pi/4)$, and $(2, -7\pi/4)$ as polar coordinates of R

Second, *if we measure r along the polar axis, then we consider r to be positive.* But *if r is measured along the opposite ray of* $\overrightarrow{OP}$ (that is, $\overrightarrow{OQ}$), *then r is negative.* Thus, in Figure 10-19, $(2, \pi/4)$, $(-2, 5\pi/4)$, and $(-2, -3\pi/4)$ are shown as coordinates of R. Notice that $\overrightarrow{OQ}$, *the opposite ray of* $\overrightarrow{OP}$, *is the initial side of the angle component of the polar coordinate when r is negative.*

Since r and θ can be both positive and negative, we recommend the following procedure to locate $R(r, \theta)$ in the polar coordinate system:

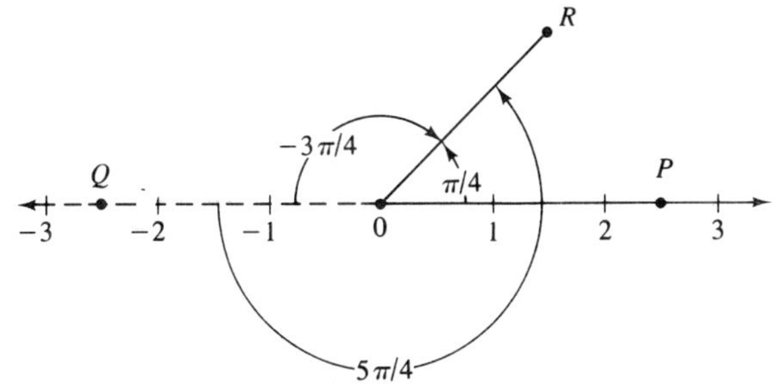

FIGURE 10-19 $(2, \pi/4)$, $(-2, 5\pi/4)$, and $(-2, -3\pi/4)$ as polar coordinates of R

Case 1. If $r > 0$ and $\theta > 0$, then move r units along the polar axis and rotate θ units counterclockwise to point R.

Case 2. If $r > 0$ and $\theta < 0$, then move r units along the polar axis and rotate $|\theta|$ units clockwise to point R.

Case 3. If $r < 0$ and $\theta > 0$, then move r units along $\overrightarrow{OQ}$, the ray opposite to the polar axis, and rotate θ units counterclockwise to point R.

Case 4. If $r < 0$ and $\theta < 0$, then move r units along the ray opposite to the polar axis and rotate $|\theta|$ units clockwise to point R.

Recall that a negative sign on an angle indicates that the rotation of the terminal side of the angle is clockwise from the initial side. *The amount of rotation is the absolute value of the measure of the angle.* Therefore, the initial side of θ is $\overrightarrow{OP}$ when $r > 0$ and the opposite ray of $\overrightarrow{OP}$ when $r < 0$. The rotation is clockwise when $\theta < 0$, and the amount of rotation is $|\theta|$.

Why are absolute value bars used on θ in cases 2 and 4?

EXAMPLE 10–10 Plot the points $A(3, 2\pi/3)$, $B(4, -\pi/3)$, $C(-2, 7\pi/6)$, and $D(-5, -\pi/6)$.

Solution: Figure 10-20 shows the polar coordinate system with several values of θ and values of r from 1 through 5 indicated.

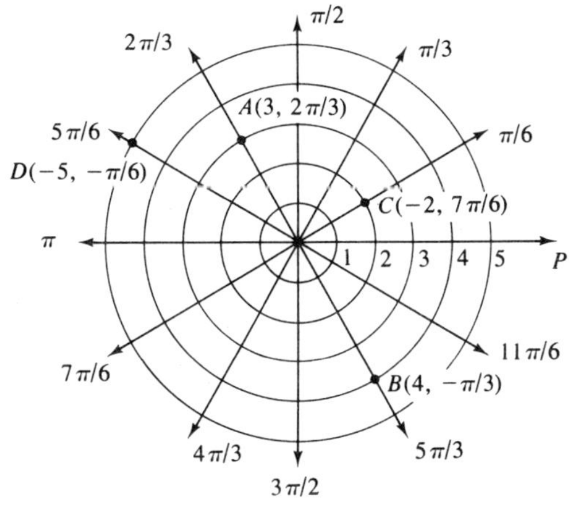

FIGURE 10-20 Graphs of points A, B, C, and D

(1) For $A(3, 2\pi/3)$, move 3 units to the right and rotate $2\pi/3$ units counterclockwise to point A.

(2) For $B(4, -\pi/3)$, move 4 units to the right and rotate $\pi/3$ units clockwise to point B.

(3) For $C(-2, 7\pi/6)$, move 2 units to the left and rotate $7\pi/6$ units counterclockwise to point C.

(4) For $D(-5, -\pi/6)$, move 5 units to the left and rotate $\pi/6$ units clockwise to point D.

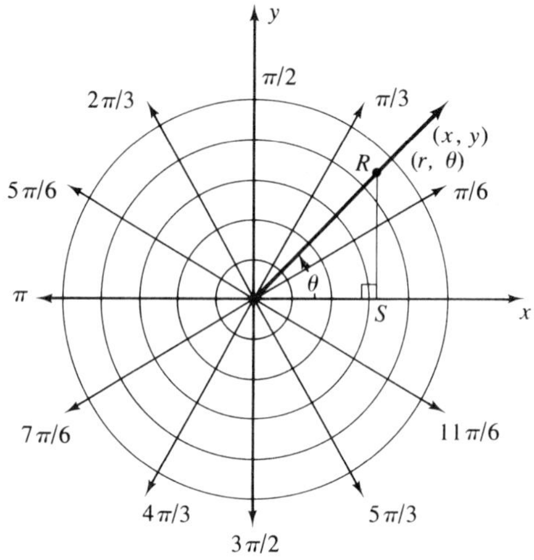

FIGURE 10-21 Point R in a rectangular and a polar system

In Figure 10-21 point R is shown as a point in both the rectangular and polar coordinate systems. The positive x-axis and the polar axis are the same ray. Triangle ORS is a right triangle with acute angle θ. The lengths of the leg adjacent, leg opposite, and hypotenuse are x, y, and r, respectively. By the alternate definitions of cosine and sine,

$$\cos\theta = \frac{x}{r} \quad \text{and} \quad \sin\theta = \frac{y}{r}$$

Thus, we obtain Equations (1) and (2).

$$x = r\cos\theta \qquad \textbf{(1)}$$

$$y = r\sin\theta \qquad \textbf{(2)}$$

Equations (1) and (2) are used to convert the coordinates of a point in polar form to coordinates in rectangular form. Referring to Figure 10-21, we use the Pythagorean Theorem and the alternate definition of tangent to obtain equations (3) and (4).

$$r = \sqrt{x^2 + y^2} \qquad \textbf{(3)}$$

$$\tan\theta = \frac{y}{x},\ x \neq 0 \qquad \textbf{(4)}$$

Equations (3) and (4) are used to convert the coordinates of a point in rectangular form to coordinates in polar form. Unless otherwise indicated, we will automatically restrict θ to values between 0 and 2π in Equation (4).

EXAMPLE 10–11 Convert the polar coordinates $(4, 7\pi/6)$ to rectangular form.

Solution: With $r = 4$, $\theta = 7\pi/6$, and Equations (1) and (2), we get

$$x = 4 \cos \frac{7\pi}{6} \quad \text{and} \quad y = 4 \sin \frac{7\pi}{6}$$

$$x = 4\left(\frac{-\sqrt{3}}{2}\right) = -2\sqrt{3} \quad \text{and} \quad y = 4\left(\frac{-1}{2}\right) = -2$$

Thus, $(-2\sqrt{3}, -2)$ are the rectangular coordinates of the point with polar coordinates $(4, 7\pi/6)$.

EXAMPLE 10–12 Convert the rectangular coordinates $(-\sqrt{3}, 1)$ to polar form.

Solution: With $x = -\sqrt{3}$, $y = 1$, and Equations (3) and (4), we get

$$r = \sqrt{(-\sqrt{3})^2 + 1^2} \quad \text{and} \quad \tan \theta = \frac{1}{-\sqrt{3}}$$

$$r = \sqrt{4} = 2 \quad \text{and} \quad \theta = 5\pi/6$$

Thus, $(2, 5\pi/6)$ are the polar coordinates of the point with rectangular coordinates $(-\sqrt{3}, 1)$.

For the given rectangular coordinates, $x < 0$ and $y > 0$; therefore, the point is in Quadrant II. Thus, θ is a Quadrant II angle whose tangent value is $-1/\sqrt{3}$, and $\theta = 5\pi/6$.

Why is θ $5\pi/6$ and not $11\pi/6$, since $\tan\frac{11\pi}{6} = \frac{-1}{\sqrt{3}}$ also?

EXAMPLE 10–13 Express the equation $r^2 = 2r \sin \theta + 1$ in rectangular form.

Solution: From Equations (3) and (2), $r^2 = x^2 + y^2$ and $y = r \sin \theta$. If $r^2 = 2r \sin \theta + 1$, then

$$x^2 + y^2 = 2y + 1$$

$$x^2 + y^2 - 2y = 1$$

EXAMPLE 10–14 Express the equation $x^2 - 4y + 3 = 0$ in polar form.

Solution: From Equations (1) and (2), $x = r \cos \theta$ and $y = r \sin \theta$. If $x^2 - 4y + 3 = 0$, then

$$(r \cos \theta)^2 - 4(r \sin \theta) + 3 = 0$$

$$r^2\cos^2\theta - 4r \sin \theta + 3 = 0$$

Exercise Set 10–3

Plot the following points.

1. $(3, 5\pi/6)$
2. $(2, 5\pi/3)$
3. $(-1, \pi/2)$
4. $(-2, 13\pi/6)$
5. $(4, -5\pi/6)$
6. $(-1, -\pi/3)$

Convert the coordinates of the following points from polar form to rectangular form.

7. $(2, 5\pi/4)$
8. $(-6, \pi/6)$
9. $(-3, -\pi/2)$
10. $(3/4, 11\pi/6)$
11. $(\sqrt{3}, \pi)$
12. $(-2, 0)$

Convert the coordinates of the following points from rectangular form to polar form.

13. $(2, 2)$
14. $(-\sqrt{3}, -1)$
15. $(2, 0)$
16. $(2, -2\sqrt{3})$
17. $(1/2, \sqrt{3}/2)$
18. $(-\sqrt{2}, \sqrt{2})$
19. For the point (1, 1) in rectangular coordinates, name the coordinates of the point in polar coordinates in four different ways.
20. For the point (0, 5) in rectangular coordinates, name the coordinates of the point in polar coordinates in four different ways.

Convert the following polar equations to rectangular form.

21. $r \sin\theta = 2$
22. $r = 2$
23. $\theta = \pi/4$
24. $r = 2\cos\theta$
25. $r^2 = 4\sin\theta$
26. $r = -2\sec\theta$
27. $r = \sin 2\theta$
28. $r = 1 + \cos\theta$

Convert the following rectangular equations to polar form.

29. $x^2 + y^2 = 16$
30. $x + y = 3$
31. $x = 4$
32. $y = -x$
33. $y^2 = 2x$
34. $\dfrac{x^2}{4} + \dfrac{y^2}{9} = 1$

10–4 GRAPHS OF POLAR EQUATIONS

In this section we will graph relations of the forms

$$(1)\ \{(r, \theta) \mid r = 1 + \cos\theta\} \qquad (2)\ \{(r, \theta) \mid r^2 = 4\sin 2\theta\}$$
$$(3)\ \{(r, \theta) \mid r = 4\} \qquad (4)\ \{(r, \theta) \mid \theta = \pi/3\}$$

The equations in these relations are polar equations. The graph of a polar equation, therefore, consists of the points in the plane whose polar coordinates satisfy the equation.

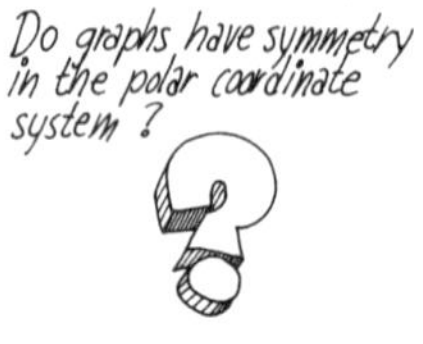

The graph of an equation in polar form may possess three types of symmetry: (1) symmetry with respect to the polar axis, (2) symmetry with respect to the pole (or origin), and (3) symmetry with respect to the line $\theta = \pi/2$. We will define each type of symmetry and provide a test to apply to an equation to determine whether the graph of the equation possesses any symmetry.

Symmetry with respect to the polar axis

> **Definition:** The graph of an equation in polar form has symmetry with respect to the polar axis if whenever $P(r_1, \theta_1)$ is a point of the graph, then $Q(r_1, -\theta_1)$ is also a point of the graph.

Test 1. If a given equation is unchanged when θ is everywhere replaced with $-\theta$, then the graph of the equation is symmetric with respect to the polar axis.

Symmetry with respect to the pole

> **Definition:** The graph of an equation in polar form has symmetry with respect to the pole if whenever $P(r_1, \theta_1)$ is a point of the graph, then $Q(-r_1, \theta_1)$ is also a point of the graph.

Test 2. If a given equation is unchanged when r is everywhere replaced with $-r$, then the graph of the equation is symmetric with respect to the pole.

Symmetry with respect to the line $\theta = \pi/2$

> **Definition:** The graph of an equation in polar form has symmetry with respect to the line $\theta = \pi/2$ if whenever $P(r_1, \theta_1)$ is a point of the graph, then $Q(-r_1, -\theta_1)$ is also a point of the graph.

Test 3. If a given equation is unchanged when r and θ are everywhere replaced with $-r$ and $-\theta$, then the graph of the equation is symmetric with respect to the line $\theta = \pi/2$.

There are some definite similarities when we compare the definitions for symmetry of graphs of polar equations with the definitions for symmetry of graphs of equations in rectangular form. Recall that symmetry for graphs of equations in rectangular form consists of three kinds: (1) symmetry with respect to the horizontal axis, (2) symmetry with respect to the vertical axis, and (3) symmetry with respect to the origin. We will not compare in detail the two sets of symmetries at this time. However, you may benefit from a quick comparison of the two sets of symmetries to determine which pairs of symmetries are alike for the two systems.

Aren't the definitions for symmetry of graphs of polar equations something like the definitions for symmetry of graphs of equations in rectangular form?

There is, however, one significant difference in symmetry in graphs of polar equations from symmetry in graphs of rectangular equations. That is, *the graph of an equation in polar form may possess symmetry, but the test of the equation for that form of symmetry may fail.* The reason for the inability of the tests to uncover symmetry in every polar equation is due to the multiple sets of coordinates that can be used for any point in the polar coordinate system.

EXAMPLE 10–15 Graph $r = 1 + \cos\theta$.

Solution:
Test 1. Replace θ with $-\theta$.

$$r = 1 + \cos(-\theta) = 1 + \cos\theta$$

Thus, the graph of the equation is symmetric with respect to the polar axis.

Test 2. Replace r with $-r$.

$$-r = 1 + \cos\theta, \text{ and } r = -1 - \cos\theta \qquad \text{(Test fails)}$$

Test 3. Replace r with $-r$ and θ with $-\theta$.

$$-r = 1 + \cos(-\theta), \text{ and } r = -1 - \cos\theta \qquad \text{(Test fails)}$$

The following table of values determines several points of the graph. By symmetry the graph is obtained for values of θ from π to 2π. The curve is called a **cardioid** and is shown in Figure 10-22.

θ	$1 + \cos\theta$
0	2
$\pi/3$	3/2
$\pi/2$	1
$2\pi/3$	1/2
π	0

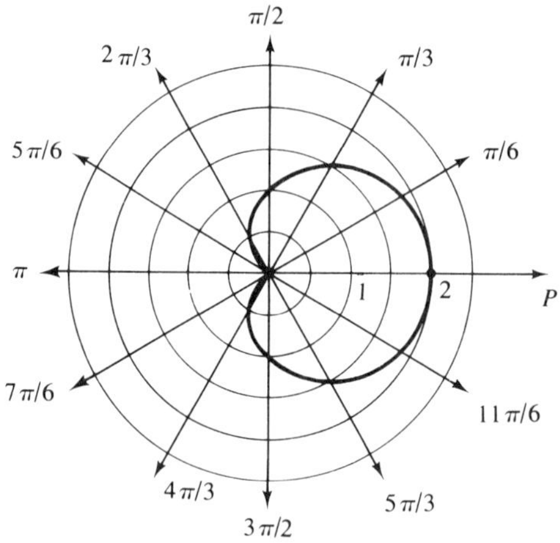

FIGURE 10-22 Graph of $r = 1 + \cos\theta$

EXAMPLE 10–16 Graph $r^2 = 4 \sin 2\theta$.

Solution:
Test 1. Replace θ with $-\theta$.

$$r^2 = 4 \sin 2(-\theta) = -4 \sin 2\theta \qquad \text{(Test fails)}$$

Test 2. Replace r with $-r$.

$$(-r)^2 = 4 \sin 2\theta, \text{ and } r^2 = 4 \sin 2\theta$$

Thus, the graph is symmetric with respect to the pole.

Test 3. Replace r with $-r$ and θ with $-\theta$.

$$(-r)^2 = 4 \sin 2(-\theta), \text{ and } r^2 = -4 \sin 2\theta \quad \text{(Test fails)}$$

The following table of values determines several points of the graph. The two values of r for each value of θ result in the two ''leaf-shaped'' curves in Figure 10-23. The figure is called a **lemniscate** or **two-leaved rose.**

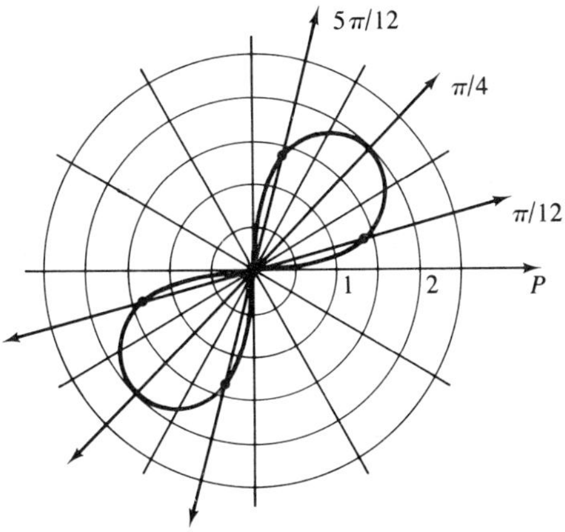

FIGURE 10-23 Graph of $r^2 = 4 \sin 2\theta$

θ	$\pm\sqrt{4 \sin 2\theta}$
0	0
$\pi/12$	$\pm\sqrt{2} \approx \pm 1.4$
$\pi/4$	± 2
$5\pi/12$	$\pm\sqrt{2} \approx \pm 1.4$
π	0

EXAMPLE 10–17 Graph $r = 4$.

Solution: The given equation can also be written as $0(\theta) + r = 4$. The given equation restricts r to the constant value 4, but θ can assume any value. The graph of the equation (Figure 10-24) is therefore a circle with radius 4. In general, *the graph of an equation of the form $r = k$, $k > 0$, is a circle with radius k.*

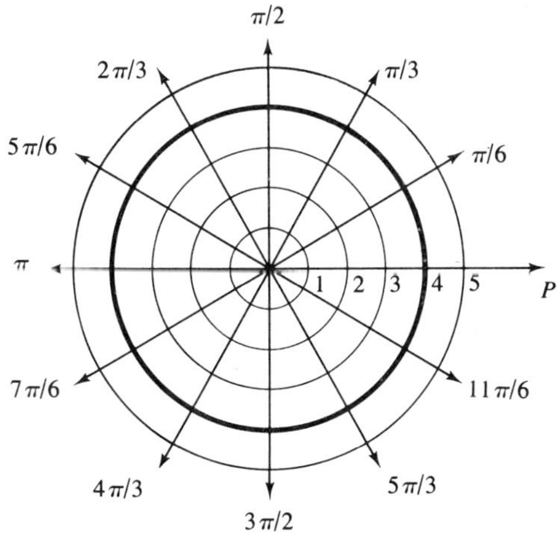

FIGURE 10-24 Graph of $r = 4$

EXAMPLE 10–18 Graph $\theta = \pi/3$.

Solution: The given equation can be written $0(r) + \theta = \pi/3$. Thus, θ is restricted to the constant value $\pi/3$, but r may assume any real number value. The graph of the equation is therefore a line, as shown in Figure 10-25. In general, *the graph of any equation of the form $\theta = k$ is a line through the pole.*

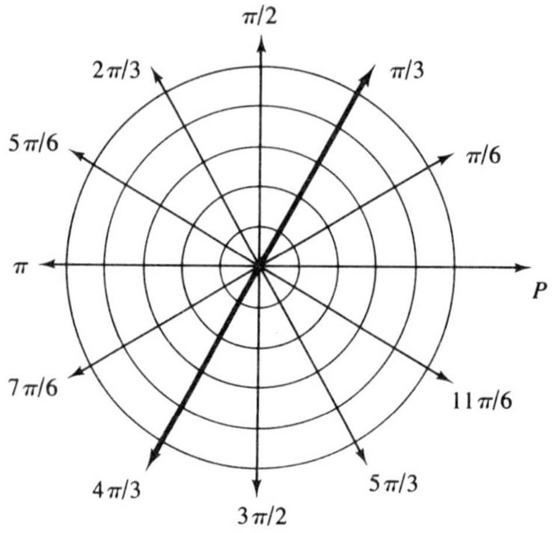

FIGURE 10-25 Graph of $\theta = \pi/3$

EXAMPLE 10–19 Graph $r = 2 \cos 2\theta$.

Solution:
Test 1. Replace θ with $-\theta$.

$$r = 2 \cos 2(-\theta) = 2 \cos 2\theta$$

The graph has symmetry with respect to the polar axis.

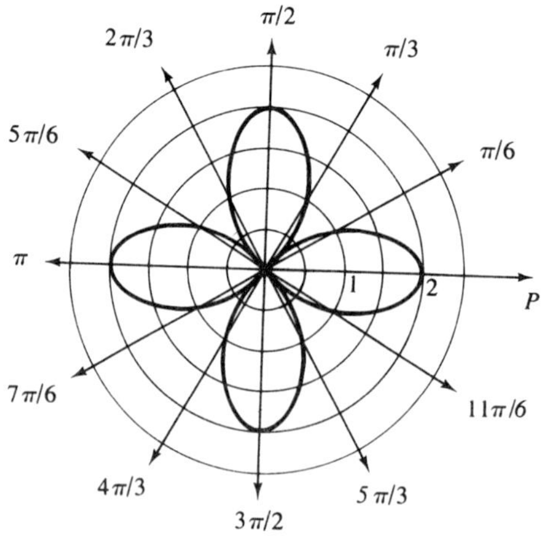

FIGURE 10-26 Graph of $r = 2 \cos 2\theta$

Test 2. Replace r with $-r$.

$$-r = 2 \cos 2\theta, \text{ and } r = -2 \cos 2\theta \qquad \text{(Test fails)}$$

Test 3. Replace r with $-r$ and θ with $-\theta$.

$$-r = 2 \cos 2(-\theta), \text{ and } r = -2 \cos 2\theta \qquad \text{(Test fails)}$$

The following table of values determines several points of the graph of the equation shown in Figure 10-26. In general, *the graph of an equation of the form $r = a \cos k\theta$ is a rose with k leaves if k is odd and 2k leaves if k is even.* Notice that the graph exhibits symmetry with respect to the pole and $\theta = \pi/2$ line even though the tests for symmetry failed.

θ	$2 \cos 2\theta$
0	2
$\pi/6$	1
$\pi/4$	0
$\pi/3$	-1
$\pi/2$	-2
$2\pi/3$	-1
$3\pi/4$	0
$5\pi/6$	1
π	2

Exercise Set 10–4

In each of the following polar equations, determine whether symmetry with respect to (a) the polar axis, (b) the pole, or (c) the line $\theta = \pi/2$ exists.

1. $r = 2 \sin \theta$
2. $r = \dfrac{1}{1 - \cos \theta}$
3. $r = 2 + \cos \theta$
4. $r = \dfrac{1}{\sin \theta + \cos \theta}$

Graph the following equations in polar form.

5. $\theta = \pi/6$ (line)
6. $\theta = -2\pi/3$ (line)
7. $r = -2$ (circle)
8. $r = 2 \cos \theta$ (circle)
9. $r = 3 \sin \theta$ (circle)
10. $r = \sin 3\theta$ (3-leaved rose)
11. $r = \cos 2\theta$ (4-leaved rose)
12. $r = 2 \sin 5\theta$ (5-leaved rose)
13. $r = 2 + 2 \cos \theta$ (cardioid)
14. $r = 1 - \sin \theta$ (cardioid)
15. $r = 2 + \cos \theta$ (limacon)
16. $r = 1 - 2 \sin \theta$ (limacon)
17. $r^2 = 16 \cos 2\theta$ (lemniscate)
18. $r^2 = 9 \sin 2\theta$ (lemniscate)
19. $r = 2\theta$ (spiral of Archimedes)
20. $r\theta = 2$ (hyperbolic spiral)
21. $r = \dfrac{1}{1 - \cos \theta}$ (parabola)
22. $r = \dfrac{5}{1 - \cos \theta}$ (parabola)

10-5 PARAMETRIC EQUATIONS

In the study of functions we stated that, whenever possible, it was desirable to express the function explicitly in terms of the independent variable. Thus,

$$f = \{(x, y) \mid y = f(x)\}$$

states that for the function f, each value of the dependent variable y is obtained for a given value of the independent variable x in the domain of the function.

However, there are times when we want to express both variables of the given function in terms of a third variable. Equations of this type are called *parametric equations*.

Parametric equations

> **Definition:** If
>
> $$x = f(t) \quad \text{and} \quad y = g(t)$$
>
> and the domains of both functions are the same, then the given equations are a pair of parametric equations, and the variable t is called the *parameter*.

Figure 10-27 shows a circle with radius of 4 units and center at the origin. For each point $P(x, y)$ on the circle, we know that $x^2 + y^2 = 16$. We also know that for any point $P(x, y)$ on the circle, $x = 4 \cos \theta$ and $y = 4 \sin \theta$, where θ is the measure in radians of a standard position angle. We can therefore represent the circle in terms of the parameter θ and the pair of parametric equations

$$x = 4 \cos \theta \quad \text{and} \quad y = 4 \sin \theta$$

For the relation C, where

$$C = \{(\theta, (x, y)) \mid x = 4 \cos \theta \text{ and } y = 4 \sin \theta\}$$

the domain $= \{\theta \mid \theta \in \text{real numbers}\}$ and range $= \{(x, y) \mid -4 \leq x \leq 4$ and $-4 \leq y \leq 4\}$. Notice that the domains of both functions $x = 4 \cos \theta$ and $y = 4 \sin \theta$ are the same. Notice also that the range of the parametric equations is a set

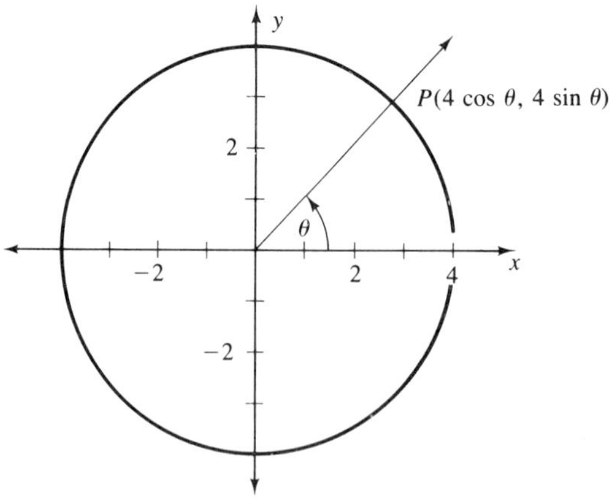

FIGURE 10-27 Circle $x^2 + y^2 = 16$, and $x = 4 \cos \theta$ and $y = 4 \sin \theta$

of ordered pairs. Thus, the range of C is the subset of the coordinate plane that is contained on the given circle.

If the equation of the circle is of the form $x^2 + y^2 = r^2$, with $r > 0$, then the parametric equations

$$x = r \cos \theta \quad \text{and} \quad y = r \sin \theta$$

also represent the circle.

Can any circle centered at the origin be represented by a similar pair of parametric equations?

To sketch the graph of an equation in parametric form, we assign values to the parameter and then compute the corresponding values for the variables in the range of the equation.

EXAMPLE 10–20 Graph $\{(t, (x, y)) \mid x = 2(t - \sin t) \text{ and } y = 2(1 - \cos t)\}$.

Solution: The following table of values is used to sketch the graph in Figure 10-28. The approximate values for x and y are easily obtained with a hand-held calculator. We have approximated these values to one decimal place.

t	$2(t - \sin t)$	$2(1 - \cos t)$
0	0	0
$\pi/4$	0.2	0.6
$\pi/2$	1.1	2
$3\pi/4$	3.3	3.4
π	6.3	4
$5\pi/4$	9.3	3.4
$3\pi/2$	11.4	2
$7\pi/4$	12.4	0.6
2π	12.6	0

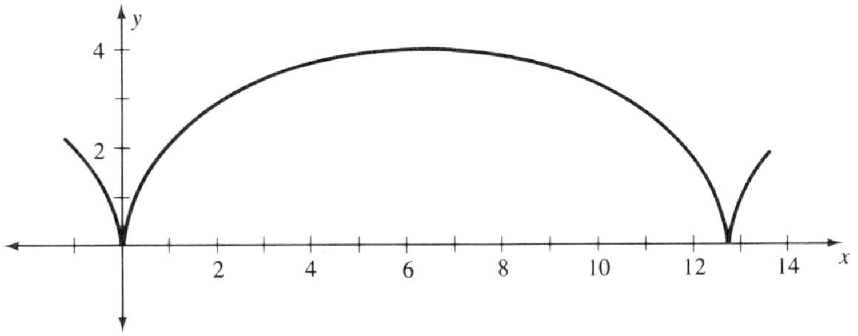

FIGURE 10-28 Graph of $x = 2(t - \sin t)$, $y = 2(1 - \cos t)$

For values of $t < 0$ or $t > 2\pi$, the curve obtained is repeated. The curve is called a **cycloid** and can be generated by tracing the location of a point on a circle with radius of 2 units as the circle rolls along the horizontal axis. In general, *if a is the radius of the circle, then the parametric equations*

$$x = a(t - \sin t) \quad \text{and} \quad y = a(1 - \cos t)$$

represent the cycloid.

An equation in parametric form can be converted to an equation in rectangular form by **eliminating the parameter.**

EXAMPLE 10–21 Eliminate the parameter in the parametric equations

$$x = 2t + 1 \quad \text{and} \quad y = 2 - t$$

Solution: We can solve both equations for t and then subtract the two equations to eliminate the parameter t.

$$\text{If } x = 2t + 1, \text{ then } t = \frac{x - 1}{2}$$

$$\text{If } y = 2 - t, \quad \text{then } t = 2 - y$$

Then we have

$$0 = \frac{x - 1}{2} - (2 - y)$$

$$x + 2y = 5$$

EXAMPLE 10–22 Eliminate the parameter in the parametric equations

$$x = 3 \cos \theta \quad \text{and} \quad y = 2 \sin \theta$$

Solution: We will use the identity $\cos^2\theta + \sin^2\theta = 1$ to eliminate the parameter θ in the given pair of equations.

$$\text{If } x = 3 \cos \theta, \text{ then } \frac{x}{3} = \cos \theta \quad \text{and} \quad \frac{x^2}{9} = \cos^2\theta$$

$$\text{If } y = 2 \sin \theta, \text{ then } \frac{y}{2} = \sin \theta \quad \text{and} \quad \frac{y^2}{4} = \sin^2\theta$$

Adding the two equations, we get

$$\frac{x^2}{9} + \frac{y^2}{4} = \cos^2\theta + \sin^2\theta = 1$$

Notice that we have obtained the equation of a **standard ellipse.** In general, *the parametric equations*

$$x = a \cos t \quad \text{and} \quad y = b \sin t$$

represent the ellipse

$$\frac{x^2}{a^2} + \frac{y^2}{b^2} = 1$$

EXAMPLE 10–23 Graph $\{(\theta, (x, y)) \mid x = 2 \sec \theta \text{ and } y = 3 \tan \theta\}$. Then express the equation in rectangular form by eliminating the parameter.

Solution: The following information is used to graph the given pair of parametric equations in Figure 10-29. We recognize the graph as a **standard hyperbola** with foci and vertices on the x-axis.

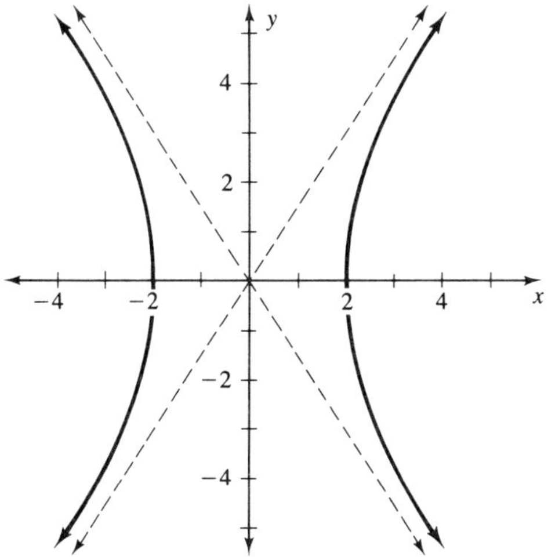

FIGURE 10-29 Graph of $x = 2 \sec \theta$, $y = 3 \tan \theta$

θ	$2 \sec \theta$	$3 \tan \theta$
0	2.0	0
$\pi/6$	2.3	1.7
$\pi/4$	2.8	3.0
$\pi/3$	4.0	5.2
$\pi/2$	undefined	undefined
$2\pi/3$	−4.0	−5.2
$3\pi/4$	−2.8	−3.0
$5\pi/6$	−2.3	−1.7
π	−2.0	0
$7\pi/6$	−2.3	1.7
$5\pi/4$	−2.8	3.0
$4\pi/3$	−4.0	5.2
$3\pi/2$	undefined	undefined
$5\pi/3$	4.0	−5.2
$7\pi/4$	2.8	−3.0
$11\pi/6$	2.3	−1.7

To eliminate the parameter θ, we use the identity $\sec^2\theta - \tan^2\theta = 1$. If $x = 2 \sec \theta$ and $y = 3 \tan \theta$, then

$$\frac{x}{2} = \sec \theta \quad \text{and} \quad \frac{y}{3} = \tan \theta$$

$$\frac{x^2}{4} = \sec^2\theta \quad \text{and} \quad \frac{y^2}{9} = \tan^2\theta$$

Then we have

$$\frac{x^2}{4} - \frac{y^2}{9} = \sec^2\theta - \tan^2\theta = 1$$

With $a^2 = 4$ and $b^2 = 9$, the equations of the asymptotes of the hyperbola are $y = \pm\frac{3}{2}x$. In general, the parametric equations $x = a \sec \theta$ and $y = b \tan \theta$

represent the hyperbola

$$\frac{x^2}{a^2} - \frac{y^2}{b^2} = 1$$

and the parametric equations $y = a \sec \theta$ and $x = b \tan \theta$ represent the hyperbola

$$\frac{y^2}{a^2} - \frac{x^2}{b^2} = 1$$

Exercise Set 10–5

Sketch the graph of the following parametric equations.

1. $x = 6t + 2, \quad y = 2t$
2. $x = t, \quad y = \cos t$
3. $x = 2t + 4, \quad y = t^2$
4. $x = \cos t, \quad y = \sin^2 t$
5. $x = 2t, \quad y = t^3$
6. $x = 1 - t, \quad y = t^2 + 2t$
7. $x = 2 \sin \theta, \quad y = 2 \cos \theta$
8. $x = 5 \sin \theta, \quad y = 5 \cos \theta$
9. $x = 4(t - \sin t), \quad y = 4(1 - \cos t)$
10. $x = 5(t - \sin t), \quad y = 5(1 - \cos t)$
11. $x = \tan \theta, y = \sec \theta$
12. $x = 1 + 3 \cos \theta, \quad y = 1 - 3 \sin \theta$
13. $x = 2 + 3 \cos \theta, \quad y = 3 \sin \theta - 3$
14. $x = 3 \sec \theta, \quad y = 2 \tan \theta$
15. $x = 4 \cos \theta, \quad y = 3 \sin \theta$
16. $x = 2 \cos \theta, \quad y = 5 \sin \theta$

Eliminate the parameter in the following parametric equations to find a single equation in rectangular coordinates.

17. $x = t^2, \quad y = t - 2$
18. $x = 2t + 5, \quad y = 3t - 7$
19. $x = \cos^2\theta, \quad y = \sin \theta$
20. $x = \cos \theta, \quad y = \cos 2\theta$
21. $x = 2 \sec \theta, \quad y = 3 \tan \theta$
22. $x = t - \frac{1}{t}, \quad y = t + \frac{1}{t}$
23. $x = t^3 + t, \quad y = t^3 - t$
24. $x = \cos \theta - \sin \theta, \quad y = \cos \theta + \sin \theta$
25. $x = t^2 + 1, \quad y = t^3 - 4$
26. $x = \frac{t}{1 + t^3}, \quad y = \frac{t^2}{1 + t^3}$
27. $x = 1.5 \sec \theta, \quad y = 2.5 \tan \theta$
28. $x = 5 \tan \theta, \quad y = 4 \sec \theta$
29. $x = 5 \cos \theta, \quad y = 3 \sin \theta$
30. $x = 3.5 \cos \theta, \quad y = 1.5 \sin \theta$

10-6 VECTORS

In science and engineering, measured quantities are divided into two general categories: scalar quantities and vector quantities. Both quantities have *magnitude*, a description of how much quantity is measured. But a vector quantity also has *direction*.

Vector quantity

> **Definition:** A vector quantity has a magnitude and a direction.

What are some examples of scalar and vector quantities?

Some common examples of scalar quantities are time, cost, temperature, and measurements of length and volume. An object's mass is also an example of a scalar quantity. Three common examples of vector quantities are force, velocity, and acceleration. An object's weight is actually a vector quantity that measures the force with which the object is pushing down on the earth's surface. Thus, the mass of an object is scalar, but the weight of the same object is a vector quantity.

A vector quantity is represented graphically as a **directed segment.** If the endpoints of the segment are A and B, then $\overrightarrow{AB}$ is the symbol for the vector quantity when the direction is from A to B. Or, if the direction of the vector quantity is from B to A, we write $\overrightarrow{BA}$. For the vector $\overrightarrow{AB}$, A is called the **initial point** and B the **terminal point.** We also use boldface letters such as **v**, **w**, and **x** to represent vector quantities.

Magnitude of $\overrightarrow{AB}$

> **Definition:** If $\overrightarrow{AB}$ is a vector quantity, then $\|\overrightarrow{AB}\|$ is the norm or magnitude of $\overrightarrow{AB}$.

Direction of $\overrightarrow{AB}$

> **Definition:** If $\overrightarrow{AB}$ is a vector quantity, then θ is the direction of $\overrightarrow{AB}$, where $-180° < \theta \le 180°$ and θ is the angle between $\overrightarrow{AB}$ and some fixed direction.

When $\overrightarrow{AB}$ is placed in a polar coordinate system with A at the origin, then the coordinates of $B(r,\ \theta)$ describe the magnitude and direction of $\overrightarrow{AB}$. That is,

$$r = \|\overrightarrow{AB}\| \qquad \textbf{(1)}$$

$$\theta = \text{the direction of } \overrightarrow{AB} \qquad \textbf{(2)}$$

and the fixed direction is the polar axis.

When $\overrightarrow{AB}$ is placed in a rectangular coordinate system with A at the origin, then the coordinates of $B(x,\ y)$ are used to compute the magnitude and direction of $\overrightarrow{AB}$ with the following equations:

$$\|\overrightarrow{AB}\| = \sqrt{x^2 + y^2} \qquad \textbf{(3)}$$

$$\tan\theta = \frac{y}{x},\quad -180° < \theta \le 180° \qquad \textbf{(4)}$$

and the fixed direction is the positive x-axis.

Standard position vector

> **Definition:** A vector **v** is in standard position if the initial point of **v** is the origin of the coordinate system in which it is drawn.

In Figure 10-30(a) and (b), the vectors **v**, **w**, and **x** are drawn as standard position vectors in a rectangular coordinate system and a polar coordinate system. Notice that in each system the coordinates of the terminal points of the vectors are unique. Thus, if **v** is a standard position vector, then $\mathbf{v} = (x, y)$ or $\mathbf{v} = (r,\ \theta)$ uniquely identifies **v** for a given pair of values $(x,\ y)$ or $(r,\ \theta)$.

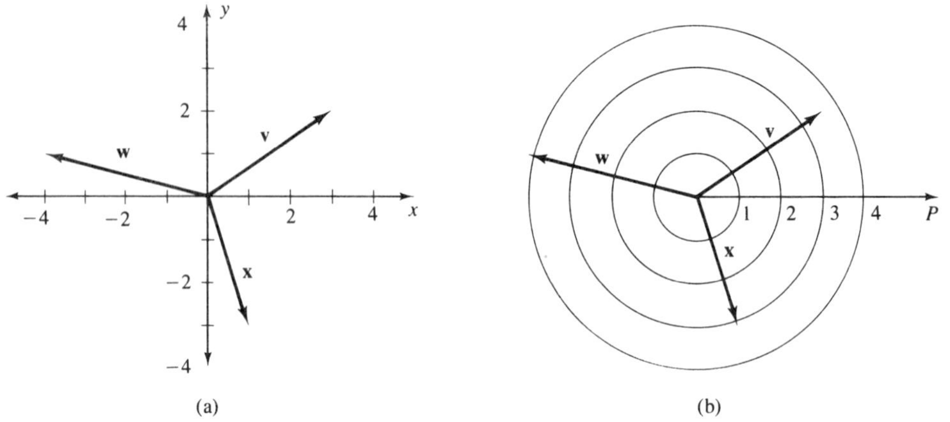

FIGURE 10-30 Vectors **v**, **w**, and **x** in standard position in (a) rectangular polar system and (b) polar coordinate system

Zero vector

> **Definition:** If **v** is the zero vector, then **v** = (0, 0).

Sum of two vectors

> **Definition:** If **u** and **v** are standard position vectors with **u** = (x_1, y_1) and **v** = (x_2, y_2), then
>
> $$\mathbf{w} = \mathbf{u} + \mathbf{v}$$
>
> or
>
> $$\mathbf{w} = ((x_1 + x_2), (y_1 + y_2))$$
>
> and **w** is the sum or resultant of **u** and **v**.

What is meant by the sum of two vectors?

To illustrate the sum of two vector quantities, let's suppose **u** and **v** are two forces applied to an object. The vector **w** = **u** + **v** is the **net force** applied to the object as a result of the initial forces **u** and **v.** If **u** = 15 newtons of force and **v** = 10 newtons of force, then the value of **w** can range from −5 to 25 newtons, depending on the directions in which **u** and **v** are applied relative to each other.

Another example of a resultant vector quantity can be seen on a windy day. Light airplanes "drift" across the sky at an angle to the heading of the plane. The direction and speed (**w**) of the plane is the result of the direction and airspeed of the plane (**u**) plus the direction and speed of the wind (**v**).

EXAMPLE 10–24 If **u** and **v** are standard position vectors with **u** = (3, 5) and **v** = (4, −2), find the magnitude and direction of **w** = **u** + **v**.

Solution: In Figure 10-31, **u** and **v** are drawn. The resultant, **w,** is the diagonal of the parallelogram with sides **u** and **v.** By the definition of the sum of two vectors, with **u** = (3, 5) and **v** = (4, −2), then

$$\mathbf{w} = ((3 + 4), (5 + (-2))) = (7, 3)$$

With Equations (3) and (4), we get

$$\|\mathbf{w}\| = \sqrt{7^2 + 3^2} = \sqrt{58}$$

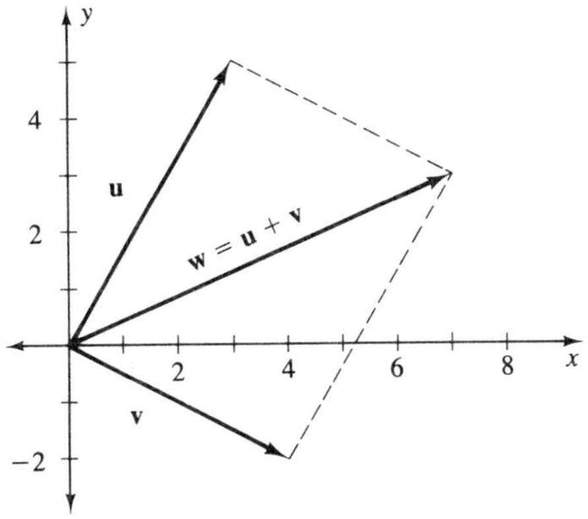

FIGURE 10-31 $\mathbf{u} = (3, 5)$, $\mathbf{v} = (4, -2)$, and $\mathbf{w} = \mathbf{u} + \mathbf{v} = (7, 3)$

and

$$\tan\theta = 3/7 = 0.4286$$
$$\theta = 23°10' \text{ to the nearest ten minutes}$$

Notice that the rectangular and polar coordinates of a vector serve two distinctly different purposes. When the standard position vector is expressed in polar coordinates, then the coordinates are the magnitude and direction of the vector. But when the sum of two or more standard position vectors is needed, then the sum of the x-coordinates and the sum of the y-coordinates of each vector are the coordinates of the resultant vector. If the magnitude and direction of the resultant are needed, we simply convert the rectangular coordinates to polar coordinates with Equations (3) and (4).

EXAMPLE 10–25 Find the resultant of $\mathbf{u} = (10.0, 30°)$ and $\mathbf{v} = (8.0, -60°)$.

Solution: If (x_1, y_1) are the rectangular coordinates of $\mathbf{u} = (10.0, 30°)$, then

$$x_1 = 10.0 \cos 30° = 8.7$$
$$y_1 = 10.0 \sin 30° = 5.0$$

If (x_2, y_2) are the rectangular coordinates of $\mathbf{v} = (8.0, -60°)$, then

$$x_2 = 8.0 \cos(-60°) = 4.0$$
$$y_2 = 8.0 \sin(-60°) = -6.9$$

Then

$$\begin{aligned}\mathbf{w} = \mathbf{u} + \mathbf{v} &= ((x_1 + x_2), (y_1 + y_2)) \\ &= ((8.7 + 4.0), (5.0 + (-6.9))) \\ &= (12.7, -1.9)\end{aligned}$$

Thus,

$$\|\mathbf{w}\| = \sqrt{12.7^2 + (-1.9)^2} = 12.8$$

and

$$\tan\theta = -1.9/12.7 = -0.1496$$
$$\theta = -9^\circ \text{ to the nearest whole degree}$$

Therefore, $\mathbf{w} = \mathbf{u} + \mathbf{v} = (12.8,\ -9^\circ)$.

EXAMPLE 10–26 A 100-pound block is placed on a board, and the board is inclined until the block starts to slide down the board. The angle of inclination of the board is 26°. What forces, to the nearest pound, are holding the block to the board and pushing the block down the board?

Solution: Let **w** be the weight of the block toward the center of the earth. Let **x** and **y** be the forces acting parallel and perpendicular to the board, respectively, and $\mathbf{w} = \mathbf{x} + \mathbf{y}$. The three vectors are shown in Figure 10-32.

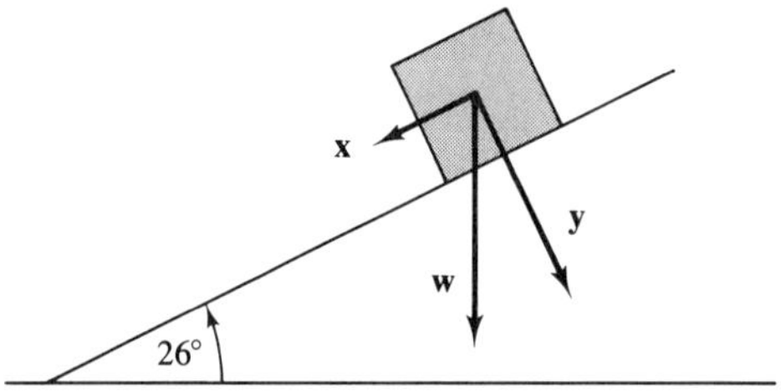

FIGURE 10-32 $|\mathbf{w}| = 100$ pounds and $\mathbf{w} = \mathbf{x} + \mathbf{y}$

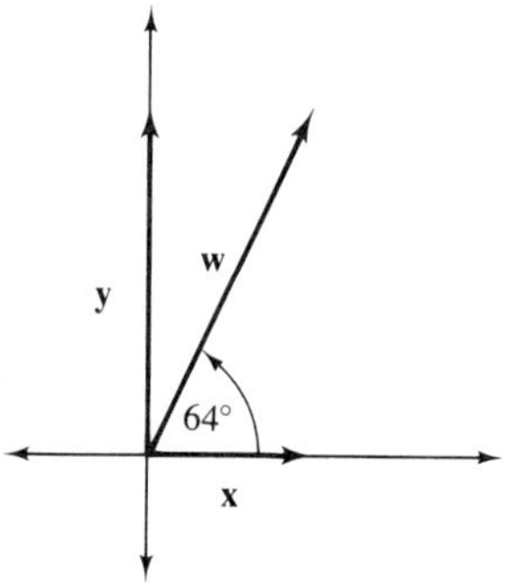

FIGURE 10-33 Vectors **w, x,** and **y** as standard position vectors

We can show that the angle between **w** and **y** is also 26°; thus, the angle between vectors **w** and **x** is 64°. In Figure 10-33, **w, x,** and **y** are shown as standard position vectors. Then we have

$$|\mathbf{x}| = 100\cos 64^\circ$$
$$= 43.84, \text{ or } 44 \text{ pounds}$$

and

$$|\mathbf{y}| = 100\sin 64^\circ$$
$$= 89.88, \text{ or } 90 \text{ pounds}$$

Therefore, the force holding the block to the plane is about 90 pounds, while the force pushing the block down the plane is about 44 pounds.

The answers in Example 10–26 clearly show that *vector quantities do not add like scalar quantities*. Since vector quantities have direction, the only time that the magnitude of the sum of two vectors is equal to the sum of their magnitudes is when the direction of the two vectors is the same.

But 90 + 44 is greater than 100, the weight of the block!

Vectors in equilibrium

> **Definition:** If $\mathbf{v}_1 = (x_1, y_1)$, $\mathbf{v}_2 = (x_2, y_2)$, . . ., $\mathbf{v}_n = (x_n, y_n)$ and $\mathbf{v}_1 + \mathbf{v}_2 + \cdots + \mathbf{v}_n = (0, 0)$, then the set of vectors is in equilibrium.

To illustrate this definition, let $\mathbf{v}_1$, $\mathbf{v}_2$, . . . , $\mathbf{v}_n$ be a set of forces acting on an object. If $\mathbf{v}_1 + \mathbf{v}_2 + \cdots + \mathbf{v}_n = (0, 0)$, then the object is at rest or moving at a constant velocity. That is, *for any object in equilibrium, the sum of the forces acting on the object is zero.*

Example 10–27 If $\mathbf{v}_1 = (10.0, 30°)$, $\mathbf{v}_2 = (20.0, 300°)$, and $\mathbf{v}_3 = (15.0, 180°)$ are three forces measured in newtons acting on an object, find the force $\mathbf{v}_4$ necessary to put the system of forces in equilibrium.

Solution:

If $\mathbf{v}_1 = (10.0, 30°)$, then $\mathbf{v}_1 = (10.0 \cos 30°, 10.0 \sin 30°) = (8.7, 5.0)$

If $\mathbf{v}_2 = (20.0, 300°)$, then $\mathbf{v}_2 = (20.0 \cos 300°, 20.0 \sin 300°) = (10.0, -17.3)$

If $\mathbf{v}_3 = (15.0, 180°)$, then $\mathbf{v}_3 = (15.0 \cos 180°, 15.0 \sin 180°) = (-15.0, 0)$

Now let $\mathbf{v}_4 = (a, b)$ such that $\mathbf{v}_1 + \mathbf{v}_2 + \mathbf{v}_3 + \mathbf{v}_4 = (0, 0)$. That is,

$$(8.7 + 10.0 + (-15.0) + a, 5.0 + (-17.3) + 0 + b) = (0, 0)$$

Then

$$3.7 + a = 0 \quad \text{and} \quad -12.3 + b = 0$$
$$a = -3.7 \quad \text{and} \quad b = 12.3$$

Thus, $\mathbf{v}_4 = (-3.7, 12.3)$. Finally,

$$\|\mathbf{v}_4\| = \sqrt{(-3.7)^2 + 12.3^2} = 12.8$$

and

$$\tan \theta = \frac{-12.3}{3.7} = -3.3243$$
$$\theta = 107° \text{ to the nearest whole degree}$$

Therefore, a force of about 12.8 newtons applied to the object at an angle of approximately 107° will cancel out the combined effect of forces $\mathbf{v}_1$, $\mathbf{v}_2$, and $\mathbf{v}_3$ and bring the object to a state of equilibrium.

Exercise Set 10–6

For the following vectors, find (a) the magnitude to the nearest tenth, and (b) the direction to the nearest ten minutes.

1. $\mathbf{v} = (5, 12)$
2. $\mathbf{w} = (-4, 3)$
3. $\mathbf{x} = (7, -24)$
4. $\mathbf{u} = (-8, -15)$
5. $\mathbf{v} = (10, -3)$
6. $\mathbf{w} = (-6, -5)$

Find the rectangular coordinates of the following vectors to the nearest tenth of a unit.

7. $\mathbf{v} = (10, 30°)$
8. $\mathbf{w} = (4, -120°)$
9. $\mathbf{x} = (5, 135°)$
10. $\mathbf{y} = (20, 90°)$
11. $\mathbf{v} = (8, 120°20')$
12. $\mathbf{w} = (15, -85°40')$

For the following, (a) sum the vectors, (b) find the magnitude of the resultant to the nearest tenth of a unit, and (c) find the direction of the resultant to the nearest ten minutes.

13. $\mathbf{u} = (2, 3)$; $\mathbf{v} = (3, 4)$
14. $\mathbf{u} = (3, -2)$; $\mathbf{v} = (2, 5)$
15. $\mathbf{u} = (-2, 5)$; $\mathbf{v} = (2, -3)$
16. $\mathbf{u} = (-5, -4)$; $\mathbf{v} = (2, -1)$
17. $\mathbf{u} = (2, 1)$; $\mathbf{v} = (3, 4)$; $\mathbf{w} = (2, -2)$
18. $\mathbf{u} = (-2, 4)$; $\mathbf{v} = (-1, -2)$; $\mathbf{w} = (1, -2)$
19. $\mathbf{u} = (4, 45°)$; $\mathbf{v} = (2, -45°)$
20. $\mathbf{u} = (1, 150°)$; $\mathbf{v} = (6, 30°)$
21. $\mathbf{u} = (3, -120°)$; $\mathbf{v} = (2, 180°)$
22. $\mathbf{u} = (2, 135°)$; $\mathbf{v} = (3, -60°)$
23. Two forces of 130 pounds and 100 pounds act on an object at an angle of 45°. What is the magnitude of the resultant of the two forces?
24. An airplane flying due north at an airspeed of 200 miles per hour is blown off course by a westerly wind of 15 miles per hour. What is the ground speed and the direction of actual flight?
25. A man pushes a wheelbarrow carrying 300 pounds up a ramp or inclined plane. If he pushes with a force of 180 pounds, what is the largest angle the ramp may make with the ground?
26. A weight of 100 pounds is hung from the middle of a rope which is stretched between two boards 40 feet apart. If the rope sags 5 feet in the middle, find the force on each half of the rope.
27. An airplane is headed on an azimuth of 135° with an airspeed of 400 miles per hour. If the plane has a ground speed of 450 miles per hour and is flying on an azimuth of 145°, find the wind direction and wind speed.
28. Find the angle of elevation of a plane if a block weighing 100 pounds needs 60 pounds of force to keep it from sliding down the plane.
29. If $\mathbf{F}_1 = (30, 40)$, $\mathbf{F}_2 = (-10, 30)$, and $\mathbf{F}_3 = (20, -10)$ are three vector forces acting on an object, find the vector force $\mathbf{F}_4$ necessary to put the system of vectors in equilibrium.
30. If $\mathbf{F}_1 = (20, 45°)$, $\mathbf{F}_2 = (30, 330°)$, and $\mathbf{F}_3 = (10, -180°)$ are three vector forces acting on an object, find the vector force $\mathbf{F}_4$ necessary to put the system of vectors in equilibrium.

ROTATION OF AXES 10-7

In Chapter 6 we studied relations of set C, where $C = \{(x, y) \mid Ax^2 + Bxy + Cy^2 + Dx + Ey + F = 0\}$. At that time we considered only equations in which $B = 0$. In this section we will graph relations of set C for which $B \neq 0$. To simplify the job, we will introduce an x^*y^*-coordinate system on which the graphs (if they exist) are standard.

To illustrate, let's consider the relation R, where

$$R = \{(x, y) \mid x^2 + xy + y^2 = 18\}$$

When we compare the equation of R with the general form of the equation in C, we see that $A = B = C = 1$. Since $B \neq 0$, the graph of R is not standard in the xy-coordinate system. However, as we shall see, the graph of R is standard in an x^*y^*-coordinate system that is related to the xy system of the variables in the equation.

To establish the relationship between the two coordinate systems, in Figure 10-34 we show a point $P(x, y)$ different from the origin in the xy-coordinate plane. We have also drawn an x^*y^*-coordinate system in which the origins of the two systems are the same point. However, the x^*-axis is rotated through an angle ϕ from the x-axis. Thus, a rotation of the x-axis about the origin by an amount equal to ϕ will cause the two axes to coincide. Such a rotation is called a **rotation transformation.** Because P is contained in two different coordinate systems, it has two ordered pairs of coordinates, (x, y) and (x^*, y^*).

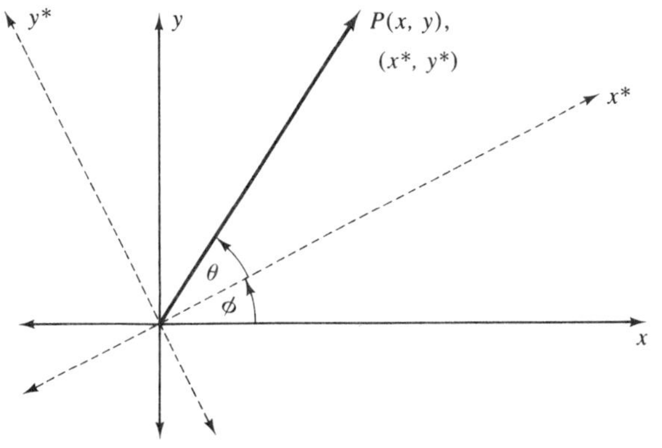

FIGURE 10-34 Coordinate axes xy and x^*y^* rotated through an angle ϕ

As noted in Section 10-3, the rectangular coordinates (x, y) of any point P can also be expressed in terms of polar coordinates (r, θ), where

$$r = \sqrt{x^2 + y^2} \quad \text{and} \quad \tan\theta = \frac{y}{x}$$

If we let $r = |OP|$ and θ be the angle between the x^*-axis and $\overline{OP}$, then

$$x^* = r\cos\theta \quad \text{and} \quad y^* = r\sin\theta$$

With ϕ the angle between the x- and x^*-axes and θ the angle between the x^*-axis and $\overline{OP}$, then

$$x = r\cos(\theta + \phi) \quad \text{and} \quad y = r\sin(\theta + \phi)$$

But

$$\cos(\theta + \phi) = \cos\theta\cos\phi - \sin\theta\sin\phi$$

and

$$\sin(\theta + \phi) = \sin\theta\cos\phi + \cos\theta\sin\phi$$

Therefore,

$$x = r\cos\theta\cos\phi - r\sin\theta\sin\phi$$

and

$$y = r\sin\theta\cos\phi + r\cos\theta\sin\phi$$

When we substitute x^* for $r\cos\theta$ and y^* for $r\sin\theta$, we obtain Equations (1) and (2) in the following theorem:

> **Theorem:** If xy and x^*y^* are the coordinate axes of two rectangular coordinate systems with the same origin, and if ϕ is the angle from the positive half of the x-axis to the positive half of the x^*-axis, then the coordinates of a point P are (x, y) and (x^*, y^*) with respect to the xy- and x^*y^*-coordinate systems, respectively. The pairs of coordinates are related by the following equations:
>
> $$x = x^*\cos\phi - y^*\sin\phi \qquad \textbf{(1)}$$
>
> $$y = y^*\cos\phi + x^*\sin\phi \qquad \textbf{(2)}$$

EXAMPLE 10–28 If $(x^*, y^*) = (4, 2)$ in an x^*y^*-coordinate system in which $\phi = 30°$, find the corresponding (x, y) coordinates.

Solution: Given that $x^* = 4$, $y^* = 2$, and $\phi = 30°$, with Equation (1) we find that

$$\begin{aligned} x &= x^*\cos\phi - y^*\sin\phi \\ &= 4\cos 30° - 2\sin 30° \\ &= 2\sqrt{3} - 1 \end{aligned}$$

With Equation (2),

$$\begin{aligned} y &= y^*\cos\phi + x^*\sin\phi \\ &= 2\cos 30° + 4\sin 30° \\ &= \sqrt{3} + 2 \end{aligned}$$

Thus, $P(4, 2)$ in the x^*y^* system is $P(2\sqrt{3} - 1, \sqrt{3} + 2)$ in the xy system. With Table 1 or a calculator, we can approximate the coordinates of P as $(2.5, 3.7)$.

With Equations (1) and (2), we can convert an equation in terms of x and y to an equation in terms of x^* and y^* for a given value of ϕ. To illustrate, we will convert the equation of the relation R identified earlier in this section.

EXAMPLE 10–29 If $\phi = 45°$, express $x^2 + xy + y^2 = 18$ in terms of x^* and y^*.

Solution: If $\phi = 45°$, then $\cos\phi = \sin\phi = 1/\sqrt{2}$. With Equations (1) and (2), we get

$$x = \frac{x^*}{\sqrt{2}} - \frac{y^*}{\sqrt{2}} \quad \text{and} \quad y = \frac{y^*}{\sqrt{2}} + \frac{x^*}{\sqrt{2}}$$

We now substitute these values for x and y in the given equation and simplify as follows:

$$\left(\frac{x^*}{\sqrt{2}} - \frac{y^*}{\sqrt{2}}\right)^2 + \left(\frac{x^*}{\sqrt{2}} - \frac{y^*}{\sqrt{2}}\right)\left(\frac{y^*}{\sqrt{2}} + \frac{x^*}{\sqrt{2}}\right) + \left(\frac{y^*}{\sqrt{2}} + \frac{x^*}{\sqrt{2}}\right)^2 = 18$$

$$\left(\frac{(x^*)^2}{2} - x^*y^* + \frac{(y^*)^2}{2}\right) + \left(\frac{x^*y^*}{2} + \frac{(x^*)^2}{2} - \frac{(y^*)^2}{2} - \frac{x^*y^*}{2}\right)$$
$$+ \left(\frac{(y^*)^2}{2} + x^*y^* + \frac{(x^*)^2}{2}\right) = 18$$

$$\frac{3(x^*)^2}{2} + \frac{(y^*)^2}{2} = 18$$

$$3(x^*)^2 + (y^*)^2 = 36$$

This equation is a member of set C^*, where

$$C^* = \{(x^*, y^*) \mid A(x^*)^2 + Bx^*y^* + C(y^*)^2 + Dx^* + Ey^* + F = 0\}$$

Since the equation can be written in the form

$$\frac{(x^*)^2}{12} + \frac{(y^*)^2}{36} = 1$$

the graph of R is a *standard ellipse in the x^*y^*-coordinate system*. The graph is shown in Figure 10-35 with vertices on the y^*-axis.

If $B \neq 0$ in the general equation of the second-degree relation, then the xy term can always be removed by rotating the xy-axis through an angle ϕ $(0° < \phi < 90°)$. The proof of this statement is not included in this text, and we accept it as true. We simply state the details in the following theorem.

Why did a rotation transformation of 45° eliminate the xy term in the original equation for R?

Theorem: If $B \neq 0$, the xy term in the equation

$$Ax^2 + Bxy + Cy^2 + Dx + Ey + F = 0$$

is removed by rotating the xy-axis through an angle ϕ $(0° < \phi < 90°)$, and

$$\cot 2\phi = \frac{A - C}{B} \tag{3}$$

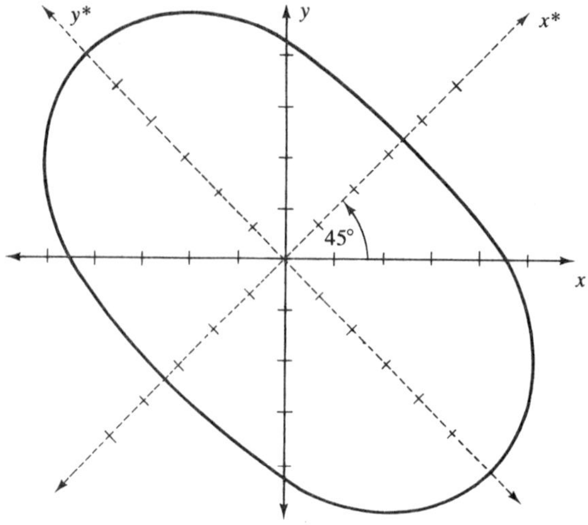

FIGURE 10-35 Graph of $x^2 + xy + y^2 = 18$ or $\dfrac{(x^*)^2}{12} + \dfrac{(y^*)^2}{36} = 1$

In the given equation for R, $x^2 + xy + y^2 = 18$, with $A = B = C = 1$. Thus, with Equation (3), we have

$$\cot 2\phi = \frac{1 - 1}{1} = 0$$
$$2\phi = 90°$$
$$\phi = 45°$$

The rotation of 45° therefore guaranteed that the resulting equation would not contain an x^*y^* term.

EXAMPLE 10–30 Determine the angle of rotation ϕ that eliminates the xy term in the equation $8x^2 - 6\sqrt{3}xy + 14y^2 = 20$.

Solution: In the given equation, $A = 8$, $B = -6\sqrt{3}$, and $C = 14$. With Equation (3), we have

$$\cot 2\phi = \frac{8 - 14}{-6\sqrt{3}} = \frac{1}{\sqrt{3}}$$

With the table of exact function values, we find

$$2\phi = 60°$$
$$\phi = 30°$$

Thus, the xy term is removed from the equation with the rotation transformation equations and $\phi = 30°$.

EXAMPLE 10–31 Graph the equation $3x^2 - 2\sqrt{3}xy + y^2 - 2x - 2\sqrt{3}y = 0$.

Solution: With $A = 3$, $B = -2\sqrt{3}$, and $C = 1$, then by Equation (3),

$$\cot 2\phi = \frac{3-1}{-2\sqrt{3}} = \frac{-1}{\sqrt{3}}$$

With the table of exact values, we find

$$2\phi = 120°$$
$$\phi = 60°$$

With $\phi = 60°$ and transformation equations (1) and (2), $3x^2 - 2\sqrt{3}xy + y^2 - 2x - 2\sqrt{3}y = 0$ becomes

$$3\left(\frac{x^*}{2} - \frac{\sqrt{3}y^*}{2}\right)^2 - 2\sqrt{3}\left(\frac{x^*}{2} - \frac{\sqrt{3}y^*}{2}\right)\left(\frac{y^*}{2} + \frac{\sqrt{3}x^*}{2}\right) + \left(\frac{y^*}{2} + \frac{\sqrt{3}x^*}{2}\right)^2$$
$$- 2\left(\frac{x^*}{2} - \frac{\sqrt{3}y^*}{2}\right) - 2\sqrt{3}\left(\frac{y^*}{2} + \frac{\sqrt{3}x^*}{2}\right) = 0$$

When we perform the indicated multiplications and combine like terms, we get the following equations:

$$4(y^*)^2 - 4x^* = 0$$
$$x^* = (y^*)^2$$

The graph of the given equation is a *parabola* that is symmetric about the x^*-axis and vertex $V(0, 0)$. The graph is shown in Figure 10-36.

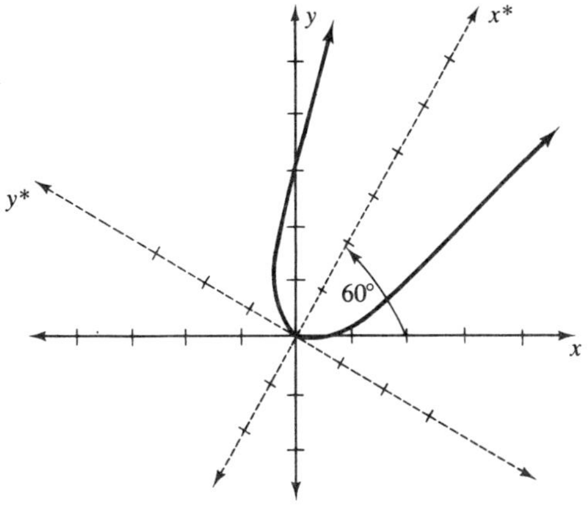

FIGURE 10-36 Graph of $3x^2 - 2\sqrt{3}xy + y^2 - 2x - 2\sqrt{3}y = 0$ or $(y^*)^2 = x^*$ with $\phi = 60°$

Exercise Set 10–7

Convert the given (x^*, y^*) coordinates to (x, y) coordinates for the indicated ϕ, the angle between the positive x-axis and positive x^*-axis. (*Optional:* With a hand-held calculator, approximate the x- and y-coordinates to one decimal place.)

1. $(x^*, y^*) = (6, 2)$; $\phi = 45°$
2. $(x^*, y^*) = (2, 4)$; $\phi = 90°$
3. $(x^*, y^*) = (-1, 3)$; $\phi = 60°$
4. $(x^*, y^*) = (-3, -4)$; $\phi = -30°$
5. $(x^*, y^*) = (10, 6)$; $\phi = \pi/6$
6. $(x^*, y^*) = (4, 8)$; $\phi = 2\pi/3$
7. $(x^*, y^*) = (-5, 3)$; $\phi = -\pi/3$
8. $(x^*, y^*) = (-4, 2)$; $\phi = -\pi/4$

In each of the following equations, (a) remove the xy term by performing a rotation of the xy-axes through an angle ϕ, (b) identify the curve defined by the equation, and (c) graph the relation, showing both the xy- and x^*y^*-axes.

9. $3x^2 - 2xy + 3y^2 = 16$
10. $xy = 4$
11. $2x^2 + \sqrt{3}xy + y^2 = 2$
12. $5x^2 + 6\sqrt{3}xy - y^2 = 1$
13. $11x^2 - 10\sqrt{3}xy + y^2 = 16$
14. $x^2 + 2\sqrt{3}xy + 3y^2 - 5 = 0$
15. $x^2 - 2xy + y^2 + 8\sqrt{2}x + 8\sqrt{2}y = 0$
16. $3x^2 - 6xy + 3y^2 + 2\sqrt{2}x + 2\sqrt{2}y = 4$

Optional

Definition: For any second-degree equation of the form $Ax^2 + Bxy + Cy^2 + Dx + Ey + F = 0$, the expression $B^2 - 4AC$ is called the **discriminant.**

If $B^2 - 4AC < 0$, the graph of the equation is an **ellipse.**
If $B^2 - 4AC = 0$, the graph of the equation is a **parabola.**
If $B^2 - 4AC > 0$, the graph of the equation is a **hyperbola.**

Use the discriminant to determine the kind of graph the following second-degree equations have, if they exist.

17. $5x^2 + y^2 - 8x + 1 = 0$
18. $9x^2 + 4y^2 - 18x + 16y - 11 = 0$
19. $3x^2 - 6xy + y^2 - 10 = 0$
20. $x^2 - 6xy + 9y^2 - 3 = 0$
21. $4x^2 + 8xy + 4y^2 - 2\sqrt{2}x + \sqrt{2}y = 4$
22. $3x^2 + xy + y^2 + 5x - 2y = 5$

SUMMARY

Law of Sines: In any triangle, if A, B, and C are angles opposite the sides of lengths a, b, and c, respectively, then

$$\frac{\sin A}{a} = \frac{\sin B}{b} = \frac{\sin C}{c}$$

Law of Cosines: If a, b, and c are the lengths of the sides of any triangle ABC, then

$$a^2 = b^2 + c^2 - 2bc \cos A$$
$$b^2 = a^2 + c^2 - 2ac \cos B$$
$$c^2 = a^2 + b^2 - 2ab \cos C$$

The **polar coordinates** of point R in a polar coordinate system with origin O and polar axis $\overrightarrow{OP}$ are (r, θ), where $r = |\overline{OR}|$ and θ is an angle measured from $\overrightarrow{OP}$ to $\overline{OR}$.

If (x, y) and (r, θ) are the coordinates of R in rectangular and polar coordinate systems, respectively, with O the origin of both systems and the positive x-axis and polar axis the same ray, then

$$x = r\cos\theta \quad (1) \qquad y = r\sin\theta \quad (2)$$
$$r = \sqrt{x^2 + y^2} \quad (3) \qquad \tan\theta = \frac{y}{x} \quad (4)$$

Symmetry with respect to the polar axis

Definition: The graph of an equation in polar form has symmetry with respect to the polar axis if whenever $P(r_1, \theta_1)$ is a point of the graph, then $Q(r_1, -\theta_1)$ is also a point of the graph.

Definition: The graph of an equation in polar form has symmetry with respect to the pole if whenever $P(r_1, \theta_1)$ is a point of the graph, then $Q(-r_1, \theta_1)$ is also a point of the graph.

Symmetry with respect to the pole

Definition: The graph of an equation in polar form has symmetry with respect to the line $\theta = \pi/2$ if whenever $P(r_1, \theta_1)$ is a point of the graph, then $Q(-r_1, -\theta_1)$ is also a point of the graph.

Symmetry with respect to the line $\theta = \pi/2$

Definition: If $x = f(t)$ and $y = g(t)$ and the domains of both functions are the same, then the equations are called a pair of parametric equations, and the variable t is called the parameter.

Parametric equations

Definition: A vector quantity has a magnitude and a direction.

Vector quantity

Definition: If $\overrightarrow{AB}$ is a vector quantity, then $\|\overrightarrow{AB}\|$ is the norm or magnitude of $\overrightarrow{AB}$.

Magnitude of $\overrightarrow{AB}$

Definition: If $\overrightarrow{AB}$ is a vector quantity, then θ is the direction of $\overrightarrow{AB}$, where $-180° < \theta \leq 180°$ and θ is the angle between $\overrightarrow{AB}$ and some fixed direction.

Direction of $\overrightarrow{AB}$

Definition: A vector **v** is in standard position if the initial point of **v** is the origin of the coordinate system in which it is drawn.

Standard position vector

Definition: If **v** is the zero vector, then **v** = (0, 0).

Zero vector

Definition: If **u** and **v** are standard position vectors with $\mathbf{u} = (x_1, y_1)$ and $\mathbf{v} = (x_2, y_2)$, then $\mathbf{w} = \mathbf{u} + \mathbf{v}$, or $\mathbf{w} = ((x_1 + x_2), (y_1 + y_2))$, and **w** is the sum or resultant of **u** and **v**.

Sum of two vectors

Definition: If $\mathbf{v}_1 = (x_1, y_1)$, $\mathbf{v}_2 = (x_2, y_2)$, . . . , $\mathbf{v}_n = (x_n, y_n)$, and $\mathbf{v}_1 + \mathbf{v}_2 + \cdots + \mathbf{v}_n = (0, 0)$, then the set of vectors is in equilibrium.

Vectors in equilibrium

Theorem: If xy and x^*y^* are the coordinate axes of two rectangular coordinate systems with the same origin, and if ϕ is the angle from the positive half of the x-axis to the positive half of the x^*-axis, then the coordinates of a point P are (x, y) and (x^*, y^*) with respect to the xy- and x^*y^*-coordinate systems, respectively. The pairs of coordinates are related by the following equations:

$$x = x^* \cos \phi - y^* \sin \phi$$
$$y = y^* \cos \phi + x^* \sin \phi$$

Theorem: If $B \neq 0$, the xy term in the equation $Ax^2 + Bxy + Cy^2 + Dx + Ey + F = 0$ is removed by rotating the x-axis through an angle ϕ $(0° < \phi < 90°)$, and $\cot 2\phi = \dfrac{A - C}{B}$.

SELF-CHECKING EXERCISE

Solve the following right triangles. **(10-1)**

1. $B = 57°30'$ and $a = 13.1$
2. $a = 9.3$ and $b = 6.5$

Solve the following triangles with the Law of Sines. **(10-1)**

3. $c = 160$; $b = 128$; $B = 40°00'$
4. $A = 81°00'$; $B = 68°00'$; $a = 3.5$

For the following ambiguous case problems, find the number of triangles determined. Do not solve the triangles. **(10-1)**

5. $a = 8.4$; $b = 10.3$; $A = 62°20'$

6. $b = 15.1$; $c = 13.5$; $C = 40°30'$

Solve the following triangles with the Law of Cosines. **(10-2)**

7. $A = 45°50'$; $b = 15.3$; $c = 19.2$

8. $a = 61.0$; $b = 57.0$; $c = 82.0$. Approximate each angle measure to the nearest 10 minutes.

Find the required part of the following triangles. **(10-2)**

9. Find a if $A = 75°20'$, $b = 3.5$, and $c = 5.2$.

10. Find c if $a = 9.8$, $b = 10.4$, and $C = 112°30'$.

Solve the following applied problems. **(10-1** and **10-2)**

11. Steve is flying a kite, using 500 feet of string. If the angle of elevation is 15°, how high is the kite flying above the ground if Steve is 4 feet tall and all the string is being used?

12. Two ships leave port at the same time, one going 28° west of North at 40 miles per hour and the other due east at 30 miles per hour. How far apart are the ships after 3 hours?

13. From two lighthouses 8.2 miles apart, a ship is sighted at sea with angles of 20° and 32°, respectively, relative to the line between the lighthouses. How far is the ship from the first lighthouse?

Give three other pairs of polar coordinates for the following points. **(10-3)**

14. $(-2, \pi/3)$

15. $(5, -2\pi/7)$

Convert the coordinates of the following points from polar form to rectangular form. **(10-3)**

16. $(6, 7\pi/6)$

17. $(-10, -3\pi/4)$

Convert the coordinates of the following points from rectangular form to polar form. If necessary, approximate θ to the nearest hundredth of a radian. **(10-3)**

18. $(4, -4)$

19. $(-5, 12)$

Rewrite the following equations in rectangular form. **(10-3)**

20. $r = 2 \sin \theta$

21. $r = \dfrac{12}{3 \cos \theta - 4 \sin \theta}$

Rewrite the following equations in polar form. **(10-3)**

22. $x^2 = 4y$

23. $3x^2 + 3y^2 + 4y = 0$

Sketch the graphs of the following equations. **(10-4)**

24. $r = 2 - \cos \theta$

25. $r^2 = \sin \theta$

Eliminate the parameter and find an equation in rectangular form for the following equations. **(10-5)**

26. $x = \sin t, \quad y = \sin 2t$

27. $x = \dfrac{2t}{t^2 + 1}, \quad y = \dfrac{4}{t^2 + 1}$

Sketch a graph of the following parametric equations. **(10-5)**

28. $x = \sin t, \quad y = \cos^2 t$

29. $x = 2 - \tan t, \quad y = 2 + 3 \tan t$

Find the magnitude and direction (to the nearest hundredth of a radian) of the sum of **v** and **w**. **(10-6)**

30. $\mathbf{v} = (4, 7)$ and $\mathbf{w} = (-5, 3)$

31. $\mathbf{v} = (-2, 6)$ and $\mathbf{w} = (7, 6)$

32. Find the magnitude and direction of the resultant of two forces of 150 newtons and 200 newtons acting on the same object if the angle between the forces is 60°. **(10-6)**

Convert the given (x^*, y^*) coordinates to (x, y) coordinates for the given angle of rotation of axes. **(10-7)**

33. $(x^*, y^*) = (10, -6); \phi = 3\pi/4$

34. $(x^*, y^*) = (-3, 4); \phi = -150°$

In the following equation, (a) remove the xy term by performing a rotation of the xy-axes through an angle ϕ, (b) identify the curve defined by the equation, and (c) graph the relation, showing both the xy- and x^*y^*-axes. **(10-7)**

35. $x^2 + 2\sqrt{3}xy + 3y^2 + 2\sqrt{3}x - 2y = 0$

Answers to Self-Checking Exercise

1. $A = 32°30'; b = 20.6; c = 24.4$
2. $c = 11.4; A = 55°00'; B = 35°00'$
3. $C = 53°30'; A = 86°30'; a = 199$
4. $C = 31°00'; b = 3.3; c = 1.8$
5. no triangle
6. two triangles
7. $a = 13.9; B = 52°10'; C = 82°00'$
8. $A = 48°00'; B = 44°00'; C = 88°00'$
9. $a = 5.5$
10. $c = 16.8$
11. 138 feet
12. 181 miles
13. 5.5 miles
14. Answers can vary. Examples: $(2, -2\pi/3), (2, 4\pi/3), (-2, -5\pi/3)$
15. Answers can vary. Examples: $(5, 12\pi/7), (-5, 5\pi/7), (-5, -9\pi/7)$
16. $(-3\sqrt{3}, -3)$
17. $(5\sqrt{2}, 5\sqrt{2})$
18. $(4\sqrt{2}, 7\pi/4)$
19. $(13, 1.97)$
20. $x^2 + y^2 - 2y = 0$
21. $3x - 4y = 12$
22. $r^2\cos^2\theta = 4r \sin \theta$
23. $3r = -4 \sin \theta$

24.

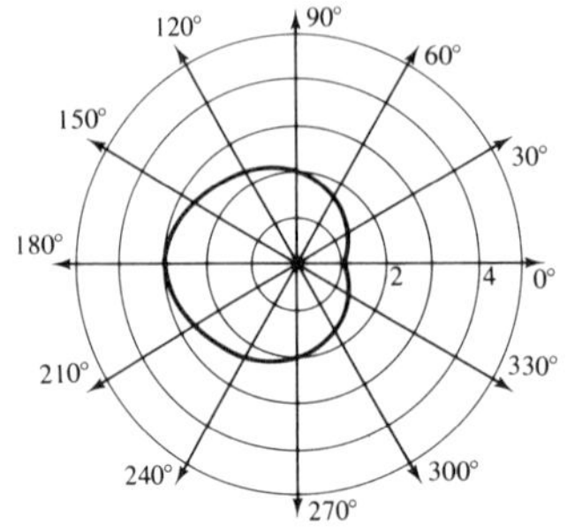

25.

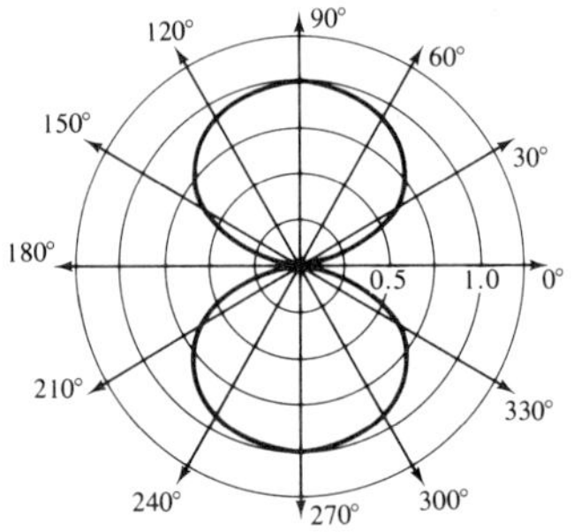

26. $y = 2x\sqrt{1 - x^2}$ or $y^2 = 4x^2 - 4x^4$

27. $4x^2 + y^2 - 4y = 0$

28.

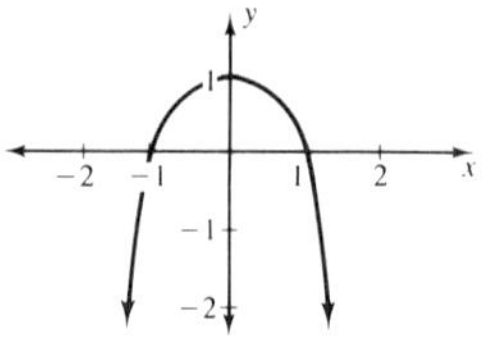

29.

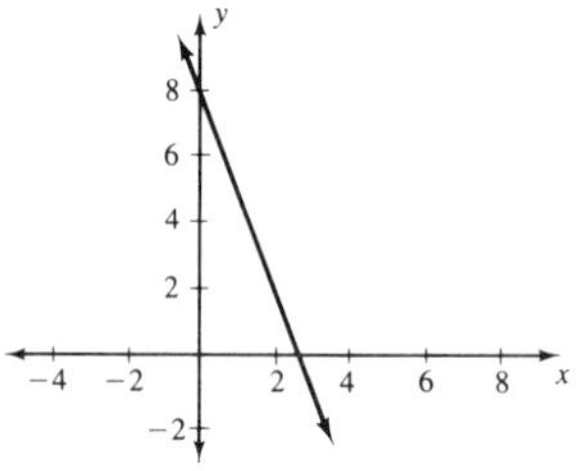

30. Magnitude $= \sqrt{101}$; direction $= 1.67$

31. Magnitude $= 13$; direction $= 1.18$

32. Magnitude = 304 newtons; direction = $25°20'$ from the 200-newton force toward the 150-newton force.

33. $(-2\sqrt{2}, 8\sqrt{2})$

34. $\left(\frac{4 + 3\sqrt{3}}{2}, \frac{3 - 4\sqrt{3}}{2}\right)$

35. (a) $y^* = (x^*)^2$ (b) parabola

(c)

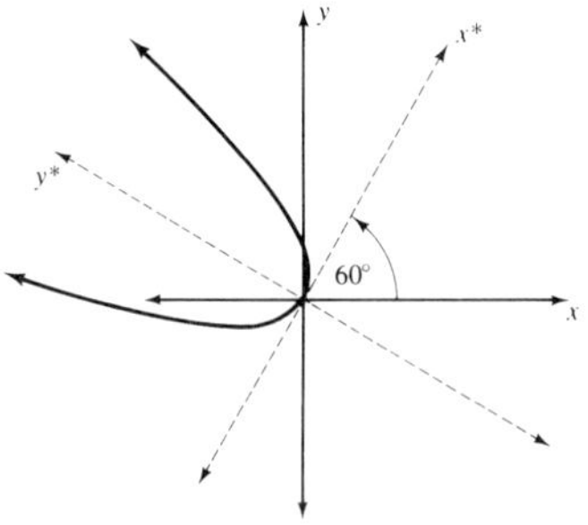

Complex Numbers 11

In Chapter 2, the rational exponent $1/m$ (m a positive integer) was defined as follows:

$$a^{1/m} = b \text{ if and only if } a = b^m$$

Since a and b could only be real numbers, the restriction was made that a and b had to be positive when m was even. The need for such a restriction is apparent when we let $a = -16$ and $m = 2$. Then, $(-16)^{1/2} = b$ if and only if $b^2 = -16$. But for any real number b, b^2 is nonnegative. Therefore, *there is no real number* b *such that* $b^2 = -16$.

In this chapter we define a **nonreal number** whose square is negative. As a result, we will have numbers that satisfy the equation $b^2 = -16$. In Section 11-1 we define this new number and then state theorems that govern the four basic operations on the new number. In Section 11-2 we study a trigonometric form of the number. In Section 11-3 we raise the number to powers and extract roots. The content of this chapter is very important to further work in mathematics, engineering, and science.

COMPLEX NUMBERS 11-1

One of the properties of real numbers states that the square of any real number is nonnegative. Thus, if $a \in$ real numbers, then $a^2 \geq 0$. We now define a number whose square is negative.

Imaginary unit i

> **Definition:** $i = \sqrt{-1}$ such that $i^2 = -1$

The number i is called the *imaginary unit.*

Imaginary number

> **Definition:** An imaginary number is any number that can be written in the form $a + bi$, where a and b are real numbers ($b \neq 0$) and i is the imaginary unit.

Examples of imaginary numbers are $1 + 4i$, $0 - 2i$, and $3 + i\sqrt{2}$. Notice that when b is an irrational number written as a radical, the i factor precedes the radical symbol.

EXAMPLE 11–1 Simplify

(a) i^3 (b) i^4 (c) i^{15}

Solutions:

(a) $i^3 = i^2(i) = -1(i) = -i$

(b) $i^4 = (i^2)^2 = (-1)^2 = 1$

(c) $i^{15} = (i^4)^3(i^3) = 1(i^3) = -i$

The results of this example suggest that any positive integral power of i can be simplified to either i, -1, $-i$, or 1. To verify this assumption as true, let n be an integer greater than 3. Then, for any i^n, we may write

$$i^n = (i^4)^j(i^k), \quad \text{where } k \in \{0, 1, 2, 3\} \text{ and } j \text{ is a positive integer}$$

That is, $n = 4j + k$, and k is the remainder when n is divided by 4. Since $i^4 = 1$, then $(i^4)^j = 1$ for any positive integer j. Furthermore, $i^0 = 1$, $i^1 = i$, $i^2 = -1$, and $i^3 = -i$. Thus,

$$i^n = 1, \quad \text{for } k = 0 \qquad i^n = i, \quad \text{for } k = 1$$
$$i^n = -1, \quad \text{for } k = 2 \qquad i^n = -i, \quad \text{for } k = 3$$

For example, $i^{32} = (i^4)^8(i^0) = 1$ and $i^{27} = (i^4)^6(i^3) = -i$.

$\sqrt{-b}$, $b > 0$

Definition: If $b > 0$, then $\sqrt{-b} = i\sqrt{b}$.

EXAMPLE 11–2 Write the following in simplified form:

(a) $\sqrt{-12}$ (b) $\sqrt{-4}\,(\sqrt{-9})$ (c) $\dfrac{\sqrt{-36}}{\sqrt{6}}$

Solutions:

(a) $\sqrt{-12} = i\sqrt{12} = i\sqrt{4(3)} = 2i\sqrt{3}$

(b) $\sqrt{-4}(\sqrt{-9}) = 2i(3i) = 6i^2 = -6$

(c) $\dfrac{\sqrt{-36}}{\sqrt{2}} = \dfrac{6i}{\sqrt{2}} = \dfrac{6i\sqrt{2}}{\sqrt{2}\sqrt{2}} = 3i\sqrt{2}$

Why was the imaginary unit used to make the radicands positive before doing any operations?

The laws of radicals do not apply to negative radicands. For example, in part (b) of Example 11–2, $\sqrt{-4}(\sqrt{-9}) \neq \sqrt{(-4)(-9)}$, since $\sqrt{(-4)(-9)} = \sqrt{36} = 6$. But $\sqrt{-4}(\sqrt{-9}) = -6$.

Complex number

Definition: A complex number is any number that can be written in the form $a + bi$, where a and b are real numbers and i is the imaginary unit.

With the definition above, if $b = 0$, then $a + bi = a$, a real number. If $a = 0$, then $a + bi = bi$, called a **pure imaginary number.** Notice that the real numbers and imaginary numbers are proper subsets of the set of complex numbers. Also, when the complex number is written in the form $a + bi$, it is said to be in **standard form.** The tree diagram in Figure 11-1 shows the proper subsets of the complex numbers as we have developed them in this text.

Equality of complex numbers

Definition: If $a + bi$ and $c + di$ are complex numbers, then $a + bi = c + di$ if and only if $a = c$ and $b = d$.

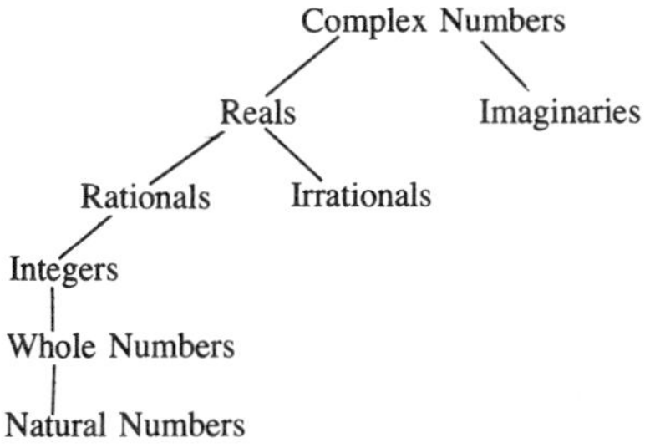

FIGURE 11-1 Subsets of the set of complex numbers

EXAMPLE 11–3 Find the values of x and y, where $2x + 3yi = 4 - 12i$.

Solution: If $2x + 3yi = 4 - 12i$, then

$$2x = 4 \quad \text{and} \quad 3y = -12$$
$$x = 2 \quad \text{and} \quad y = -4$$

Conjugate of a complex number

> **Definition:** The conjugate of the complex number $a + bi$ is the complex number $a - bi$.

EXAMPLE 11–4 State the conjugates of

(a) $5 - 2i$ (b) $-3 + 4i$

Solutions:

(a) By definition, the conjugate of $5 - 2i$ is $5 + 2i$.
(b) The conjugate of $-3 + 4i$ is $-3 - 4i$.

The following definitions state the procedure for adding, subtracting, multiplying, and dividing complex numbers.

Sum of complex numbers

> **Definition:** If $a + bi$ and $c + di$ are complex numbers, then
> $$(a + bi) + (c + di) = (a + c) + (b + d)i$$

Difference of complex numbers

> **Definition:** If $a + bi$ and $c + di$ are complex numbers, then
> $$(a + bi) - (c + di) = (a - c) + (b - d)i$$

Multiplication of complex numbers

> **Definition:** If $a + bi$ and $c + di$ are complex numbers, then
> $$(a + bi)(c + di) = (ac - bd) + (ad + bc)i$$

Division of complex numbers

> **Definition:** If $a + bi$ and $c + di$ are complex numbers, then
> $$\frac{a + bi}{c + di} = \left(\frac{ac + bd}{c^2 + d^2}\right) + \left(\frac{bc - ad}{c^2 + d^2}\right)i$$

To divide by the complex number $c + di$, we multiply numerator and denominator by $c - di$, the conjugate of $c + di$. Thus,

$$\frac{a + bi}{c + di} = \left(\frac{a + bi}{c + di}\right)\left(\frac{c - di}{c - di}\right)$$

$$= \frac{ac - adi + bci - bdi^2}{c^2 - cdi + cdi - d^2i^2}$$

$$= \frac{(ac + bd) + (bc - ad)i}{c^2 + d^2} \qquad \text{(since } -bdi^2 = bd \text{ and } -d^2i^2 = d^2\text{)}$$

$$\frac{a + bi}{c + di} = \left(\frac{ac + bd}{c^2 + d^2}\right) + \left(\frac{bc - ad}{c^2 + d^2}\right)i$$

EXAMPLE 11–5 Perform the indicated operations.

(a) $(3 + 4i) + (-1 + 2i)$ (b) $(12 - 5i) - (6 + 2i)$

Solutions:

(a) With Theorem 1,

$$(3 + 4i) + (-1 + 2i) = (3 + (-1)) + (4 + 2)i$$
$$= 2 + 6i$$

(b) With Theorem 2,

$$(12 - 5i) - (6 + 2i) = (12 - 6) + (-5 - 2)i$$
$$= 6 - 7i$$

EXAMPLE 11–6 Perform the indicated operations.

(a) $(3 + 2i)(3 - 2i)$ (b) $\dfrac{-3 + 2i}{1 + i}$

Solutions:

(a) By Theorem 3,

$$(3 + 2i)(3 - 2i) = (9 - (-4)) + (-6 + 6)i$$
$$= 13 + 0i = 13$$

Notice that $3 + 2i$ and $3 - 2i$ are conjugates, and their product is the real number $3^2 + 2^2 = 13$. In general,

$$(a + bi)(a - bi) = a^2 + b^2, \text{ a real number}$$

(b) To divide, we multiply numerator and denominator by the conjugate of the denominator:

$$\frac{-3 + 2i}{1 + i} = \left(\frac{-3 + 2i}{1 + i}\right)\left(\frac{1 - i}{1 - i}\right) = \left(\frac{-3 + 2}{1 + 1}\right) + \left(\frac{2 + 3}{1 + 1}\right)i$$
$$= \frac{-1}{2} + \frac{5}{2}i$$

The identity elements for the operations of addition and multiplication for complex numbers are 0 and 1. However, we write 0 and 1 in the standard form of the complex number in Theorems 1 and 2. The proofs of these theorems are included in the exercise set.

Are 0 and 1 the identity elements for addition and multiplication in the set of complex numbers?

Theorem 1: For the set of complex numbers, $0 + 0i$ is the **identity element for addition.**

Theorem 2: For the set of complex numbers, $1 + 0i$ is the **identity element for multiplication.**

Recall that the **closure property** was used to show the need for a new number. For example, if a and b are integers and $b \neq 0$, then a/b is not always an integer; that is, the integers are not closed under the operation of division. Hence we define the set of rational numbers. In the exercises for this section we are asked to prove that the sets of irrational numbers and imaginary numbers are not closed under the operations of addition, subtraction, multiplication, and division. To prove that the sets are not closed, we need to find but one example for which the sets fail to be closed under a given operation. For example, $\sqrt{2}$ and $-\sqrt{2}$ are both irrational numbers, but $\sqrt{2} + (-\sqrt{2}) = 0$, and 0 is a rational number. Thus, *the irrational numbers are not closed under the operation of addition.* The one exception we stated is sufficient proof for such an assertion.

Exercise Set 11–1

Simplify the following.

1. i^7
2. i^5
3. i^{18}
4. $i^9(i^{34})$
5. i^{-8}
6. $\sqrt{-144}$
7. $\sqrt{-32}$
8. $\sqrt{-7}(\sqrt{-7})$
9. $\sqrt{13}(\sqrt{-13})$
10. $\sqrt{-8}\,(\sqrt{-2})$
11. $(i\sqrt{50})(i\sqrt{3})$
12. $(i\sqrt{200})(\sqrt{-10})(i\sqrt{3})^2$
13. $(3i)(7i)^2$
14. $(-i\sqrt{98})(i\sqrt{2})$
15. $(\sqrt{-48})(\sqrt{-3})(i\sqrt{2})^3$

Perform the indicated operations. Express all complex numbers in the $(a + bi)$ form.

16. $(3 + 2i) + (3 - 2i)$
17. $(8 + 5i) - (2 - 6i)$
18. $(\sqrt{-13} + \sqrt{9}) + (\sqrt{16} + \sqrt{-52})$
19. $(6i - \sqrt{32}) - (\sqrt{32} + 2i)$
20. $(2 + 3i)(2 - 7i)$
21. $(17 - i)(6 + i)$
22. $(7 - 4i)(7 + 4i)$
23. $(2 + 5i)^2$
24. $(\sqrt{-49} + 3)(\sqrt{-64} + 2)$
25. $(1 + i\sqrt{3})(1 + \sqrt{-3})$
26. $\dfrac{2 + i}{7 + 5i}$
27. $\dfrac{2 - i}{2 + i}$
28. $\dfrac{6}{3 - 2i}$
29. $\dfrac{2i}{8 - i}$
30. $\dfrac{1 - 3i}{3 - \sqrt{-16}}$
31. $\dfrac{(2 + i)(7 - i)}{3 + i}$

32. $(3 + \sqrt{-1})[(2 + \sqrt{-3}) - (3 - \sqrt{-3})]$

33. $\dfrac{(\sqrt{-8} + \sqrt{3}) - (-\sqrt{-2} - \sqrt{3})}{i}$

Find the conjugates of the following complex numbers.

34. $2 - 3i$ 35. $\sqrt{3} - \sqrt{-2}$ 36. $\dfrac{1}{1 + i}$

Solve the following equations (that is, find real numbers x and y such that the complex numbers are equal).

37. $x + yi = 3 - 2i$

38. $2x + 3yi = -4 - 12i$

39. $(x + y) + (2x - 3y)i = 4 + 3i$

40. $(x + 2yi) + (y - xi) = 3$

41. Prove that $1 + 0i$ is the identity element for multiplication in the set of complex numbers.

42. Prove that $0 + 0i$ is the identity element for addition in the set of complex numbers.

43. Show that the set of irrational numbers is not closed under division, multiplication, and subtraction.

44. Show that the set of imaginary numbers is not closed under multiplication and addition.

11-2 THE TRIGONOMETRIC FORM OF THE COMPLEX NUMBER

In the preceding section we identified the standard form of the complex number as $a + bi$, where a and b are real numbers and i is the imaginary unit. The $a + bi$ form of the complex number is also called the *rectangular form* of the number.

We graph a complex number $a + bi$ in a rectangular coordinate system, using the following interpretations:

1. The horizontal axis is labeled the **real axis.**
2. The vertical axis is labeled the **imaginary axis.**
3. For point P, the graph of $a + bi$, the **abscissa is a** and the **ordinate is b.**

Thus, $P(a, b)$ is the graph of the complex number $a + bi$. There now exists a one-to-one correspondence between the set of points in the plane and the set of complex numbers. We refer to the plane on which the numbers are graphed as the **complex plane.**

EXAMPLE 11–7 Graph the following numbers on a rectangular coordinate system:

(a) $3 + 2i$ (b) $-5 - 4i$ (c) $3i$
(d) -2 (e) $2 - 3i$

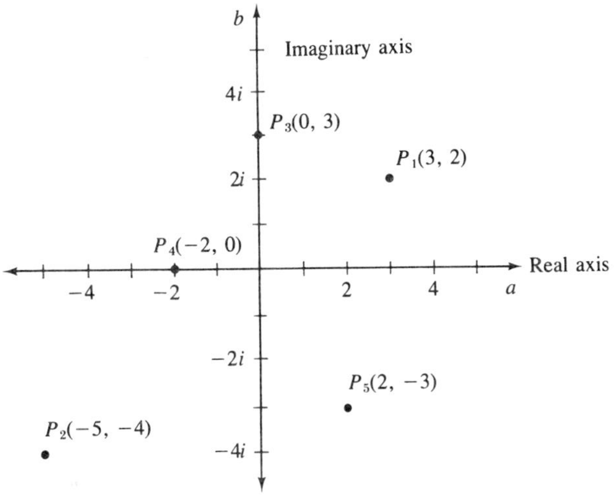

FIGURE 11-2 Graph of five complex numbers

Solutions: In Figure 11-2 a rectangular coordinate system is drawn with the horizontal axis labeled *real* and the vertical axis labeled *imaginary*.
(a) For the number $3 + 2i$, $a = 3$ and $b = 2$. Thus, $P_1(3, 2)$ is the graph of the complex number $3 + 2i$.
(b) With $a = -5$ and $b = -4$, $P_2(-5, -4)$ is the graph of $-5 - 4i$.
(c) $P_3(0, 3)$ is the graph of the pure imaginary number $3i$.
(d) $P_4(-2, 0)$ is the graph of the real number -2.
(e) $P_5(2, -3)$ is the graph of $2 - 3i$.

The following definition identifies the trigonometric form of the complex number $a + bi$.

Trigonometric form of a complex number

Definition: The trigonometric form of the complex number $a + bi$ is

$$r(\cos \theta + i \sin \theta)$$

where $r = \sqrt{a^2 + b^2}$, $\cos \theta = a/r$, and $\sin \theta = b/r$.

The expression $r(\cos \theta + i \sin \theta)$ can also be written in the abbreviated form $r \text{ cis } \theta$. In the trigonometric form of the complex number, r is called the **modulus** and θ is called the **argument** or **amplitude** of the number.

We can use degree or radian measure for θ in the trigonometric form of the number. In the following examples and exercises we will use both units of angle measure.

Is θ expressed in degree or radian measure?

EXAMPLE 11–8 Express the following complex numbers in trigonometric form, and identify the modulus and agrument of each number.
(a) $3 + 3i$ (b) $5 - 12i$

Solutions:
(a) With $a = 3$ and $b = 3$, $r = \sqrt{3^2 + 3^2} = 3\sqrt{2}$. Thus,

$$\cos\theta = \frac{3}{3\sqrt{2}} = \frac{1}{\sqrt{2}} \quad \text{and} \quad \sin\theta = \frac{3}{3\sqrt{2}} = \frac{1}{\sqrt{2}}$$

Since $\cos\theta > 0$ and $\sin\theta > 0$, θ is in Quadrant I; thus, $\theta = 45°$.

$$3 + 3i = 3\sqrt{2}(\cos 45° + i \sin 45°), \text{ or } 3\sqrt{2} \text{ cis } 45°$$

The modulus is $3\sqrt{2}$ and the argument is 45°.
(b) With $a = 5$ and $b = -12$, $r = \sqrt{25 + 144} = 13$. Thus, $\cos\theta = 5/13$, or 0.3846, and $\sin\theta = -12/13$, or -0.9231. Since $\cos\theta > 0$ and $\sin\theta < 0$, θ is in Quadrant IV. With Table 4 or a hand calculator, $\theta = 292°40'$ to the nearest ten minutes, and

$$\begin{aligned} 5 - 12i &= 13(\cos 292°40' + i \sin 292°40') \\ &= 13 \text{ cis } 292°40' \end{aligned}$$

The modulus is 13 and the argument is 292°40'.

Couldn't -67°20' also be used for the argument of that complex number in Example 11-8?

There are infinitely many values of θ that can be used for the argument, since $\cos(\theta + 360°k) = \cos\theta$ and $\sin(\theta + 360°k) = \sin\theta$ for any integer k. Thus, *the argument of the trigonometric form of any complex number is not unique*. Let us therefore agree that, unless otherwise stated, we will use the smallest positive value for θ as the argument of any complex number. Furthermore, we will round all radian measures of θ to two decimal places (unless the value is exact) and approximate all degree measures of θ to the nearest ten minutes.

EXAMPLE 11–9 Change the form of the complex number to standard form.

(a) $2\left(\cos\frac{4\pi}{3} + i \sin\frac{4\pi}{3}\right)$ (b) $3 \text{ cis } 316°30'$

Solutions:

(a) With $a = r\cos\theta$ and $b = r\sin\theta$,

$$a = 2\cos\frac{4\pi}{3} = 2\left(\frac{-1}{2}\right) = -1$$

and

$$b = 2\sin\frac{4\pi}{3} = 2\left(\frac{-\sqrt{3}}{2}\right) = -\sqrt{3}$$

Thus, $a + bi = -1 - i\sqrt{3}$.
(b)

$$a = 3\cos 316°30' = 3(0.7254) \approx 2.2$$

and

$$b = 3 \sin 316°30' = 3(-0.6884) \approx -2.1$$

Thus, $a + bi = 2.2 - 2.1i$.

We realize an immediate advantage for writing complex numbers in trigonometric form when complex numbers are multiplied or divided. That is, the multiplication and division are easily performed.

Theorem 1: If A and B are complex numbers with $A = r \text{ cis } \alpha$ and $B = s \text{ cis } \beta$, then

$$AB = rs \text{ cis}(\alpha + \beta)$$

Proof. If $A = r \text{ cis } \alpha$ and $B = s \text{ cis } \beta$, then

$$\begin{aligned} AB &= r(\cos\alpha + i \sin\alpha) \cdot s(\cos\beta + i \sin\beta) \\ &= rs(\cos\alpha \cos\beta + i \cos\alpha \sin\beta + i \sin\alpha \cos\beta + i^2 \sin\alpha \sin\beta) \\ &= rs((\cos\alpha \cos\beta - \sin\alpha \sin\beta) + i(\sin\alpha \cos\beta + \cos\alpha \sin\beta)) \end{aligned}$$

But $\cos\alpha \cos\beta - \sin\alpha \sin\beta = \cos(\alpha + \beta)$ and $\sin\alpha \cos\beta + \cos\alpha \sin\beta = \sin(\alpha + \beta)$. Therefore, $AB = rs\,(\cos(\alpha + \beta) + i \sin(\alpha + \beta))$, or $rs \text{ cis}(\alpha + \beta)$.

In words the theorem states that *the modulus of a product of two complex numbers is the product of the moduli, and the argument of the product is the sum of the arguments.*

Theorem 2: If A and B are two complex numbers with $A = r \text{ cis } \alpha$ and $B = s \text{ cis } \beta$, then

$$\frac{A}{B} = \frac{r}{s} \text{ cis}(\alpha - \beta)$$

The proof of Theorem 2 is included in the exercise set for this section.

EXAMPLE 11–10 If $A = 1 + i\sqrt{3}$ and $B = -2\sqrt{3} + 2i$,
(a) convert A and B to trigonometric form.
(b) compute AB with Theorem 1.

Solutions:
(a) For A, $a = 1$ and $b = \sqrt{3}$; thus, $r = 2$. Then we have $\cos\alpha = 1/2$ and $\sin\alpha = \sqrt{3}/2$. Therefore, $\alpha = 60°$ and $1 + i\sqrt{3} = 2 \text{ cis } 60°$.

For B, $a = -2\sqrt{3}$ and $b = 2$; thus, $r = 4$. Then we have $\cos\beta = -\sqrt{3}/2$ and $\sin\beta = 1/2$. Therefore, $\beta = 150°$ and $-2\sqrt{3} + 2i = 4 \text{ cis } 150°$.
(b) With $A = 2 \text{ cis } 60°$ and $B = 4 \text{ cis } 150°$,

$$\begin{aligned} AB &= 2(4)\text{cis}(60° + 150°) \\ &= 8 \text{ cis } 210° = -4\sqrt{3} - 4i \end{aligned}$$

EXAMPLE 11–11 If $A = 1 + i\sqrt{3}$ and $B = -2\sqrt{3} + 2i$, use the trigonometric forms of A and B obtained in Example 11–10 to compute B/A.

Solution: If $A = 2 \text{ cis } 60°$ and $B = 4 \text{ cis } 150°$, then

$$\frac{B}{A} = \frac{4 \text{ cis } 150°}{2 \text{ cis } 60°} = \frac{4}{2}(\cos(150° - 60°) + i \sin(150° - 60°))$$
$$= 2(\cos 90° + i \sin 90°) = 2i$$

Exercise Set 11–2

Graph the following complex numbers in a rectangular coordinate system.

1. $3 + i$
2. $-2 - 3i$
3. $2 - 4i$
4. $-5 - i$
5. $-3i$
6. 3
7. $5 + 4i$
8. $4i$
9. -4
10. $-2 + 2i$

Write the following complex numbers in trigonometric form. Use the exact value of r. Approximate θ to the nearest ten minutes and $0° \leq \theta < 360°$.

11. $1 + i\sqrt{3}$
12. $2 - 2i$
13. -2
14. $-3\sqrt{3} + 3i$
15. $5 + 4i$
16. $5i$
17. $-\sqrt{7} - i\sqrt{21}$
18. $3 - 10i$
19. $-3 - 4i$
20. $-2 + i$

Express the following complex numbers in $a + bi$ form. Where necessary, approximate a and b to one decimal place.

21. $3\left(\cos \frac{\pi}{6} + i \sin \frac{\pi}{6}\right)$
22. $7\left(\cos \frac{3\pi}{2} + i \sin \frac{3\pi}{2}\right)$
23. $8(\cos 120° + i \sin 120°)$
24. $\sqrt{2}(\cos 225° + i \sin 225°)$
25. $3 \text{ cis } 2\pi$
26. $\text{cis } 313°$
27. $10 \text{ cis } 130°$
28. $5 \text{ cis } 2.8973$
29. $5 \text{ cis } \frac{7\pi}{2}$
30. $3 \text{ cis } \frac{19\pi}{6}$

Do the indicated operations. Leave answers in $a + bi$ form. Where necessary, approximate a and b to one decimal place.

31. $\left(2 \text{ cis } \frac{\pi}{2}\right)\left(3 \text{ cis } \frac{\pi}{4}\right)$
32. $(4 \text{ cis } 71°)(0.5 \text{ cis } 109°)$
33. $\dfrac{10 \text{ cis } 5\pi/6}{5 \text{ cis } 5\pi/6}$
34. $\dfrac{30 \text{ cis } 7\pi/6}{3 \text{ cis } \pi}$
35. $\dfrac{8(\cos 135° + i \sin 135°)}{\sqrt{2} - i\sqrt{2}}$
36. $(7 \text{ cis } 14°)(13 \text{ cis } 191°)$
37. $\dfrac{5 \text{ cis } 95°}{2 \text{ cis } 37°}$
38. $\dfrac{\cos 110° + i \sin 110°}{\cos 30° + i \sin 30°}$
39. $(39 \text{ cis } 43°)(3 \text{ cis } 127°)$
40. $(3 \text{ cis } 15°)^2$

Do the indicated operations. Leave answers in trigonometric form. Use the exact value of r and approximate θ to the nearest hundredth of a radian.

41. $\dfrac{2 + 7i}{7 - 2i}$

42. $(3 + i)(1 - 2i)$

43. $(2 + 2i)\left(3 \text{ cis } \dfrac{5\pi}{4}\right)$

44. $(3 - 3i)\left(2 \text{ cis } \dfrac{\pi}{4}\right)$

45. $\dfrac{2\sqrt{3} - 2i}{4 \text{ cis } 45°}$

46. $(-3 + 3i)^3$

47. Prove Theorem 2.

DE MOIVRE'S THEOREM 11–3

This section explains two theorems on complex numbers. The first theorem gives an equation for raising a complex number in trigonometric form to any positive integral power. The second theorem gives an equation for extracting roots of complex numbers.

> **Theorem 3: De Moivre's Theorem** If A is a complex number and $A = r \text{ cis } \theta$, then
>
> $$A^n = r^n \text{ cis } n\theta$$
>
> where n is any positive integer.

In words the theorem states that the nth power of A equals the nth power of the modulus and the product of the argument and n.

EXAMPLE 11–12 If $A = 2 \text{ cis } \dfrac{\pi}{4}$, find A^5.

Solution: Using De Moivre's Theorem with $n = 5$, we get

$$A^5 = 2^5 \text{cis } 5\left(\frac{\pi}{4}\right)$$
$$= 32 \text{ cis } \frac{5\pi}{4}$$

De Moivre's Theorem is also valid for any integer n — positive or negative. However, we will restrict our examples and exercises to only positive integers n.

Does De Moivre's theorem only apply to positive integers n?

EXAMPLE 11–13 If $B = \sqrt{3} + i$,
(a) convert B to the trigonometric form.
(b) find B^4, using De Moivre's Theorem.

Solutions:

(a) With $a = \sqrt{3}$ and $b = 1$, then $r = \sqrt{3+1} = 2$.
Now, $\cos\theta = \sqrt{3}/2$ and $\sin\theta = 1/2$. Therefore, $\theta = 30°$ and $\sqrt{3} + i = 2 \text{ cis } 30°$.

(b) If $B = 2 \text{ cis } 30°$, then

$$B^4 = 2^4 \text{cis } 4(30°) = 16 \text{ cis } 120° = -8 + 8i\sqrt{3}$$

EXAMPLE 11–14 Find $\left(\frac{-\sqrt{3}}{2} + \frac{1}{2}i\right)^{12}$.

Solution: With $a = -\sqrt{3}/2$ and $b = 1/2$, then $r = \sqrt{3/4 + 1/4} = 1$. Now $\cos\theta = -\sqrt{3}/2$ and $\sin\theta = 1/2$. Therefore, $\theta = 5\pi/6$ and

$$\begin{aligned}\left(\frac{-\sqrt{3}}{2} + \frac{1}{2}i\right)^{12} &= \left(\text{cis } \frac{5\pi}{6}\right)^{12} \\ &= \cos 10\pi + i \sin 10\pi \\ &= 1 + 0i = 1\end{aligned}$$

Do the results of Example 11-14 imply that a twelfth root of 1 is $\frac{-\sqrt{3}}{2} + \frac{1}{2}i$?

As we shall see, a complex number has n distinct nth roots. For example, since 1 is a complex number (namely, $1 + 0i$), 1 has two square roots, three cube roots, four fourth roots, and twelve distinct twelfth roots. Thus, there exist eleven complex numbers distinct from $\frac{-\sqrt{3}}{2} + \frac{1}{2}i$ that are twelfth roots of $1 + 0i$.

When we convert a complex number from the $a + bi$ form to the $r(\cos\theta + i\sin\theta)$ form, we obtain one value for the modulus r. However, we have selected for the amplitude the smallest positive number θ such that $\cos\theta = a/r$ and $\sin\theta = b/r$; thus, $0 \le \theta < 2\pi$. But any number ϕ, such that $\phi = \theta + 2k\pi$ (k an integer), can also be used for an amplitude of $a + bi$. We therefore have an infinite number of expressions for the trigonometric form of a complex number, such that $a + bi = r \text{ cis}(\theta + 2k\pi)$, $k \in$ integers.

Equality of complex numbers in trigonometric form

Definition: Two complex numbers $r \text{ cis } \theta$ and $s \text{ cis } \phi$ are equal if and only if $r = s$ and $\phi = \theta + 2k\pi$, $k \in$ integers.

To compute an nth root of the complex number $A = r \text{ cis } \theta$, we look for a complex number $W = s \text{ cis } \phi$ such that $W^n = A$. Since $W^n = s^n \text{cis } n\phi$, we conclude that W is an nth root of A if and only if

$$s^n \text{cis } n\phi = r \text{ cis}(\theta + 2k\pi)$$

and $s = \sqrt[n]{r}$ and $n\phi = \theta + 2k\pi$, or $\phi = \dfrac{\theta + 2k\pi}{n}$.

> **Theorem 4:** If n is a positive integer and r is a positive real number, then the n roots of r cis θ are of the form
>
> $$\sqrt[n]{r}\left(\cos\frac{\theta + 2k\pi}{n} + i\sin\frac{\theta + 2k\pi}{n}\right)$$
>
> where $k \in \{0, 1, 2, \ldots, n - 1\}$.

EXAMPLE 11–15 Find all complex numbers W such that $W^5 = 32$.

Solution: We convert $32 + 0i$ to trigonometric form, with $r = 32$ and $\theta = 0$. Thus, $32 + 0i = 32 \text{ cis}(0 + 2k\pi)$. Now $W = s \text{ cis } \phi$ is a fifth root of $32 + 0i$ if and only if $s = \sqrt[5]{32}$ or $s = 2$ and $\phi = \dfrac{0 + 2k\pi}{5}$, $k \in \{0, 1, 2, 3, 4\}$. Therefore,

$$\phi_1 = 0 \quad \text{for } k = 0 \qquad \phi_2 = 2\pi/5 \quad \text{for } k = 1$$
$$\phi_3 = 4\pi/5 \quad \text{for } k = 2 \qquad \phi_4 = 6\pi/5 \quad \text{for } k = 3$$
$$\phi_5 = 8\pi/5 \quad \text{for } k = 4$$

Thus, we get

$$W_1 = 2 \text{ cis } 0 \qquad W_2 = 2 \text{ cis } \frac{2\pi}{5}$$
$$W_3 = 2 \text{ cis } \frac{4\pi}{5} \qquad W_4 = 2 \text{ cis } \frac{6\pi}{5}$$
$$W_5 = 2 \text{ cis } \frac{8\pi}{5}$$

Notice that if $k = 5$, then $\phi_6 = 10\pi/5 = 2\pi$, and ϕ_6 is coterminal with ϕ_1. Thus, 2 cis 2π is not a new root, but the same root as W_1, except with a larger amplitude. The five roots are graphed in Figure 11-3. Notice that the five points divide the circumference of a circle of radius 2 into five arcs of equal lengths.

Why don't we let k = 5, 6, 7, and even higher values?

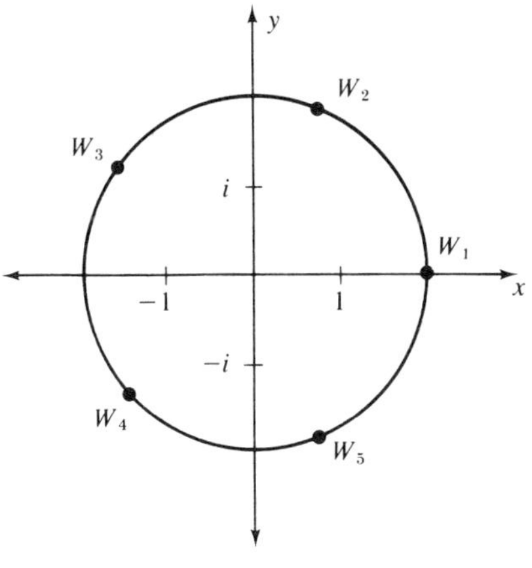

FIGURE 11-3 Graph of the five roots of 32 cis 0

EXAMPLE 11–16 Find the fourth roots of $A = 8 + 8i\sqrt{3}$. Express the answers in $a + bi$ form.

Solution: With $a = 8$ and $b = 8\sqrt{3}$, then $r = \sqrt{64 + 192} = 16$. Since $\cos \theta = 1/2$ and $\sin \theta = \sqrt{3}/2$, then $\theta = \pi/3$. Thus, $8 + 8i\sqrt{3} = 16 \text{ cis } \pi/3$. $W = s \text{ cis } \phi$ is a fourth root of $16 \text{ cis } \pi/3$ if and only if $s^4 = 16$ and $\phi = \dfrac{\pi/3 + 2k\pi}{4}$, $k \in \{0, 1, 2, 3\}$. Therefore, $s = 2$ and

$$\begin{aligned} \phi_1 &= \pi/12 && \text{for } k = 0 \\ \phi_2 &= 7\pi/12 && \text{for } k = 1 \\ \phi_3 &= 13\pi/12 && \text{for } k = 2 \\ \phi_4 &= 19\pi/12 && \text{for } k = 3 \end{aligned}$$

Therefore, the four fourth roots of A are

$$\begin{aligned} W_1 &= 2 \text{ cis } \frac{\pi}{12}, && \text{or} \quad 1.9 + 0.5i \\ W_2 &= 2 \text{ cis } \frac{7\pi}{12}, && \text{or} \quad -0.5 + 1.9i \\ W_3 &= 2 \text{ cis } \frac{13\pi}{12}, && \text{or} \quad -1.9 - 0.5i \\ W_4 &= 2 \text{ cis } \frac{19\pi}{12}, && \text{or} \quad 0.5 - 1.9i \end{aligned}$$

The four roots are shown in Figure 11-4 as points on a circle of radius 2. Notice that the four points divide the circumference of the circle into four arcs of equal lengths.

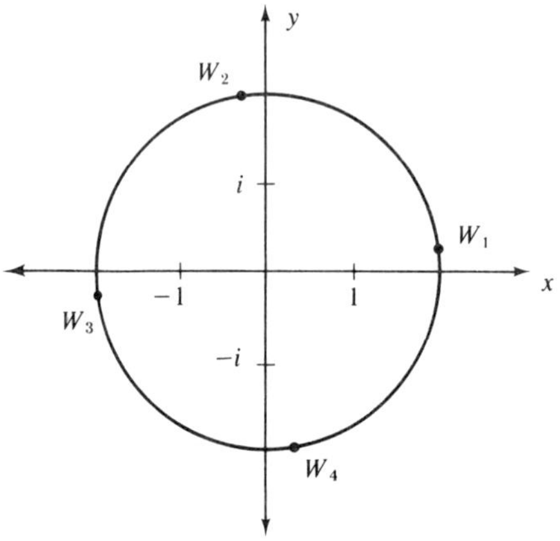

FIGURE 11-4 Graph of the four roots of $8 + 8i\sqrt{3}$

Exercise Set 11–3

Find the following powers of complex numbers, using De Moivre's Theorem. Leave each answer in the same form of the number in which it is given. Reduce the amplitude to the smallest positive value possible. Use $\pi = 3.14$.

1. $(2 \text{ cis } 60°)^3$
2. $(3 \text{ cis } 150°)^5$
3. $(\sqrt{3} \text{ cis } 200°)^4$
4. $(\sqrt[3]{2} \text{ cis } 315°)^6$

5. $(\sqrt{5} \text{ cis } 2.90)^6$

6. $\left(2 \text{ cis } \frac{5\pi}{4}\right)^5$

7. $\left(\sqrt[3]{3} \text{ cis } \frac{9\pi}{5}\right)^9$

8. $(\sqrt[4]{4} \text{ cis } 3.80)^8$

9. $(1 + i)^3$

10. $(-1 + i\sqrt{3})^4$

11. $(-\sqrt{3} + i)^5$

12. $(2 - 2i)^6$

13. $(3 - i\sqrt{3})^2$

14. $(1 - i)^5$

15. $(-2 - 2i)^4$

16. $(-\sqrt{3} - i)^{10}$

Find all the indicated roots. Express all answers in r cis θ form and graph the roots. Approximate roots in $a + bi$ form to the nearest tenth.

17. The square roots of 9 cis 30°

18. The cube roots of 8 cis 36°

19. The fourth roots of 16 cis 200°

20. The cube roots of 125 cis 150°

21. The cube roots of $-32\sqrt{3} - 4i$

22. The fifth roots of $1 + i$

23. The sixth roots of 64

24. The eighth roots of 1

Solve the following equations with solutions from the set of complex numbers.

25. $x^4 - 1 = 0$

26. $x^4 - i = 0$

27. $x^5 + 32 = 0$

28. $x^3 + 8 = 0$

29. $x^6 + 1 = 0$

30. $x^5 - 243 = 0$

SUMMARY

Definition: $i = \sqrt{-1}$ such that $i^2 = -1$ — **Imaginary unit i**

Definition: An imaginary number is any number that can be written in the form $a + bi$, where a and b are real numbers ($b \neq 0$) and i is the imaginary unit. — **Imaginary number**

Definition: If $b > 0$, then $\sqrt{-b} = i\sqrt{b}$. — **$\sqrt{-b}$, $b > 0$**

Definition: A complex number is any number that can be written in the form $a + bi$, where a and b are real numbers and i is the imaginary unit. — **Complex number**

Definition: If $a + bi$ and $c + di$ are complex numbers, then $a + bi = c + di$ if and only if $a = c$ and $b = d$. — **Equality of complex numbers**

Definition: The conjugate of the complex number $a + bi$ is the complex number $a - bi$. — **Conjugate of a complex number**

Definition of Complex Numbers in Rectangular Form If A and B are complex numbers, with $A = a + bi$ and $B = c + di$, then

Definition: $A + B = (a + c) + (b + d)i$

Definition: $A - B = (a - c) + (b - d)i$

Definition: $AB = (ac - bd) + (ad + bc)i$

Definition: $\frac{A}{B} = \left(\frac{ac + bd}{c^2 + d^2}\right) + \left(\frac{bc - ad}{c^2 + d^2}\right)i$

Theorem 1: $0 + 0i$ is the **identity element for addition.**

Theorem 2: $1 + 0i$ is the **identity element for multiplication.**

Trigonometric form of a complex number

Definition: The trigonometric form of the complex number $a + bi$ is $r(\cos\theta + i\sin\theta)$ where $r = \sqrt{a^2 + b^2}$, $\cos\theta = a/r$, and $\sin\theta = b/r$.

Theorems on Complex Numbers in Trigonometric Form If A and B are complex numbers, with $A = r \text{ cis } \alpha$ and $B = s \text{ cis } \beta$, then

Theorem 1: $AB = rs \text{ cis}(\alpha + \beta)$

Theorem 2: $\frac{A}{B} = \frac{r}{s} \text{ cis}(\alpha - \beta)$

Theorem 3: **De Moivre's Theorem** $A^n = r^n \text{ cis } n\theta$

Theorem 4: If n is a positive integer, then

$$\sqrt[n]{A} = \sqrt[n]{r}\left(\cos\frac{\theta + 2k\pi}{n} + i\sin\frac{\theta + 2k\pi}{n}\right)$$

where $k \in \{0, 1, 2, \ldots, n - 1\}$.

Equality of complex numbers in trigonometric form

Definition: Two complex numbers $r \text{ cis } \theta$ and $s \text{ cis } \phi$ are equal if and only if $r = s$ and $\phi = \theta + 2k\pi$, $k \in \{\text{integers}\}$.

SELF-CHECKING EXERCISE

Simplify. **(11-1)**

1. i^{13}
2. i^{99}
3. $\sqrt{-27}$
4. $\sqrt{8}\sqrt{-18}$
5. $\frac{\sqrt{-10}}{\sqrt{6}}$
6. $\frac{i\sqrt{-30}}{\sqrt{5}}$
7. $(\sqrt{-7})(i\sqrt{2})$
8. $(2i)^2(\sqrt{-12})$

Perform the following operations. Leave answers in $a + bi$ form. **(11-1)**

9. $(-3 + 10i) + (5 - 3i)$
10. $(1 + 5i) - (-4 + i)$
11. $(18 - 7i)(3 + 2i)$
12. $(2 - 3i)^2$
13. $\frac{4 - 4i\sqrt{3}}{1 + i}$
14. $\frac{(3 + i)(-1 - 2i)}{4 - i}$

Solve the following equations. **(11-1)**

15. $3x + 8i = -12 - 4yi$
16. $(x + y) + (2x - 3y)i = 5 - 5i$

Express the following complex numbers in trigonometric form. **(11-2)**

17. $-3 + 3i$
18. $8 - 15i$

Express the following complex numbers in rectangular form. **(11-2)**

19. $10 \text{ cis } \frac{5\pi}{6}$

20. $20 \text{ cis } 195°$

Perform the following operations. Leave answers in r cis θ form. **(11-2)**

21. $(2 \text{ cis } 102°)(3 \text{ cis } 108°)$

22. $\frac{14 \text{ cis } 3.71}{2 \text{ cis } 0.45}$

23. $(\sqrt{6} \text{ cis } 3.45)(\sqrt{10} \text{ cis } 5.05)$

24. $\frac{48 \text{ cis } 75°10'}{32 \text{ cis}(-42°30')}$

25. $(\sqrt[3]{10} \text{ cis } 83°)^3$

26. $\frac{(2 \text{ cis } 4.15)^3}{3 \text{ cis } 5.05}$

Find the indicated power of the given complex numbers. Express answers in the form in which the number is given. **(11-3)**

27. $(\cos 1.15 + i \sin 1.15)^4$

28. $(2 \text{ cis } 17°)^6$

29. $(5 - 5i)^3$

30. $\left(\frac{-\sqrt{3}}{2} - \frac{1}{2}i\right)^{12}$

31. Find the fourth roots of 81. **(11-3)**

32. Find the fifth roots of $4\sqrt{2} + 4i\sqrt{2}$. **(11-3)**

Answers to Self-Checking Exercise

1. i
2. $-i$
3. $3i\sqrt{3}$
4. $12i$
5. $i\sqrt{15}/3$
6. $-\sqrt{6}$
7. $-\sqrt{14}$
8. $-8i\sqrt{3}$
9. $2 + 7i$
10. $5 + 4i$
11. $68 + 15i$
12. $-5 - 12i$
13. $(2 - 2\sqrt{3}) + (-2 - 2\sqrt{3})i$
14. $3/17 - 29i/17$
15. $(x, y) = (-4, -2)$
16. $(x, y) = (2, 3)$
17. $3\sqrt{2} \text{ cis } \frac{3\pi}{4}$
18. $17 \text{ cis } 298°$, to the nearest 10 minutes
19. $-5\sqrt{3} + 5i$
20. $-19.3 - 5.2i$, to the nearest tenth
21. $6 \text{ cis } 210°$
22. $7 \text{ cis } 3.26$
23. $2\sqrt{15} \text{ cis } 8.50$ or $2\sqrt{15} \text{ cis } 2.22$
24. $\frac{3}{2} \text{ cis } 117°40'$
25. $10 \text{ cis } 249°$
26. $\frac{8}{3} \text{ cis } 7.40$ or $\frac{8}{3} \text{ cis } 1.12$
27. $\cos 4.60 + i \sin 4.60$
28. $64 \text{ cis } 102°$
29. $-250 - 250i$
30. 1
31. $3, 3i, -3, -3i$
32. $\sqrt{2} \text{ cis } \frac{\pi}{20}, \sqrt{2} \text{ cis } \frac{9\pi}{20}, \sqrt{2} \text{ cis } \frac{17\pi}{20}, \sqrt{2} \text{ cis } \frac{25\pi}{20}, \sqrt{2} \text{ cis } \frac{33\pi}{20}$

12 Sequences and Series

We sometimes find it difficult to be motivated to study topics whose relevance is based entirely on the theme "we will need these topics at some later time." Nevertheless, we must admit that the motivation for studying the topics in this chapter basically centers on this theme. That is, sequences and series are important tools in certain areas of higher mathematics. Because this text is designed to prepare you for advanced courses in mathematics, we therefore study these topics in this chapter.

In Section 12-1 we define and study a sequence. The series is then defined in Section 12-2. The special series known as arithmetic and geometric progressions are developed in Sections 12-3, 12-4, and 12-5.

SEQUENCES 12-1

In past chapters we have studied different kinds of functions. As examples we can name the linear, quadratic, trigonometric, logarithmic, and exponential functions. If x and y are the variables used to define a function, we designate x as the **independent variable** and y as the **dependent variable.** To emphasize the functional relationship we use the notation

$$y = f(x)$$

An important part of any functional relationship is the identification of the **domain** and **range** of the function. Recall that for the function f, where

$$f = \{(x, y) \mid y = f(x)\}$$

the domain of f is the set of values assigned to x. The range of f is the set of all values of y for every value of x in the domain. In this section we study the function in the following definition.

Sequence function

> **Definition:** A sequence function is a function whose domain is a set of consecutive positive integers.

Does every sequence function have the same domain?

Every sequence function does not have the same domain because every sequence function does not have the same set of positive integers for the domain. To illustrate, let S_1, S_2, and S_3 be sequence functions, such that

$$\text{the domain of } S_1 = \{1, 2, 3, 4, 5\}$$
$$\text{the domain of } S_2 = \{1, 2, 3, \ldots, n\}$$
$$\text{the domain of } S_3 = \{1, 2, 3, \ldots\}$$

Notice that for each sequence, the domain is a set of consecutive positive integers. However, for S_1, the domain is finite with 5 elements. For S_2, the domain is finite with n elements, $n \in \{\text{positive integers}\}$. For S_3, the domain is an infinite number of positive integers. As a result of this example, we find it useful to make a distinction between sequences on the basis of the domains of the sequences.

Finite sequence

> **Definition:** A sequence is finite if the domain of the sequence function is finite.

Infinite sequence

> **Definition:** A sequence is infinite if the domain of the sequence function is infinite.

We now turn our attention to the elements in the range of a sequence function. Recall that for most of the functions we have studied, the range has been given in terms of an equation. We can then generate specific values in the range by substituting elements from the domain into the equation. For example, let f, g, and h be functions with domains the set of real numbers, with

$$f = \{(x, f(x)) \mid f(x) = 3x + 1\}$$
$$g = \{(x, g(x)) \mid g(x) = x^2 - 1\}$$
$$h = \{(x, h(x)) \mid h(x) = 2^x - 5\}$$

We can list several ordered pairs in the three functions by assigning any real number for x. For example,

$$f = \{(1, 4), (2, 7), (3, 10), (4, 13), (5, 16), \ldots\}$$
$$g = \{(1, 0), (2, 3), (3, 8), (4, 15), (5, 24), \ldots\}$$
$$h = \{(1, -3), (2, -1), (3, 3), (4, 11), (5, 27), \ldots\}$$

The ordered pairs listed for f, g, and h make them look like sequence functions.

For convenience, we used positive integers to generate some of the ordered pairs for f, g, and h. However, -1 is not a positive integer; therefore, $(-1, -2)$, $(-1, 0)$, and $(-1, -4.5)$ are also ordered pairs of f, g, and h, respectively, when the functions are real valued. But the same ordered pairs would not be elements of sequence functions. We can convert f, g, and h to sequence functions by simply restricting the domains to the set of positive integers. With this restriction, each of the functions defined above is then a sequence function.

Since the domains of all sequence functions are the same, symbolism associated with the function is modified according to the following agreements:

1. We will usually exhibit the function by simply listing the range elements in order. For example, for f, g, and h defined above as sequence functions,

$$f = \{4, 7, 10, 13, 16, 19, 22, \ldots\}$$
$$g = \{0, 3, 8, 15, 24, 35, 48, \ldots\}$$
$$h = \{-3, -1, 3, 11, 27, 59, 123, \ldots\}$$

2. We call the range elements the *terms* of the sequence. For the function f, the first term is $f(1) = 4$, the second term is $f(2) = 7$, the third term is $f(3) = 10$, and the nth term, or *general term,* is $f(n) = 3n + 1$.
3. The notation $a_1, a_2, a_3, \ldots, a_n, \ldots$ is a preferred form for the terms of the sequence rather than $f(1), f(2), f(3), \ldots, f(n), \ldots$. For example, for the function g defined above, $a_1 = 0$, $a_2 = 3$, $a_3 = 8$, and the general term $a_n = n^2 - 1$. We can also use other letters such as $b_1, b_2, b_3, \ldots$ or $c_1, c_2, c_3, \ldots$ to represent the terms of a sequence.

EXAMPLE 12–1 Find the terms of the sequence defined by $a_n = 3^n - 10$, $n \in \{1, 2, 3, 4, 5\}$.

Solution: Notice that the given sequence is finite since the domain of the function is the positive integers 1 through 5. The terms of the sequence are

$$a_1 = 3^1 - 10 = -7$$
$$a_2 = 3^2 - 10 = -1$$
$$a_3 = 3^3 - 10 = 17$$
$$a_4 = 3^4 - 10 = 71$$
$$a_5 = 3^5 - 10 = 233$$

The sequence is therefore $-7, -1, 17, 71, 233$.

EXAMPLE 12–2 Find the terms of the sequence defined by $a_n = \dfrac{n - 1}{n + 1}$, $n \in \{1, 2, 3, \ldots, 20\}$.

Solution: The given sequence is finite with 20 terms.

$$a_1 = \frac{1 - 1}{1 + 1} = 0$$
$$a_2 = \frac{2 - 1}{2 + 1} = \frac{1}{3}$$
$$a_3 = \frac{3 - 1}{3 + 1} = \frac{2}{4}$$
$$a_{20} = \frac{20 - 1}{20 + 1} = \frac{19}{21}$$

Thus, the sequence is $0, 1/3, 2/4, \ldots, 19/21$. Notice that a sequence can usually be stated by listing the first 3 or 4 terms and then the last term in the sequence.

EXAMPLE 12–3 Find the terms of the sequence defined by $a_n = \frac{(-1)^n}{3n}$, $n \in \{\text{positive integers}\}$.

Solution: Since the domain of the sequence function is infinite, the sequence is also infinite.

$$a_1 = \frac{(-1)^1}{3(1)} = \frac{-1}{3}$$

$$a_2 = \frac{(-1)^2}{3(2)} = \frac{1}{6}$$

$$a_3 = \frac{(-1)^3}{3(3)} = \frac{-1}{9}$$

$$a_4 = \frac{(-1)^4}{3(4)} = \frac{1}{12}$$

Therefore, the sequence is $-1/3$, $1/6$, $-1/9$, $1/12$, . . . , $\frac{(-1)^n}{3n}$, ... Notice that for an infinite sequence we list 3 or 4 terms and then insert the general term a_n.

We sometimes are interested in finding a general term for a sequence when several terms of the sequence are given. We should note first that *for most sequences, the general term is not unique*. That is, frequently we can find more than one formula that will generate the terms given for the sequence. However, there is usually one expression that is more obvious than any other.

The task of finding an a_n for a given sequence is not always simple. The object of our search is a pattern that is common to each of the given terms of the sequence. When a pattern yields a formula for a_n that satisfies each of the given terms, then our task is done.

EXAMPLE 12–4 Find a general term for the sequence 2, 7, 12, 17, 22, . . . , a_n, ...

Solution: We need to answer the following question: "What expression in n will yield 2 when $n = 1$, 7 when $n = 2$, 12 when $n = 3$, 17 when $n = 4$, and 22 when $n = 5$?"

Notice that successive terms in the sequence differ by 5. That is, $7 - 2 = 5$, $12 - 7 = 5$, $17 - 12 = 5$, and so on. We might "guess," therefore, that a_n contains a factor of 5. Since $2 = 5(1) - 3$, $7 = 5(2) - 3$, $12 = 5(3) - 3$, $17 = 5(4) - 3$, and finally $22 = 5(5) - 3$, our guess is accurate. Thus,

$$a_n = 5n - 3$$

is an expression for the general term of the given sequence.

EXAMPLE 12–5 Find a general term for the sequence 1, $-1/2$, $1/3$, $-1/4$, $1/5$, . . . , b_n, ...

Solution: Notice that the given terms of the sequence alternate in sign with the odd-numbered terms negative and the even-numbered terms positive. Therefore, b_n must contain a factor $(-1)^n$. Since the denominators of each term correspond to the number of the term in the sequence, then

$$b_n = \frac{(-1)^n}{n}$$

is an expression for the given sequence.

Exercise Set 12–1

Write the first five terms of the sequence with the given general term.

1. $a_n = 2n + 3$
2. $b_n = n^2 + 2$
3. $c_n = 4n - 1$
4. $a_n = n^3 - 8$
5. $b_n = (-2)^n$
6. $c_n = (-3)^{n-1}$
7. $a_n = \dfrac{2n + 1}{3n - 2}$
8. $b_n = \dfrac{n}{n^2 + 1}$
9. $c_n = \dfrac{(-1)^n(n - 1)}{2n + 1}$
10. $a_n = \dfrac{(-1)^n n^2}{3n + 5}$

Find the terms of the following sequences with the given general term and $n \in \{\text{positive integers}\}$.

11. $a_n = \dfrac{1 - 2n}{1 + n}$
12. $b_n = \dfrac{n^2 - 1}{n}$
13. $c_n = n^2 + 2n$
14. $a_n = 8 - 2^n$
15. $b_n = \dfrac{(-1)^n}{n^2}$
16. $c_n = \dfrac{(-1)^n}{n^3 + 1}$
17. $a_n = \sqrt{n + 2}$
18. $b^n = \sqrt[3]{2n}$

Find a general term for each of the following sequences.

19. 6, 10, 14, 18, 22, . . .
20. 3, 6, 9, 12, 15, . . .
21. 1, 2, 4, 8, 16, . . .
22. 5, 25, 125, 625, 3125, . . .
23. $-1/2$, $1/4$, $-1/8$, $1/16$, $-1/32$, . . .
24. 0, -1, $\sqrt{2}$, $-\sqrt{3}$, 2, . . .
25. 2, 3/2, 4/3, 5/4, 6/5, . . .
26. 1/2, 4/3, 9/4, 16/5, 25/6, . . .
27. -3, -9, -15, -21, -27, . . .
28. 4/3, $-4/9$, 4/27, $-4/81$, 4/243, . . .

SERIES 12-2

In Section 12-1 we defined a sequence function that has a subset of the positive integers for a domain. In this section we study a series associated with a sequence and the symbolism used to express a series.

Series

> **Definition:** A series is the sum of the terms of a sequence.

From the definition we can see that the series associated with a given sequence is obtained by adding all of the terms in the sequence.

EXAMPLE 12–6 Find the series associated with the indicated sequences.

(a) 1, 3, 5, 7, 9, 11 (b) 1, 4, 9, 16, 25, 36, 49

Solutions:

(a) For the sequence 1, 3, 5, 7, 9, 11, the corresponding series is $1 + 3 + 5 + 7 + 9 + 11 = 36$. Notice that we find the series by simply summing the terms of the sequence.

(b) For the sequence, 1, 4, 9, 16, 25, 36, 49, the corresponding series is $1 + 4 + 9 + 16 + 25 + 36 + 49 = 140$.

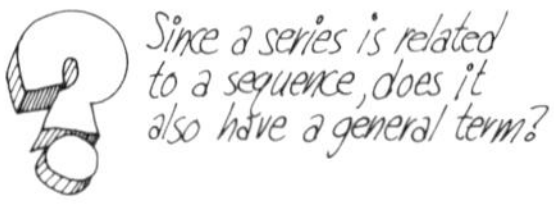

Like a sequence, a series also has a general term. As stated in the following definition, the nth term of a series is the same as the general term of the corresponding sequence.

General term of a series

> **Definition:** The general term, or nth term, of a series is the general term of the corresponding sequence.

Therefore, if a_n is the general term of a sequence, then a_n is also the general term of the corresponding series of the sequence. With sigma notation we can express the series with the nth term.

> **Notation:** If a_n is the general term of a series, then
>
> $$\sum_{i=1}^{n} a_i \quad \text{or} \quad \textstyle\sum_{i=1}^{n} a_i$$
>
> is the sum of n terms of the series. The letter i is called the *index of summation*. The upper index n is a positive integer in the domain of the sequence and indicates the number of terms in the series.

To illustrate sigma notation, let's use the sequence $a_n = 2n - 3$ with $n = 7$. That is, the sequence is $-1, 1, 3, 5, 7, 9, 11$. The general term of the sum of the sequence is also $2n - 3$, and the series is written $-1 + 1 + 3 + 5 + 7 + 9 + 11$. In summation notation we write the series as follows:

$$\sum_{i=1}^{7} (2i - 3) \quad \text{or} \quad \textstyle\sum_{i=1}^{7} (2i - 3)$$

Note 1. We have replaced n in the nth term of the series with the variable i, called the **index of summation.** The variables j and k are also commonly used as indices of summation. Thus,

$$\textstyle\sum_{i=1}^{7} (2i - 3) = \sum_{j=1}^{7} (2j - 3) = \sum_{k=1}^{7} (2k - 3)$$

Note 2. The lower and upper limits of the summation symbol are related to the domain of the sequence function from which the series is obtained. That is, for the given sequence with $a_n = 2n - 3$, $n \in \{1, 2, 3, \ldots, 7\}$. Each term of the corresponding series is obtained when i in the expression $2i - 3$ is replaced with $1, 2, 3, \ldots, 7$.

Note 3. We call the expression

$$-1 + 1 + 3 + 5 + 7 + 9 + 11$$

the **expanded form** of $\sum_{i=1}^{7} (2i - 3)$.

EXAMPLE 12–7 Write the following series in expanded form and find the indicated sum:

$$\sum_{i=1}^{5} (i^2 - 3)$$

Solution: We obtain the 5 terms of the expanded form of the stated series by replacing i in the expression $i^2 - 3$ with 1 through 5.

$$\text{For } i = 1,\ i^2 - 3 = -2$$
$$\text{For } i = 2,\ i^2 - 3 = 1$$
$$\text{For } i = 3,\ i^2 - 3 = 6$$
$$\text{For } i = 4,\ i^2 - 3 = 13$$
$$\text{For } i = 5,\ i^2 - 3 = 22$$

Thus,

$$\sum_{i=1}^{5} (i^2 - 3) = -2 + 1 + 6 + 13 + 22$$

and the indicated sum is 40.

What do we use for the upper limit of the index when the number of terms in the sequence is not known?

Recall that there are two instances in which the number of terms in a sequence is not a fixed value:

1. The sequence function has a domain $= \{1, 2, 3, \ldots, n\}$.
2. The sequence function has a domain $= \{1, 2, 3, \ldots\}$.

Suppose that a_n is the general term of a sequence function with domain $\{1, 2, 3, \ldots, n\}$. Then the corresponding series is written in summation form as

$$\sum_{i=1}^{n} a_i$$

and in expanded form we write

$$\sum_{i=1}^{n} a_i = a_1 + a_2 + a_3 + \cdots + a_n$$

Furthermore, if b_n is the general term of a sequence function with domain $\{1, 2, 3, \ldots\}$, then the corresponding series is written in summation form as

$$\sum_{i=1}^{\infty} b_i$$

and in expanded form we write

$$\sum_{i=1}^{\infty} b_i = b_1 + b_2 + b_3 + \cdots + b_n + \cdots$$

The symbol ∞ as the upper limit in the last two equations is used to indicate that *the domain of the sequence function is infinite.*

EXAMPLE 12–8 Write the following series in expanded form:

(a) $\sum_{i=1}^{n} (-2)^i$ (b) $\sum_{k=1}^{\infty} (-1)^k(3 - k)$

Solutions:

(a)

For $i = 1$, $(-2)^1 = -2$

For $i = 2$, $(-2)^2 = 4$

For $i = 3$, $(-2)^3 = -8$

For $i = 4$, $(-2)^4 = 16$

Therefore,

$$\sum_{i=1}^{n} (-2)^i = -2 + 4 - 8 + 16 + \cdots + (-2)^n$$

Notice that the value of the indicated sum cannot be determined until the specific value of n is known.

(b) When the expression $(-1)^k(3 - k)$ is evaluated for $k = 1, 2, 3, 4$, and n, we obtain the following expanded form:

$$\sum_{k=1}^{\infty} (-1)^k(3 - k) = -2 + 1 - 0 - 1 + \cdots + (-1)^n(3 - n) + \cdots$$

In Section 12-1 we determined a general term for a sequence for which a few terms were listed. If a series is stated in expanded form, we can write the series in sigma notation by determining a general term for the series. We should note, however, that the nth term of a series is not necessarily unique. We will therefore look for an "obvious" general term for the given series.

EXAMPLE 12–9 Write the following in sigma notation:

$$\frac{1}{2} + \frac{4}{3} + \frac{9}{4} + \frac{16}{5} + \frac{25}{6} + \frac{36}{7} + \frac{49}{8}$$

Solution: The given series contains 7 terms; thus, the lower and upper limits of the index of summation are 1 and 7, respectively. The numerator of each term of the series is a perfect square, that is, 1, 4, 9, 16, 25, 36, and 49 are $1^2, 2^2, 3^2, 4^2, 5^2, 6^2$, and 7^2, respectively. Furthermore, the denominator of each term is one more than the number of the term in the series. Thus, for the 1st term, the denominator is $1 + 1 = 2$; for the 2nd term, the denominator is $2 + 1 = 3$; and for the 7th term, the denominator is $7 + 1 = 8$. For the general term, the numerator is n^2 and the denominator is $n + 1$, and $a_n = \dfrac{n^2}{n + 1}$. In sigma notation,

$$\frac{1}{2} + \frac{4}{3} + \frac{9}{4} + \frac{16}{5} + \frac{25}{6} + \frac{36}{7} + \frac{49}{8} = \sum_{i=1}^{7} \left(\frac{i^2}{i + 1}\right)$$

Exercise Set 12–2

Write the given series in expanded form and find the indicated sum.

1. $\sum_{i=1}^{5} (i^2 - 2)$
2. $\sum_{i=1}^{5} \frac{i(i+1)}{2}$
3. $\sum_{j=1}^{6} (-1)^j j^3$
4. $\sum_{j=1}^{5} \frac{(-1)^j}{2^j}$
5. $\sum_{k=1}^{5} \log k$
6. $\sum_{k=1}^{6} 2^k$
7. $\sum_{j=1}^{4} [(-3)^j + j]$
8. $\sum_{i=1}^{5} (3i^2 - 2i)$
9. $\sum_{j=1}^{10} j(j-1)$
10. $\sum_{j=1}^{20} j$

Write the following series in expanded notation.

11. $\sum_{k=1}^{n} (k^3 - 2k)$
12. $\sum_{k=1}^{n} \frac{k+1}{\log(k+1)}$
13. $\sum_{i=1}^{n} \frac{i^2}{5i-1}$
14. $\sum_{i=1}^{n} \frac{(-5)^i}{\ln 2i}$
15. $\sum_{j=1}^{\infty} \frac{2x^j}{j}$
16. $\sum_{j=1}^{\infty} \frac{(-2x)^{j-1}}{3j}, \quad x \neq 0$
17. $\sum_{k=1}^{\infty} (3^k - 2^k)$
18. $\sum_{k=1}^{\infty} (x+1)^k$
19. $\sum_{i=1}^{\infty} \frac{i}{i+1}$
20. $\sum_{i=1}^{\infty} \frac{1}{i^2}$

Write the following series in sigma notation.

21. $2 + 4 + 8 + 16 + 32 + 64$
22. $1 + \frac{1}{3} + \frac{1}{9} + \frac{1}{27} + \frac{1}{81}$
23. $1 + 4 + 7 + 10 + 13 + 16$
24. $\frac{-1}{2} + \frac{1}{3} - \frac{1}{4} + \frac{1}{5} - \frac{1}{6} + \frac{1}{7}$
25. $-3 + 9 - 27 + 81 - 243 + 729$
26. $0 + \frac{1}{3} + \frac{2}{4} + \frac{3}{5} + \frac{4}{6} + \frac{5}{7} + \frac{6}{8}$
27. $1 + 2x + 4x^2 + 8x^3 + 16x^4$
28. $4xy^2 + 16x^2y^4 + 64x^3y^6 + 256x^4y^8 + 1024x^5y^{10}$

ARITHMETIC PROGRESSIONS 12–3

In the study of sequences we find two types that have special significance — the **arithmetic sequence** and the **geometric sequence.** These sequences are of such importance that mathematicians have given them the special name of **progressions.** In this section we define an arithmetic progression and then study the corresponding series. In Sections 12-4 and 12-5 a similar treatment is given to the geometric progression.

Arithmetic progression

> **Definition:** An arithmetic progression is a sequence in which the difference between any two adjacent terms is a constant.

We call the difference between any two adjacent terms the **common difference,** for which we use the symbol d.

EXAMPLE **12–10** Determine whether the following sequences are arithmetic progressions.
(a) 3, 7, 11, 15, 19 (b) 5, 9, 14, 20

Solutions:
(a) For the given sequence, $a_1 = 3, a_2 = 7, a_3 = 11, a_4 = 15$, and $a_5 = 19$. We compute the difference between adjacent terms to determine whether there is a common difference d:

$$\begin{aligned} a_2 - a_1 &= 7 - 3 = 4 \\ a_3 - a_2 &= 11 - 7 = 4 \\ a_4 - a_3 &= 15 - 11 = 4 \\ a_5 - a_4 &= 19 - 15 = 4 \end{aligned}$$

Since the difference between adjacent pairs of terms is the constant 4, the given sequence is an arithmetic progression in which $d = 4$.
(b) If $a_1 = 5$, $a_2 = 9$, $a_3 = 14$, and $a_4 = 20$, then

$$\begin{aligned} a_2 - a_1 &= 9 - 5 = 4 \\ a_3 - a_2 &= 14 - 9 = 5 \\ a_4 - a_3 &= 20 - 14 = 6 \end{aligned}$$

This sequence does not have a common difference between adjacent terms; thus, it is not an arithmetic progression.

We now refer to the arithmetic progression in Example 12–10(a), with a common difference $d = 4$. The first term in the progression is $a_1 = 3$. Notice the following set of equations:

$$\begin{aligned} a_2 &= 7 = 3 + 4 = 3 + 1(4) = a_1 + d \\ a_3 &= 11 = 3 + 8 = 3 + 2(4) = a_1 + 2d \\ a_4 &= 15 = 3 + 12 = 3 + 3(4) = a_1 + 3d \\ a_5 &= 19 = 3 + 16 = 3 + 4(4) = a_1 + 4d \end{aligned}$$

For each term in the progression, the numerical coefficient of d is one less than the number of the term in the sequence. For example, the coefficient of d is $3 - 1 = 2$ for a_3. Using the pattern established above, we identify an equation for determining the general term of an arithmetic progression in the following theorem.

> **Theorem:** If a_1 is the first term of an arithmetic progression in which the common difference is d, then the **general term** of the progression, a_n, is given by the equation
>
> $$a_n = a_1 + (n - 1)d \qquad \textbf{(1)}$$

Equation (1) contains three variables: (1) the first term of the progression, a_1, (2) the common difference in the progression, d, and (3) the value of the nth term in the progression, a_n. When any two of the variables in Equation (1) are known, then the value of the third variable can be calculated.

EXAMPLE 12–11 Find the value of a_{10} for the sequence $-3, 2, 7, 12, \ldots$

Solution: The given sequence is an arithmetic progression with $a_1 = -3$ and $d = 5$. (Notice that $2 - (-3) = 7 - 2 = 12 - 7 = 5$.) For a_{10}, $n = 10$. With Equation (1), $a_{10} = -3 + (10 - 1)5 = 42$.

EXAMPLE 12–12 Find the first term of the arithmetic progression in which the 10th term is -12 and the common difference is -2.

Solution: With $n = 10$, $a_{10} = -12$, $d = -2$, and with Equation (1),

$$-12 = a_1 + (10 - 1)(-2)$$
$$a_1 = 6$$

An arithmetic progression can have either a positive or a negative difference. If $d > 0$, then $a_{n+1} > a_n$. That is, the values of successive terms increase, if $d < 0$, then $a_{n+1} < a_n$, and the values of successive terms decrease.

According to Example 12-12, the value of d can be negative as well as positive !

EXAMPLE 12–13 If the 1st and 35th terms of an arithmetic progression are 15 and -2, respectively, find the common difference.

Solution: If $a_1 = 15$ and $a_{35} = -2$, then with Equation (1), $-2 = 15 + (35 - 1)d$. Therefore, $d = -17/34 = -1/2$. Notice that since $a_{35} < a_1$, we would expect a value of d less than 0. Thus, $d = -1/2$ is consistent with our expectation.

We now turn our attention to the **series** associated with an arithmetic progression.

> **Notation:** If a_n is the general term of an arithmetic progression, then
>
> $$S_n = \Sigma_{i=1}^{n} a_i$$
>
> and S_n is the sum of the first n terms of the progression.

The following equations are expanded forms of S_n:

$$\text{(i) } S_n = a_1 + (a_1 + d) + (a_1 + 2d) + \cdots + (a_1 + (n-1)d)$$
$$\text{(ii) } S_n = a_n + (a_n - d) + (a_n - 2d) + \cdots + (a_n - (n-1)d)$$

Equation (i) is obtained by expressing each term in the series after a_1 as the sum of a_1 and a multiple of d. For example, $a_2 = a_1 + d$, $a_3 = a_1 + 2d$, and so on, with $a_n = a_1 + (n-1)d$.

Equation (ii) is obtained by expressing each term in the series before a_n (the last term) as the difference between a_n and a multiple of d. For example, $a_{n-1} = a_n - d$, $a_{n-2} = a_n - 2d$, and so on, with $a_1 = a_n - (n-1)d$.

When Equations (i) and (ii) are added, each term with a factor of d cancels. That is, $d + (-d) = 0$, $2d + (-2d) = 0$, and $(n-1)d + [-(n-1)d] = 0$. The remaining terms in the sum of (i) and (ii) can be paired as follows:

$$2S_n = (a_1 + a_n) + (a_1 + a_n) + \cdots + (a_1 + a_n)$$

There are n pairs of $a_1 + a_n$; thus, we have

$$2S_n = n(a_1 + a_n)$$
$$S_n = \frac{n}{2}(a_1 + a_n)$$

Theorem: If a_1 and a_n are the first and last terms, respectively, of a series formed from an arithmetic progression of n terms, then the **sum** of the series is S_n, and

$$S_n = \frac{n}{2}(a_1 + a_n) \qquad \textbf{(2)}$$

To illustrate Equation (2), we have displayed a few of the terms of the series $S_n = \sum_{i=1}^{100} i$ in Figure 12-1. That is, $S_{100} = 1 + 2 + 3 + \cdots + 100$, the sum of the first 100 positive integers. Notice that $a_1 + a_{100} = 1 + 100 = 101$. Furthermore, $a_2 + a_{99} = 2 + 99 = 101$, and $a_3 + a_{98} = 3 + 98 = 101$. In the center of the series, $a_{50} + a_{51} = 50 + 51 = 101$. Since $n = 100$, then $n/2 = 50$; and there are 50 pairs, each with a sum of 101. Therefore,

$$S_{100} = 50(101) = 5050$$

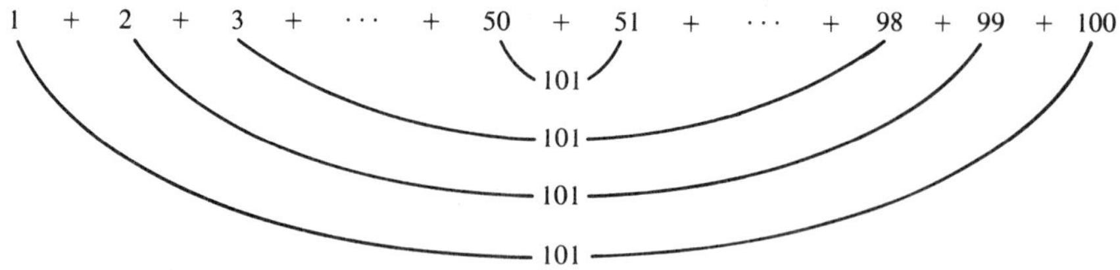

FIGURE 12-1 $S_n = \sum_{i=1}^{100} i$

EXAMPLE 12–14 Find the sum of the arithmetic progression 1, 4, 7, . . . , 58.

Solution: To use Equation (2) to find S_n, we need to find n, the number of terms in the progression. For the given sequence, $a_1 = 1$, $a_n = 58$, and $d = 4 - 1 = 3$. With Equation (1),

$$58 = 1 + (n - 1)(3)$$
$$n = 20$$

Substituting $a_1 = 1$, $a_{20} = 58$, and $n = 20$ in Equation (2), we get

$$S_{20} = \frac{20}{2}(1 + 58) = 590$$

Thus, the sum of the given progression is 590.

EXAMPLE 12–15 Evaluate $\sum_{i=1}^{30} (23 - 4i)$.

Solution: In expanded form, we write

$$\sum_{i=1}^{30} (23 - 4i) = 19 + 15 + 11 + \cdots + (-97)$$

Thus, $a_1 = 19$, $a_{30} = -97$, and $d = 15 - 19 = -4$. With Equation (2) and $n = 30$, we get

$$S_{30} = \frac{30}{2}(19 - 97) = -1170$$

Exercise Set 12–3

For the stated conditions, write the first five terms of the following arithmetic progressions.

1. $a_1 = 5$; $d = 3$
2. $a_1 = -8$; $d = 5$
3. $a_1 = 20$; $d = -7$
4. $a_1 = 2$; $d = -0.6$
5. $a_5 = 30$; $d = 6$
6. $a_5 = -8$; $d = -4$
7. $a_3 = 15$; $d = 2.5$
8. $a_3 = -3$; $d = 3.2$

Find the indicated term of the following arithmetic progressions.

9. 16th term of 1, 6, 11, . . .
10. 21st term of −5, −1, 3, . . .
11. 32nd term of 1/2, 3/4, 1, 5/4, . . .
12. 42nd term of 2, 8/3, 10/3, 4, . . .
13. 24th term of 1.5, 2.7, 3.9, . . .
14. 34th term of 8.5, 6.2, 3.9, . . .

Find the number of terms in the following arithmetic progressions.

15. 1, 8, 15, . . . , 141
16. 3, 1, −1, . . . , −61
17. $\pi/5, 2\pi/5, \ldots, 4\pi$
18. $\sqrt{3}, 5\sqrt{3}, 9\sqrt{3}, \ldots, 137\sqrt{3}$
19. 3, 23/8, 11/4, . . . , −9/8
20. 101, 95, 89, . . . , −133

Find the sum of the following arithmetic progressions.

21. −8, −3, 2, . . . , 87
22. 3, 6, 9, . . . , 75
23. 35, 28, 21, . . . , −56
24. −19, −15, −11, . . . , 61
25. $3x, 7x, 11x, \ldots, 79x$
26. 0.1, 1.2, 2.3, . . . , 24.3

27. 11/2, 5, 9/2, . . . , −43/2

28. 9.7y, 8.4y, 7.1y, . . . , −9.8y

Find the following sums.

29. $\Sigma_{i=1}^{15} (2i - 5)$

30. $\Sigma_{i=1}^{20} (50 - 3i)$

31. $\Sigma_{j=1}^{12} (0.3j + 1)$

32. $\Sigma_{j=1}^{18} \left(3 + \frac{j}{2}\right)$

33. $\Sigma_{k=1}^{21} (63 - 3.2k)$

34. $\Sigma_{k=1}^{16} \left(\frac{k}{3} + 3\right)$

12-4 FINITE GEOMETRIC PROGRESSIONS

In Section 12-3 we studied the arithmetic progression, a sequence in which successive terms have a common difference, d. In this section we will study a sequence in which successive terms have a **common ratio,** r. This sequence is called a geometric progression.

Geometric progression

> **Definition:** A geometric progression is a sequence in which each term in the sequence after the first term is the product of the preceding term and a single constant.

The single constant is called the common ratio, r. The following sequences are geometric progressions:

(i) 1/8, 1/4, 1/2, 1, 2, 4, 8, 16, 32, 64, 128, 256

(ii) 27, 9, 3, 1, 1/3, 1/9, 1/27, 1/81, 1/243

(iii) 3/64, −3/16, 3/4, −3, 12, −48, 192, −768

In (i), the value of r is 2. That is, after the first term 1/8, each term in the sequence is the product of 2 and the preceding term. In (ii), the common ratio is 1/3. In (iii), $r = -4$. Therefore, the terms of the sequence alternate in sign as well as in the common ratio 4.

Is it difficult to determine the value of r ?

It is very easy to determine the common ratio for a given geometric sequence. We simply compute the ratio of any two successive terms in the sequence to obtain r. That is, if a_k is the kth term of a geometric progression of n terms $(1 \leq k < n)$, then

$$r = \frac{a_{k+1}}{a_k} \tag{1}$$

EXAMPLE 12–16 Find the common ratio for the geometric progression 9, 6, 4, 8/3, 16/9.

Solution: Since $n = 5$, then $1 \leq k < 5$. If we arbitrarily select $k = 2$, then $k + 1 = 3$. Thus, $a_2 = 6$ and $a_3 = 4$ for the given sequence, and with Equation (1),

$$r = \frac{a_3}{a_2} = \frac{4}{6} = \frac{2}{3}$$

To check $r = 2/3$ for the given sequence, we arbitrarily choose a_4 and a_5 to compare $2/3(a_4)$ with the stated value for a_5. With $a_4 = 8/3$, then

$$\frac{2}{3}\left(\frac{8}{3}\right) = \frac{16}{9} = a_5$$

If we let a_1 be the first term of a geometric progression with a common ratio r, then the sequence can be written as

$$a_1, a_1r, a_1r^2, a_1r^3, \ldots, a_n$$

Notice that each term in the sequence is the product of a_1 and a power of r. Furthermore, *the exponent on r is always one less than the numbered term in the sequence*. That is:

For the 1st term, the exponent on r is $1 - 1 = 0$ and $r^0 = 1$.

For the 2nd term, the exponent on r is $2 - 1 = 1$.

For the nth term, the exponent on r is $n - 1$.

> **Theorem.** The **general** or ***n*th term** of a geometric progression with first term a_1 and common ratio r is given by
>
> $$a_n = a_1r^{n-1} \qquad \textbf{(2)}$$

EXAMPLE 12–17 Find the 6th term of a geometric progression with 1st term 12 and common ratio 1/2.

Solution: For the given information, $n = 6$; thus, $n - 1 = 5$. Furthermore, $a_1 = 12$ and $r = 1/2$. With Equation (2),

$$a_6 = 12\left(\frac{1}{2}\right)^5 = \frac{3}{8}$$

Therefore, the 6th term in the sequence is 3/8.

EXAMPLE 12–18 Find the common ratio of a geometric progression in which $a_1 = 81/8$ and $a_7 = 8/9$.

Solution: From the given information, $n = 7$, $a_1 = 81/8$, and $a_7 = 8/9$. With Equation (2),

$$\frac{8}{9} = \frac{81}{8}r^6$$

Thus, we get

$$\begin{aligned} r^6 &= \left(\frac{8}{9}\right) \div \left(\frac{81}{8}\right) \\ &= \frac{64}{729} \text{ or } \left(\frac{2}{3}\right)^6 \\ r &= \frac{2}{3} \end{aligned}$$

Let S_n be the series formed by summing the first n terms of a finite geometric progression. Then

$$S_n = a_1 + a_1r + a_1r^2 + \cdots + a_1r^{n-1}$$

We now multiply the equation by r and by -1 to form the following two equations:

$$rS_n = a_1r + a_1r^2 + a_1r^3 + \cdots + a_1r^{n-1} + a_1r^n$$
$$-S_n = -a_1 - a_1r - a_1r^2 - a_1r^3 - \cdots - a_1r^{n-1}$$

When we add these two equations, all terms on the right side of the equation cancel except for $-a_1$ and a_1r^n. Thus,

$$rS_n - S_n = -a_1 + a_1r^n$$
$$(r - 1)S_n = a_1(r^n - 1)$$
$$S_n = \frac{a_1(r^n - 1)}{r - 1}$$

> **Theorem:** The **sum** of the first n terms of a finite geometric progression with first term a_1 and common ratio r is given by
>
> $$S_n = \frac{a_1(r^n - 1)}{r - 1} \qquad \textbf{(3)}$$

EXAMPLE 12–19 Find the sum of the first 6 terms of the geometric progression with first term 3 and common ratio 2.

Solution: From the given information, $n = 6$, $a_1 = 3$, and $r = 2$. With Equation (3),

$$S_6 = \frac{3(2^6 - 1)}{2 - 1} = 3(63) = 189$$

EXAMPLE 12–20 Find $\sum_{i=1}^{5} 3(4^i)$.

Solution: First, let's write the series in expanded form:

$$\sum_{i=1}^{5} 3(4^i) = 3(4) + 3(4^2) + 3(4^3) + 3(4^4) + 3(4^5)$$

Each term of the series is a term of a geometric progression in which $a_1 = 3(4) = 12$ and $r = 4$. We can therefore use Equation (3) to find the indicated sum:

$$S_5 = \frac{12(4^5 - 1)}{4 - 1} = 4092$$

Notice that the index of summation is the exponent on the factor 4 in the given expression. Thus, as i takes on the values from 1 through 5, each term of the series increases by a factor of 4. The result is the series for a geometric progression with common ratio 4.

EXAMPLE 12–21 A sheet of paper is approximately 0.003 inch thick. When the paper is divided in half and stacked, the stack of paper is doubled. When this stack is divided in half and stacked, the stack of paper is again doubled. The dividing and stacking can be expressed as a geometric progression with $a_1 = 0.003$ inch and $r = 2$. How high is the stack of papers after 50 such doublings?

Solution: With $a_1 = 0.003$, $n = 51$, and $r = 2$, we get

$$a_{51} = 0.003(2^{50})$$

We can use an electronic calculator with a y^x key to compute 2^{50}. Enter 2, enter 50 and depress the y^x key, and display reads $1.125899909 \times 10^{15}$. Now multiply by 0.003, and display reads $3.377699727 \times 10^{12}$ inches. Now enter 12 and divide to convert the answer to feet; the display reads $2.814749773 \times 10^{11}$. Now enter 5280 and divide to convert the answer to miles; the display reads 5.330965479×10^7 or approximately 53,309,655 miles. The stack of paper would reach over halfway to the sun.

Exercise Set 12–4

Use Equation (1) to determine the common ratio for each of the following geometric progressions. Check the computed value of r with any other pair of consecutive terms of the sequence.

1. 3/25, 3/5, 3, 15, 75
2. 5/8, −5/4, 5/2, −5, 10, −20
3. 32, 8, 2, 1/2, 1/8, 1/32
4. 49, 35, 25, 125/7, 625/49
5. 1000, 300, 90, 27, 8.1, 2.43
6. 10, −12, 14.4, −17.28, 20.736
7. 12/125, 6/25, 3/5, 3/2, 15/4
8. 27/64, −9/16, 3/4, −1, 4/3, −16/9
9. $a^{1/2}, a^2, a^{7/2}, a^5, a^{13/2}$
10. $a^3/16,\ a^2/8,\ a/4,\ 1/2,\ 1/a,\ 2/a^2\ (a \neq 0)$

Find the indicated term for each of the following geometric progressions.

11. a_8 with $a_1 = 1/8$ and $r = -2$
12. a_5 with $a_1 = 75$ and $r = 1/5$
13. a_6 with $a_1 = 100$ and $r = 0.3$
14. a_4 with $a_1 = 10$ and $r = 9.8$
15. a_{10} with $a_1 = 160$ and $r = -1/2$
16. a_6 with $a_1 = 27x$ and $r = x/3$

Find the indicated term of the following geometric progressions.

17. 7th term of 640, −320, 160, . . .
18. 10th term of 1/12, 1/6, 1/3, . . .
19. 8th term of 135/56, 45/28, 15/14, . . .
20. 6th term of 2, −10, 50, . . .
21. 5th term of 2000, −400, 80, . . .
22. 12th term of 1, 2, 4, . . .

Find r for the geometric progression with the given first and nth terms.

23. $a_1 = 1/9;\ a_5 = 9$
24. $a_1 = 16;\ a_8 = -1/8$
25. $a_1 = 300;\ a_6 = 0.003$
26. $a_1 = 4;\ a_4 = 27/2$
27. $a_1 = 45b;\ a_5 = 5b^3/9$
28. $a_1 = 5t;\ a_4 = 5000t^7$

Find the sums of the first n terms of each of the following geometric progressions.

29. $a_1 = 1; r = 4; n = 6$
30. $a_1 = 1; r = 1/2; n = 10$
31. $a_1 = 81; r = -1/3; n = 8$
32. $a_1 = 125; r = -1/5; n = 7$
33. $a_1 = 0.009; r = 10; n = 6$
34. $a_1 = 1/8; r = 1/4; n = 8$
35. $a_1 = 4/3; r = 3/2; n = 7$
36. $a_1 = 2; r = 3; n = 5$

Compute the following.

37. $\Sigma_{i=1}^{5} 4\left(\frac{1}{2}\right)^i$
38. $\Sigma_{i=1}^{4} 2592\left(\frac{1}{6}\right)^i$
39. $\Sigma_{j=1}^{8} 4\left(\frac{3}{2}\right)^j$
40. $\Sigma_{j=1}^{6} \left(\frac{1}{4}\right)^j$
41. $\Sigma_{k=1}^{7} 5(2^k)$
42. $\Sigma_{k=1}^{5} 7(-2)^k$
43. A "super ball" rebounds to 95 percent of the height from which it fell. The total distance the ball travels vertically is the sum of a geometric progression where $r = 0.95$ and a_1 is the initial height from which the ball falls. If $a_1 = 10$ meters, how far does the ball travel in (a) 5 round trips and (b) 8 round trips?
44. Suppose a worker is offered wages of 1 cent for the first day's work, 2 cents for the second day's work, 4 cents for the third day's work, and so on. That is, a day's wages is double the previous day's pay. How much does the worker earn the first month on the job if she works 22 days?

12-5 INFINITE GEOMETRIC PROGRESSIONS

In Section 12-4 we used the equation

$$S_n = \frac{a_1(r^n - 1)}{r - 1}$$

to find the sum of the first n terms of a geometric progression with first term a_1 and common ratio r. In this section we want to consider the sums of geometric progressions in which the number of terms in the progression is infinite.

Let's begin by considering two infinite geometric progressions (A) and (B) identified below:

(A) 1, 2, 4, 8, 16, 32, 64, 128, . . .

(B) 1, 1/2, 1/4, 1/8, 1/16, 1/32, 1/64, 1/128, . . .

For (A), $a_1 = 1$ and $r = 2$, and for (B), $a_1 = 1$ and $r = 1/2$. Let S_n and T_n represent the sums of n terms of (A) and (B), respectively, and $n \in \{1, 2, 3, 4, 5, 6, 7, 8\}$:

$$S_1 = 1 \qquad T_1 = 1$$
$$S_2 = 3 \qquad T_2 = 3/2$$
$$S_3 = 7 \qquad T_3 = 7/4$$

$S_4 = 15$ $T_4 = 15/8$
$S_5 = 31$ $T_5 = 31/16$
$S_6 = 63$ $T_6 = 63/32$
$S_7 = 127$ $T_7 = 127/64$
$S_8 = 255$ $T_8 = 255/128$

For (A), the values of S_n are obviously getting larger as n increases. Intuitively, we might guess that there is no limit to how large S_n can get. We can reasonably expect S_n to become infinitely large by simply adding more terms of the progression. After all, the number of terms in (A) is infinite, and each term is twice the value of the preceding term. Thus, we can make S_n as large as we want to by adding enough terms in the progression. Our expectation is quite accurate, and to express this fact we write

$$\lim_{n \to \infty} S_n = \infty$$

which we read as "the limt of S_n is infinitely large as n, the number of terms in the series, becomes infinitely large." Notice that ∞ is not a number, but a symbol that indicates that no finite number exists that is large enough to express S_n as n becomes infinitely large.

Consider now the set of values given for T_n. Similar to S_n, the values of T_n are increasing as n increases. But unlike S_n, we might question the ability of T_n to have *unlimited growth*. Notice that the given values of T_n are all less than 2! As n increases, we always increase the value of T_n by adding another term in the sequence. However, the size of the term added is always 1/2 the value of the preceding term. Thus, we are always adding smaller numbers to the value of T_n as n increases. As we shall see, the value of T_n will never equal 2 nor exceed it, no matter how many terms of the progression are added. We express this fact as

$$\lim_{n \to \infty} T_n = 2$$

which we read as "the limit of T_n is 2 as n becomes infinitely large." That is, no matter how many terms of the geometric progression with first term 1 and common ratio 1/2 are added, the sum will never equal 2. But — and this is important — we can get a sum as close to 2 as we want by simply adding enough terms of the progression.

Does that mean that the sum of an infinite number of terms of (B) will never equal 2?

To explain it another way, suppose we are required to fill a 2-liter container with water. However, we are instructed to first pour in 1 liter of water ($a_1 = 1$). Then pour in 1/2 that amount ($a_2 = 1/2$), then 1/2 that amount ($a_3 = 1/4$), then 1/2 that amount ($a_4 = 1/8$), and so on. We are therefore adding water according to a geometric progression with $a_1 = 1$ and $r = 1/2$. Since $\lim_{n\to\infty} T_n = 2$, we can conclude that we will never completely fill the 2-liter container with the scheme described. However, we can fill the container to as close to capacity as we choose by simply continuing the process long enough.

Notice that (A) and (B) are both infinite geometric progressions, but the corresponding series for the two progressions are distinctly different. That is,

$$\lim_{n\to\infty} S_n = \infty \quad \text{and} \quad \lim_{n\to\infty} T_n = 2$$

The reason that T_n has a finite value for a limit while S_n does not is the result of the difference in the common ratios of the two sequences. For (A), $r = 2$, and for

(B), $r = 1/2$. To observe that the difference in the values of r is the reason for the difference in the limits of S_n and T_n, consider Equation (3) from Section 12-4 for the sum of the first n terms of any geometric progression:

$$S_n = \frac{a_1(r^n - 1)}{r - 1}$$

When we multiply the numerator and denominator by (-1), we get

$$S_n = \frac{a_1(1 - r^n)}{1 - r} = \frac{a_1 - a_1r^n}{1 - r}$$

We now write the ratio as two fractions:

$$S_n = \frac{a_1}{1 - r} - \frac{a_1r^n}{1 - r}$$

For any geometric progression, a_1 and r are constant; thus, the term $\frac{a_1}{1 - r}$ is also a constant. But the term $\frac{a_1r^n}{1 - r}$ varies as n varies. Consider the effect on geometric progressions (A) and (B) for $n \in \{1, 2, 3, 4, 5\}$:

(A)	(B)
$a_1 = 1$ and $r = 2$	$a_1 = 1$ and $r = 1/2$
$S_n = \frac{1}{1 - 2} - \frac{1(2)^n}{1 - 2}$	$T_n = \frac{1}{1 - \frac{1}{2}} - \frac{1\left(\frac{1}{2}\right)^n}{1 - \frac{1}{2}}$
$= -1 + 2^n$	$= 2 - 2\left(\frac{1}{2}\right)^n$
For $n = 1$, $S_1 = -1 + 2 = 1$	$T_1 = 2 - 1 = 1$
For $n = 2$, $S_2 = -1 + 4 = 3$	$T_2 = 2 - \frac{1}{2} = \frac{3}{2}$
For $n = 3$, $S_3 = -1 + 8 = 7$	$T_3 = 2 - \frac{1}{4} = \frac{7}{4}$
For $n = 4$, $S_4 = -1 + 16 = 15$	$T_4 = 2 - \frac{1}{8} = \frac{15}{8}$
For $n = 5$, $S_5 = -1 + 32 = 31$	$T_5 = 2 - \frac{1}{16} = \frac{31}{16}$, and so on

For the series S_n, the term 2^n increases as n increases; therefore, S_n increases. But for the series T_n, the term $2(1/2)^n$ decreases as n increases. Specifically,

$$\lim_{n\to\infty} 2\left(\frac{1}{2}\right)^n = 0$$

That is, as n becomes infinitely large, the term $\frac{a_1r^n}{1 - r}$ "approaches" the value of 0. Hence,

$$\lim_{n\to\infty} T_n = \frac{a_1}{1 - r} = 2$$

for the given values of a_1 and r. In general, if $-1 < r < 1$, then any geometric progression will have a corresponding series whose sum is defined to be the ratio $\frac{a_1}{1-r}$.

Definition: If S is the sum of an infinite geometric progression with first term a_1 and common ratio r with $-1 < r < 1$, and $S = a_1 + a_1 r + a_1 r^2 + \cdots$, then

$$\lim_{n\to\infty} S_n = \frac{a_1}{1-r} \qquad (1)$$

$\lim_{n\to\infty} S_n$

Notice that $\frac{a_1}{1-r}$ is defined as the sum of S_n as $n \to \infty$. We also use the notation

$$S_\infty \quad \text{or} \quad \textstyle\sum_{i=1}^{\infty} a_i$$

for the sum of an infinite geometric progression.

EXAMPLE 12–22 Find the sum $2 + \frac{2}{3} + \frac{2}{9} + \cdots$.

Solution: With $a_1 = 2$ and $r = 1/3$ and Equation (1),

$$\lim_{n\to\infty} S_n = \frac{2}{1 - \frac{1}{3}} = 3$$

Thus, $S_\infty = 3$, and the indicated series will never exceed the value 3, no matter how many terms of the sequence are added.

EXAMPLE 12–23 Find the sum $2 + \frac{12}{5} + \frac{72}{25} + \frac{432}{125} + \cdots$.

Solution: Since $r = 6/5$, and $6/5 > 1$, the given series does not have a sum, and we write $S_\infty = \infty$.

EXAMPLE 12–24 Evaluate $\sum_{i=1}^{\infty} 5\left(\frac{-1}{2}\right)^i$.

Solution: We first write a few terms of the given series:

$$\textstyle\sum_{i=1}^{\infty} 5\left(\frac{-1}{2}\right)^i = \frac{-5}{2} + \frac{5}{4} - \frac{5}{8} + \frac{5}{16} - \cdots$$

Substituting $a_1 = -5/2$ and $r = -1/2$ into Equation (1), we get

$$\textstyle\sum_{i=1}^{\infty} 5\left(\frac{-1}{2}\right)^i = \frac{-5/2}{1 - (-1/2)} = \frac{-5}{3}$$

Example 12–25 Express 3.151515 . . . as the ratio of two integers.

Solution: The given rational number can be written as

$$3 + 0.15 + 0.0015 + 0.000015 + \cdots$$

If we let $a_1 = 0.15$, then each successive term in the series is 0.01 times the preceding term. Thus, we have a geometric series $S = 0.15 + 0.0015 + 0.000015 + \cdots$, with $r = 0.01$. Since $-1 < 0.01 < 1$, then

$$S_\infty = \frac{0.15}{1 - 0.01} = \frac{0.15}{0.99} = \frac{5}{33}$$

We add the 3 to the value for S_∞, and

$$3.151515\ldots = 3\frac{5}{33} = \frac{104}{33}$$

Exercise Set 12–5

Find the sum, if it exists, of the following infinite geometric series. If the sum does not exist, write ∞.

1. $2 + 4/5 + 8/25 + 16/125 + \cdots$
2. $1 + 3/7 + 9/49 + 27/343 + \cdots$
3. $-3 + 3/2 - 3/4 + 3/8 - \cdots$
4. $12 - 10 + 25/3 - 125/18 + \cdots$
5. $6 + 8 + 32/3 + 128/9 + \cdots$
6. $3 - 10/3 + 100/27 - 1000/243 + \cdots$
7. $-8 - 6 - 9/2 - 27/8 - \cdots$
8. $-10 - 2 - 2/5 - 2/25 - \cdots$
9. $12 + 1.2 + 0.12 + 0.012 + \cdots$
10. $7 + 2.1 + 0.63 + 0.189 + \cdots$

Find the following sums. If a sum does not exist, write ∞.

11. $\sum_{i=1}^{\infty} (0.6)^i$
12. $\sum_{i=1}^{\infty} 4\left(\frac{5}{7}\right)^i$
13. $\sum_{j=1}^{\infty} 15\left(\frac{-2}{3}\right)^j$
14. $\sum_{j=1}^{\infty} 12\left(\frac{-5}{6}\right)^j$
15. $\sum_{k=1}^{\infty} 20(-0.2)^k$
16. $\sum_{k=1}^{\infty} 8(0.9)^k$
17. $\sum_{i=1}^{\infty} 3(1.1)^i$
18. $\sum_{i=1}^{\infty} -10\left(\frac{-9}{8}\right)^i$
19. $\sum_{j=1}^{\infty} -10(0.15)^j$
20. $\sum_{j=1}^{\infty} 7.5(-0.33)^j$

Express the following rational numbers as the ratio of two integers.

21. 0.4444 . . .
22. 0.6666 . . .
23. 0.060606 . . .
24. 0.191919 . . .
25. 5.272727 . . .
26. 9.434343 . . .
27. 4.072072072 . . .
28. 7.301301301 . . .
29. 0.9999 . . .
30. 1.342342342 . . .

SUMMARY

Sequence function

Definition: A sequence function is a function whose domain is a set of consecutive positive integers.

Finite sequence

Definition: A sequence is finite if the domain of the sequence function is finite.

Infinite sequence

Definition: A sequence is infinite if the domain of the sequence function is infinite.

Series

Definition: A series is the sum of the terms of a sequence.

General term of a series

Definition: The general term, or nth term, of a series is the general term of the corresponding sequence.

Arithmetic progression

Definition: An arithmetic progression is a sequence in which the difference between any two adjacent terms is a constant.

Theorem: If a_1 is the first term of an arithmetic progression in which the common difference is d, then the **general term** of the progression, a_n, is given by the equation $a_n = a_1 + (n - 1)d$.

Theorem: If a_1 and a_n are the first and last terms, respectively, of a series formed from an arithmetic progression of n terms, then the **sum** of the series is S_n, and $S_n = \frac{n}{2}(a_1 + a_n)$.

Geometric progression

Definition: A geometric progression is a sequence in which each term in the sequence after the first term is the product of the preceding term and a single constant.

Theorem: The **general** or ***n*th term** of a geometric progression with first term a_1 and common ratio r is given by the equation $a_n = a_1 r^{n-1}$.

Theorem: The **sum** of the first n terms of a finite geometric progression with first term a_1 and common ratio r is given by the equation $S_n = \dfrac{a_1(r^n - 1)}{r - 1}$.

$\lim_{n\to\infty} S_n$

Definition: If S is the sum of an infinite geometric progression with first term a_1 and common ratio r with $-1 < r < 1$, and $S = a_1 + a_1 r + a_1 r^2 + \cdots$, then

$$\lim_{n\to\infty} S_n = \frac{a_1}{1 - r}$$

SELF-CHECKING EXERCISE

Write the first five terms of the sequence with the given general term. **(12-1)**

1. $a_n = \dfrac{n^2 + 1}{n}$

2. $b_n = 2^n - 5$

Write the terms of the following sequences, where $n \in \{\text{positive integers}\}$. **(12-1)**

3. $a_n = \dfrac{2n + 1}{3n}$

4. $b_n = \sqrt[3]{n^2 + 1}$

Find a general term for each of the following sequences. **(12-1)**

5. 7, 12, 17, 22, . . .
6. 1, 4/3, 2, 16/5, . . .

Find the sum of the given series. **(12-2)**

7. $\sum_{i=1}^{6} (3i - 1)$
8. $\sum_{j=1}^{10} (j + 5)$

Write the following series in expanded notation. **(12-2)**

9. $\sum_{k=1}^{n} \dfrac{k^2}{k+3}$
10. $\sum_{i=1}^{\infty} 4(-x)^i$

Write the following series in sigma notation. **(12-2)**

11. $3 + 6 + 9 + 12 + 15 + 18 + 21$
12. $-3 + 9/2 - 27/3 + 81/4 - 243/5$

Write the first five terms of the following arithmetic progressions. **(12-3)**

13. $a_5 = 43; d = -7$
14. $a_3 = 10; d = 4.2$

Find the indicated term of the following arithmetic progressions. **(12-3)**

15. 12th term of 0.6, 1.3, 2.0, . . .
16. 34th term of 5, 11/3, 7/3, 1, . . .

Find the sum of the following arithmetic progressions. **(12-3)**

17. −5, 1, 7, . . . , 109
18. 1, 3.4, 5.8, . . . , 27.4

Find the following sums. **(12-3)**

19. $\sum_{i=1}^{10} (3i + 1)$
20. $\sum_{j=1}^{15} \dfrac{j-1}{j+1}$

Determine the common ratio for the following geometric progressions. **(12-4)**

21. 16/25, 4/5, 1, 5/4, 25/16
22. 10, −14, 19.6, −27.44

Find the indicated term for the following geometric progressions. **(12-4)**

23. a_5 with $a_1 = 72$ and $r = 1/6$
24. a_6 with $a_1 = 100$ and $r = 0.2$

Find the sum of the first n terms of each of the following geometric progressions. **(12-4)**

25. $a_1 = 5; r = 2; n = 7$
26. $a_1 = -18; r = -1/3; n = 5$

Find the sum of the following infinite geometric series. If the sum does not exist, write ∞. **(12-5)**

27. $3 + 0.9 + 0.27 + \cdots$
28. $16 + 20 + 25 + \cdots$

Find the following sums. If a sum does not exist, write ∞. **(12-5)**

29. $\sum_{i=1}^{\infty} (0.4)^i$
30. $\sum_{j=1}^{\infty} 5\left(\dfrac{-1}{3}\right)^j$

Express the following rational numbers as the ratio of two integers. **(12-5)**

31. 5.818181 . . .
32. 3.225225225 . . .

Answers to Self-Checking Exercise

1. 2, 5/2, 10/3, 17/4, 26/5
2. −3, −1, 3, 11, 27
3. 1, 5/6, 7/9, . . . , $\dfrac{2n + 1}{3n}$, . . .
4. $\sqrt[3]{2}$, $\sqrt[3]{5}$, $\sqrt[3]{10}$, . . . ,$\sqrt[3]{n^2 + 1}$, . . .
5. $a_n = 5n + 2$
6. $b_n = \dfrac{2^n}{n + 1}$
7. 57
8. 105
9. $\dfrac{1}{4} + \dfrac{4}{5} + \dfrac{9}{6} + \cdots + \dfrac{n^2}{n + 3}$
10. $-4x + 4x^2 - 4x^3 + \cdots + 4(-x)^n + \cdots$
11. $\sum_{i=1}^{7} 3i$
12. $\sum_{i=1}^{5} \dfrac{(-3)^i}{i}$
13. 71, 64, 57, 50, 43
14. 1.6, 5.8, 10, 14.2, 18.4

15. 8.3	16. −39	17. 1040	18. 170.4
19. 175	20. 105/16	21. 5/4	22. −1.4
23. 1/18	24. 0.032	24. 635	26. −122/9
27. 30/7	28. ∞	29. 2/3	30. −5/4
31. 64/11	32. 358/111		

13 Combinatorics

In our modern and complex society, decisions or inferences are made on what statisticians call "incomplete information." For example, a decision to accept or reject a large number (or **population**) of items is frequently made by determining the number of defective or faulty items in a part (or **sample**) of the items. The inference is made that the percentage of defectives in the sample is the same as the percentage of defectives in the total population.

Similarly, opinion polls state that a certain proportion of the public has attitudes on a certain issue of concern. Actually, the stated percentage is true for the sample of the public interviewed by the people conducting the poll. The inference is made that the stated percentage also applies to the public at large.

When decisions or inferences are made with incomplete information, there exists a chance that an error has been made. That is, a population of items can be rejected for having too high a percentage of defectives when in fact the percentage is within allowable limits; or the population of items can be accepted when in fact it has a higher percentage of defectives than is permissible. In the study of statistics, we learn methods to quantify or measure the probability of making such errors. It is for this reason that elementary statistics is becoming a required course in many college degree programs.

Therefore, in this chapter we will study **probability theory.** We will develop several important theorems associated with probability and apply the theory to as many "real life" situations as possible.

To help us compute probabilities, we need a method for counting the number of ways a series of occurrences can take place. We therefore begin our study in this chapter with sections on permutations and combinations.

PERMUTATIONS 13-1

Every state government needs to provide license plates for each car registered in that state. The task usually requires some technique of assigning letters and numbers to various places on the license plate. Suppose that a certain state uses a pattern of two letters of the English alphabet followed by three digits. In Figure

13-1 each box represents a space on a given license plate that is filled with a letter or a digit. The question we want to answer is "How many *different* license plates can be made with this technique?"

1	2	3	4	5

FIGURE 13-1 Letter in boxes 1 and 2; digit in boxes 3, 4, and 5

Under what conditions would two license plates be different?

There are two ways in which two license plates can be different. First, if two plates differ by either a letter or a digit, then we say they are different plates. For example, the plates AB 123, AC 123, and AB 124 are all different. Second — and this point is important — if we use the same set of letters and digits but change the order in which they appear, then the plates are different. For example, the plates AB 123, BA 123, and AB 321 are all different, even though they each have the same letters and digits.

To find out how many different plates can be formed, we must consider the number of ways in which each rectangle in Figure 13-1 can be filled. Boxes 1 and 2 are filled with letters. There are 26 letters in the English alphabet; thus, we have 26 choices of a letter for box 1. Assuming that any letter may be used more than once, we also have 26 choices of a letter for box 2.

Boxes 3, 4, and 5 are each filled with a digit, and there are ten digits from 0 through 9. Assuming that any digit may be used more than once, we have 10 choices of a digit for each of the boxes 3 through 5. In Figure 13-2 the 5 boxes are shown with the number of choices of letters or digits indicated for each box.

26	26	10	10	10
1	2	3	4	5

FIGURE 13-2 Number of choices indicated for each box

The total number of different plates is the product of the numbers given in Figure 13-2. That is, there are

$$(26)(26)(10)(10)(10) = 676{,}000$$

different license plates that can be made with the scheme described. Furthermore, each license plate differs from any other plate by at least one letter, one digit, or the order in which the same letters and digits appear. The generalization of this example is the following axiom.

> **Counting Principle:** If task A can be performed in m ways and task B can be performed in n ways, then tasks A and B can be jointly performed in $(m)(n)$ ways, provided the performance of task A does not influence the performance of task B.

As in the case of the license plate example, if more than two tasks are involved, then the total number of ways in which all of the tasks are performed is the product of the number of ways in which each of the tasks is performed.

EXAMPLE 13–1 A math quiz consists of 5 multiple choice questions. Each question has 4 possible answers, a, b, c, or d. In how many different ways can a student "guess" the answers to the quiz?

Solution: In answering all the questions to the quiz, the student performs 5 tasks:

Task A. "Guess" the answer to question 1.

Task B. "Guess" the answer to question 2.

Task C. "Guess" the answer to question 3.

Task D. "Guess" the answer to question 4.

Task E. "Guess" the answer to question 5.

There are 4 ways of performing each of the tasks; thus, the total number of ways in which all 5 tasks can be performed is found with the counting principle:

$$(4)(4)(4)(4)(4) = 1024$$

Thus, there are 1024 different sets of answers that can be submitted for this quiz.

EXAMPLE 13–2 How many 3-digit numbers can be formed with the set of digits $\{0, 1, 2, 3, 4, 5\}$ if (a) each digit can be used more than once and (b) each digit can be used only once in a given number?

Solutions: For both parts (a) and (b) we have 3 tasks to perform:

Task A. Select the first digit in the number.

Task B. Select the second digit in the number.

Task C. Select the third digit in the number.

(a) Since 0 cannot be used for the first digit (that is, 012 is the 2-digit number 12 and not a 3-digit number), then task A can be performed in 5 ways. A digit may be used more than once; thus, tasks B and C can each be performed in 6 ways. Thus, the total number of 3-digit numbers that can be formed by repeating digits is $5(6)(6) = 180$.

(b) Task A can be performed in 5 ways, since 0 cannot be used. Task B can also be performed in 5 ways since 0 may now be used, but the digit used for task A cannot be used for task B. (Notice that each digit may be used only once in a given number.) Task C can be performed in 4 ways since the digits used in A and B cannot be used in C. Thus, $5(5)(4) = 100$, and 100 3-digit numbers can be formed with 3 different digits from the given set.

A special application of the counting principle, called a **permutation,** requires that we do not use the elements of a given set more than once. We use the following example to illustrate the permutation of the members of a set of 4 letters, using 2 letters at a time.

EXAMPLE 13–3 If $S = \{a, b, c, d\}$,

(a) Compute the number of 2-letter "words" that can be formed with the letters in S if no letter can be used more than once.

(b) List the 2-letter "words" that can be formed with the letters in S.

Solutions:

(a) We have two tasks to perform, A and B. For task A we select the first letter of the word, and for task B we select a second letter. Task A can be performed in 4 ways. Since letters may not be used more than once, task B can be performed in 3 ways. Thus, $4(3) = 12$ 2-letter "words" can be formed with the elements of S.

(b) We list the 12 2-letter words below. Notice that because letters cannot be repeated, words such as aa, bb, and cc do not appear in the list. Furthermore, ab and ba are different words because of the different order of the letters.

ab	ba	ca	da
ac	bc	cb	db
ad	bd	cd	dc

The following theorem, stated without proof, provides an equation for computing the permutation of r elements of a set containing n elements.

Theorem: If set S contains n distinct elements, then the number of **permutations** of r of these elements (written ${}_nP_r$) is given with the equation

$$ {}_nP_r = n(n-1)(n-2)\cdots(n-r+1), \text{ and } r \leq n $$

EXAMPLE 13–4 Compute the following:

(a) ${}_8P_3$ (b) ${}_6P_4$

Solutions:

(a) With $n = 8$ and $r = 3$, then $n - r + 1 = 8 - 3 + 1 = 6$. Thus, ${}_8P_3 = 8(7)(6) = 336$.

(b) With $n = 6$ and $r = 4$, then $n - r + 1 = 6 - 4 + 1 = 3$. Therefore, ${}_6P_4 = 6(5)(4)(3) = 360$.

A useful symbol in mathematics is the factorial notation.

***n*!**

Definition: If n is a positive integer, then

$$ n! = n(n-1)(n-2)\cdots(3)(2)(1) $$

We read $n!$ as "n factorial."

0!

Definition: $0! = 1$

EXAMPLE 13–5 Compute the following:

(a) $4!$ (b) $\frac{8!}{5!}$ (c) $\frac{6!}{2!}$

Solutions:

(a) By definition, $4! = 4(3)(2)(1) = 24$.

(b) $\frac{8!}{5!} = \frac{8(7)(6)(5)(4)(3)(2)(1)}{(5)(4)(3)(2)(1)} = 336$

(c) $\frac{6!}{2!} = \frac{6(5)(4)(3)(2)(1)}{(2)(1)} = 360$

Notice that ${}_8P_3 = 336$, and $\frac{8!}{5!} = 336$. Also, ${}_6P_4 = 360$ and $\frac{6!}{2!} = 360$. The result is not surprising when we expand the equation for ${}_nP_r$ as follows:

$$\begin{aligned} {}_nP_r &= \frac{n(n-1)(n-2)\cdots(n-r+1)(n-r)(n-r-1)\cdots(3)(2)(1)}{(n-r)(n-r-1)\cdots(3)(2)(1)} \\ &= \frac{n!}{(n-r)!} \end{aligned}$$

Therefore, if $n = 8$ and $r = 3$, then

$${}_8P_3 = \frac{8!}{(8-3)!} = \frac{8!}{5!} = 336$$

And if $n = 6$ and $r = 4$, then

$${}_6P_4 = \frac{6!}{(6-4)!} = \frac{6!}{2!} = 360$$

The equation for calculating the permutation of n things taken r at a time in terms of factorials is especially useful for calculators that have an $n!$ key.

EXAMPLE 13–6 Use a hand-held calculator to compute ${}_{15}P_7$.

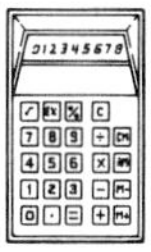

Solution: Enter 15 and depress $n!$ key. The display reads $1.307674368 \times 10^{12}$. Since $15 - 7 = 8$, enter 8 and depress $n!$ key. The display now reads 40320. Depress division key and display reads 32,432,400, the value of ${}_{15}P_7$.

EXAMPLE 13–7 Linda has 7 plants in the solarium that are suitable for moving inside the house. A new wrought iron plant stand will hold 4 plants. In how many ways can Linda arrange 4 of the 7 plants in the stand?

Solution: The word *arrange* implies that the given problem is a permutation. With $n = 7$ and $r = 4$, then

$${}_7P_4 = \frac{7!}{(7-4)!} = \frac{7(6)(5)(4)(3!)}{(3!)} = 840$$

Thus, there are 840 different plant arrangements possible. Each arrangement will differ from all other arrangements by at least one plant or the order in which the plants are placed.

Exercise Set 13–1

Use the counting principle to solve the following problems.

1. A United States social security number consists of 3 digits, then 2 digits, and finally a block of 4 digits. Assuming that any digit can be used in any spot and digits can be repeated, how many different numbers can be formed?

2. A telephone number consists of a 3-digit area code, a 3-digit prefix, and a 4-digit number. If 0 and 1 cannot be used for the first digits in the area code or the prefix, how many phone numbers can be made with this scheme of forming them?

3. The United States Postal Service uses a 5-digit zip code. Assuming that any digit can be used in any position and digits can be repeated, how many zip codes can be formed?

4. The State of California uses a pattern of 3 digits followed by 3 letters on license plates for cars. If any digit or letter may appear in any of the designated positions and digits and letters may be repeated, how many different license plates can be formed with this scheme?

5. A major department store chain uses a pattern of a letter and 7 digits on its credit cards. The letters O and Q may not be used, and 0 may not be used as the first digit. Repetition of digits is allowed. How many different credit cards can be issued by the store?

6. The medical cards of a medical group consists of 6 digits and 2 letters. The digit 0 may not be used as the first number, and digits may not be repeated. Furthermore, a given letter may not be used more than once. How many different medical identifications can be made with this scheme?

7. A baseball team has on its roster 3 catchers, 8 pitchers, 5 outfielders, and 7 infielders. How many different 9-person teams can the roster provide?

8. Kimberley S. has 3 pair of shorts, 4 halter tops, and 2 pair of sandals. Assuming any combination of shorts, tops, and sandals may be worn together, how many different outfits can she wear?

9. A restaurant offers 3 kinds of soups, 4 salad dressings, 12 entrees, 6 desserts, and 5 beverages on its dinner menu. How many different complete meals can one order?

10. In making out his fall schedule of classes, Tim must select from 5 biology 1A classes, 3 college algebra sections, 4 English 1A classes, and 2 statistics classes. Assuming that no time conflicts occur among these classes, how many different programs can he set up?

Compute the following.

11. $6!$ 12. $8!$ 13. $\frac{10!}{7!}$ 14. $\frac{5!}{0!}$

15. $\frac{12!}{8!}$ 16. $\frac{10!}{4!6!}$ 17. $\frac{52!}{5!47!}$ 18. $\frac{13!}{10!3!}$

Compute the following.

19. ${}_5P_2$ 20. ${}_{20}P_1$ 21. ${}_{52}P_0$ 22. ${}_7P_3$

23. ${}_{10}P_4$ 24. ${}_{12}P_2$ 25. ${}_5P_5$ 26. ${}_6P_6$

Compute the following permutations.

27. In how many ways can 5 books be selected and arranged from a collection of 9 books?

28. A color code chart contains 45 colors. Two different colors are to be selected for stripes on a new Datsun 280Z. How many color permutations can be formed?

29. Six different obstacles are to be used for a horse jumping contest. If all obstacles are to be used in the contest, how many different courses can be laid out if each obstacle is jumped exactly once?

30. In how many ways can five cards be selected and arranged from a deck of 52 playing cards?

31. A family portrait of Mom, Dad, and 3 children is to be taken. How many different arrangements of the family can be taken in the photograph?

32. A class of 40 students is to elect a president, secretary, and social chairperson. How many different slates of officers can be elected from the members of the class?

33. Ten pacers are competing in the feature race at Golden Bear Raceway. In how many different ways can the horses finish win, place, or show?

34. There are 7 teams in the American League East division. In how many different ways can the 7 teams finish the current baseball season?

COMBINATIONS 13–2

When performing a series of tasks, we are not always concerned with the order of the elements that results when the tasks are performed. To illustrate, suppose that in Example 13–7 Linda is concerned only with the *number of selections* of 4 plants that can be taken from the set of 7 plants available. She is unconcerned with the *number of arrangements* that can be made when the plants are placed on the plant stand. In other words, in how many ways can 4 different plants be selected from the set of 7 plants so that each *selection* will differ by at least 1 plant?

Let A, B, C, D, E, F, and G represent the 7 plants in the solarium. Then A, B, C, and D represent one selection of four plants that Linda can make. The set of plants A, B, C, and D is the same set, no matter which plant Linda selects first, which one she selects second, and so on. The only way this particular set of plants can be changed is to replace one of the plants in the group. Thus, A, B, C, E is a different selection; B, C, D, E is a different selection; and so on. But — and this point is important — D, C, B, A is the *same set of plants* as A, B, C, D.

Notice that once a specific set of plants has been selected, they can be arranged on the plant stand in 4!, or 24, different ways. (Recall that ${}_4P_4 = 4(3)(2)(1) = 24$.) However, each arrangement consists of the same 4 plants.

When r elements are selected *and* arranged from a set of n elements, we call the resulting subsets permutations. When r elements are selected from a set of n elements, we call the resulting subsets **combinations.** The relationship between the numbers of permutations and combinations of r things taken from a set of n things is given in the following theorem.

> **Theorem:** If set S contains n distinct elements, then the number of **combinations** of r of these elements (written ${}_nC_r$) is given with the equation
>
> $${}_nC_r = \frac{n!}{r!(n-r)!}, \quad \text{and} \quad r \le n$$

EXAMPLE 13–8 How many different selections of 4 plants can Linda make from the 7 plants in the solarium?

Solution: With $n = 7$, $r = 4$, and $7 - 4 = 3$, then

$$ {}_7C_4 = \frac{7!}{4!3!} = \frac{7(6)(5)(4)(3!)}{4(3)(2)(1)(3!)} = 35 $$

Thus, Linda can make 35 different selections from the set of 7 plants, using 4 at a time. Notice that each selection differs from any other selection by at least one plant.

How are the number of permutations and the number of combinations related to each other?

Recall that ${}_nP_r = \frac{n!}{(n-r)!}$. Since

$$ {}_nC_r = \frac{n!}{r!(n-r)!} = \frac{1}{r!}\left(\frac{n!}{(n-r)!}\right) $$

we can write the following equations:

$$ {}_nC_r = \frac{{}_nP_r}{r!} $$

or

$$ {}_nP_r = {}_nC_r \times r! \qquad \textbf{(1)} $$

To illustrate Equation (1), consider $S = \{a, b, c, d, e\}$. If we select 3 letters at a time from S, then

$$ {}_5C_3 = \frac{5!}{3!(5-3)!} = \frac{5(4)(3)(2)(1)}{3(2)(1)(2)(1)} = 10 $$

If we select and arrange 3 letters at a time from S, then

$$ {}_5P_3 = \frac{5!}{(5-3)!} = \frac{5(4)(3)(2)(1)}{(2)(1)} = 60 $$

In Figure 13-3 we list the 10 combinations and 60 permutations of the 5 letters of S taken 3 at a time. Notice that a given 3-letter combination is ordered in $3! = 6$ different ways. In general, *each combination of n things taken r at a time can be permuted in r! ways.*

EXAMPLE 13–9 Compute the following:

(a) ${}_5C_0$ (b) ${}_5C_1$ (c) ${}_5C_4$ (d) ${}_5C_5$

Solutions:

(a) With $n = 5$ and $r = 0$, ${}_5C_0 = \frac{5!}{0!(5-0)!} = 1$.

(b) With $n = 5$ and $r = 1$, ${}_5C_1 = \frac{5!}{1!(5-1)!} = 5$.

(c) With $n = 5$ and $r = 4$, ${}_5C_4 = \frac{5!}{4!(5-4)!} = 5$.

(d) With $n = 5$ and $r = 5$, ${}_5C_5 = \frac{5!}{5!(5-5)!} = 1$.

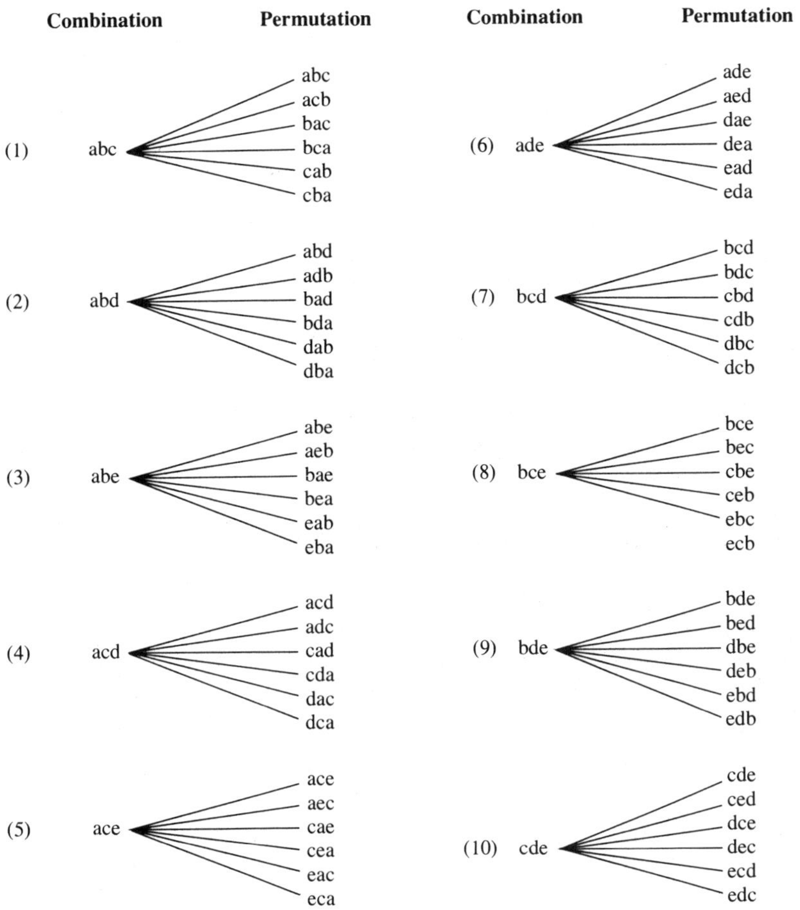

FIGURE 13-3

When we generalize the results of parts (a) and (d) of this example, we obtain the following theorem.

> **Theorem:** For any nonnegative integer n, ${}_nC_0 = 1$ and ${}_nC_n = 1$.

Notice that in parts (b) and (c) of the same example, $5 - 1 = 4$ and $5 - 4 = 1$, and ${}_5C_1 = {}_5C_4 = 5$. When we generalize these results, we obtain the following theorem:

> **Theorem:** For any nonnegative integers n and r $(r \leq n)$, ${}_nC_r = {}_nC_{n-r}$.

EXAMPLE 13–10 If $S = \{u, v, w, x, y, z\}$, how many subsets of S contain:

(a) 0 elements? (b) 1 element? (c) 2 elements?
(d) 3 elements? (e) 4 elements? (f) 5 elements?
(g) 6 elements? (h) What is the total number of subsets of S?

Solutions: Set S contains $n = 6$ elements.
(a) Substituting $r = 0$ in the preceding theorem, we get ${}_6C_0 = 1$. Recall that a set with 0 elements is the null set, and the null set is a subset of every set.
(b) With $r = 1$,

$$ {}_6C_1 = \frac{6!}{1!(6-1)!} = \frac{6(5!)}{1(5!)} = 6 $$

In general, for any positive integer n, ${}_nC_1 = n$. Any set with n elements has n subsets with exactly 1 element each.
(c) With $r = 2$,

$$ {}_6C_2 = \frac{6!}{2!(6-2)!} = \frac{6(5)(4!)}{2(1)(4!)} = 15 $$

Thus, S has 15 subsets of exactly 2 elements each.
(d) With $r = 3$,

$$ {}_6C_3 = \frac{6!}{3!(6-3)!} = \frac{6(5)(4)(3!)}{3(2)(1)(3!)} = 20 $$

(e) With $r = 4$, then $n - r = 6 - 4 = 2$. Using the preceding theorem, we get ${}_6C_4 = {}_6C_2 = 15$.
(f) With $r = 5$, then $n - r = 6 - 5 = 1$. Using the preceding theorem, we get ${}_6C_5 = {}_6C_1 = 6$.
(g) Substituting $r = 6$ in the preceding theorem, we get ${}_6C_6 = 1$. Recall that any given set is an improper subset of itself.
(h) From the results of parts (a) through (g), the total number of subsets of S = $1 + 6 + 15 + 20 + 15 + 6 + 1 = 64$. Notice that set S has 6 elements, and $64 = 2^6$. In general, for any set with n elements, the set has 2^n subsets.

Why are subsets of a set a combination of n things taken r at a time and not a permutation?

In listing the elements of a set, we do not order the elements in the set. Rearranging the elements within a given set does not create a different set. Thus, subsets are combinations and not permutations.

EXAMPLE 13–11 The faculty of a high school has 8 male members and 12 female members. A committee of 3 is to be selected to attend a national teachers' organization meeting in Washington, D.C. If the names are drawn by lots, how many different committees can be formed with (a) only female faculty members, (b) two female and one male faculty members, (c) one female and two male faculty members, and (d) only male faculty members?

Solutions:
(a) There are 12 female faculty members. If the committee of 3 consists only of women, then $n = 12$, $r = 3$, and

$$ {}_{12}C_3 = \frac{12!}{3!(12-3)!} = \frac{12(11)(10)(9!)}{3(2)(1)(9!)} = 220 $$

(b) There are 12 female faculty members and 8 male members. If the committee is composed of 2 women and 1 man, then

$$ {}_{12}C_2 = \frac{12!}{2!(12-2)!} = \frac{12(11)(10!)}{2(1)(10!)} = 66 $$

means 66 different committees of 2 women can be made from the 12 available; and ${}_8C_1 = 8$ different committees of 1 man can be made from the 8 available. With the counting principle, we get

$$ {}_{12}C_2 \times {}_8C_1 = 66(8) = 528 $$

Thus, 528 committees of 2 women and 1 man can be formed.
(c) For a committee of 1 woman and 2 men,

$$ {}_{12}C_1 = 12 \quad \text{and} \quad {}_8C_2 = \frac{8!}{2!(8-2)!} = 28 $$

Thus, we get ${}_{12}C_1 \times {}_8C_2 = 12(28) = 336$ different committees of 1 woman and 2 men.
(d) For a committee of 3 men, we get

$$ {}_8C_3 = \frac{8!}{3!(8-3)!} = \frac{8(7)(6)(5!)}{3(2)(1)(5!)} = 56 $$

different committees of 3 men.

Notice that from the results of parts (a) through (d), there are $220 + 528 + 336 + 56 = 1140$ different committees that can be formed from the $12 + 8 = 20$ faculty members. To check the results, we can compute ${}_{20}C_3$ as follows:

$$ {}_{20}C_3 = \frac{20!}{3!(20-3)!} = \frac{20(19)(18)(17!)}{3(2)(1)(17!)} = 1140 $$

Exercise Set 13–2

Compute the following combinations.

1. ${}_{99}C_0$
2. ${}_{28}C_1$
3. ${}_{40}C_2$
4. ${}_{15}C_3$
5. ${}_{10}C_4$
6. ${}_{20}C_{19}$
7. ${}_{85}C_{85}$
8. ${}_{14}C_3$
9. ${}_8C_5$
10. ${}_{75}C_{74}$
11. ${}_{12}C_6$
12. ${}_9C_5$

13. Set T contains 5 elements. Compute the number of subsets of T that contain exactly
(a) 0 elements? (b) 1 element? (c) 2 elements?
(d) 3 elements? (e) 4 elements? (f) 5 elements?
(g) What is the total number of subsets of set T?

14. Set U contains 7 elements. How many subsets of U contain exactly
(a) 0 elements? (b) 1 element? (c) 2 elements?
(d) 3 elements? (e) 4 elements? (f) 5 elements?
(g) 6 elements? (h) 7 elements?
(i) How many subsets does U have?

15. A Boy Scout troup of 12 boys is on an overnight outing. How many different "volunteers" of 3 boys can the scoutmaster select to get wood for the camp fire?

16. A bin of electrical switches in a hardware store contains 20 switches, of which 3

are faulty. A customer randomly selects 2 switches from the bin. In how many different ways can the customer select

(a) 2 switches from the bin?
(b) 2 switches with 0 faulty switches in the selection?
(c) 2 switches with 1 faulty and 1 good switch?
(d) both switches faulty?

17. A form of poker can be played in which 5 cards only are dealt to each player.
    (a) How many 5-card hands can be dealt from a deck of 52 playing cards?
    (b) How many such hands contain the ace, king, queen, jack, and ten of spades?
    (c) How many such hands contain the ace and five of hearts, the ten of spades, the two of diamonds, and the seven of clubs?

18. A chef at a large cafeteria decides to make stew with some of the leftovers in the cooler. There are 5 different meats and 8 different vegetables available. The chef wants to use 3 different meats and 5 different vegetables in the stew. How many different kinds of stew can he make?

19. A bank contains 10 dimes and 6 quarters. The bank is shaken until 3 coins fall out.
    (a) How many combinations of 3 coins each are possible?
    (b) How many of the combinations in (a) contain all dimes?
    (c) How many of the combinations in (a) contain 2 dimes and 1 quarter?
    (d) How many of the combinations in (a) contain 1 dime and 2 quarters?
    (e) How many of the combinations in (a) contain 3 quarters?

20. In the game of bridge, each player is dealt 13 cards from a deck of 52 playing cards. How many different hands are possible?

21. An extension class of a state college at a nearby military installation has 25 students, of which 16 are military personnel and 9 are civilians. Two people are randomly selected from the class. In how many different ways can the two people
    (a) be selected?
    (b) be selected and both be military people?
    (c) be selected and one be military and one civilian?
    (d) be selected and both be civilians?

22. A horse race has 10 horses competing.
    (a) How many different combinations of horses can finish first, second, and third?
    (b) How many different combinations of horses can finish first, second, and third if the favored horse wins the race?

23. The local pizza restaurant offers 10 different items that can be added to the tomato base on a pizza. How many different pizzas can be offered with
    (a) 2 additional items?
    (b) 4 additional items?
    (c) 5 different items?
    (d) all 10 items?

24. Prove that for any $n \geq 0$, ${}_nC_0 = 1$.

25. Prove that for any $n \geq 0$, ${}_nC_1 = n$.

26. Prove that for any $n \geq 0$, ${}_nC_r = {}_nC_{n-r}$.

27. Prove that for any $n \geq 0$, ${}_nC_n = 1$.

PROBABILITY OF A SINGLE EVENT 13–3

In statistics an item of information concerning an activity in which we are interested is called a **datum.** The set of two or more such items is then called **data.** The activity that generates the data is called an **experiment,** and each result of the experiment is called an **outcome.**

To illustrate the terms identified above, consider the simple activity of shaking 4 coins and then placing them on a table. Suppose we are interested in the number of heads and tails displayed on the 4 coins. The record of the number of heads and tails in a given display is called a datum. The experiment is the actual shaking and displaying of the coins. An outcome of the experiment can be a display of 4 heads and no tails or no heads and 4 tails. The repetition of the experiment generates a set of data. The set of all outcomes of the experiment is called the **sample space.** If the number of outcomes in the experiment is finite, we can often display the outcomes in a table.

EXAMPLE 13–12 Make a table of the outcomes when 4 coins are shaken and displayed.

Solution: If H is the symbol for a head displayed on a coin and T is the symbol for a tail, then the table of outcomes is shown in Figure 13-4. Notice that the counting principle verifies that 16 outcomes are possible when 4 coins are tossed. That is, the outcome on each coin is a task, each task can be performed in 2 ways (head or tail), and there are 4 tasks. Thus, there are $2(2)(2)(2) = 16$ different ways in which the 4 tasks can be performed.

	First coin	Second coin	Third coin	Fourth coin
(1)	H	H	H	H
(2)	H	H	H	T
(3)	H	H	T	H
(4)	H	H	T	T
(5)	H	T	H	H
(6)	H	T	H	T
(7)	H	T	T	H
(8)	H	T	T	T
(9)	T	H	H	H
(10)	T	H	H	T
(11)	T	H	T	H
(12)	T	H	T	T
(13)	T	T	H	H
(14)	T	T	H	T
(15)	T	T	T	H
(16)	T	T	T	T

FIGURE 13-4 Table of outcomes when four coins are tossed

Event

> **Definition:** Any subset of the set of outcomes of a given experiment is called an event.

What events can be defined for the experiment in Example 13-12?

For the experiment of displaying 4 coins, we can define the following events. We use E to designate an event, then subscript the E to distinguish between the different events for a given experiment.

E_1 = {outcomes with 4 heads}, (outcome (1) in Figure 13-4)

E_2 = {outcomes with 3 heads}, (outcomes (2), (3), (5), and (9))

E_3 = {outcomes with 2 heads}, (outcomes (4), (6), (7), (10), (11), and (13))

E_4 = {outcomes with 1 head}, (outcomes (8), (12), (14), and (15))

E_5 = {outcomes with 0 heads}, (outcome (16))

E_6 = {outcomes with 10 heads}, (no outcome satisfies this event)

Notice that an event can be defined for which the experiment yields no data. That is, we defined E_6 as the event that contains 10 heads, even though only 4 coins are used in the experiment. Thus, no outcome of the experiment will satisfy the requirements of E_6.

> **Symbol:** Let $n(E_1)$ = the number of outcomes in event E_1 and $n(S)$ = the number of outcomes in the sample space.

In the experiment of tossing 4 coins, let $n = 4$ and r = the number of coins showing a head. Then we have

$$
\begin{aligned}
{}_4C_4 &= 1 = n(E_1) \\
{}_4C_3 &= 4 = n(E_2) \\
{}_4C_2 &= 6 = n(E_3) \\
{}_4C_1 &= 4 = n(E_4) \\
{}_4C_0 &= 1 = n(E_5)
\end{aligned}
$$

Example 13–13 A bank contains 5 dimes and 3 nickels. Three coins are shaken from the bank.

(a) Compute the total number of possible outcomes.

(b) For the following events, compute $n(E)$:

(i) E_1 = {outcomes with 3 dimes}

(ii) E_2 = {outcomes with 2 dimes and 1 nickel}

(iii) E_3 = {outcomes with 1 dime and 2 nickels}

(iv) E_4 = {outcomes with 3 nickels}

Solutions:

(a) With $n = 5 + 3 = 8$ coins, and $r = 3$, then

$${}_8C_3 = \frac{8!}{3!(8-3)!} = 56$$

There are 56 different combinations of 3 coins each that can fall from the bank.

(b) (i) E_1 requires that 3 dimes be obtained from the 5 dimes in the bank. With $n = 5$ and $r = 3$, we have

$$_5C_3 = \frac{5!}{3!(5-3)!} = 10 = n(E_1)$$

Thus, 10 of the 56 outcomes will consist of exactly 3 dimes.

(ii) E_2 requires that 2 dimes be obtained from the 5 dimes and 1 nickel be obtained from the 3 nickels in the bank. We therefore use the counting principle as follows:

$$_5C_2 \times {}_3C_1 = 10(3) = 30 = n(E_2)$$

(iii) E_3 requires 1 dime and 2 nickels. Thus,

$$_5C_1 \times {}_3C_2 = 5(3) = 15 = n(E_3)$$

and 15 outcomes contain 1 dime and 2 nickels.

(iv) E_4 requires all 3 coins to be nickels. With $n = 3$ and $r = 3$, we have $_3C_3 = 1 = n(E_4)$.

Notice that we have defined every possible combination of coins for the experiment in events E_1 through E_4. That is, when 3 coins are obtained from the bank, we will always get either 3 dimes, 2 dimes and 1 nickel, 1 dime and 2 nickels, or 3 nickels. To further verify this fact, we notice that

$$n(E_1) + n(E_2) + n(E_3) + n(E_4) = 10 + 30 + 15 + 1 = 56$$

Since the total number of outcomes for the experiment is 56, we know that whenever the experiment is conducted, one of the 4 events defined must occur.

Probability of E_1

Definition: If E_1 is an event in an experiment, then the probability that E_1 will occur whenever the experiment is carried out is

$$P(E_1) = \frac{f}{t} \qquad (1)$$

where f is the number of outcomes favorable to E_1 and t is the total number of outcomes possible for the experiment.

Notice that the probability that E_1 will occur whenever the experiment is conducted requires a counting of outcomes. That is, we must somehow compute t, the total number of outcomes for the experiment. We must also determine f, the total number of the outcomes that satisfy E_1. The ratio of the whole numbers f and t is the probability that E_1 will occur. We can write $f = n(E_1)$ and $t = n(S)$, where S is the sample space for the experiment. Thus,

$$P(E_1) = \frac{n(E_1)}{n(S)} \qquad (2)$$

EXAMPLE 13–14 An experiment consists of flipping one coin and rolling one die.

(a) List the outcomes.

(b) Find $P(E_1)$, where E_1 is a head on the coin and a 3 on the die.

(c) Find $P(E_2)$, where E_2 is a head and an even number.
(d) Find $P(E_3)$, where E_3 is a tail and a number less than 5.

Solutions:
(a) With 2 outcomes on the coin (H or T) and 6 outcomes on the die (1, 2, 3, 4, 5, 6), the sample space of 12 outcomes ($n(S) = 12$) is shown in Figure 13-5.
(b) One outcome in S shows a head and 3; thus, $n(E_1) = 1$, and $P(E_1) = 1/12$.
(c) Three outcomes in S show a head and an even number, namely H–2, H–4, and H–6. Thus, $n(E_2) = 3$, $P(E_2) = 3/12$ or $1/4$.
(d) Four outcomes of S pair a tail with a number less than 5, namely T–1, T–2, T–3, and T–4. Therefore, $n(E_3) = 4$ and $P(E_4) = 4/12$ or $1/3$.

H–1	T–1
H–2	T–2
H–3	T–3
H–4	T–4
H–5	T–5
H–6	T–6

FIGURE 13-5 Display of outcomes when a coin is flipped and a die is rolled

If some event E_1 is defined such that $n(E_1) = n(S)$, then

$$P(E_1) = \frac{n(S)}{n(S)} = 1$$

An event with a probability of 1 will occur whenever the experiment is conducted.

If some event E_2 is defined such that $n(E_2) = 0$, then

$$P(E_2) = \frac{0}{n(S)} = 0$$

An event with a probability of 0 will never occur whenever the experiment is conducted.

> **Theorem:** For any event E_1 defined for some experiment, $0 \leq P(E_1) \leq 1$.

The inequality in this theorem states that the probability of any E_1 must be between 0 and 1 inclusive. We can never be more than 100 percent certain that some event will occur.

EXAMPLE 13–15 Each of the letters of the word PROBABILITY is painted on a disc, and the discs are placed in a box. You select a disc from the box without looking. Find the probability that the letter on the disc is (a) a Y, (b) a B, (c) a vowel, (d) a C, and (e) precedes the letter Z in the alphabet.

Solutions:
(a) PROBABILITY contains 11 letters; thus, $n(S) = 11$. One disc has the letter Y; thus, $P(\text{Y}) = 1/11$.

(b) Two discs have a letter B; thus, $P(\text{B}) = 2/11$.
(c) The letters A, I, and O are vowels; therefore, 4 discs contain vowels, and $P(\text{vowel}) = 4/11$.
(d) No disc has the letter C; thus, $P(\text{C}) = 0/11 = 0$.
(e) All discs have letters that precede Z in the alphabet; thus,

$$P(\text{the letter precedes Z}) = 11/11 = 1$$

In many "real life experiments" the probability of an event is obtained empirically. That is, a given experiment is observed over a period of time and the number of outcomes that satisfy an event E_1 are counted. The definition of probability is then applied to the given data to compute $P(E_1)$ for any event E_1.

How can $n(E)$ and $n(S)$ be computed when the number of outcomes is too large to be listed, or is not obvious as in the last two examples?

For example, insurance companies use the empirical method to determine rates for a policy on an individual's life, automobile, or home. In manufacturing, quality control inspections are based on empirical probabilities that predict a given number of defective items in a sample of the product. Weather forecasting is based on the data, gathered over many years, of observing what weather conditions follow from a set of meteorological conditions. For our study of probability, we will restrict our experiments to those for which $n(S)$ can be calculated directly.

If E_1 is a set of outcomes of an experiment, then $\overline{E}_1$ is the set of outcomes of the experiment that do not satisfy E_1. We call $\overline{E}_1$ the **complement** of E_1. For any event E_1,

$$n(E_1) + n(\overline{E}_1) = n(S) \qquad \textbf{(3)}$$

That is, every outcome of the experiment is either in E_1 or $\overline{E}_1$, the complement of E_1. When we divide each term of Equation (3) by $n(S)$, we obtain

$$\frac{n(E_1)}{n(S)} + \frac{n(\overline{E}_1)}{n(S)} = \frac{n(S)}{n(S)}$$

or

$$P(E_1) + P(\overline{E}_1) = 1 \qquad \textbf{(4)}$$

Equation (4) asserts that we are certain (a probability of 1) that whenever a given experiment is conducted, then *for any event E_1, either E_1 or $\overline{E}_1$ will occur every time*. We state this result in the following theorem:

> **Theorem:** If E_1 and $\overline{E}_1$ are **complementary events,** then $P(E_1) + P(\overline{E}_1) = 1$.

EXAMPLE 13–16 What is the probability of obtaining at least one tail when 5 coins are flipped?

Solution: When 5 coins are displayed, then 0, 1, 2, 3, 4, or 5 tails are displayed. If E_1 is the event that at least one tail is showing, then the outcomes in which 1, 2, 3, 4, or 5 tails are showing satisfies E_1. The complement of E_1 is the event that 0 tails are showing. Thus, $\overline{E}_1 = \{\text{outcomes showing 0 tails}\}$. When 5 coins are flipped, $n(S) = 2(2)(2)(2)(2) = 32$. For $\overline{E}_1$, $n(\overline{E}_1) = 1$, and $P(\overline{E}_1) =$

1/32. Since $P(E_1) + P(\bar{E}_1) = 1$, then $P(E_1) = 1 - P(\bar{E}_1)$; and for this experiment,

$$P(E_1) = 1 - \frac{1}{32} = \frac{31}{32}$$

Therefore, the probability that at least 1 tail is showing when 5 coins are flipped is 31/32.

Exercise Set 13–3

For the following experiments, (a) list the elements of the sample space, and (b) find the probability of the defined events.

1. A box contains 3 red balls (r_1, r_2, r_3) and 2 blue balls (b_1, b_2). Two balls are selected from the box.

   E_1 = {outcomes with 2 red balls}
   E_2 = {outcomes with 1 red and 1 blue ball}
   E_3 = {outcomes with 2 blue balls}

2. A solid with 4 triangle-shaped faces numbered 1 through 4 is tossed twice. An outcome is the sum of the two numbers obtained.

   E_1 = {outcomes with sum 3}
   E_2 = {outcomes with sum 5}
   E_3 = {outcomes with sum 8}
   E_4 = {outcomes with sum 10}

3. A family has 4 children. Let g and b represent a girl and a boy, respectively. Assume that the probabilities are the same that an individual child is a girl or a boy.

   E_1 is a family of 4 girls.
   E_2 is a family of 3 girls and 1 boy.
   E_3 is a family of 2 girls and 2 boys.
   E_4 is a family of 1 girl and 3 boys.
   E_5 is a family of 4 boys.

4. A bag contains 4 yellow tulip bulbs (y_1, y_2, y_3, y_4) and 2 red tulip bulbs (r_1, r_2). Two bulbs are selected and planted in a clay pot.

   E_1 is 2 yellow tulips.
   E_2 is 1 yellow and 1 red tulip.
   E_3 is 2 red tulips
   E_4 is 2 white tulips.

5. A coin purse has 3 dimes (d_1, d_2, d_3) and 3 nickels (n_1, n_2, n_3). Two coins are selected from the purse.

   E_1 is 2 dimes.
   E_2 is 1 dime and 1 nickel.

E_3 is a value of more than 25 cents.

E_4 is a value of less than a dollar.

6. The letters of the word EVENT are painted on discs, and the discs are placed in a bag. The two E letters are labeled E_1 and E_2. Two discs are selected.

$$E_1 = \{\text{outcomes with 2 vowels}\}$$

$$E_2 = \{\text{outcomes with 2 consonants}\}$$

7. A political subcommittee consists of 3 independents (i_1, i_2, i_3), 2 Democrats (d_1, d_2), and 1 Republican. Two members are randomly selected to head a special task force.

$$E_1 = \{\text{task forces with two independents}\}$$

$$E_2 = \{\text{task forces with two Democrats}\}$$

$$E_3 = \{\text{task forces with two Republicans}\}$$

8. A baseball player batted 4 times in last night's game. Each time at bat he either got a hit (H) or did not get a hit ($\overline{\text{H}}$). For this game, assume that his chances of getting a hit or not getting a hit were the same for each time at bat.

$$E_1 = \{\text{outcomes in which he got 4 hits}\}$$

$$E_2 = \{\text{outcomes in which he got 3 hits}\}$$

9. A deck of playing cards consists of 52 cards, of which 4 are aces, 4 are kings, and so on. There are also 4 suits — spades, hearts, clubs, and diamonds. One card is selected from the deck. Do not list the sample space, but compute the following probabilities:

(a) $P(10)$ (b) $P(\text{spade})$ (c) $P(\text{red card})$
(d) $P(\text{even-numbered card})$ (e) $P(\text{face card})$

10. List the sample space of outcomes when a pair of dice is rolled. Compute the following probabilities:
(a) $P(7)$ (b) $P(5)$ (c) $P(13)$
(d) $P(\text{odd-numbered sum})$ (e) $P(7 \text{ or } 11)$

11. At the close of trading on the New York Stock Exchange yesterday, data revealed the following information:

950 stock prices advanced

800 stock prices declined

250 stock prices remained unchanged

A stock is randomly selected from the list on the N.Y.S.E. What is the probability that
(a) it advanced in price?
(b) it remained unchanged in price?
(c) it did not advance in price?
(d) it did not remain unchanged in price?

12. A charity benefit was attended by 300 individuals, consisting of 80 men, 125 women, and 95 children. Each person's ticket was placed in a box and one was drawn for a door prize. What is the probability that the prize was won by
(a) a woman? (b) a child?
(c) not a woman? (c) not a man?

13-4 PROBABILITY OF THE UNION OF EVENTS

When an experiment is conducted and data are generated, we can define an event E_1 that contains all, some, or none of the outcomes of the experiment. In the last section we defined the probability of a single event E_1. In this section we will develop the rules of probability that enable us to compute the probability of the union of two or more events E_1, E_2, E_3, and so on. That is, we undertake the task of formulating an equation for $P(E_1 \text{ or } E_2)$, where E_1 and E_2 are defined events for the same experiment.

What is meant by the union of events?

In set theory, if A and B are two sets, then the union of A and B, written $A \cup B$, is a set that contains the elements of set A or set B. For example, if $A = \{a, b, c\}$ and $B = \{c, d, e\}$, then $A \cup B = \{a, b, c, d, e\}$.

Recall that an event E_1 is a set of outcomes favorable to E_1. Thus, the union of events E_1 and E_2 contains the outcomes of the experiment favorable to E_1 or E_2.

However, we must first recognize that our task is complicated by the fact that the word *or* has a dual purpose in the English language. Specifically, the same connective *or* is used to join two sentences with an *inclusive intent* and an *exclusive intent*. To illustrate, consider the following compound sentences:

(1) Tonight I will take my wife to dinner or to the theater.
(2) Tomorrow I will join the Army or the Navy.

In sentence (1) the inclusive intent of the word *or* is used, since it is possible that I will take my wife to dinner and also take her to the theater. Frequently we emphasize the inclusive intent by adding the phrase "or both." Thus, I might say "Tonight I will take my wife to dinner or to the theater, or both."

In sentence (2) the exclusive intent of the word is used since it is legally impossible to join both the Army and the Navy at the same time. Frequently we emphasize the exclusive intent of the word *or* by adding the phrase "but not both." Thus, I might say "I will join the Army or the Navy, but not both."

For events E_1 and E_2 of some experiment, to calculate $P(E_1 \text{ or } E_2)$ we must first determine whether E_1 and E_2 are inclusive events or exclusive events. The meanings of the terms are identified in the following definitions.

Exclusive events

> **Definition:** If E_1 and E_2 are events of some experiment, then E_1 and E_2 are exclusive events if no outcome of the experiment satisfies both events at the same time.

Inclusive events

> **Definition:** If E_1 and E_2 are events of some experiment, E_1 and E_2 are inclusive events if at least one outcome of the experiment satisfies both events at the same time.

A Venn diagram can be used to show graphically the difference between exclusive and inclusive events. In a Venn diagram we imagine that all of the outcomes of an experiment are contained in a rectangle. A closed curve is then used to enclose those outcomes that are favorable to a particular event. Figure 13-6 shows a diagram for an experiment in which the outcomes favorable to E_1 are inside the indicated circle.

If E_1 and E_2 are both events in the same experiment, then the way the closed curves are drawn indicates whether E_1 and E_2 are inclusive or exclusive. In

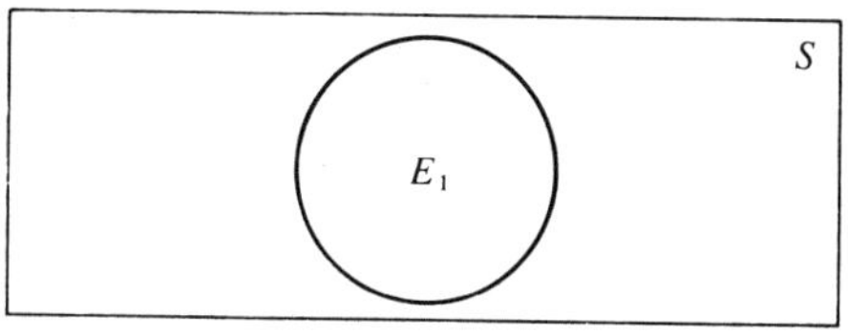

FIGURE 13-6 Venn diagram of event E_1 in a sample space S

Figure 13-7 E_1 and E_2 are exclusive events because the closed curves for E_1 and E_2 are **disjoint.** Thus, no outcome in the sample space is included in both events. We say that E_1 and E_2 are **mutually exclusive events.** In Figure 13-8 E_1 and E_2 are inclusive events because the closed curves overlap. Therefore, there is at least one outcome in the sample space that simultaneously satisfies E_1 and E_2.

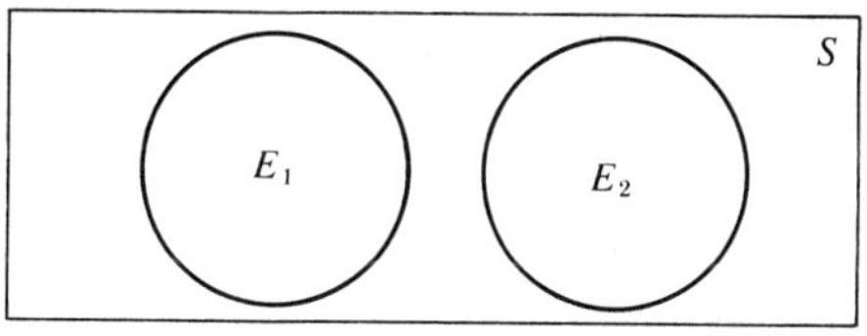

FIGURE 13-7 Venn diagram of mutually exclusive events E_1 and E_2 in S

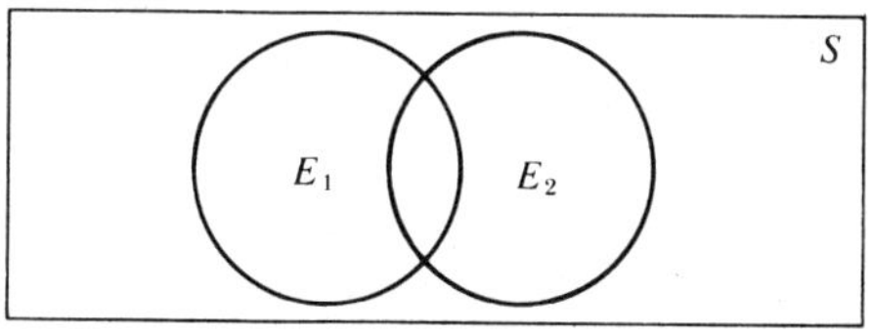

FIGURE 13-8 Venn diagram of inclusive events E_1 and E_2 in S

We can determine whether E_1 and E_2 are exclusive or inclusive by answering the question, "Can E_1 and E_2 both occur at the same time?" If the answer is no, then E_1 and E_2 are mutually exclusive events. If the answer is yes, then E_1 and E_2 are inclusive events.

How can a decision be made as to whether two events are exclusive or inclusive ?

EXAMPLE 13–17 Determine whether the following pairs of events are inclusive or exclusive.

(a) E_1: it rained today; E_2: it snowed today.
(b) E_3: the sum of 2 dice is 7; E_4: the sum of 2 dice is 11.
(c) E_5: a drawn card is a spade; E_6: a drawn card is a king.

Solutions:

(a) Since it is possible for both rain and snow to occur on a given day, E_1 and E_2 are inclusive events.
(b) A given roll of a pair of dice cannot yield a sum of 7 and 11 at the same time; thus, E_3 and E_4 are exclusive events.
(c) A given card drawn from the deck can simultaneously be a spade and a king; thus, E_5 and E_6 are inclusive events.

$P(E_1$ or $E_2)$

> **Definition:** If E_1 and E_2 are events for some experiment, then
>
> $$P(E_1 \text{ or } E_2) = \frac{n(E_1 \cup E_2)}{n(S)} \qquad \textbf{(1)}$$

In the theory of sets it is shown that

$$n(E_1 \cup E_2) = n(E_1) + n(E_2) - n(E_1 \cap E_2) \qquad \textbf{(2)}$$

where $n(E_1 \cap E_2)$ is the number of outcomes in the intersection of E_1 and E_2. In the Venn diagram, the intersection of two events is the overlapping portion of the two curves.

When we substitute Equation (2) into Equation (1), we get the following equality:

$$P(E_1 \text{ or } E_2) = \frac{n(E_1) + n(E_2) - n(E_1 \cap E_2)}{n(S)}$$
$$= \frac{n(E_1)}{n(S)} + \frac{n(E_2)}{n(S)} - \frac{n(E_1 \cap E_2)}{n(S)}$$

But $\frac{n(E_1)}{n(S)} = P(E_1)$ and $\frac{n(E_2)}{n(S)} = P(E_2)$; and we write $P(E_1$ and $E_2)$ for $\frac{n(E_1 \cap E_2)}{n(S)}$. When these substitutions are made, we obtain Equations (3) and (4) in the following theorem:

> **Theorem:** If E_1 and E_2 are events for some experiment, then
>
> $$P(E_1 \text{ or } E_2) = P(E_1) + P(E_2) - P(E_1 \text{ and } E_2) \qquad \textbf{(3)}$$
>
> for any events E_1 and E_2. If E_1 and E_2 are mutually exclusive, then
>
> $$P(E_1 \text{ or } E_2) = P(E_1) + P(E_2) \qquad \textbf{(4)}$$

Notice that Equation (4) follows from the fact that for mutually exclusive events, $E_1 \cap E_2 = \varnothing$. That is, no outcome satisfies both events simultaneously when the events are exclusive. Thus, $n(E_1 \cap E_2) = 0$, and Equation (3) reduces to the simplified form of Equation (4).

EXAMPLE 13–18 Each of the 10 digits 0 through 9 is painted on similar discs, and the discs are placed in a box. A disc is selected from the box. Find the probability that the digit on the disc is

(a) 5 or 7.
(b) an odd number or a number less than 5.
(c) a multiple of 3 or an even number.

Solutions:

(a) A given disc cannot have both a 5 and a 7 painted on it; thus, the events "the disc displays a 5" and "the disc displays a 7" are mutually exclusive. With $n(S) = 10$ and Equation (4), we get

$$P(5 \text{ or } 7) = \frac{1}{10} + \frac{1}{10} = \frac{2}{10}, \text{ or } \frac{1}{5}$$

(b) There are 5 odd numbers (1, 3, 5, 7, and 9) and 5 numbers less than 5 (0, 1, 2, 3, and 4). Since 1 and 3 are both odd numbers and less than 5, the events "the disc displays an odd number" and "the disc displays a number less than 5" are inclusive events. Furthermore, the number of outcomes in the intersection is 2. With Equation (3), we get

$$P(\text{odd number or a number less than 5}) = \frac{5}{10} + \frac{5}{10} - \frac{2}{10} = \frac{8}{10}$$

(c) The numbers 3, 6, and 9 are multiples of 3, and the numbers 0, 2, 4, 6, and 8 are even. The events are inclusive because 6 is contained in both events. With Equation (3), we get

$$P(\text{multiple of 3 or an even number}) = \frac{3}{10} + \frac{5}{10} - \frac{1}{10} = \frac{7}{10}$$

EXAMPLE 13–19 A box contains 9 discs of similar size and shape. Five of the discs are blue and 4 are red. An X is stamped on 3 blue discs and 2 red discs. A Y is stamped on 2 blue discs and 2 red discs. One disc is selected from the box. Find the following probabilities:
(a) P(blue disc or a disc with an X)
(b) P(disc with an X or a disc with a Y)

Solutions: A diagram of the sample space is shown in Figure 13-9.
(a) From the diagram we can see that

$$P(\text{blue disc}) = 5/9$$
$$P(\text{disc with an X}) = 5/9$$
$$P(\text{blue disc and disc with an X}) = 3/9$$

With Equation (3), we get

$$P(\text{blue disc or a disc with an X}) = \frac{5}{9} + \frac{5}{9} - \frac{3}{9} = \frac{7}{9}$$

Thus, the probability that a disc selected from the box is either blue (marked X or Y) or marked with an X (colored blue or red) is 7/9.
(b) The events "the disc has an X" and "the disc has a Y" are mutually exclusive. With Equation (4), we get

$$P(\text{disc with an X or a disc with a Y}) = \frac{5}{9} + \frac{4}{9} - 1$$

We are certain that each time a disc is selected, it will have an X or Y stamped on it; thus, the probability is 1.

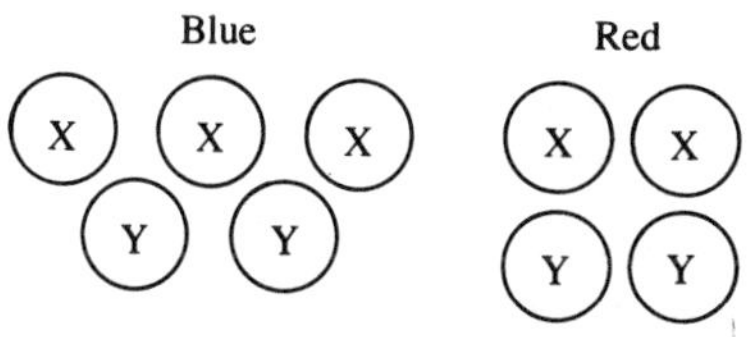

FIGURE 13-9 Display of discs for Example 13-19

EXAMPLE 13–20 In Figure 13-10, a Venn diagram displays the probabilities of inclusive events E_1 and E_2. For reference purposes the regions are labled ①, ②, ③, and ④. Use the values in the diagram to compute the following probabilities:

(a) $P(E_1)$ (b) $P(E_2)$ (c) $P(E_1 \text{ and } E_2)$
(d) $P(E_1 \text{ or } E_2)$ (e) $P(\bar{E}_1)$

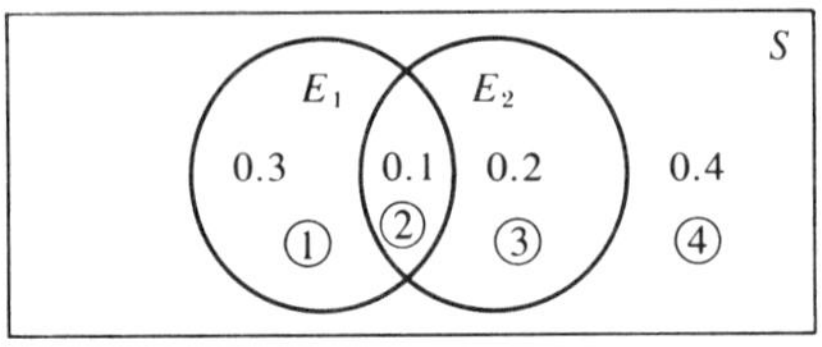

FIGURE 13-10 Display of probabilities for Example 13–20

Solutions:

(a) The outcomes of E_1 are included in regions ① and ②. Thus, $P(E_1) = 0.3 + 0.1 = 0.4$.
(b) The outcomes of E_2 are included in regions ② and ③. Therefore, $P(E_2) = 0.1 + 0.2 = 0.3$.
(c) The outcomes of E_1 and E_2 are in the intersection of the two events, namely region ②. Thus, $P(E_1 \text{ and } E_2) = 0.1$.
(d) The outcomes of E_1 or E_2 are in the union of the two events, that is, ①, ②, and ③. Hence, $P(E_1 \text{ or } E_2) = 0.3 + 0.1 + 0.2 = 0.6$.
(e) The outcomes of $\bar{E}_1$ are in the complement of E_1, that is, regions ③ and ④. Therefore, $P(\bar{E}_1) = 0.2 + 0.4 = 0.6$. Notice that $P(\bar{E}_1) = 1 - P(E_1) = 1 - 0.4 = 0.6$.

Exercise Set 13–4

Determine whether the following pairs of events are inclusive or exclusive.

1. E_1: My final grade in statistics is an A.
   E_2: My final grade in statistics is a B.

2. E_3: Ballplayer A got 2 hits in last night's game.
   E_4: Ballplayer A got 3 hits in last night's game.

3. E_5: The next 3 people I meet will all be women.
   E_6: The next 3 people I meet will all be college students.

4. E_7: Of the 10 tulip bulbs I planted, 3 will have red flowers.
   E_8: Of the 10 tulip bulbs I planted, 8 will grow.

5. E_9: The state legislature will raise the sales tax this year.
   E_{10}: The county will raise the property tax this year.

6. E_{11}: A card drawn from a deck of cards is a face card.
   E_{12}: A card drawn from a deck of cards is a heart.

7. E_{13}: A real estate salesperson sells 2 houses today.
   E_{14}: The same real estate salesperson sells 1 house today.

8. E_{15}: The flip of 3 coins yields exactly 2 heads.
   E_{16}: The same flip of 3 coins yields exactly 1 head.

9. A pair of fair dice is rolled. Find the probability that the sum of the numbers on the dice is
   (a) 5 or 9 (b) an odd number or an even number
   (c) less than 5 or greater than 9 (d) less than 5 or an even number
   (e) a multiple of 3 or an odd number

10. A card is drawn from a deck of playing cards. Find the probability that the drawn card is
    (a) a spade or a club (b) a ten or an ace
    (c) a heart or a face card (d) an even-numbered card or a club

11. A bank contains 2 quarters, 5 dimes, and 8 nickels. A coin is randomly selected from the bank. Find the probability that the coin is
    (a) a quarter or a dime (b) a quarter or a nickel
    (c) a coin worth less than 15 cents or a dime
    (d) a dime or not a quarter

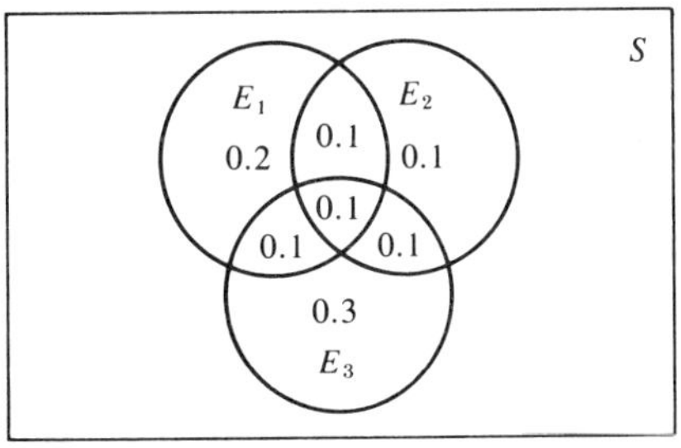

FIGURE 13-11

12. Figure 13-11 gives the probabilities for events E_1, E_2, and E_3. Use the values in the diagram to compute the following probabilities.
    (a) $P(E_1)$ (b) $P(E_2)$ (c) $P(E_3)$
    (d) $P(E_1 \text{ or } E_2)$ (e) $P(E_1 \text{ or } E_3)$ (f) $P(E_2 \text{ or } E_3)$
    (g) $P(\bar{E}_1)$ (h) $P(\bar{E}_2)$ (i) $P(E_1 \text{ or } \bar{E}_2)$
    (j) $P(E_3 \text{ or } \bar{E}_1)$ (k) $P(\bar{E}_1 \text{ or } \bar{E}_2)$

13. Figure 13-12 gives the probabilities for events E_4, E_5, and E_6. Use the values in the diagram to compute the following probabilities.
    (a) $P(E_4)$ (b) $P(E_5)$ (c) $P(E_6)$
    (d) $P(E_4 \text{ or } E_5)$ (e) $P(E_4 \text{ or } E_6)$ (f) $P(E_5 \text{ or } E_6)$
    (g) $P(\bar{E}_4)$ (h) $P(\bar{E}_5)$ (i) $P(E_4 \text{ or } \bar{E}_5)$
    (j) $P(E_6 \text{ or } \bar{E}_4)$ (k) $P(\bar{E}_4 \text{ or } \bar{E}_5)$

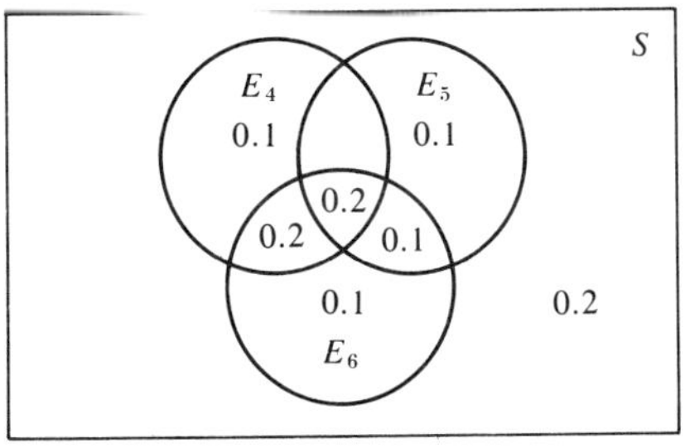

FIGURE 13-12

14. Linda's phone number and address are 654–1014 and 279 14th Street, respectively. A box contains 10 discs, and on each disc is painted a digit. If one disc is randomly selected from the box, what is the probability the disc contains a digit from
(a) Linda's phone number?
(b) Linda's address?
(c) Linda's phone number or address?
(d) Linda's phone number or not in her address?
(e) the digits not in her phone number or not in her address?

15. A disc is selected from a box that contains 10 similar discs, with one digit from 0 through 9 painted on each disc. What is the probability that the disc contains a digit
(a) that is labeled 5 or 9?
(b) that is labeled 2, 3, 5, or 7?
(c) that is labeled an even number or a number less than 4?
(d) that is labeled a multiple of 3 or an odd number?

16. The 11 letters of the word MATHEMATICS are each placed on a separate wooden disc. A disc is randomly selected from the box containing the discs. What is the probability that the letter on the disc is
(a) an M or an A?
(b) a T or a vowel?
(c) a consonant or a P?
(d) a letter in the word HAPPY or a letter in the word STATISTICS?
(e) a letter in the word ACTIVE or a letter in the word APATHETIC?

17. A mathematics class of 30 students consists of 12 women, of which 7 are white and 5 are black, and 18 men, of which 10 are white and 8 are black. One student is selected at random from the class. What is the probability that the student selected is
(a) a woman or black?
(b) a woman or white?
(c) a man or a woman?
(d) a man or black?

18. Thirty-five people who are members of the American Work Force were surveyed. It was determined that 20 were classified as "head of the household," 8 were classified "underemployed," and 3 of the "underemployed" were also "head of the household." One of these individuals is selected at random. What is the probability that the individual is
(a) head of the household or underemployed?
(b) head of the household or not underemployed?
(c) underemployed or not a head of the household?
(*Hint:* Draw a Venn diagram of the sample space of the 35 people surveyed.)

13-5 JOINT PROBABILITY

In Section 13-4 we developed Equation (3),

$$P(E_1 \text{ or } E_2) = P(E_1) + P(E_2) - P(E_1 \text{ and } E_2)$$

To compute $P(E_1 \text{ and } E_2)$, we determined $n(E_1 \cap E_2)$, the number of outcomes in the intersection of the two events. In this section we will develop an equation for computing $P(E_1 \text{ and } E_2)$ in terms of the probabilities of E_1 and E_2. Therefore, it will not be necessary to count the number of outcomes in the intersection of E_1 and E_2.

To begin, consider the experiment of flipping two legal coins A and B. What is the probability of obtaining a head on both coins? We first define the following events:

$$E_1 = \{\text{outcomes with a head on coin A}\}$$
$$E_2 = \{\text{outcomes with a head on coin B}\}$$

Figure 13-13 lists the outcomes of the experiment. We see that $n(E_1 \text{ and } E_2) = 1$; that is, only one outcome has a head on A and a head on B. Since $n(S) = 4$, then $P(E_1 \text{ and } E_2) = 1/4$.

Coin A	Coin B
H	H
H	T
T	H
T	T

FIGURE 13-13 Sample space for outcomes of coins *A* and *B*

We now make the following observations:

1. $P(E_1) = P(\text{head on coin A}) = 1/2$ and $P(E_2) = P(\text{head on coin B}) = 1/2$. Furthermore,

$$\frac{1}{2}\left(\frac{1}{2}\right) = \frac{1}{4} = P(E_1 \text{ and } E_2)$$

Thus, $P(E_1 \text{ and } E_2) = [P(E_1)][P(E_2)]$ for the experiment defined.

2. $P(E_2) = 1/2$ whether a head or a tail occurs on coin A.

Since $P(E_2)$ for coin B is the same no matter what outcome occurs on coin A, we say that E_1 and E_2 are **independent events.** In general, two events are independent if the probability of the second event (E_2) is the same whether or not the first event (E_1) has occurred.

Independent events

Definition: If E_1 and E_2 are two events for the same experiment, then E_1 and E_2 are independent events if and only if

$$P(E_2 \mid E_1) = P(E_2) \qquad \textbf{(5)}$$

where $P(E_2 \mid E_1)$ is read "the probability of event E_2 given that E_1 has already occurred."

EXAMPLE 13–21 Determine whether the following pairs of events are independent for the stated experiment.

(a) A pair of dice A and B is rolled.
$E_1 = \{\text{outcomes of a 4 on die A}\}$
$E_2 = \{\text{outcomes of a 4 on die B}\}$

(b) Individuals A and B are randomly selected in the U.S.
$E_3 = \{\text{outcomes for which A is overweight}\}$
$E_4 = \{\text{outcomes for which B is overweight}\}$

(c) An individual is randomly selected in the U.S.

$E_5 =$ {outcomes for which the individual is under 25 and has a driver's license}

$E_6 =$ {outcomes for which the individual has a driver's license and has an accident this year}

Solutions:

(a) The outcomes on die B are independent of the outcomes on die A. That is, the number obtained on die B does not change because a certain number does or does not occur on die A. Each number on B has the same probability of occurring, independent of any number that occurs on A. Thus, $P(4 \mid 4) = P(4) = 1/6$.

(b) If A and B are randomly selected, then the probability that B is overweight is independent of whether or not A is overweight. Therefore, E_3 and E_4 are independent events, and $P(\text{B is overweight} \mid \text{A is overweight}) = P(B \text{ is overweight})$.

(c) According to the empirical data collected by insurance companies, events E_5 and E_6 are dependent. That is, P(an individual has an accident this year | the person is under 25 years of age) $\neq P$(the individual has an accident this year). To verify that these events are dependent, we need only compare the cost of car insurance for people under 25 years of age with that for people over 25 years of age.

Consider now the experiment of simultaneously selecting two cards X and Y from a deck of playing cards. If

$$E_1 = \{\text{outcomes for which X is a spade}\}$$
$$E_2 = \{\text{outcomes for which Y is a spade}\}$$

what is $P(E_1 \text{ and } E_2)$? That is, what is the probability that both cards drawn are spades? We cannot list the outcomes of the sample space for E_1 and E_2 since $n(S) = {}_{52}C_2 = 1326$. However, we know that a deck of playing cards has 52 cards and 13 are spades. Thus, $P(E_1) = 13/52$, and the probability that X is a spade is 13/52.

One card is removed when E_1 occurs; therefore, only 51 cards remain when X is taken. Furthermore,

$$P(E_2) = \frac{12}{51}, \quad \text{if X is a spade}$$

$$P(E_2) = \frac{13}{51}, \quad \text{if X is not a spade}$$

That is, the probability that Y is a spade is different, depending on whether or not X is a spade. Thus, for the stated experiment, E_1 and E_2 are dependent events.

What probability value is used for E_2 when there can be two values?

For our experiment we want both X and Y to be spades. If X, the first card drawn, is not a spade, then we simply replace the card and keep drawing until X is a spade. Then, of the remaining 51 cards, 12 are spades and $P(E_2) = 12/51$.

For the given experiment,

$$P(\text{spade and spade}) = P(\text{X is a spade})\,[P(\text{Y is a spade} \mid \text{X is a spade})]$$
$$= \frac{13}{52}\left(\frac{12}{51}\right) = \frac{156}{2652}, \text{ or } \frac{1}{17}$$

This example illustrates the **product formula** for computing the joint probability of two events.

> **Theorem: Multiplication Rule of Probability** If E_1 and E_2 are two events for the same experiment, then the joint probability of E_1 and E_2 is given by the equation
>
> $$P(E_1 \text{ and } E_2) = [P(E_1)][P(E_2 \mid E_1)] \qquad \textbf{(6)}$$

In the definition of independent events, we stated that $P(E_2 \mid E_1) = P(E_2)$. Therefore, for independent events E_1 and E_2, Equation (6) is simplified to the form of Equation (7).

> **Theorem:** If E_1 and E_2 are independent events for the same experiment, then the joint probability of E_1 and E_2 is given by the equation
>
> $$P(E_1 \text{ and } E_2) = [P(E_1)][P(E_2)] \qquad \textbf{(7)}$$

There is a way in which the two events in the experiment of drawing two cards can be made independent. If we replace card X before we draw card Y, then E_1 and E_2 are independent events. That is, replacing the first card changes the probability that the second card is also a spade, and $P(\text{Y is a spade}) = 13/52$. The effect of sampling *with replacement* and *without replacement* on the probabilities of successive events is studied in the next examples.

Can the experiment of drawing two cards be conducted so that the events are independent?

EXAMPLE 13–22 A bank contains 1 quarter (q), 2 dimes (d_1, d_2), and 3 nickels (n_1, n_2, n_3). Two coins are selected without replacement from the bank. Find the following probabilities:

(a) $P(\text{2 dimes})$ (b) $P(\text{2 nickels})$ (c) $P(\text{a quarter and a dime})$

Solutions: For a visual aid, the 30 outcomes of the sample space are shown in Figure 13-14.

(1)	qd_1	(6)	d_1q	(11)	d_2q	(16)	n_1q	(21)	n_2q	(26)	n_3q
(2)	qd_2	(7)	d_1d_2	(12)	d_2d_1	(17)	n_1d_1	(22)	n_2d_1	(27)	n_3d_1
(3)	qn_1	(8)	d_1n_1	(13)	d_2n_1	(18)	n_1d_2	(23)	n_2d_2	(28)	n_3d_2
(4)	qn_2	(9)	d_1n_2	(14)	d_2n_2	(19)	n_1n_2	(24)	n_2n_1	(29)	n_3n_1
(5)	qn_3	(10)	d_1n_3	(15)	d_2n_3	(20)	n_1n_3	(25)	n_2n_3	(30)	n_3n_2

FIGURE 13-14 Outcomes of successive draws of two coins from box containing 1 quarter, 2 dimes, and 3 nickels

(a)
$$\begin{aligned} P(\text{2 dimes}) &= P(\text{a dime and a dime}) \\ &= P(\text{dime})[P(\text{dime} \mid \text{dime})] \\ &= \frac{2}{6}\left(\frac{1}{5}\right) = \frac{1}{15} \end{aligned}$$

There are 2 dimes in the 6 coins, thus $P(\text{dime}) = 2/6$. When 1 dime is removed, then 1 of the 5 remaining coins is a dime, and $P(\text{dime} \mid \text{dime}) = 1/5$. By the multiplication rule, $P(\text{2 dimes})$ is the product of the two probabilities. In Figure 13-14 we see that outcomes (7) and (12) consist of 2 dimes, and $2/30 = 1/15$ as computed.

(b)
$$\begin{aligned} P(\text{2 nickels}) &= P(\text{a nickel and a nickel}) \\ &= P(\text{nickel})[P(\text{nickel} \mid \text{nickel})] \\ &= \frac{3}{6}\left(\frac{2}{5}\right) = \frac{1}{5} \end{aligned}$$

Since 3 of the 6 coins are nickels, then $P(\text{nickel}) = 3/6$. When 1 nickel is removed, then 2 of the remaining coins are nickels, and $P(\text{nickel} \mid \text{nickel}) = 2/5$. Thus, $P(\text{2 nickels}) = \frac{3}{6}\left(\frac{2}{5}\right) = \frac{1}{5}$. In Figure 13-14, outcomes (19), (20), (24), (25), (29), and 30 contain 2 nickels, and $P(\text{2 nickels}) = 1/5$ as computed.

(c) For the selected coins, let q and d represent a quarter and a dime, respectively. Then $P(\text{a quarter and a dime}) = P(\text{q and d})$. We must first recognize that there are *two* events that satisfy the given condition: E_1: the first coin is a quarter and the second coin is a dime; and E_2: the first coin is a dime and the second coin is a quarter. Thus,

$$\begin{aligned} P(\text{q and d}) &= P(E_1 \text{ or } E_2) \\ &= P(E_1) + P(E_2) \end{aligned}$$

Notice that E_1 and E_2 are mutually exclusive events since only one of the two sequences of selecting the coins can occur at any one time. Now

$$P(E_1) = P(\text{q})[P(\text{d} \mid \text{q})] = \left(\frac{1}{6}\right)\left(\frac{2}{5}\right) = \frac{1}{15}$$

$$P(E_2) = P(\text{d})[P(\text{q} \mid \text{d})] = \left(\frac{2}{6}\right)\left(\frac{1}{5}\right) = \frac{1}{15}$$

Therefore,

$$P(\text{q and d}) = \frac{1}{15} + \frac{1}{15} = \frac{2}{15}$$

In Figure 13-14, outcomes (1), (2), (6), and (11) contain a quarter and a dime, and $4/30 = 2/15$ as computed. It helps to think of the given probability problem as

$$\begin{aligned} P(\text{q and d}) &= P((\text{q and d}) \text{ or } (\text{d and q})) \\ &= P(\text{q})[P(\text{d} \mid \text{q})] + P(\text{d})[P(\text{q} \mid \text{d})] \end{aligned}$$

When the values of the four indicated probabilities are substituted in the equation, then

$$P(\text{q and d}) = \left(\frac{1}{6}\right)\left(\frac{2}{5}\right) + \left(\frac{2}{6}\right)\left(\frac{1}{5}\right) = \frac{2}{15}$$

EXAMPLE 13–23 A bank contains 1 quarter (q), 2 dimes (d_1, d_2), and 3 nickels (n_1, n_2, n_3). Two coins are selected *with replacement*. Find the following probabilities:

(a) $P(\text{2 dimes})$ (b) $P(\text{2 nickels})$ (c) $P(\text{a quarter and a dime})$

Solutions: Since the coins are selected with replacement (the first coin is returned to the bank before the second one is selected), then $n(S) = 6(6) = 36$. Furthermore, the two events are now independent. That is, $P(\text{d} \mid \text{d}) = P(\text{d})$, $P(\text{d} \mid \text{q}) = P(\text{d})$, and so on.

(a) $P(\text{a dime and a dime}) = P(\text{dime})[P(\text{dime})] = \left(\frac{2}{6}\right)\left(\frac{2}{6}\right) = \frac{1}{9}$

(b) $P(\text{a nickel and a nickel}) = P(\text{nickel})[P(\text{nickel})] = \left(\frac{3}{6}\right)\left(\frac{3}{6}\right) = \frac{1}{4}$

(c) $P(\text{a quarter and a dime}) = P(\text{q and d}) = P(\text{q})[P(\text{d})] + P(\text{d})[P(\text{q})]$

$$= \left(\frac{1}{6}\right)\left(\frac{2}{6}\right) + \left(\frac{2}{6}\right)\left(\frac{1}{6}\right) = \frac{2}{18}, \text{ or } \frac{1}{9}$$

We can use combinations to compute the probabilities in Example 13–22, since the coins are selected without replacement. Recall that the bank has 1 quarter, 2 dimes, and 3 nickels. For each probability, 2 coins are selected. With $n = 6$ and $r = 2$,

Can combinations be used to compute the probabilities of Examples 13-22 and 13-23?

$${}_6C_2 = \frac{6!}{2!(6-2)!} = \frac{6(5)}{2(1)} = 15$$

(a) $P(\text{2 dimes}) = \frac{{}_2C_2}{{}_6C_2} = \frac{1}{15}$, since ${}_2C_2 = 1$

(b) $P(\text{2 nickels}) = \frac{{}_3C_2}{{}_6C_2} = \frac{3}{15}$ or $\frac{1}{5}$, since ${}_3C_2 = 3$

(c) $P(\text{a quarter and a dime}) = \frac{{}_1C_1 \times {}_2C_1}{{}_6C_2} = \frac{1(2)}{15} = \frac{2}{15}$

Notice that we want 1 quarter out of 1 quarter (thus, ${}_1C_1$) and 1 dime out of 2 dimes (thus, ${}_2C_1$). The counting principle is used to multiply the values of the two combinations.

EXAMPLE 13–24 A bin at a hardware store contains 20 electrical outlet devices. Four of the devices are faulty. If a customer randomly selects 3 to buy, what is the probability that in the purchase

(a) none of the devices are faulty?
(b) one device is faulty?
(c) two devices are faulty?
(d) all three devices are faulty?
(e) What is the sum of the probabilities (a) through (d)?

Solutions: The customer buys 3 outlet devices from 20 available. The number of different ways in which the 3 items can be selected is

$${}_{20}C_3 = \frac{20!}{3!(20-3)!} = 1140$$

For the following probabilities, let g represent a good device and f represent a faulty device.

(a) $P(\text{0 faulty devices}) = P(\text{3 good devices}) = P(\text{g + g + g})$. There are 16 good devices in the bin; thus,

$$P(\text{g and g and g}) = \left(\frac{16}{20}\right)\left(\frac{15}{19}\right)\left(\frac{14}{18}\right) = \frac{28}{57}$$

Notice that the selections are made without replacement. Furthermore,

$$P(\text{3 good devices}) = \frac{{}_{16}C_3}{{}_{20}C_3} = \frac{560}{1140}, \text{ or } \frac{28}{57}$$

(b) $P(\text{1 faulty device}) = P(\text{1 faulty and 2 good})$
$= P((\text{f} + \text{g} + \text{g})\ or\ (\text{g} + \text{f} + \text{g})\ or\ (\text{g} + \text{g} + \text{f}))$

There are 16 good and 4 faulty devices in the bin; thus,

$$P(\text{1 faulty and 2 good}) = \left(\frac{4}{20}\right)\left(\frac{16}{19}\right)\left(\frac{15}{18}\right) + \left(\frac{16}{20}\right)\left(\frac{4}{19}\right)\left(\frac{15}{18}\right) + \left(\frac{16}{20}\right)\left(\frac{15}{19}\right)\left(\frac{4}{18}\right)$$
$$= \frac{8}{57} + \frac{8}{57} + \frac{8}{57} = \frac{24}{57}$$

Notice that each sequence of faulty and good devices qualify as mutually exclusive events. Furthermore,

$$P(\text{1 faulty and 2 good}) = \frac{{}_4C_1 \times {}_{16}C_2}{{}_{20}C_3} = \frac{4(120)}{1140} = \frac{24}{57}$$

(c) $P(\text{2 faulty devices}) = P(\text{2 faulty and 1 good})$
$= P((\text{f} + \text{f} + \text{g})\ or\ (\text{f} + \text{g} + \text{f})\ or\ (\text{g} + \text{f} + \text{f}))$

Thus,

$$P(\text{2 faulty and 1 good}) = \left(\frac{4}{20}\right)\left(\frac{3}{19}\right)\left(\frac{16}{18}\right) + \left(\frac{4}{20}\right)\left(\frac{16}{19}\right)\left(\frac{3}{18}\right) + \left(\frac{16}{20}\right)\left(\frac{4}{19}\right)\left(\frac{3}{18}\right)$$
$$= \frac{8}{285} + \frac{8}{285} + \frac{8}{285} = \frac{24}{285}$$

Also,

$$P(\text{2 faulty and 1 good}) = \frac{{}_4C_2 \times {}_{16}C_1}{{}_{20}C_3} = \frac{6(16)}{1140} = \frac{24}{285}$$

(d) $$P(\text{3 faulty devices}) = P(\text{f} + \text{f} + \text{f}) = \left(\frac{4}{20}\right)\left(\frac{3}{19}\right)\left(\frac{2}{18}\right) = \frac{1}{285}$$

Also,

$$P(\text{3 faulty devices}) = \frac{{}_4C_3}{{}_{20}C_3} = \frac{4}{1140} = \frac{1}{285}$$

(e) The sum of the probabilities (a) through (d) is

$$\frac{28}{57} + \frac{24}{57} + \frac{24}{285} + \frac{1}{285} = \frac{285}{285},\ \text{or } 1$$

The probability of 1 is not surprising, since we are certain that for any selection of 3 devices from the bin, any customer will get either (a) 3 good devices, (b) 2 good and 1 faulty device, (c) 1 good and 2 faulty devices, or (d) 3 faulty devices. Notice also that computing the probabilities with combinations and the counting principle is easier than using the different sequences in which a given outcome can occur. Either method can be used, and we are free to use whichever method is easier for us to comprehend.

Exercise Set 13–5

Determine which of the following pairs of events are independent.

1. E_1: A student gets A grades in English classes.
   E_2: A student gets A grades in mathematics classes.

2. E_3: A student driver has a grade point average over 3.0.
   E_4: A student driver has a chargeable accident this year.

3. E_5: A male over 30 years of age smokes cigarettes.
   E_6: A male over 30 years of age has a heart attack.

4. E_7: A ballplayer gets a base hit the first time at bat.
   E_8: The same ballplayer gets a base hit the second time at bat.

5. E_9: A randomly selected digit is an even number.
   E_{10}: A randomly selected digit is greater than 4.

6. E_{11}: A letter selected from the set $\{a, b, c, d, e\}$ is a letter in the word *candy*.
   E_{12}: A letter selected from the set $\{a, b, c, d, e\}$ is a letter in the word *glad*.

7. E_{13}: The sum of the numbers on a pair of dice is 7.
   E_{14}: The sum of the numbers on a pair of dice is 7 when one of the dice is known to be showing a 5.

8. E_{15}: A vehicle stopped at an inspection station has a faulty emission system.
   E_{16}: A second vehicle stopped at the same inspection station has a faulty emission system.

9. An experiment requires that a number be selected from the set $S = \{0, 1, 2, 3, 4, 5\}$. Compute the following probabilities.
   (a) P(even number) (b) P(number less than 3)
   (c) P(even number and a number less than 3)
   (d) Are the two defined events independent?

10. One card is selected from a deck of playing cards. Find the probability that the given card is
    (a) an ace and a spade (b) a heart and a face card
    (c) a club and a card numbered less than 5
    (d) a club and a diamond

11. Two cards are selected without replacement from a deck of playing cards. Find the probability that both cards are
    (a) spades (b) face cards
    (c) aces (d) face cards and hearts
    (e) ten of hearts

12. A mathematics class has 8 girls and 7 boys. Two students are randomly selected. What is the probability that
    (a) both students are girls?
    (b) one student is a girl and one is a boy?
    (c) both students are boys?
    (d) What is the sum of the probabilities of parts (a) through (c)?

13. A neighborhood organization contains 20 registered Democrats and 15 registered Republicans. Three people are randomly selected to serve on a committee. What is the probability that
    (a) all three people are Democrats?
    (b) two are Democrats and one is a Republican?
    (c) one is a Democrat and two are Republicans?
    (d) all three people are Republicans?
    (e) What is the sum of the probabilities of parts (a) through (d)?

14. A salesperson has a record of making a sale to 30 percent of the individuals she contacts. She contacts 3 people. What is the probability that

(a) 3 sales are made? (b) 2 sales are made?
(c) 1 sale is made? (d) no sales are made?
(e) What is the sum of the probabilities of (a) through (d)?

15. The probabilities for events E_1, E_2, and E_3 are displayed in Figure 13-15. Use the indicated probabilities to compute the following:
(a) $P(E_1)$ (b) $P(E_2)$ (c) $P(E_3)$
(d) $P(E_1 \text{ or } E_2)$ (e) $P(E_1 \text{ or } E_3)$ (f) $P(E_2 \text{ or } E_3)$
(g) $P(E_1 \text{ and } E_2)$ (h) $P(E_1 \text{ and } E_3)$ (i) $P(E_2 \text{ and } E_3)$
(j) Which, if any, of the events are independent? (Recall that if E_1 and E_2 are independent, then $P(E_1 \text{ and } E_2) = P(E_1)[P(E_2)]$.)

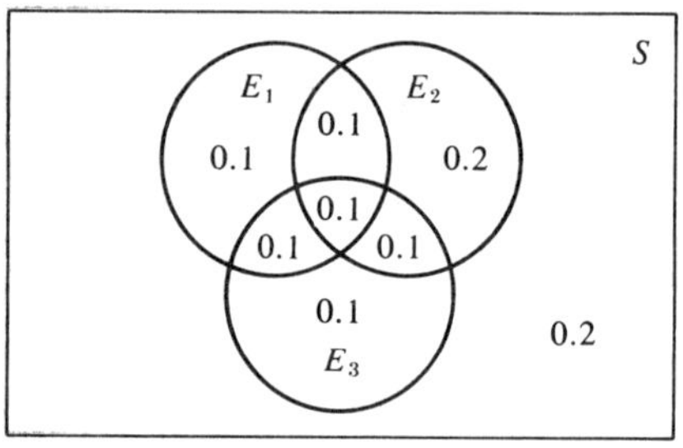

FIGURE 13-15

16. The probabilities for events E_4, E_5, and E_6 are displayed in Figure 13-16. Use the indicated probabilities to compute the following:
(a) $P(E_4)$ (b) $P(E_5)$ (c) $P(E_6)$
(d) $P(E_4 \text{ or } E_5)$ (e) $P(E_4 \text{ or } E_6)$ (f) $P(E_5 \text{ or } E_6)$
(g) $P(E_4 \text{ and } E_5)$ (h) $P(E_5 \text{ and } E_6)$ (i) $P(E_4 \text{ and } E_6)$
(j) Which, if any, of the events are independent?

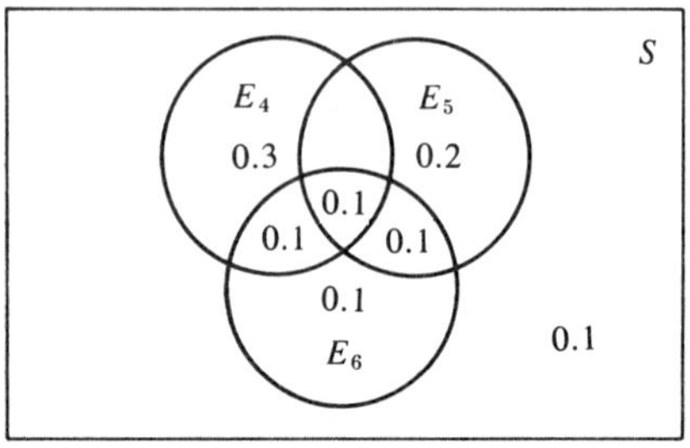

FIGURE 13-16

17. Eric's bank contains 3 quarters, 7 dimes, and 10 nickels. He shakes 2 coins from the bank. What is the probability that the 2 coins are
(a) both quarters? (b) both dimes?
(c) both nickels? (d) a quarter and a dime?
(e) a quarter and a nickel? (f) a dime and a nickel?
(g) What is the sum of the probabilities of (a) through (f)?

18. Of the issues traded on the New York Stock Exchange one day, it was reported that 800 stock prices advanced, 500 stock prices declined, and 200 stock prices remained unchanged. Two stocks are randomly selected from the list. What is the probability that the prices of the stocks selected
(a) both advanced? (b) both declined?
(c) both remained unchanged? (d) one advanced and one declined?

(e) one advanced and one remained unchanged?
(f) one declined and one remained unchanged?
(g) What is the sum of the probabilities of (a) through (f)?

19. A shipment of 30 television sets to a local retail outlet has 8 sets with some defect. A delivery person randomly selects 3 sets for delivery to customers. What is the probability that the delivery person selects
(a) 3 nondefective sets? (b) 2 nondefective sets?
(c) 1 nondefective set? (d) all defective sets?
(e) What is the sum of the probabilities (a) through (d)?

20. An avid bowler has computed the probability of getting a strike on a given ball (that is, knocking all ten pins down with the first ball) as 0.35. Compute to three decimal places the probability that in the next 3 frames the bowler has
(a) 3 strikes (b) 2 strikes (c) 1 strike (d) 0 strikes
(e) What is the sum of the probabilities (a) through (d)?

THE BINOMIAL THEOREM 13–6

Frequently in mathematics we need to expand a **bionomial expression.** By definition, a binomial expression contains two terms. If we let a and b be the first and second terms, respectively, then $a + b$ represents any general binomial expression.

In this section we have two goals: (1) to develop a formula for writing the expanded form of $(a + b)^n$ for any positive integer n, and (2) to develop an equation to write the rth term in the expansion for any positive integer $r \leq n$.

To help us achieve the first goal, we write the expanded form of $(a + b)^n$ for $n = 1, 2, 3, 4$, and 5. We can verify each of the five equations by simply multiplying $(a + b)$ by itself the indicated number of times. To establish patterns for the coefficients and exponents in the expanded forms, we insert the coefficient 1 and exponent 1 where needed:

$$(a + b)^1 = 1a^1 + 1b^1$$
$$(a + b)^2 = 1a^2 + 2a^1b^1 + 1b^2$$
$$(a + b)^3 = 1a^3 + 3a^2b^1 + 3a^1b^2 + 1b^3$$
$$(a + b)^4 = 1a^4 + 4a^3b^1 + 6a^2b^2 + 4a^1b^3 + 1b^4$$
$$(a + b)^5 = 1a^5 + 5a^4b^1 + 10a^3b^2 + 10a^2b^3 + 5a^1b^4 + 1b^5$$

We now make the following observations from the above expansions:

Note 1. The first and last terms in each expansion are a^n and b^n, respectively. For example, the first and last terms in $(a + b)^4$ are a^4 and b^4, respectively.

Note 2. For $a \neq 0$ and $b \neq 0$, $a^0 = b^0 = 1$. Therefore, in each expansion the first and last terms can be written as $a^n(b^0)$ and $a^0(b^n)$, respectively. The sums of the exponents on a and b in each term of the expansions are therefore n. For example, in $(a + b)^4$, the sums of the exponents are $4 + 0 = 4$, $3 + 1 = 4$, $2 + 2 = 4$, $1 + 3 = 4$, and $0 + 4 = 4$, respectively.

Note 3. Each expansion of $(a + b)^n$ contains $n + 1$ terms. For example, when $n = 3$, the expansion contains $3 + 1 = 4$ terms.

Note 4. Each term in the expansions contains 3 factors: (1) a power of a, called the **a factor;** (2) a power of b, called the **b factor;** and (3) a numerical coefficient.

Note 5. For successive terms, the exponents on the a factors decrease according to the pattern

$$n, n - 1, n - 2, \ldots, 0$$

For example, the 5 terms in the expanded form of $(a + b)^4$ have a factors a^4, a^3, a^2, a^1, and a^0, respectively.

Note 6. For successive terms, the exponents on the b factors increase according to the pattern

$$0, 1, 2, \ldots, n$$

For example, the 5 terms in the expanded form of $(a + b)^4$ have b factors b^0, b^1, b^2, b^3, and b^4, respectively.

Note 7. The numerical coefficients in the expanded forms are obtained with the formula for computing the combination of n things taken r at a time, written ${}_nC_r$.

To illustrate, let k represent the *position* of any term in the expanded form of $(a + b)^n$; thus, $k \in \{1, 2, 3, \ldots, n + 1\}$. Consider now the six coefficients in the expanded form of $(a + b)^5$, that is, $n = 5$:

In term 1, $k = 1$ and ${}_5C_0 = 1$; thus, $r = k - 1 = 0$.
In term 2, $k = 2$ and ${}_5C_1 = 5$; thus, $r = k - 1 = 1$.
In term 3, $k = 3$ and ${}_5C_2 = 10$; thus, $r = k - 1 = 2$.
In term 4, $k = 4$ and ${}_5C_3 = 10$; thus, $r = k - 1 = 3$.
In term 5, $k = 5$ and ${}_5C_4 = 5$; thus, $r = k - 1 = 4$.
In term 6, $k = 6$ and ${}_5C_5 = 1$; thus, $r = k - 1 = 5$.

In each instance, the value of r in ${}_5C_r$ is one less than k, the number of the position of the term in the expanded form of $(a + b)^5$. In general, to obtain the numerical coefficient of the kth term in the expanded form of $(a + b)^n$ and $1 \leq k \leq n + 1$, we compute ${}_nC_r$ and $r = k - 1$.

When we summarize the details of these comments, we obtain the following theorem.

Binomial Theorem: For any positive integer n,

$$(a + b)^n = {}_nC_0(a^n) + {}_nC_1(a^{n-1})(b) + \cdots + {}_nC_{n-1}(a)(b^{n-1}) + {}_nC_n(b^n)$$

Is the value of r in ${}_nC_r$ and the exponent on the b factor the same for each term of the expansion?

As we just observed, for the kth term in the expansion, $r = k - 1$. Furthermore, the exponent on the b factor of the kth term is also $k - 1$; thus, the two values are the same for each term. Since ${}_nC_0 = {}_nC_n = 1$, we can eliminate these factors to simplify the theorem to the following form:

$$(a + b)^n = a^n + {}_nC_1(a^{n-1})(b) + \cdots + {}_nC_{n-1}(a)(b^{n-1}) + b^n$$

EXAMPLE 13–26 Expand and simplify $(x + 2)^6$.

Solution: With $a = x$, $b = 2$, and the Binomial Theorem, we have

$$(x + 2)^6 = x^6 + {}_6C_1(x^5)(2) + {}_6C_2(x^4)(2^2) + {}_6C_3(x^3)(2^3) + {}_6C_4(x^2)(2^4) + {}_6C_5(x)(2^5) + 2^6$$

Therefore,

$$\begin{aligned}(x + 2)^6 &= x^6 + 6(x^5)(2) + 15(x^4)(4) + 20(x^3)(8) + 15(x^2)(16) + 6(x)(32) + 64 \\ &= x^6 + 12x^5 + 60x^4 + 160x^3 + 240x^2 + 192x + 64\end{aligned}$$

EXAMPLE 13–27 Expand and simplify $(3x - 2y)^4$.

Solution: With $a = 3x$, $b = (-2y)$, and the Binomial Theorem, we have

$$\begin{aligned}(3x - 2y)^4 &= (3x)^4 + {}_4C_1(3x)^3(-2y) + {}_4C_2(3x)^2(-2y)^2 + {}_4C_3(3x)(-2y)^3 + (-2y)^4 \\ &= 81x^4 - 4(27x^3)(2y) + 6(9x^2)(4y^2) - 4(3x)(8y^3) + 16y^4 \\ &= 81x^4 - 216x^3y + 216x^2y^2 - 96xy^3 + 16y^4\end{aligned}$$

We can quickly check the exponents on x and y in each term to see that their sum in each case is $n = 4$. Also, a quick check indicates that the expansion contains $n + 1 = 5$ terms.

> **Theorem:** The kth term of the expanded form of $(a + b)^n$, where $1 \le k \le n + 1$, has the form
>
> $${}_nC_{k-1}(a^{n-k+1})(b^{k-1})$$

Why is the exponent on the a factor $n-k+1$?

We have made a couple of observations that lead to the fact that the exponent on the a factor is $n - k + 1$. First, we noted that the value of r on ${}_nC_r$ in the kth term of the expanded form is always $k - 1$. Thus, for the kth term, we write ${}_nC_{k-1}$.

We noted further that the exponent on the b factor in each term is the value of r. For the kth term, $r = k - 1$ and the exponent on the b factor is $k - 1$. Finally, we have noted that the sum of the exponents for the a and b factors for each term must be n. If for the kth term we let the exponent of the a factor be z, then

$$\begin{aligned}z + (k - 1) &= n \\ z &= n - k + 1\end{aligned}$$

Thus, the exponent on the a factor of the kth term is $n - k + 1$.

EXAMPLE 13–28 Find the seventh term in the expanded form of $(3s - t/2)^{10}$.

Solution: For the given binomial expression, $a = 3s$, $b = -t/2$, and $n = 10$. If $k = 7$, then $k - 1 = 6$ and $n - k + 1 = 4$. The following equality follows from the preceding theorem:

$$
\begin{aligned}
{}_nC_{k-1}(a^{n-k+1})(b^{k-1}) &= {}_{10}C_6(3s)^4\left(\frac{-t}{2}\right)^6 \\
&= 210(81s^4)\left(\frac{t^6}{64}\right) \\
&= \frac{8505s^4t^6}{32}
\end{aligned}
$$

A certain triangular array of positive integers is called **Pascal's Triangle.** The first 7 rows in the array are shown in Figure 13-17. Notice that the array can be generated by using the following rules:

1. Begin and end each row n with a 1.
2. An interior number is the sum of the numbers to the immediate left and right of the number in the preceding row.

1
1 1
1 2 1
1 3 3 1
1 4 6 4 1
1 5 10 10 5 1
1 6 15 20 15 6 1

FIGURE 13-17 The first seven rows in Pascal's Triangle

To illustrate these rules, we obtain the 8 numbers in the next row of the triangle in Figure 13-18 as follows:

First number is 1

Second number is $1 + 6 = 7$

Third number is $6 + 15 = 21$

Fourth number is $15 + 20 = 35$

Fifth number is $20 + 15 = 35$

Sixth number is $15 + 6 = 21$

Seventh number is $6 + 1 = 7$

Eighth number is 1

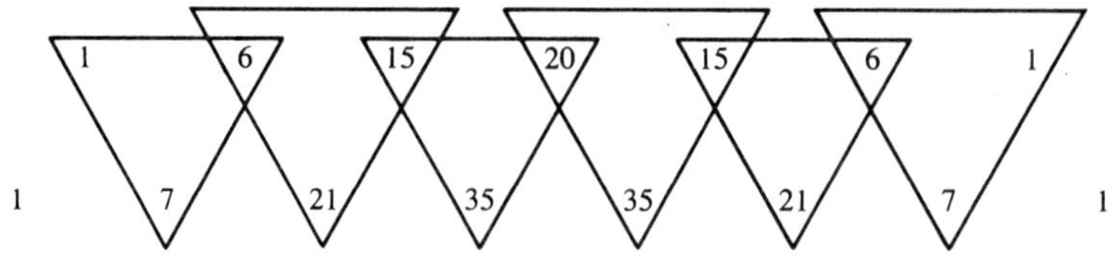

FIGURE 13-18 The numbers in row 8 of Pascal's Triangle

Also, for $n = 7$ and $r \in \{0, 1, 2, \ldots, 7\}$, we obtain the following values:

$$_7C_0 = 1$$
$$_7C_1 = 7$$
$$_7C_2 = 21$$
$$_7C_3 = 35$$
$$_7C_4 = 35$$
$$_7C_5 = 21$$
$$_7C_6 = 7$$
$$_7C_7 = 1$$

When the values of the combinations are compared with the numbers in row 8 of Pascal's Triangle, we see that they are the same. Consequently, each number in the triangle can also be determined with the formula for $_nC_r$.

For row 1 in the triangle, $n = 0$. For row 2, $n = 1$; and, in general, for row c, $n = c - 1$. Notice also that after row 1, the second number from the left in any row is the value of n.

What determines the value of n to use to generate the numbers in a given row of the triangle?

Therefore, for the 15th row of the triangle, $c = 15$ and $n = 15 - 1 = 14$. There are 15 numbers in the row, and each number is obtained as follows:

$$_{14}C_0 \quad _{14}C_1 \quad _{14}C_2 \quad \cdots \quad _{14}C_{14}$$

In the expanded form of $(a + b)^n$, we need to determine the **numerical coefficient factor** of each term in the expression. *The values of the numerical coefficients are the numbers in Pascal's Triangle.* Therefore, we can use the array to write the numerical coefficients for the expansion of $(a + b)^n$ within reasonable values of n. We use the following example to illustrate.

EXAMPLE 13–29 Write the expanded form of $(x + y)^{10}$

Solution: In Figure 13-19, the values of the triangle for $n = 10$ are shown. Since $a = x$ and $b = y$, the expanded form of $(x + y)^{10}$ can be written by using the entries in the table and by observing the pattern of exponents on x and y. That is,

$$(x + y)^{10} = x^{10} + 10x^9y + 45x^8y^2 + 120x^7y^3 + 210x^6y^4 + 252x^5y^5 + 210x^4y^6 + 120x^3y^7 + 45x^2y^8 + 10xy^9 + y^{10}$$

1 7 21 35 35 21 7 1
1 8 28 56 70 56 28 8 1
1 9 36 84 126 126 84 36 9 1
1 10 45 120 210 252 210 120 45 10 1

FIGURE 13-19 Pascal's Triangle for $n = 10$

Exercise Set 13–6

Write the expanded form of the following binomials and simplify each term.

1. $(2x + 1)^5$
2. $(x + 3)^5$
3. $(3y - 1)^4$
4. $(y - 4)^4$
5. $(x + 2y)^6$
6. $(2x - y)^6$

7. $\left(m + \frac{n}{5}\right)^5$

8. $\left(\frac{m}{2} - \frac{n}{3}\right)^6$

9. $(m^2 + 2n^3)^7$

10. $\left(3m^2 - \frac{n}{2}\right)^7$

Find the indicated term in the expanded form of the following binomials.

11. Third term of $(2r - s)^{10}$

12. Fourth term of $(r - 3s)^{12}$

13. Fifth term of $\left(x - \frac{y}{2}\right)^{14}$

14. Sixth term of $\left(\frac{x}{2} + 3y\right)^{15}$

15. Seventh term of $(x^2 - y^3)^{16}$

16. Eighth term of $(2k^2 + 3l)^{10}$

17. Ninth term of $\left(3k^2 - \frac{1}{2}\right)^{11}$

18. Fifth term of $\left(\frac{x^2}{3} + \frac{2y^2}{5}\right)^7$

19. Use Figure 13-19 to write the next three rows to Pascal's Triangle.

Use the results of Exercise 19 to write the expanded forms of the following binomials.

20. $(k + 1)^{11}$

21. $(x - 1)^{12}$

22. $(p + q)^{13}$

SUMMARY

Counting Principle: If task A can be performed in m ways and task B can be performed in n ways, then tasks A and B can be jointly performed in $(m)(n)$ ways, provided the performance of task A does not influence the performance of task B.

Theorem: If set S contains n distinct elements, then the number of **permutations** of r of these elements (written ${}_nP_r$) is given with the equation

$${}_nP_r = n(n-1)(n-2)\cdots(n-r+1),\ r \le n \quad \text{or} \quad {}_nP_r = \frac{n!}{(n-r)!}$$

n!

Definition: If n is a positive integer, then $n! = n(n-1)(n-2)\cdots(3)(2)(1)$.

0!

Definition: $0! = 1$

Theorem: If set S contains n distinct elements, then the number of **combinations** of r of these elements (written ${}_nC_r$) is given with the equation

$${}_nC_r = \frac{n!}{r!(n-r)!},\ \text{and } r \le n$$

Theorem: For any nonnegative integer n, ${}_nC_0 = 1$ and ${}_nC_n = 1$.

Theorem: For any nonnegative integers n and r, $r \le n$, ${}_nC_r = {}_nC_{n-r}$.

Event

Definition: Any subset of the set of outcomes of a given experiment is called an event.

Symbol: Let $n(E)$ = the number of outcomes in event E.

Probability

Definition: If E_1 is an event in an experiment, then the probability that E_1 will occur whenever the experiment is conducted is $P(E_1) = \frac{f}{t}$, where f is the number of outcomes favorable to E_1 and t is the total number of outcomes possible for the experiment.

Theorem: For any event E_1 defined for some experiment, $0 \le P(E_1) \le 1$.

Theorem: If E_1 and $\bar{E}_1$ are **complementary events,** then $P(E_1) + P(\bar{E}_1) = 1$.

Exclusive events

Definition: If E_1 and E_2 are defined events for some experiment, then E_1 and E_2 are exclusive events if no outcome of the experiment satisfies both events at the same time.

Inclusive events

Definition: If E_1 and E_2 are defined events for some experiment, then E_1 and E_2 are inclusive events if at least one outcome of the experiment satisfies both events at the same time.

$P(E_1 \text{ or } E_2)$

Definition: If E_1 and E_2 are events for some experiment, then

$$P(E_1 \text{ or } E_2) = \frac{n(E_1 \cup E_2)}{n(S)}$$

Theorem: If E_1 and E_2 are events for some experiment, then $P(E_1 \text{ or } E_2) = P(E_1) + P(E_2) - P(E_1 \text{ and } E_2)$ for any events E_1 and E_2. If E_1 and E_2 are mutually exclusive, then $P(E_1 \text{ or } E_2) = P(E_1) + P(E_2)$.

Independent events

Definition: If E_1 and E_2 are two events for the same experiment, then E_1 and E_2 are independent events if and only if $P(E_2 \mid E_1) = P(E_2)$.

Theorem: Multiplication Rule of Probability If E_1 and E_2 are two events for the same experiment, then the joint probability of E_1 and E_2 is given by the equation

$$P(E_1 \text{ and } E_2) = [P(E_1)][P(E_2 \mid E_1)]$$

Theorem: If E_1 and E_2 are independent events for the same experiment, then the joint probability of E_1 and E_2 is given by the equation $P(E_1 \text{ and } E_2) = [P(E_1)][P(E_2)]$.

Binomial Theorem: For any positive integer n,

$$(a + b)^n = a^n + {}_nC_1(a^{n-1})(b) + \cdots + {}_nC_{n-1}(a)(b^{n-1}) + b^n$$

Theorem: The kth term of the expanded form of $(a + b)^n$, where $1 \le k \le n + 1$, has the form ${}_nC_{k-1}(a^{n-k+1})(b^{k-1})$.

SELF-CHECKING EXERCISE

Compute the following. **(13-1)**

1. $7!$
2. $\frac{9!}{5!}$
3. $\frac{20!}{14!6!}$
4. $\frac{26!}{4!22!}$

Compute the following. **(13-1)**

5. ${}_6P_3$
6. ${}_{13}P_4$
7. ${}_{40}P_0$
8. ${}_8P_8$

Compute the following, using the counting principle and permutations. **(13-1)**

9. How many different 3-digit numbers can be formed with the digits 1, 2, 3, 4, and 5 if
   (a) the digits can be used more than once in a number?
   (b) a digit cannot be used more than once in a number?

10. How many different 4-letter ''words'' can be formed with the letters a, b, c, d, e, and f if
    (a) the letters can be used more than once in a ''word''?
    (b) a letter may not be used more than once in a ''word''?

11. A delivery person has 10 scheduled stops to make on his route on a given day. He wants to make 5 deliveries before lunch. In how many different ways can he arrange his route for the 5 morning deliveries?

12. The membership of a local chapter of the Friends of Violets consists of 18 women and 15 men. According to the club's charter, the president and social chairperson must be women and the secretary and treasurer must be men. How many different slates of officers are possible with the current members?

Compute the following. **(13-2)**

13. ${}_{18}C_{14}$ 14. ${}_{12}C_6$ 15. ${}_{30}C_{30}$ 16. ${}_{50}C_1$

Compute the following combinations. **(13-2)**

17. Set S = {Alice, Beverly, Clair, Doug, Ethel, Frank}. How many subsets of S can be formed that contain
    (a) nobody? (b) one person? (c) two people?
    (d) three people? (e) four people? (f) five people?
    (g) six people?

18. Tommy has enough money to buy 6 different kinds of fireworks of a particular price. The stand has 18 different fireworks at that price. How many different kinds of firework displays can Tommy buy?

19. A fish tank in the aquarium store on the corner contains 32 goldfish. Kimberley wants to buy 3 of them for a gift. How many different selections are possible?

20. A class contains 8 right-handed people and 6 left-handed people. Three students are randomly selected from the class.
    (a) How many different selections are possible?
    (b) How many selections contain all right-handed people?
    (c) How many selections contain 1 left-handed and 2 right-handed people?
    (d) How many selections contain 1 right-handed and 2 left-handed people?
    (e) How many selections contain 3 left-handed people?

21. A sportswoman is skeet shooting clay pigeons. If H represents a hit and M a miss, list the sample space of outcomes for the next four discs she shoots at. **(13-3)**

Compute the following probabilities. **(13-3), (13-4), (13-5)**

22. A card is drawn from a standard deck of 52 cards. What is the probability that the card drawn is
    (a) red? (b) a spade or a heart?
    (c) a spade or a king? (d) a spade and an ace?
    (e) an ace or a king? (f) an ace and a king?

23. Three cards are drawn without replacement from a standard deck of 52 cards. What is the probability that the cards drawn are
    (a) all red?
    (b) all hearts?
    (c) all aces?
    (d) all face cards?
    (e) two spades and one club?
    (f) two kings and one queen?

24. The Open Hunter Jumping Competition at the last horse show had 18 competitors. In the competition, 6 riders were under 17 years of age, 4 were between 17 and 25 years, and 8 were over 25 years old. Assume all riders were of equal ability. What is the probability that of the 5 ribbons awarded,
    (a) all were won by riders over 25 years of age?
    (b) 3 were won by riders over 25 and 2 were won by riders under 17?
    (c) 3 were won by riders under 17 and 2 were won by riders between 17 and 25?

25. An urn contains 7 blue, 5 yellow, and 3 pink marbles. Two marbles are simultaneously drawn at random. What is the probability that
    (a) both are pink?
    (b) one is blue and one is yellow?
    (c) one is pink and one is blue?

26. Three of the college football teams ranked as the top ten in the nation by Associated Press were defeated last weekend. If 3 teams are picked at random from the list, what is the probability that of the teams selected
    (a) none were defeated last weekend?
    (b) exactly one of the teams was defeated last weekend?
    (c) exactly two of the teams were defeated last weekend?
    (d) all three of the teams were defeated?
    (e) What is the sum of the probabilities (a) through (d)?

Write the following binomials in expanded form. **(13-6)**

27. $(2x^2 + y)^4$

28. $(T^3 - 2u)^5$

Determine the value of the indicated term in each of the following binomial expansions. **(13-6)**

29. Third term of $(5m - 4n)^5$

30. Fourth term of $(3c + 2d)^6$

Answers to Self-Checking Exercise

1. 5040
2. 3024
3. 38,760
4. 14,950
5. 120
6. 17,160
7. 1
8. 40,320
9. (a) 125 (b) 60
10. (a) 1296 (b) 360
11. 30,240
12. 64,260
13. 3060
14. 924
15. 1
16. 1
17. (a) 1 (b) 6 (c) 15 (d) 20 (e) 15 (f) 6 (g) 1
18. 18,564
19. 4960
20. (a) 364 (b) 56 (c) 168 (d) 120 (e) 20

21.

H	H	H	H
H	H	H	M
H	H	M	H
H	H	M	M
H	M	H	H
H	M	H	M
H	M	M	H
H	M	M	M
M	H	H	H
M	H	H	M
M	H	M	H
M	H	M	M
M	M	H	H
M	M	H	M
M	M	M	H
M	M	M	M

22. (a) 1/2 (b) 1/2 (c) 16/52 (d) 1/52 (e) 8/52 (f) 0

23. (a) 2/17 (b) 11/850 (c) 1/5525 (d) 11/1105
(e) 39/850 (f) 6/5525

24. (a) 1/153 (b) 5/51 (c) 5/357

25. (a) 1/35 (b) 1/3 (c) 1/5

26. (a) 35/120 (b) 63/120 (c) 21/120 (d) 1/120 (e) 1

27. $16x^8 + 32x^6y + 24x^4y^2 + 8x^2y^3 + y^4$

28. $T^{15} - 10T^{12}u + 40T^9u^2 - 80T^6u^3 + 80T^3u^4 - 32u^5$

29. $20{,}000m^3n^2$

30. $4320c^3d^3$

Appendix

TABLE 1 Powers and Roots

n	n^2	n^3	$\sqrt{n}$	$\sqrt[3]{n}$
0	0	0	0.000	0.000
1	1	1	1.000	1.000
2	4	8	1.414	1.260
3	9	27	1.732	1.442
4	16	64	2.000	1.587
5	25	125	2.236	1.710
6	36	216	2.449	1.817
7	49	343	2.646	1.913
8	64	512	2.828	2.000
9	81	729	3.000	2.080
10	100	1 000	3.162	2.154
11	121	1 331	3.317	2.224
12	144	1 728	3.464	2.289
13	169	2 197	3.606	2.351
14	196	2 744	3.742	2.410
15	225	3 375	3.873	2.466
16	256	4 096	4.000	2.520
17	289	4 913	4.123	2.571
18	324	5 832	4.243	2.621
19	361	6 859	4.359	2.668
20	400	8 000	4.472	2.714
21	441	9 261	4.583	2.759
22	484	10 648	4.690	2.802
23	529	12 167	4.796	2.844
24	576	13 824	4.899	2.884
25	625	15 625	5.000	2.924
26	676	17 576	5.099	2.962
27	729	19 683	5.196	3.000
28	784	21 952	5.292	3.037
29	841	24 389	5.385	3.072
30	900	27 000	5.477	3.107
31	961	29 791	5.568	3.141
32	1 024	32 768	5.657	3.175
33	1 089	35 937	5.745	3.208
34	1 156	39 304	5.831	3.240
35	1 225	42 875	5.916	3.271
36	1 296	46 656	6.000	3.302
37	1 369	50 653	6.083	3.332
38	1 444	54 872	6.164	3.362
39	1 521	59 319	6.245	3.391
40	1 600	64 000	6.325	3.420
41	1 681	68 921	6.403	3.448
42	1 764	74 088	6.481	3.476
43	1 849	79 507	6.557	3.503
44	1 936	85 184	6.633	3.530
45	2 025	91 125	6.708	3.557
46	2 116	97 336	6.782	3.583
47	2 209	103 823	6.856	3.609
48	2 304	110 592	6.928	3.634
49	2 401	117 649	7.000	3.659
50	2 500	125 000	7.071	3.684
51	2 601	132 651	7.141	3.708
52	2 704	140 608	7.211	3.733
53	2 809	148 877	7.280	3.756
54	2 916	157 464	7.348	3.780
55	3 025	166 375	7.416	3.803
56	3 136	175 616	7.483	3.826
57	3 249	185 193	7.550	3.849
58	3 364	195 112	7.616	3.871
59	3 481	205 379	7.681	3.893
60	3 600	216 000	7.746	3.915
61	3 721	226 981	7.810	3.936
62	3 844	238 328	7.874	3.958
63	3 969	250 047	7.937	3.979
64	4 096	262 144	8.000	4.000
65	4 225	274 625	8.062	4.021
66	4 356	287 496	8.124	4.041
67	4 489	300 763	8.185	4.062
68	4 624	314 432	8.246	4.082
69	4 761	328 509	8.307	4.102
70	4 900	343 000	8.367	4.121
71	5 041	357 911	8.426	4.141
72	5 184	373 248	8.485	4.160
73	5 329	389 017	8.544	4.179
74	5 476	405 224	8.602	4.198
75	5 625	421 875	8.660	4.217
76	5 776	438 976	8.718	4.236
77	5 929	456 533	8.775	4.254
78	6 084	474 552	8.832	4.273
79	6 241	493 039	8.888	4.291
80	6 400	512 000	8.944	4.309
81	6 561	531 441	9.000	4.327
82	6 724	551 368	9.055	4.344
83	6 889	571 787	9.110	4.362
84	7 056	592 704	9.165	4.380
85	7 225	614 125	9.220	4.397
86	7 396	636 056	9.274	4.414
87	7 569	658 503	9.327	4.431
88	7 744	681 472	9.381	4.448
89	7 921	704 969	9.434	4.465
90	8 100	729 000	9.487	4.481
91	8 281	753 571	9.539	4.498
92	8 464	778 688	9.592	4.514
93	8 649	804 357	9.644	4.531
94	8 836	830 584	9.695	4.547
95	9 025	857 375	9.747	4.563
96	9 216	884 736	9.798	4.579
97	9 409	912 673	9.849	4.595
98	9 604	941 192	9.899	4.610
99	9 801	970 299	9.950	4.626
100	10 000	1 000 000	10.000	4.642

TABLE 2 Logarithms of Numbers

x	0	1	2	3	4	5	6	7	8	9
1.0	.0000	.0043	.0086	.0128	.0170	.0212	.0253	.0294	.0334	.0374
1.1	.0414	.0453	.0492	.0531	.0569	.0607	.0645	.0682	.0719	.0755
1.2	.0792	.0828	.0864	.0899	.0934	.0969	.1004	.1038	.1072	.1106
1.3	.1139	.1173	.1206	.1239	.1271	.1303	.1335	.1367	.1399	.1430
1.4	.1461	.1492	.1523	.1553	.1584	.1614	.1644	.1673	.1703	.1732
1.5	.1761	.1790	.1818	.1847	.1875	.1903	.1931	.1959	.1987	.2014
1.6	.2041	.2068	.2095	.2122	.2148	.2175	.2201	.2227	.2253	.2279
1.7	.2304	.2330	.2355	.2380	.2405	.2430	.2455	.2480	.2504	.2529
1.8	.2553	.2577	.2601	.2625	.2648	.2672	.2695	.2718	.2742	.2765
1.9	.2788	.2810	.2833	.2856	.2878	.2900	.2923	.2945	.2967	.2989
2.0	.3010	.3032	.3054	.3075	.3096	.3118	.3139	.3160	.3181	.3201
2.1	.3222	.3243	.3263	.3284	.3304	.3324	.3345	.3365	.3385	.3404
2.2	.3424	.3444	.3464	.3483	.3502	.3522	.3541	.3560	.3579	.3598
2.3	.3617	.3636	.3655	.3674	.3692	.3711	.3729	.3747	.3766	.3784
2.4	.3802	.3820	.3838	.3856	.3874	.3892	.3909	.3927	.3945	.3962
2.5	.3979	.3997	.4014	.4031	.4048	.4065	.4082	.4099	.4116	.4133
2.6	.4150	.4166	.4183	.4200	.4216	.4232	.4249	.4265	.4281	.4298
2.7	.4314	.4330	.4346	.4362	.4378	.4393	.4409	.4425	.4440	.4456
2.8	.4472	.4487	.4502	.4518	.4533	.4548	.4564	.4579	.4594	.4609
2.9	.4624	.4639	.4654	.4669	.4683	.4698	.4713	.4728	.4742	.4757
3.0	.4771	.4786	.4800	.4814	.4829	.4843	.4857	.4871	.4886	.4900
3.1	.4914	.4928	.4942	.4955	.4969	.4983	.4997	.5011	.5024	.5038
3.2	.5051	.5065	.5079	.5092	.5105	.5119	.5132	.5145	.5159	.5172
3.3	.5185	.5198	.5211	.5224	.5237	.5250	.5263	.5276	.5289	.5302
3.4	.5315	.5328	.5340	.5353	.5366	.5378	.5391	.5403	.5416	.5428
3.5	.5441	.5453	.5465	.5478	.5490	.5502	.5514	.5527	.5539	.5551
3.6	.5563	.5575	.5587	.5599	.5611	.5623	.5635	.5647	.5658	.5670
3.7	.5682	.5694	.5705	.5717	.5729	.5740	.5752	.5763	.5775	.5786
3.8	.5798	.5809	.5821	.5832	.5843	.5855	.5866	.5877	.5888	.5899
3.9	.5911	.5922	.5933	.5944	.5955	.5966	.5977	.5988	.5999	.6010
4.0	.6021	.6031	.6042	.6053	.6064	.6075	.6085	.6096	.6107	.6117
4.1	.6128	.6138	.6149	.6160	.6170	.6180	.6191	.6201	.6212	.6222
4.2	.6232	.6243	.6253	.6263	.6274	.6284	.6294	.6304	.6314	.6325
4.3	.6335	.6345	.6355	.6365	.6375	.6385	.6395	.6405	.6415	.6425
4.4	.6435	.6444	.6454	.6464	.6474	.6484	.6493	.6503	.6513	.6522
4.5	.6532	.6542	.6551	.6561	.6571	.6580	.6590	.6599	.6609	.6618
4.6	.6628	.6637	.6646	.6656	.6665	.6675	.6684	.6693	.6702	.6712
4.7	.6721	.6730	.6739	.6749	.6758	.6767	.6776	.6785	.6794	.6803
4.8	.6812	.6821	.6830	.6839	.6848	.6857	.6866	.6875	.6884	.6893
4.9	.6902	.6911	.6920	.6928	.6937	.6946	.6955	.6964	.6972	.6981
5.0	.6990	.6998	.7007	.7016	.7024	.7033	.7042	.7050	.7059	.7067
5.1	.7076	.7084	.7093	.7101	.7110	.7118	.7126	.7135	.7143	.7152
5.2	.7160	.7168	.7177	.7185	.7193	.7202	.7210	.7218	.7226	.7235
5.3	.7243	.7251	.7259	.7267	.7275	.7284	.7292	.7300	.7308	.7316
5.4	.7324	.7332	.7340	.7348	.7356	.7364	.7372	.7380	.7388	.7396
5.5	.7404	.7412	.7419	.7427	.7435	.7443	.7451	.7459	.7466	.7474
5.6	.7482	.7490	.7497	.7505	.7513	.7520	.7528	.7536	.7543	.7551
5.7	.7559	.7566	.7574	.7582	.7589	.7597	.7604	.7612	.7619	.7627
5.8	.7634	.7642	.7649	.7657	.7664	.7672	.7679	.7686	.7694	.7701
5.9	.7709	.7716	.7723	.7731	.7738	.7745	.7752	.7760	.7767	.7774

TABLE 2 (cont'd)

x	0	1	2	3	4	5	6	7	8	9
6.0	.7782	.7789	.7796	.7803	.7810	.7818	.7825	.7832	.7839	.7846
6.1	.7853	.7860	.7868	.7875	.7882	.7889	.7896	.7903	.7910	.7917
6.2	.7924	.7931	.7938	.7945	.7952	.7959	.7966	.7973	.7980	.7987
6.3	.7993	.8000	.8007	.8014	.8021	.8028	.8035	.8041	.8048	.8055
6.4	.8062	.8069	.8075	.8082	.8089	.8096	.8102	.8109	.8116	.8122
6.5	.8129	.8136	.8142	.8149	.8156	.8162	.8169	.8176	.8182	.8189
6.6	.8195	.8202	.8209	.8215	.8222	.8228	.8235	.8241	.8248	.8254
6.7	.8261	.8267	.8274	.8280	.8287	.8293	.8299	.8306	.8312	.8319
6.8	.8325	.8331	.8338	.8344	.8351	.8357	.8363	.8370	.8376	.8382
6.9	.8388	.8395	.8401	.8407	.8414	.8420	.8426	.8432	.8439	.8445
7.0	.8451	.8457	.8463	.8470	.8476	.8482	.8488	.8494	.8500	.8506
7.1	.8513	.8519	.8525	.8531	.8537	.8543	.8549	.8555	.8561	.8567
7.2	.8573	.8579	.8585	.8591	.8597	.8603	.8609	.8615	.8621	.8627
7.3	.8633	.8639	.8645	.8651	.8657	.8663	.8669	.8675	.8681	.8686
7.4	.8692	.8698	.8704	.8710	.8716	.8722	.8727	.8733	.8739	.8745
7.5	.8751	.8756	.8762	.8768	.8774	.8779	.8785	.8791	.8797	.8802
7.6	.8808	.8814	.8820	.8825	.8831	.8837	.8842	.8848	.8854	.8859
7.7	.8865	.8871	.8876	.8882	.8887	.8893	.8899	.8904	.8910	.8915
7.8	.8921	.8927	.8932	.8938	.8943	.8949	.8954	.8960	.8965	.8971
7.9	.8976	.8982	.8987	.8993	.8998	.9004	.9009	.9015	.9020	.9025
8.0	.9031	.9036	.9042	.9047	.9053	.9058	.9063	.9069	.9074	.9079
8.1	.9085	.9090	.9096	.9101	.9106	.9112	.9117	.9122	.9128	.9133
8.2	.9138	.9143	.9149	.9154	.9159	.9165	.9170	.9175	.9180	.9186
8.3	.9191	.9196	.9201	.9206	.9212	.9217	.9222	.9227	.9232	.9238
8.4	.9243	.9248	.9253	.9258	.9263	.9269	.9274	.9279	.9284	.9289
8.5	.9294	.9299	.9304	.9309	.9315	.9320	.9325	.9330	.9335	.9340
8.6	.9345	.9350	.9355	.9360	.9365	.9370	.9375	.9380	.9385	.9390
8.7	.9395	.9400	.9405	.9410	.9415	.9420	.9425	.9430	.9435	.9440
8.8	.9445	.9450	.9455	.9460	.9465	.9469	.9474	.9479	.9484	.9489
8.9	.9494	.9499	.9504	.9509	.9513	.9518	.9523	.9528	.9533	.9538
9.0	.9542	.9547	.9552	.9557	.9562	.9566	.9571	.9576	.9581	.9586
9.1	.9590	.9595	.9600	.9605	.9609	.9614	.9619	.9624	.9628	.9633
9.2	.9638	.9643	.9647	.9652	.9657	.9661	.9666	.9671	.9675	.9680
9.3	.9685	.9689	.9694	.9699	.9703	.9708	.9713	.9717	.9722	.9727
9.4	.9731	.9736	.9741	.9745	.9750	.9754	.9759	.9763	.9768	.9773
9.5	.9777	.9782	.9786	.9791	.9795	.9800	.9805	.9809	.9814	.9818
9.6	.9823	.9827	.9832	.9836	.9841	.9845	.9850	.9854	.9859	.9863
9.7	.9868	.9872	.9877	.9881	.9886	.9890	.9894	.9899	.9903	.9908
9.8	.9912	.9917	.9921	.9926	.9930	.9934	.9939	.9943	.9948	.9952
9.9	.9956	.9961	.9965	.9969	.9974	.9978	.9983	.9987	.9991	.9996

TABLE 3 Natural Logarithms of Numbers

a	$\ln a$	a	$\ln a$
1.0	0.0000	5.5	1.7047
1.1	0.0953	5.6	1.7228
1.2	0.1823	5.7	1.7405
1.3	0.2624	5.8	1.7579
1.4	0.3365	5.9	1.7750
1.5	0.4055	6.0	1.7918
1.6	0.4700	6.1	1.8083
1.7	0.5306	6.2	1.8245
1.8	0.5878	6.3	1.8405
1.9	0.6419	6.4	1.8563
2.0	0.6931	6.5	1.8718
2.1	0.7419	6.6	1.8871
2.2	0.7885	6.7	1.9021
2.3	0.8329	6.8	1.9169
2.4	0.8755	6.9	1.9315
2.5	0.9163	7.0	1.9459
2.6	0.9555	7.1	1.9601
2.7	0.9933	7.2	1.9741
2.8	1.0296	7.3	1.9879
2.9	1.0647	7.4	2.0015
3.0	1.0986	7.5	2.0149
3.1	1.1314	7.6	2.0281
3.2	1.1632	7.7	2.0412
3.3	1.1939	7.8	2.0541
3.4	1.2238	7.9	2.0669
3.5	1.2528	8.0	2.0794
3.6	1.2809	8.1	2.0919
3.7	1.3083	8.2	2.1041
3.8	1.3350	8.3	2.1163
3.9	1.3610	8.4	2.1282
4.0	1.3863	8.5	2.1401
4.1	1.4110	8.6	2.1518
4.2	1.4351	8.7	2.1633
4.3	1.4586	8.8	2.1748
4.4	1.4816	8.9	2.1861
4.5	1.5041	9.0	2.1972
4.6	1.5261	9.1	2.2083
4.7	1.5476	9.2	2.2192
4.8	1.5686	9.3	2.2300
4.9	1.5892	9.4	2.2407
5.0	1.6094	9.5	2.2513
5.1	1.6292	9.6	2.2618
5.2	1.6487	9.7	2.2721
5.3	1.6677	9.8	2.2824
5.4	1.6864	9.9	2.2925
		10.0	2.3026

TABLE 4 Values of the Trigonometric Functions

Angle θ									
Degrees	Radians	sin θ	cos θ	tan θ	cot θ	sec θ	csc θ		
0° 00′	.0000	.0000	1.0000	.0000	No value	1.000	No value	1.5708	90° 00′
10	029	029	000	029	343.8	000	343.8	679	50
20	058	058	000	058	171.9	000	171.9	650	40
30	087	087	1.0000	087	114.6	000	114.6	621	30
40	116	116	.9999	116	85.94	000	85.95	592	20
50	145	145	999	145	68.75	000	68.76	563	10
1° 00′	.0175	.0175	.9998	.0175	57.29	1.000	57.30	1.5533	89° 00′
10	204	204	998	204	49.10	000	49.11	504	50
20	233	233	997	233	42.96	000	42.98	475	40
30	262	262	997	262	38.19	000	38.20	446	30
40	291	291	996	291	34.37	000	34.38	417	20
50	320	320	995	320	31.24	001	31.26	388	10
2° 00′	.0349	.0349	.9994	.0349	28.64	1.001	28.65	1.5359	88° 00′
10	378	378	993	378	26.43	001	26.45	330	50
20	407	407	992	407	24.54	001	24.56	301	40
30	436	436	990	437	22.90	001	22.93	272	30
40	465	465	989	466	21.47	001	21.49	243	20
50	495	494	988	495	20.21	001	20.23	213	10
3° 00′	.0524	.0523	.9986	.0524	19.08	1.001	19.11	1.5184	87° 00′
10	553	552	985	553	18.07	002	18.10	155	50
20	582	581	983	582	17.17	002	17.20	126	40
30	611	610	981	612	16.35	002	16.38	097	30
40	640	640	980	641	15.60	002	15.64	068	20
50	669	669	978	670	14.92	002	14.96	039	10
4° 00′	.0698	.0698	.9976	.0699	14.30	1.002	14.34	1.5010	86° 00′
10	727	727	974	729	13.73	003	13.76	981	50
20	756	756	971	758	13.20	003	13.23	952	40
30	785	785	969	787	12.71	003	12.75	923	30
40	814	814	967	816	12.25	003	12.29	893	20
50	844	843	964	846	11.83	004	11.87	864	10
5° 00′	.0873	.0872	.9962	.0875	11.43	1.004	11.47	1.4835	85° 00′
10	902	901	959	904	11.06	004	11.10	806	50
20	931	929	957	934	10.71	004	10.76	777	40
30	960	958	954	963	10.39	005	10.43	748	30
40	.0989	.0987	951	.0992	10.08	005	10.13	719	20
50	.1018	.1016	948	.1022	9.788	005	9.839	690	10
6° 00′	.1047	.1045	.9945	.1051	9.514	1.006	9.567	1.4661	84° 00′
10	076	074	942	080	9.255	006	9.309	632	50
20	105	103	939	110	9.010	006	9.065	603	40
30	134	132	936	139	8.777	006	8.834	573	30
40	164	161	932	169	8.556	007	8.614	544	20
50	193	190	929	198	8.345	007	8.405	515	10
7° 00′	.1222	.1219	.9925	.1228	8.144	1.008	8.206	1.4486	83° 00′
10	251	248	922	257	7.953	008	8.016	457	50
20	280	276	918	287	7.770	008	7.834	428	40
30	309	305	914	317	7.596	009	7.661	399	30
40	338	334	911	346	7.429	009	7.496	370	20
50	367	363	907	376	7.269	009	7.337	341	10
8° 00′	.1396	.1392	.9903	.1405	7.115	1.010	7.185	1.4312	82° 00′
		cos θ	sin θ	cot θ	tan θ	csc θ	sec θ	Radians	Degrees
								Angle θ	

Angle θ									
Degrees	Radians	$\sin\theta$	$\cos\theta$	$\tan\theta$	$\cot\theta$	$\sec\theta$	$\csc\theta$		
8° 00′	.1396	.1392	.9903	.1405	7.115	1.010	7.185	1.4312	82° 00′
10	425	421	899	435	6.968	010	7.040	283	50
20	454	449	894	465	827	011	6.900	254	40
30	484	478	890	495	691	011	765	224	30
40	513	507	886	524	561	012	636	195	20
50	542	536	881	554	435	012	512	166	10
9° 00′	.1571	.1564	.9877	.1584	6.314	1.012	6.392	1.4137	81° 00′
10	600	593	872	614	197	013	277	108	50
20	629	622	868	644	6.084	013	166	079	40
30	658	650	863	673	5.976	014	6.059	050	30
40	687	679	858	703	871	014	5.955	1.4021	20
50	716	708	853	733	769	015	855	1.3992	10
10° 00′	.1745	.1736	.9848	.1763	5.671	1.015	5.759	1.3963	80° 00′
10	774	765	843	793	576	016	665	934	50
20	804	794	838	823	485	016	575	904	40
30	833	822	833	853	396	017	487	875	30
40	862	851	827	883	309	018	403	846	20
50	891	880	822	914	226	018	320	817	10
11° 00′	.1920	.1908	.9816	.1944	5.145	1.019	5.241	1.3788	79° 00′
10	949	937	811	.1974	5.066	019	164	759	50
20	.1978	965	805	.2004	4.989	020	089	730	40
30	.2007	.1994	799	035	915	020	5.016	701	30
40	036	.2022	793	065	843	021	4.945	672	20
50	065	051	787	095	773	022	876	643	10
12° 00′	.2094	.2079	.9781	.2126	4.705	1.022	4.810	1.3614	78° 00′
10	123	108	775	156	638	023	745	584	50
20	153	136	769	186	574	024	682	555	40
30	182	164	763	217	511	024	620	526	30
40	211	193	757	247	449	025	560	497	20
50	240	221	750	278	390	026	502	468	10
13° 00′	.2269	.2250	.9744	.2309	4.331	1.026	4.445	1.3439	77° 00′
10	298	278	737	339	275	027	390	410	50
20	327	306	730	370	219	028	336	381	40
30	356	334	724	401	165	028	284	352	30
40	385	363	717	432	113	029	232	323	20
50	414	391	710	462	061	030	182	294	10
14° 00′	.2443	.2419	.9703	.2493	4.011	1.031	4.134	1.3265	76° 00′
10	473	447	696	524	3.962	031	086	235	50
20	502	476	689	555	914	032	4.039	206	40
30	531	504	681	586	867	033	3.994	177	30
40	560	532	674	617	821	034	950	148	20
50	589	560	667	648	776	034	906	119	10
15° 00′	.2618	.2588	.9659	.2679	3.732	1.035	3.864	1.3090	75° 00′
10	647	616	652	711	689	036	822	061	50
20	676	644	644	742	647	037	782	032	40
30	705	672	636	773	606	038	742	1.3003	30
40	734	700	628	805	566	039	703	1.2974	20
50	763	728	621	836	526	039	665	945	10
16° 00′	.2793	.2756	.9613	.2867	3.487	1.040	3.628	1.2915	74° 00′
		$\cos\theta$	$\sin\theta$	$\cot\theta$	$\tan\theta$	$\csc\theta$	$\sec\theta$	Radians	Degrees
								Angle θ	

TABLE 4 (cont'd)

Angle θ									
Degrees	Radians	sin θ	cos θ	tan θ	cot θ	sec θ	csc θ		
16° 00′	.2793	.2756	.9613	.2867	3.487	1.040	3.628	1.2915	74° 00′
10	822	784	605	899	450	041	592	886	50
20	851	812	596	931	412	042	556	857	40
30	880	840	588	962	376	043	521	828	30
40	909	868	580	.2994	340	044	487	799	20
50	938	896	572	.3026	305	045	453	770	10
17° 00′	.2967	.2924	.9563	.3057	3.271	1.046	3.420	1.2741	73° 00′
10	.2996	952	555	089	237	047	388	712	50
20	.3025	.2979	546	121	204	048	357	683	40
30	054	.3007	537	153	172	048	326	654	30
40	083	035	528	185	140	049	295	625	20
50	113	062	520	217	108	050	265	595	10
18° 00′	.3142	.3090	.9511	.3249	3.078	1.051	3.236	1.2566	72° 00′
10	171	118	502	281	047	052	207	537	50
20	200	145	492	314	3.018	053	179	508	40
30	229	173	483	346	2.989	054	152	479	30
40	258	201	474	378	960	056	124	450	20
50	287	228	465	411	932	057	098	421	10
19° 00′	.3316	.3256	.9455	.3443	2.904	1.058	3.072	1.2392	71° 00′
10	345	283	446	476	877	059	046	363	50
20	374	311	436	508	850	060	3.021	334	40
30	403	338	426	541	824	061	2.996	305	30
40	432	365	417	574	798	062	971	275	20
50	462	393	407	607	773	063	947	246	10
20° 00′	.3491	.3420	.9397	.3640	2.747	1.064	2.924	1.2217	70° 00′
10	520	448	387	673	723	065	901	188	50
20	549	475	377	706	699	066	878	159	40
30	578	502	367	739	675	068	855	130	30
40	607	529	356	772	651	069	833	101	20
50	636	557	346	805	628	070	812	072	10
21° 00′	.3665	.3584	.9336	.3839	2.605	1.071	2.790	1.2043	69° 00′
10	694	611	325	872	583	072	769	1.2014	50
20	723	638	315	906	560	074	749	1.1985	40
30	752	665	304	939	539	075	729	956	30
40	782	692	293	.3973	517	076	709	926	20
50	811	719	283	.4006	496	077	689	897	10
22° 00′	.3840	.3746	.9272	.4040	2.475	1.079	2.669	1.1868	68° 00′
10	869	773	261	074	455	080	650	839	50
20	898	800	250	108	434	081	632	810	40
30	927	827	239	142	414	082	613	781	30
40	956	854	228	176	394	084	595	752	20
50	985	881	216	210	375	085	577	723	10
23° 00′	.4014	.3907	.9205	4245	2.356	1.086	2.559	1.1694	67° 00′
10	043	934	194	279	337	088	542	665	50
20	072	961	182	314	318	089	525	636	40
30	102	.3987	171	348	300	090	508	606	30
40	131	.4014	159	383	282	092	491	577	20
50	160	041	147	417	264	093	475	548	10
24° 00′	.4189	.4067	.9135	.4452	2.246	1.095	2.459	1.1519	66° 00′
		cos θ	sin θ	cot θ	tan θ	csc θ	sec θ	Radians	Degrees
								Angle θ	

Angle θ									
Degrees	Radians	sin θ	cos θ	tan θ	cot θ	sec θ	csc θ		
24° 00′	.4189	.4067	.9135	.4452	2.246	1.095	2.459	1.1519	66° 00′
10	218	094	124	487	229	096	443	490	50
20	247	120	112	522	211	097	427	461	40
30	276	147	100	557	194	099	411	432	30
40	305	173	088	592	177	100	396	403	20
50	334	200	075	628	161	102	381	374	10
25° 00′	.4363	.4226	.9063	.4663	2.145	1.103	2.366	1.1345	65° 00′
10	392	253	051	699	128	105	352	316	50
20	422	279	038	734	112	106	337	286	40
30	451	305	026	770	097	108	323	257	30
40	480	331	013	806	081	109	309	228	20
50	509	358	.9001	841	066	111	295	199	10
26° 00′	.4538	.4384	.8988	.4877	2.050	1.113	2.281	1.1170	64° 00′
10	567	410	975	913	035	114	268	141	50
20	596	436	962	950	020	116	254	112	40
30	625	462	949	.4986	2.006	117	241	083	30
40	654	488	936	.5022	1.991	119	228	054	20
50	683	514	923	059	977	121	215	1.1025	10
27° 00′	.4712	.4540	.8910	.5095	1.963	1.122	2.203	1.0996	63° 00′
10	741	566	897	132	949	124	190	966	50
20	771	592	884	169	935	126	178	937	40
30	800	617	870	206	921	127	166	908	30
40	829	643	857	243	907	129	154	879	20
50	858	669	843	280	894	131	142	850	10
28° 00′	.4887	.4695	.8829	.5317	1.881	1.133	2.130	1.0821	62° 00′
10	916	720	816	354	868	134	118	792	50
20	945	746	802	392	855	136	107	763	40
30	.4974	772	788	430	842	138	096	734	30
40	.5003	797	774	467	829	140	085	705	20
50	032	823	760	505	816	142	074	676	10
29° 00′	.5061	.4848	.8746	.5543	1.804	1.143	2.063	1.0647	61° 00′
10	091	874	732	581	792	145	052	617	50
20	120	899	718	619	780	147	041	588	40
30	149	924	704	658	767	149	031	559	30
40	178	950	689	696	756	151	020	530	20
50	207	.4975	675	735	744	153	010	501	10
30° 00′	.5236	.5000	.8660	.5774	1.732	1.155	2.000	1.0472	60° 00′
10	265	025	646	812	720	157	1.990	443	50
20	294	050	631	851	709	159	980	414	40
30	323	075	616	890	698	161	970	385	30
40	352	100	601	930	686	163	961	356	20
50	381	125	587	.5969	675	165	951	327	10
31° 00′	.5411	.5150	.8572	.6009	1.664	1.167	1.942	1.0297	59° 00′
10	440	175	557	048	653	169	932	268	50
20	469	200	542	088	643	171	923	239	40
30	498	225	526	128	632	173	914	210	30
40	527	250	511	168	621	175	905	181	20
50	556	275	496	208	611	177	896	152	10
32° 00′	.5585	.5299	.8480	.6249	1.600	1.179	1.887	1.0123	58° 00′
		cos θ	sin θ	cot θ	tan θ	csc θ	sec θ	Radians	Degrees
								Angle θ	

TABLE 4 (cont'd)

Angle θ									
Degrees	Radians	sin θ	cos θ	tan θ	cot θ	sec θ	csc θ		
32° 00′	.5585	.5299	.8480	.6249	1.600	1.179	1.887	1.0123	58° 00′
10	614	324	465	289	590	181	878	094	50
20	643	348	450	330	580	184	870	065	40
30	672	373	434	371	570	186	861	036	30
40	701	398	418	412	560	188	853	1.0007	20
50	730	422	403	453	550	190	844	.9977	10
33° 00′	.5760	.5446	.8387	.6494	1.540	1.192	1.836	.9948	57° 00′
10	789	471	371	536	530	195	828	919	50
20	818	495	355	577	520	197	820	890	40
30	847	519	339	619	511	199	812	861	30
40	876	544	323	661	501	202	804	832	20
50	905	568	307	703	492	204	796	803	10
34° 00′	.5934	.5592	.8290	.6745	1.483	1.206	1.788	.9774	56° 00′
10	963	616	274	787	473	209	781	745	50
20	.5992	640	258	830	464	211	773	716	40
30	.6021	664	241	873	455	213	766	687	30
40	050	688	225	916	446	216	758	657	20
50	080	712	208	.6959	437	218	751	628	10
35° 00′	.6109	.5736	.8192	.7002	1.428	1.221	1.743	.9599	55° 00′
10	138	760	175	046	419	223	736	570	50
20	167	783	158	089	411	226	729	541	40
30	196	807	141	133	402	228	722	512	30
40	225	831	124	177	393	231	715	483	20
50	254	854	107	221	385	233	708	454	10
36° 00′	.6283	.5878	.8090	.7265	1.376	1.236	1.701	.9425	54° 00′
10	312	901	073	310	368	239	695	396	50
20	341	925	056	355	360	241	688	367	40
30	370	948	039	400	351	244	681	338	30
40	400	972	021	445	343	247	675	308	20
50	429	.5995	.8004	490	335	249	668	279	10
37° 00′	.6458	.6018	.7986	.7536	1.327	1.252	1.662	.9250	53° 00′
10	487	041	969	581	319	255	655	221	50
20	516	065	951	627	311	258	649	192	40
30	545	088	934	673	303	260	643	163	30
40	574	111	916	720	295	263	636	134	20
50	603	134	898	766	288	266	630	105	10
38° 00′	.6632	.6157	.7880	.7813	1.280	1.269	1.624	.9076	52° 00′
10	661	180	862	860	272	272	618	047	50
20	690	202	844	907	265	275	612	.9018	40
30	720	225	826	.7954	257	278	606	.8988	30
40	749	248	808	.8002	250	281	601	959	20
50	778	271	790	050	242	284	595	930	10
39° 00′	.6807	.6293	.7771	.8098	1.235	1.287	1.589	.8901	51° 00′
10	836	316	753	146	228	290	583	872	50
20	865	338	735	195	220	293	578	843	40
30	894	361	715	243	213	296	572	814	30
40	923	383	698	292	206	299	567	785	20
50	952	406	679	342	199	302	561	756	10
40° 00′	.6981	.6428	.7660	.8391	1.192	1.305	1.556	.8727	50° 00′
		cos θ	sin θ	cot θ	tan θ	csc θ	sec θ	Radians	Degrees
								Angle θ	

Angle θ									
Degrees	Radians	sin θ	cos θ	tan θ	cot θ	sec θ	csc θ		
40° 00′	.6981	.6428	.7660	.8391	1.192	1.305	1.556	.8727	50° 00′
10	.7010	450	642	441	185	309	550	698	50
20	039	472	623	491	178	312	545	668	40
30	069	494	604	541	171	315	540	639	30
40	098	517	585	591	164	318	535	610	20
50	127	539	566	642	157	322	529	581	10
41° 00′	.7156	.6561	.7547	.8693	1.150	1.325	1.524	.8552	49° 00′
10	185	583	528	744	144	328	519	523	50
20	214	604	509	796	137	332	514	494	40
30	243	626	490	847	130	335	509	465	30
40	272	648	470	899	124	339	504	436	20
50	301	670	451	.8952	117	342	499	407	10
42° 00′	.7330	.6691	.7431	.9004	1.111	1.346	1.494	.8378	48° 00′
10	359	713	412	057	104	349	490	348	50
20	389	734	392	110	098	353	485	319	40
30	418	756	373	163	091	356	480	290	30
40	447	777	353	217	085	360	476	261	20
50	476	799	333	271	079	364	471	232	10
43° 00′	.7505	.6820	.7314	.9325	1.072	1.367	1.466	.8203	47° 00′
10	534	841	294	380	066	371	462	174	50
20	563	862	274	435	060	375	457	145	40
30	592	884	254	490	054	379	453	116	30
40	621	905	234	545	048	382	448	087	20
50	650	926	214	601	042	386	444	058	10
44° 00′	.7679	.6947	.7193	.9657	1.036	1.390	1.440	.8029	46° 00′
10	709	967	173	713	030	394	435	.7999	50
20	738	.6988	153	770	024	398	431	970	40
30	767	.7009	133	827	018	402	427	941	30
40	796	030	112	884	012	406	423	912	20
50	825	050	092	.9942	006	410	418	883	10
45° 00′	.7854	.7071	.7071	1.000	1.000	1.414	1.414	.7854	45° 00′
		cos θ	sin θ	cot θ	tan θ	csc θ	sec θ	Radians	Degrees
								Angle θ	

Answers to Selected Exercises

CHAPTER 1

Exercise Set 1–1

1. -7 is an integer, rational, and real; 18 is an integer, natural, whole, rational, and real; $\sqrt{5}$ is irrational and real; 0 is an integer, whole, rational, and real; 9/2 is rational and real; $\sqrt{-4}$ is not real; -0.03 is rational and real; 0.13113 . . . is irrational and real; $\sqrt[3]{8}$ is an integer, natural, whole, rational, and real; 0.3535 . . . is rational and real. **3.** N **5.** N **7.** $\mathcal{R}$ **9.** J **11.** Division by any integer k can have at most $k - 1$ different digits for a remainder. **13.** $0.\overline{3}$ **15.** $0.\overline{857142}$ **17.** $5.\overline{6}$ **19.** 3.375 **21.** 13/99 **23.** 473/111 **25.** 1/30

Exercise Set 1–2

1. Trichotomy **3.** Transitive; inequality **5.** Commutative; multiplication **7.** Commutative; multiplication **9.** Closure; multiplication **11.** Closed under multiplication **13.** Closed under addition and multiplication **15.** $-3/7$ **17.** 0 **19.** -8 **21.** 26/19 **23.** (a) Associative; addition (b) Commutative; addition (c) Associative; addition (d) Distributive (e) Addition $(7 + 5 = 12)$ **25.** (a) Commutative; addition (b) Associative; addition (c) Additive inverse (d) Identity element; addition **27.** $<$ **29.** $>$ **31.** $<$ **33.** $-a < 0$ **35.** $-a < -b$ **37.** (a) No (b) Yes (c) No **39.** (a) No (b) No (c) Yes **41.** Yes **43.** (a) Yes (b) Yes (c) Yes (d) a (e) Yes; inverse of a is a; inverse of b is b; inverse of c is c; inverse of d is d

Exercise Set 1–3

1. $\{24\}$ **3.** $\{18\}$ **5.** $\{-1\}$ **7.** $\{4\}$ **9.** $\{-76/15\}$ **11.** $\varnothing$ **13.** $[-6, \infty)$ **15.** $[-3.3, \infty)$ **17.** $[0, \infty)$ **19.** $x \in$ reals **21.** $(2, 6)$ **23.** $(7, 8]$ **25.** $(-\infty, -1) \cup [3, \infty)$ **27.** $x \in$ reals

Exercise Set 1–4

1. 1/3 **3.** -1 **5.** $\dfrac{x+2}{3x+1}$ **7.** $\dfrac{2x-3}{2(2x+3)}$ **9.** $\dfrac{x(3x+4)}{x-5}$ **11.** $x/6y^2$ **13.** $49x^3/256y^4$ **15.** $2(x^2 + x + 1)$ **17.** $4/9x^3y^2$ **19.** (a) $\dfrac{(x+2)^2}{6}$ (b) $\dfrac{3(x-2)^2}{2(x+2)^2}$ (c) $\dfrac{2(x+2)^2}{3(x-2)^2}$ **21.** $\dfrac{5}{x-1}$ **23.** $\dfrac{-6x-4}{x^2-9}$ **25.** $\dfrac{x^2+1}{x+1}$ **27.** $\dfrac{3x+1}{x^2-x+2}$ **29.** $\dfrac{-x-2}{x^2}$

Exercise Set 1–5

1. 7/12 **3.** $\dfrac{3y - x^2}{1 + x^2}$ **5.** -1 **7.** $\dfrac{x + 1}{x}$ **9.** $\dfrac{(x + 4)(x + 6)}{(x + 5)(x + 1)}$ **11.** $\dfrac{x}{x + 1}$
13. $\dfrac{1 + 6x + 2x^2}{1 - 4x - 2x^2}$ **15.** $\dfrac{(x^2 + y^2)(x + y)}{x^2y}$ **17.** $\dfrac{x - 2}{2x}$ **19.** $\dfrac{x}{x - 1}$

CHAPTER 2

Exercise Set 2–1

1. $1/2^3$ **3.** $1/5^2$ **5.** $1/36b^2$ **7.** z^9/x^3y^6 **9.** $\dfrac{y^2 + x^2}{xy}$ **11.** $\dfrac{y + x}{xy}$ **13.** $\dfrac{(y - x)(y^2 + x^2)}{x^3y^3}$
15. 3.73×10^2 **17.** 3.73×10^5 **19.** 2.4×10^5 **21.** 243 **23.** 1300 **25.** 0.0204
27. 0.0123 **29.** 0.00000188 **31.** 0.0778 **33.** 10.5 **35.** 23.7

Exercise Set 2–2

1. 3 **3.** 3 **5.** 1/3 **7.** 7/10 **9.** 1/3 **11.** 64 **13.** -4 **15.** 8/343 **17.** 7
19. 4 **21.** 32 **23.** 8 **25.** $x^{1/6}$ **27.** $x^{9/2}$ **29.** $x - y$ **31.** $x^{13y/6}$ **33.** 4
35. 125 **37.** 2.11 **39.** 1.83 **41.** 0.00032 **43.** 0.959

Exercise Set 2–3

1. $\sqrt[3]{16}$ **3.** $\sqrt[5]{x^4y^6}$ **5.** $2^{1/2}x^{1/2}y^{1/2}$ or $(2xy)^{1/2}$ **7.** $\dfrac{x^{1/4}}{y^{3/4}}$ or $\left(\dfrac{x}{y^3}\right)^{1/4}$ **9.** $2\sqrt{2}$ **11.** $3\sqrt[3]{2}$
13. $3\,|x|\,\sqrt{3x}$ **15.** $17\sqrt{2}$ **17.** $2\sqrt{x}$ **19.** $\sqrt[3]{12}/3$ **21.** $\sqrt[5]{a^2}/a$ **23.** $\dfrac{\sqrt{x} - 1}{x - 1}$
25. $\dfrac{2\sqrt{x} + 3}{4x - 9}$ **29.** $2x\sqrt[5]{2y}/y$ **35.** $4\sqrt[3]{2} + 16\sqrt[3]{3}$ **37.** $-4x^2\sqrt[4]{x}$
39. $-\sqrt{6}/2$ **41.** 144 **43.** $\sqrt{3} - 1$ **45.** $5\sqrt{2} - 210$ **47.** $\sqrt{73}/24$ **49.** $6\sqrt{5}$

CHAPTER 3

Exercise Set 3–1

1. $\left\{\dfrac{-1 \pm \sqrt{13}}{2}\right\}$ **3.** None **5.** $\{3 \pm \sqrt{15}\}$ **7.** $\{-1, 1/3\}$ **9.** $\{4, -3/2)$ **11.** $\{0, 5\}$
13. $\{2 \pm \sqrt{7}\}$ **15.** None **17.** $\{-2, 8\}$ **19.** $\left\{\dfrac{\sqrt{6} \pm \sqrt{2}}{2}\right\}$ **21.** Real and equal
23. Real and unequal **25.** Real and equal **27.** $x^2 - 5x + 6 = 0$ **29.** $x^2 - 4x + 1 = 0$
31. $6x^2 - 59x + 68 = 0$ **33.** 9 days **35.** 5 ft **37.** 40 and 60 mph **39.** 28 mph

Exercise Set 3–2

1. $\{-1/4, 1/2\}$ **3.** $\{-1/2\}$ **5.** $\left\{\dfrac{1 \pm \sqrt{37}}{2}\right\}$ **7.** $\{11/8\}$ **11.** $\{11\}$ **13.** $\varnothing$
15. $\{6\}$ **17.** $\{80\}$ **19.** $\{-31\}$ **21.** $\{7\}$ **23.** $\varnothing$ **25.** $\{1, 5\}$ **27.** $\{1\}$
29. $\{-1, 27\}$ **31.** $\{-32, -1\}$

Exercise Set 3–3

1. $(-\infty, -2) \cup (2, \infty)$ **3.** $(3, \infty)$

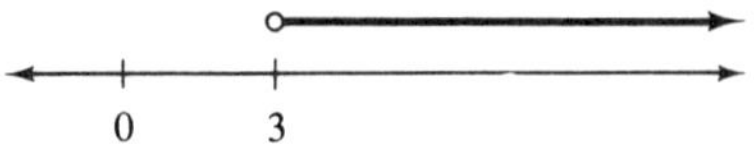

5. $(-\infty, -5] \cup [5, \infty)$ **7.** $(-\infty, -1) \cup (4, \infty)$ **9.** $(-3, 5/2)$

11. $(-\infty, -2) \cup (4, \infty)$ 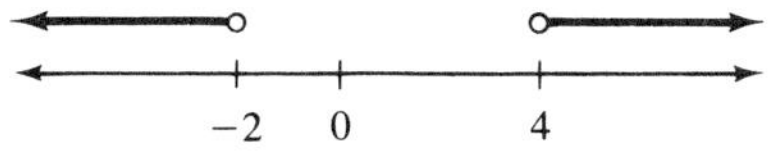

13. $(-\infty, -1/2) \cup (2, \infty)$ **15.** $[-3, 4]$ 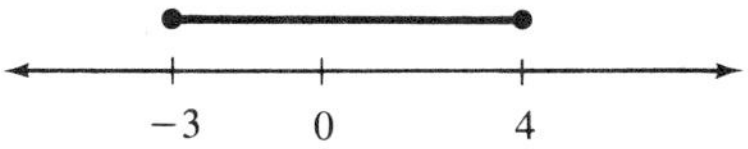

17. $(-\infty, 1) \cup (4, \infty)$ **19.** $x \in$ reals 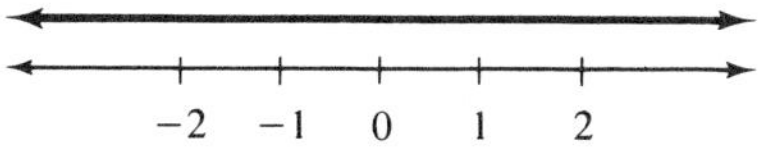

21. $\varnothing$ 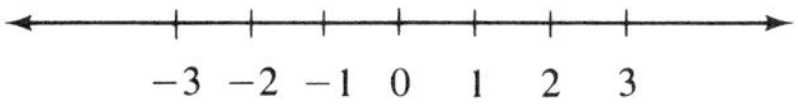

23. $2 \le t \le 4$

Exercise Set 3–4

1. $(-1, 2)$ **3.** $(0, 1/2] \cup (2, \infty)$ 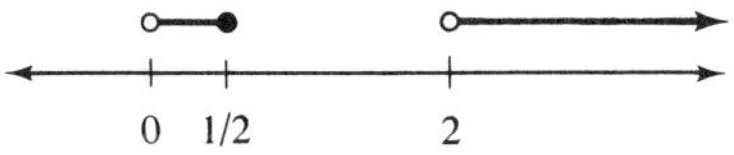

5. $(-\infty, -21] \cup (-1, 3)$ **7.** $(-\infty, -3) \cup (1, 2)$ **9.** $(-\infty, -1) \cup (0, 2)$ **11.** $[3/2, \infty)$

13. $[-3/2, 1]$ **15.** $(-\infty, 1.6) \cup (4.4, \infty)$ 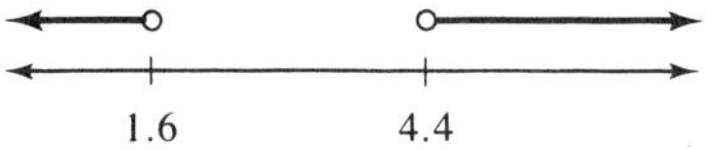

17. $(0, \infty)$ **19.** $[-2.4, -1.7] \cup [1.7, 2.4]$ 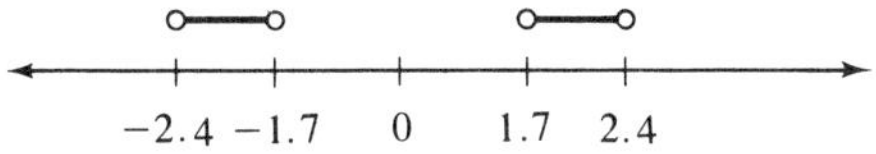

21. (a) $k = 4$ (b) $\varnothing$ **23.** (a) $k = -2$ (b) All real numbers **25.** (a) $k = 6$ or -6 (b) $\varnothing$
27. $(-1, 0) \cup (7, \infty)$ **29.** (40 meters, 60 meters)

Exercise Set 3–5

1. 3 **3.** 2 **5.** x^2; for all x **7.** $x; x \ge 0$ and $-x; x < 0$ **9.** 7 **11.** 13 **13.** 2
15. $\{-2, 4\}$ **17.** $\{-6, 6\}$ **19.** $\{x \mid x \ge 0\}$ **21.** $\{-11/4, -1/4\}$ **23.** $\{x \mid x \ge 3\}$

25. $\{-1, 5\}$ **27.** $\{x \mid 1 < x < 5\}$ 

29. $\{x \mid x < -5 \text{ or } x > -1\}$ **31.** $\{x \mid -1/2 < x < 2\}$ **33.** $\{x \mid x \in \text{reals}\}$ **35.** $\{x \mid 3 < x < 4\}$

37. $\{x \mid x \le -4 \text{ or } x \ge 3\}$ 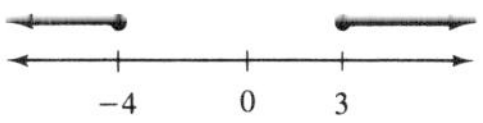

39. $|x| \le 2$ **41.** $|x - 1| \ge 4$ **43.** $|x - 2| < 5$ **45.** All real numbers x and y
47. If x and y have the same sign **49.** All real numbers x and y **51.** (a) 1 (b) -1

Exercise Set 3–6

1. 8 **3.** 0 **5.** $\{-1, 1, 2\}$ **7.** $\left\{5, \dfrac{1 \pm \sqrt{21}}{2}\right\}$ **9.** $\{1\}$ **11.** $\{-2, 3\}$
13. $\{-1, 1/2, 1, 2\}$ **15.** $(x - 5)(x + 1)(x + 3)$ **17.** $(x - 7)(2x - 1)(x + 1)(x + 3)$
19. $(x - 7)(x - \sqrt{2})(x + \sqrt{2})(x + 5)$ **21.** $k = 63$ **23.** $k = -1$ **25.** (a) $\{1/2, 2, 4\}$
(b) $(1/2, 2) \cup (4, \infty)$ **27.** (a) \$21,000 (b) (t_3, t_7); between March 31 and July 31 (c) \$585,000

Exercise Set 3–7

1. $\{-1, 1, 3\}$ **3.** $\{-4, 2, 3\}$ **5.** $\{-10, -3, 3\}$ **7.** $\{-2, 1, 2\}$ **9.** $\{-1, 2/3, 3/2\}$ **11.** $\{2, 3\}$ **13.** $\{-3, -2, 0, 1/2, 2/3\}$ **15.** $\{1\}$ **17.** $\{-2, 1\}$ **19.** $\{-1\}$ **21.** $(-3, 3)$ **23.** $(-1, 4)$ **25.** $(-2, 5)$ **27.** 1, 2, and 5 **29.** $3 \times 6 \times 8$

Exercise Set 3–8

1. $\pm 1, \pm 1/2, \pm 3/2, \pm 5/2, \pm 3, \pm 5, \pm 15/2, \pm 15$ **3.** $\pm 1/3, \pm 2/3, \pm 1, \pm 2, \pm 3, \pm 6$ **5.** $\pm 1/5, \pm 3/5, \pm 1, \pm 3, \pm 5, \pm 15$ **7.** $\pm 1/8, \pm 1/4, \pm 1/2, \pm 1$ **9.** $\{-2, -1/2, 1/2\}$ **11.** $\{5/2\}$ **13.** $\{1/7, 1 \pm \sqrt{2}\}$ **15.** $\{3/2\}$ **17.** $\{-1/2, 1/4, 5\}$ **19.** $\{-5, 3, 4\}$ **21.** $\{5/6, 1, 7\}$ **23.** $\{-1, 1/2, \pm\sqrt{5}\}$ **25.** $\{1/3, 1/2, \pm\sqrt{3}\}$ **27.** $\{-\sqrt{3}, \sqrt{3}, 2\}$ **29.** $\{-3, -1, 2, 3\}$

CHAPTER 4

Exercise Set 4–1

1. (a) Function (b) $\{1, 2, 3, 4\}$; $\{0, 2, 3, 4\}$ **3.** (a) Function (b) $\{-1/2, 0, 1/2, 2\}$; $\{3/4\}$ **5.** (a) Function (b) $\{-2, 1, 3, 4\}$; $\{-2, 1, 3, 4\}$ **7.** (a) Function (b) $\{-2, -1, 0, 1, 2, 3\}$; $\{-2, -1, 0, 1\}$ **9.** (a) Function (b) $-3 \leq x \leq 3$; $0 \leq y \leq 3$ **11.** (a) Function (b) All real numbers; $\{-2, 0, 2\}$ **13.** (a) Not a function (b) $x \geq 0$; all real numbers **15.** (a) Function (b) $x > 0$; all real numbers **17.** (a) All real numbers, $y \geq 0$ (b) Function **19.** (a) $x \geq 1$ or $x \leq -1$; all real numbers (b) Not a function **21.** (a) All real numbers except ± 2; all real numbers (b) Function **23.** (a) $-2 \leq x \leq 2$; $-3 \leq y \leq 3$ (b) Not a function **25.** (a) All real numbers; $0 < y \leq 1$ (b) Function **27.** $\sqrt{3}/3$ **29.** $\sqrt{15}/3$ **31.** 1

33. $\dfrac{a^2}{2-a}$ **35.** $\dfrac{4ah + 2h^2 - a^2h - ah^2}{(2-a-h)(2-a)}$ **37.** $\dfrac{(2-a)^2}{a}$ **39.** $x \leq -2$ or $x > 4$; $h(x) \geq 0$, $h(x) \neq 1$

Exercise Set 4–2

1. x-intercept is 2; y-intercept is -1 **3.** x-intercepts are $-1, 0, 1$; y-intercept is 0 **5.** x-intercept is 4; y-intercept is 2 **7.** $y = 2$; $x = 2$ **9.** $y = 0$; $x = -1$ **11.** $y = 3$ and $y = -3$; $x = 0$ **13.** Yes **15.** No **17.** Yes

19.

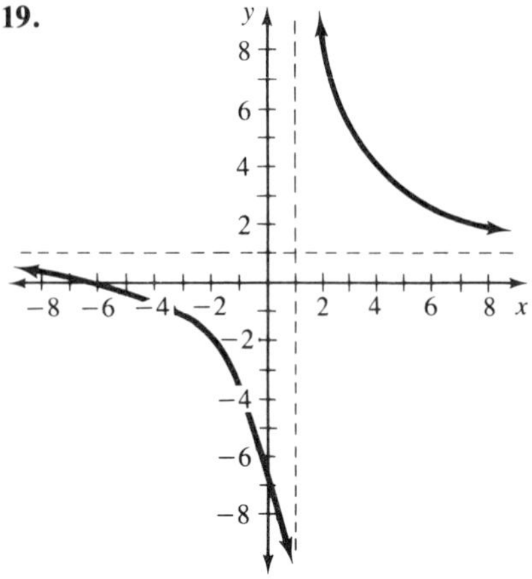

21.

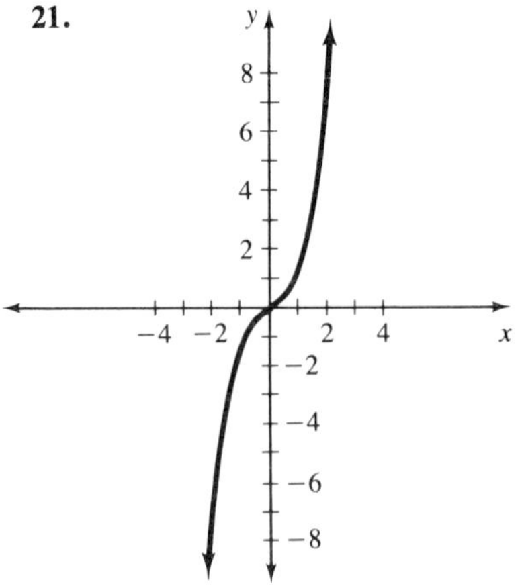

23.

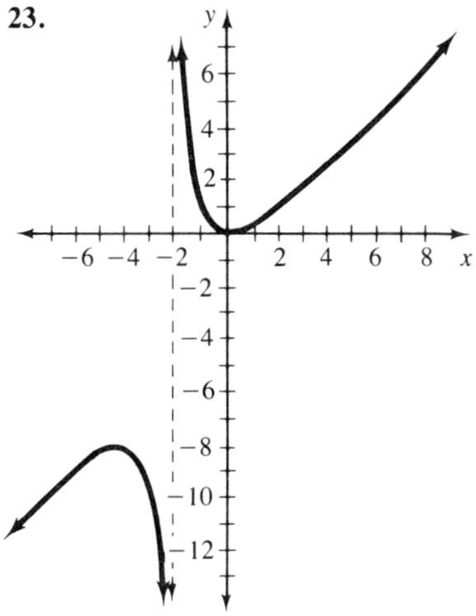

25.

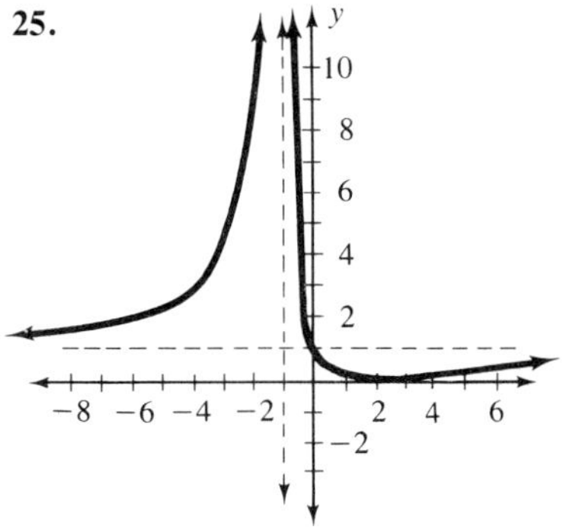

27.

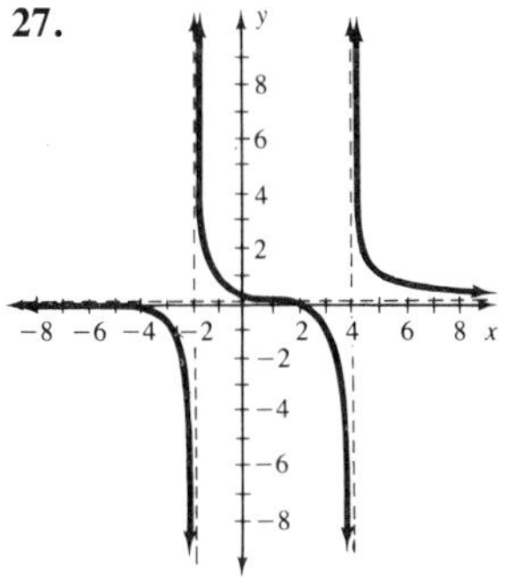

29.

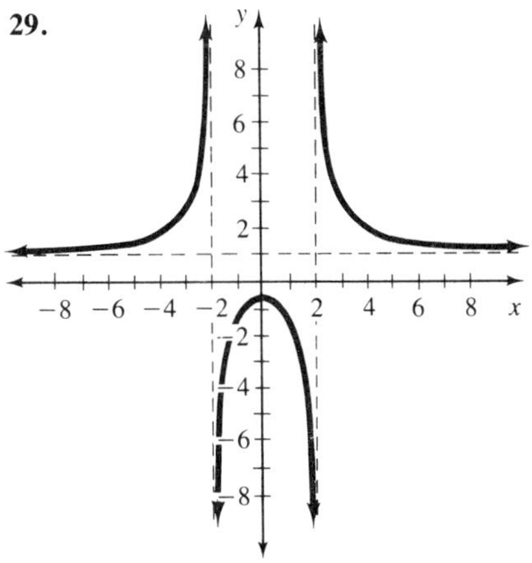

31. $y = \dfrac{-1}{x}$ **33.** $y = \dfrac{x^2}{(x^2 - 1)^2}$ **35.** $y = \dfrac{x}{4 - x^2}$

Exercise Set 4–3

1. No **3.** Yes **5.** Yes **7.** x-intercept is 6; y-intercept is 4 **9.** x-intercept is $-9/2$; y-intercept is 3 **11.** 5/2 **13.** 1 **15.** 1 **17.** $-1/3$ **19.** 0 **21.** 3/4; 2 **23.** 8/7; $-24/7$ **31.** 9/2 **33.** $6x - y + 1 = 0$ **35.** $3x - 4y + 16 = 0$ **37.** $y = 3x + 4$ **39.** $2x + 3y = 0$ **41.** $y - 6 = 0$ **43.** $y = 6$ **45.** $6x + 6y - 5 = 0$ **47.** Approximately $y = 0.9x + 12$

Exericse Set 4–4

1. $\{(9, 1)\}$ **3.** $\{(x, y) \mid x + 2y = 5\}$ **5.** $\{(-1/7, 3/7)\}$ **7.** $\{(15/4, 15/2)\}$ **9.** $\{(25/26, 9/13)\}$ **11.** $\{(1, 6)\}$ **13.** $\{(5/4, -3/8)\}$ **15.** $\{(0, -2/3)\}$ **17.** $\{13/5, -9/5)\}$ **19.** $\{(0, 0)\}$ **21.** $\{(3/29, -3/14)\}$ **23.** $\{(1/5, -1/2)\}$ **25.** $\{(5/26, -7/26, 2/26)\}$ **27.** $\{(7, -6, -2\}$ **29.** 37 nickels; 154 pennies

Exercise Set 4–5

1. 2×3 **3.** 3×2 **5.** 4×2 **7.** -6 **9.** 9 **11.** -3 **13.** 27 **15.** 19 **17.** 0 **19.** 10 **21.** -5 **23.** 0 **25.** $\{(-5, 11/2)\}$ **27.** $\{(-1/5, -7/5)\}$ **29.** $\{(9/8, -11/16)\}$ **31.** $\varnothing$ **33.** $\{(4, 10)\}$ **35.** $\varnothing$ **37.** $\{(2/7, 31/42, -5/7)\}$ **39.** $\{(0, 1, 0)\}$

Exercise Set 4–6

1.

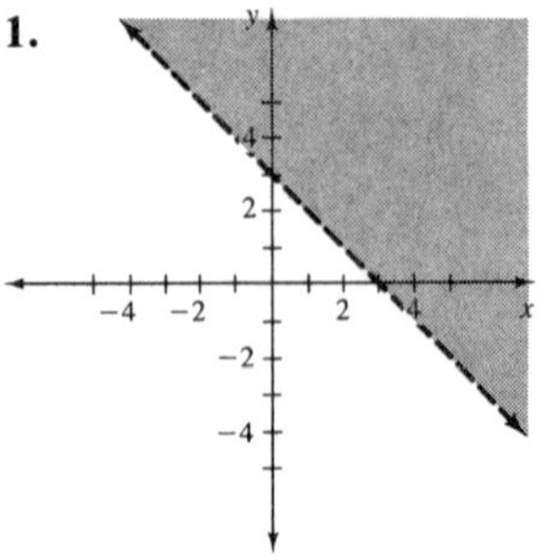

5.

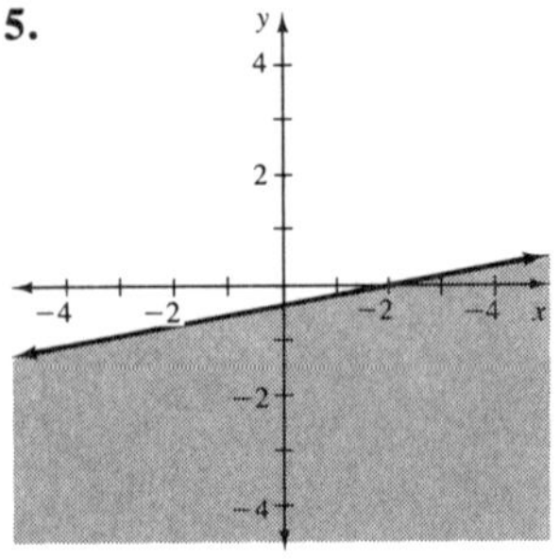

9.

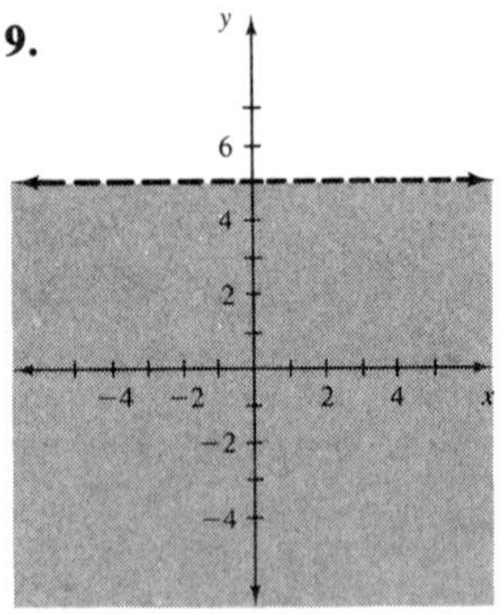

13.

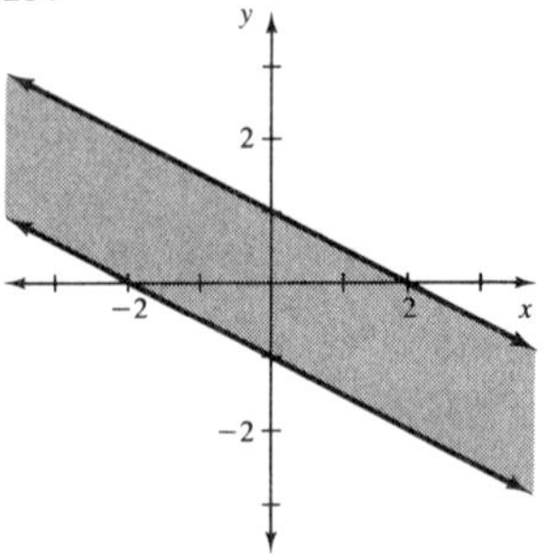

17.

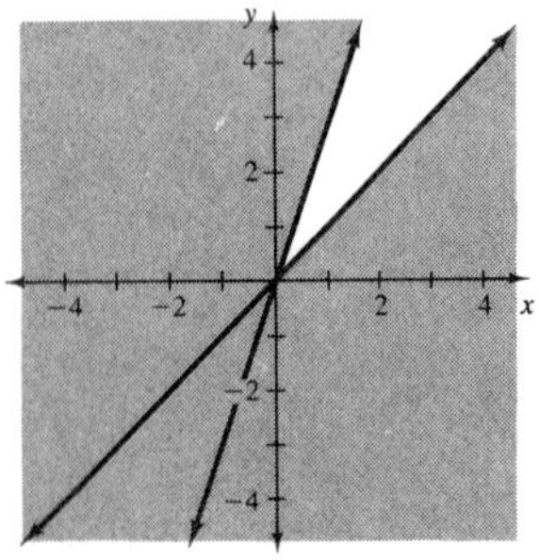

19.

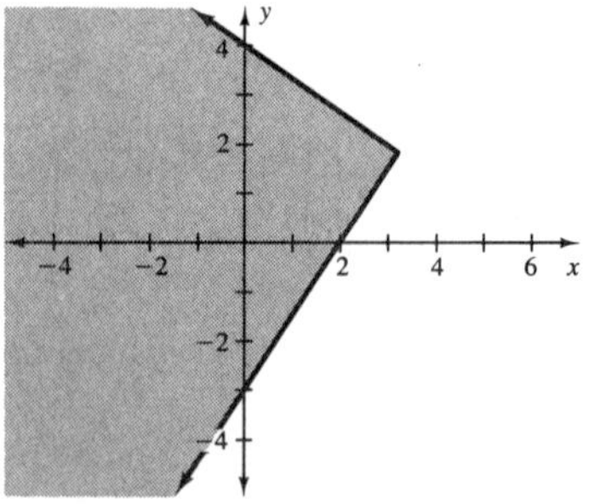

Exercise Set 4–7

5.

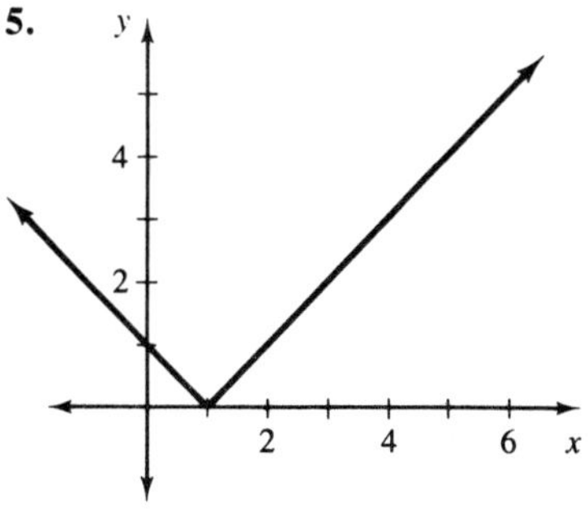

15.

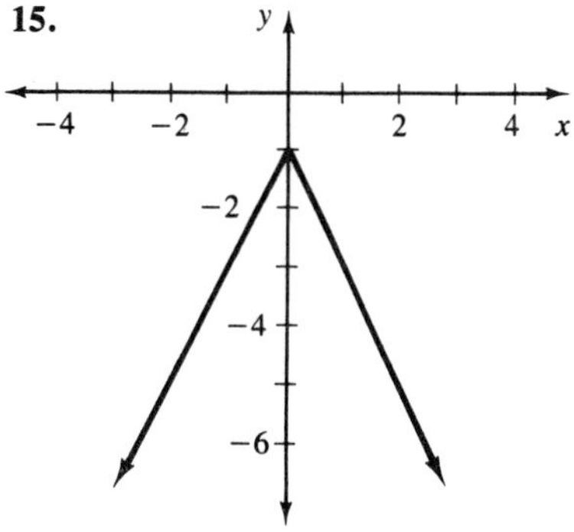

19.

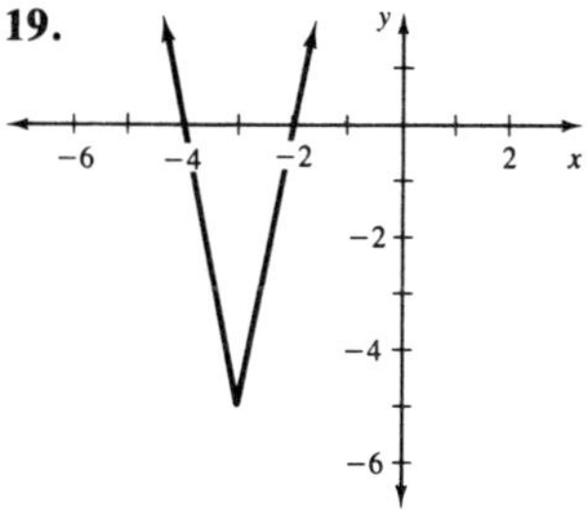

23.

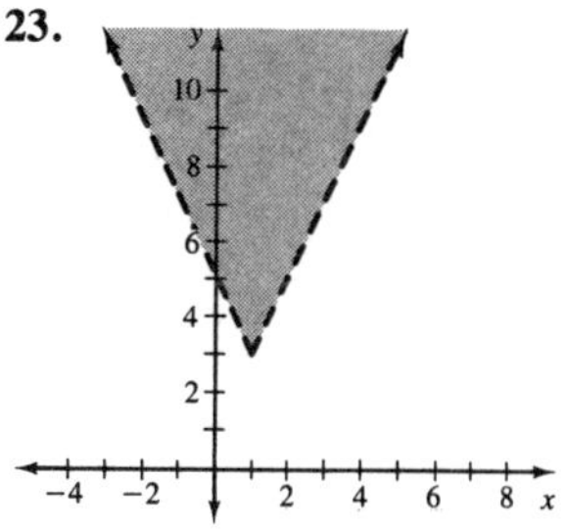

25.

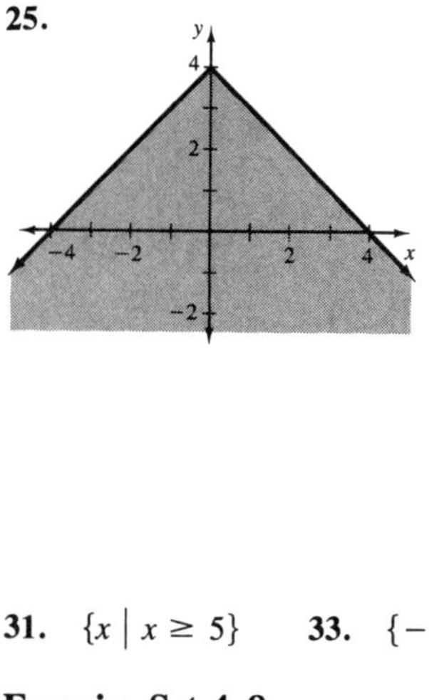

29.

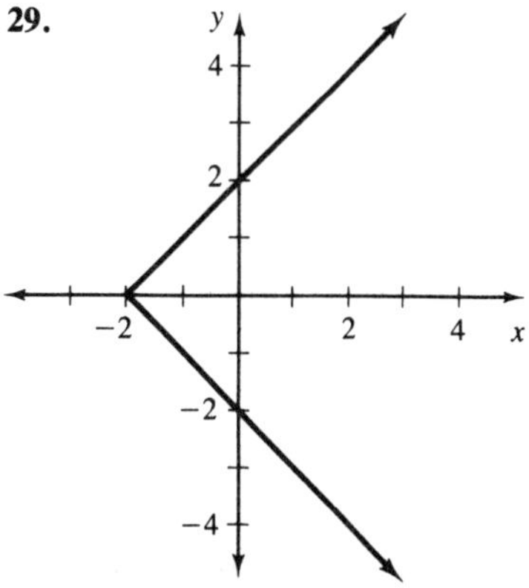

31. $\{x \mid x \geq 5\}$ **33.** $\{-2, 4/3\}$ **35.** $\{x \mid x < -3\}$ **37.** $\{x \mid x \leq 1\}$

Exercise Set 4–8

1. (a) $\{x \mid x \in \text{reals}\}$, and $x \neq 4$ for f/g (b) $(f + g)(x) = 2x + 3$ (c) $(f - g)(x) = 4x - 5$ (d) $(f \cdot g)(x) = -3x^2 + 13x - 4$ (e) $\left(\dfrac{f}{g}\right)(x) = \dfrac{3x - 1}{4 - x}$

3. (a) $\{x \mid x \in \text{reals}; x \neq -2, 0, 2\}$ (b) $(f + g)(x) = \dfrac{3x^2 - 4}{x(x^2 - 4)}$ (c) $(f - g)(x) = \dfrac{x^2 + 4}{x(x^2 - 4)}$ (d) $(f \cdot g)(x) = \dfrac{2}{x^2 - 4}$ (e) $\left(\dfrac{f}{g}\right)(x) = \dfrac{2x^2}{x^2 - 4}$

5. (a) $\{x \mid x \in \text{reals}; x \neq -1, 1, 2\}$ (b) $(f + g)(x) = \dfrac{2x - 1}{(x + 1)(x - 1)(x - 2)}$

(c) $(f - g)(x) = \dfrac{-3}{(x + 1)(x - 1)(x - 2)}$ (d) $(f \cdot g)(x) = \dfrac{1}{(x + 1)(x - 2)(x - 1)^2}$ (e) $\left(\dfrac{f}{g}\right)(x) = \dfrac{x - 2}{x + 1}$

7. (a) $\{x \mid x > 0\}$ (b) $(f + g)(x) = \dfrac{\sqrt{x^2 + x} + 1}{\sqrt{x}}$ (c) $(f - g)(x) = \dfrac{\sqrt{x^2 + x} - 1}{\sqrt{x}}$ (d) $(f \cdot g)(x) = \dfrac{\sqrt{x + 1}}{\sqrt{x}}$

(e) $\left(\dfrac{f}{g}\right)(x) = \sqrt{x^2 + x}$ **9.** (a) $\{x \mid x > 1\}$ (b) $(f + g)(x) = \dfrac{\sqrt{x - 1} + \sqrt{x + 1}}{\sqrt{x^2 - 1}}$

(c) $(f - g)(x) = \dfrac{\sqrt{x - 1} - \sqrt{x + 1}}{\sqrt{x^2 - 1}}$ (d) $(f \cdot g)(x) = \dfrac{1}{\sqrt{x^2 - 1}}$ (e) $\left(\dfrac{f}{g}\right)(x) = \dfrac{\sqrt{x - 1}}{\sqrt{x + 1}}$

11. (a) $|x|$ (b) $\{x \mid x \in \text{reals}\}$ **13.** (a) $\dfrac{x^2 - 1}{x^2}$ (b) $\{x \mid x \in \text{reals}, x \neq 0\}$

15. (a) x (b) $\{x \mid x \in \text{reals}\}$ **17.** (a) x (b) $\{x \mid x \geq 0\}$ **19.** (a) $\dfrac{(x - 1)^2}{x^2}$ (b) $\{x \mid x \in \text{reals}, x \neq 0\}$

21. (a) x (b) $\{x \mid x \in \text{reals}\}$

Exercise Set 4–9

1. Yes **3.** No **5.** No **7.** Yes **9.** Yes **11.** Yes **13.** $f^{-1} = \left\{(x, y) \mid y = \dfrac{x - 1}{2}\right\}$

15. $f^{-1} = \{(x, y) \mid y = 1/x\}$ **17.** $f^{-1} = \left\{(x, y) \mid y = \dfrac{x}{2 - 3x}\right\}$

19. (a) $f^{-1} = \{(x, y) \mid y = \sqrt{x - 2}\}$

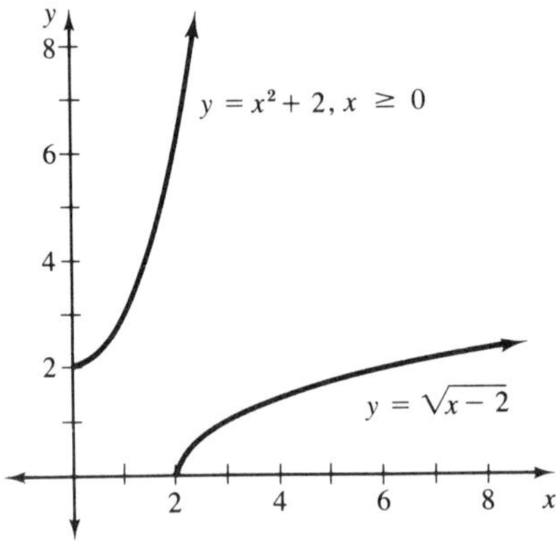

CHAPTER 5

Exercise Set 5–1

7. 1 **9.** 16 **11.** $5^{2\sqrt{3}}$ **13.** 1/8 **15.** $\{3\}$ **17.** $\{-2\}$ **19.** $\{9/5\}$ **21.** $\varnothing$
23. \$252.50

Exercise Set 5–2

1. $\log_3 9 = 2$ **3.** $\log_{11} 1 = 0$ **5.** $\log_{27} 9 = 2/3$ **7.** $5^3 = 125$ **9.** $3^1 = 3$ **11.** $2^5 = 32$
13. $\{2\}$ **15.** $\{4/3\}$ **17.** $\{3\}$ **19.** $\{1/4\}$ **21.** $\{\sqrt{5}\}$ **23.** $\{8\}$ **25.** $\{0.1\}$

Exercise Set 5–3

1. $\log_b 2 + \log_b 3$ **3.** $\log_b 2 + \log_b 3 + \log_b 5$ **5.** $1/3\ (\log_b 7 + \log_b 11)$ **7.** $\log_b 14$
9. $\log_b(6/7)$ **11.** $\log_b(2/3)$ **13.** $2m$ **15.** $j + \log_b k$ **17.** k
19. $\dfrac{\log 7}{\log 4}$ **21.** $\dfrac{2 \log 5}{\log 3}$ **23.** $\{125\}$ **25.** $\{1/5\}$ **27.** $\{68\}$ **29.** $\{2\}$

Exercise Set 5–4

1. 1.2923 **3.** −1.7077 **5.** 7.7185 **7.** 2.8280 **9.** 6.8451 **11.** 4.52 **13.** 0.00452 **15.** 205,000 **17.** 0.000998 **19.** 39,900,000 **21.** 2.1163 **23.** −0.1863 **25.** 8.8819 **27.** 1.2 **29.** 12 **31.** 0.12 **37.** ln 5 **39.** ln 2 **41.** 4 **43.** 38,400 **45.** 2.17 **47.** 0.0769 **49.** (a) 4.295×10^9 (b) Approximately 136 years

Exercise Set 5–5

1. $\dfrac{\log 5}{\log 2}$; 2.32 **3.** $\dfrac{\log 151}{\log 4} + 1$; 4.62 **5.** $\dfrac{1}{2}\left(\dfrac{2}{\log 3.5} - 1\right)$; 1.34 **7.** $\dfrac{12 \log 3 + \log 2}{3 \log 3 - \log 2}$; 5.33 **9.** $\dfrac{\log 0.567}{\log 8.11}$; −0.271 **11.** ln 2.36; 0.859 **13.** $\dfrac{\log 2}{\log 1.06}$; 11.9 years **15.** 7.34 years **17.** 4.2 years **19.** $\dfrac{\ln 3}{30}$, or about 0.037 percent **21.** $k = 0.000347$

CHAPTER 6

Exercise Set 6–1

1. $\sqrt{5}$ **3.** $\sqrt{194}$ **5.** $\sqrt{31}$ **7.** $\sqrt{a^2 + 16b^2}$ **9.** (2, 4) **11.** (5/2, 15/2) **13.** (2, 2) **15.** (1, 0) **17.** (6, 20) **19.** (a) No (b) No (c) Yes (d) $6 + 2\sqrt{5}$ **21.** (a) Yes (b) No (c) No (d) $2 + 2\sqrt{37}$ **23.** $x - y + 6 = 0$ **25.** $2\sqrt{2} + 3\sqrt{2} = 5\sqrt{2}$ **27.** 6 **29.** 5 **31.** $\dfrac{4\sqrt{58}}{29}$

Exercise Set 6–2

1. $x^2 + y^2 - 4x - 6y - 12 = 0$ **3.** $x^2 + y^2 + 10x + 4y + 20 = 0$ **5.** 19/144 **7.** $(x - a)^2 + (y - 3a)^2 = 3$ **9.** (0, 0); 6 **11.** (1, 2); 4 **13.** (0, −7); 7 **15.** (−1/2, −1/2); $\sqrt{6}/2$ **17.** (5/4, 1); $\sqrt{41}/4$ **19.** $x^2 + y^2 + 2x - 4y - 319 = 0$ **21.** $x^2 + y^2 - 4x - 6y + 11 = 0$ **23.** $x^2 + y^2 + 2x + 4y - 8 = 0$

27.

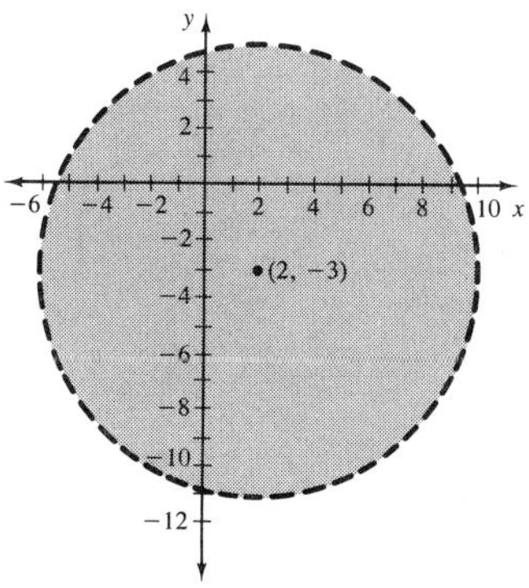

31.

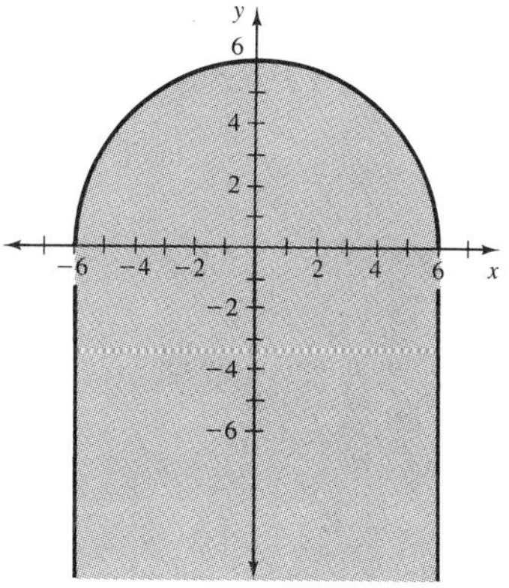

33. $x^2 + y^2 - 6x - 8y = 0$ **35.** $x^2 + y^2 - 12x - 8y + 27 = 0$

Exercise Set 6–3

1. Foci: $F_1(2\sqrt{3}, 0)$; $F_2(-2\sqrt{3}, 0)$

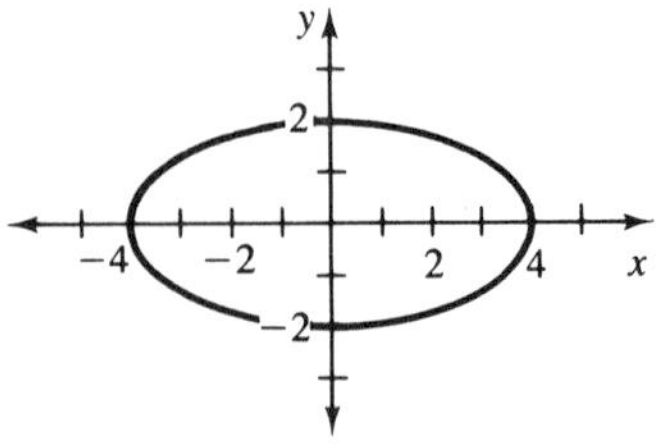

7. Foci: $F_1(0, 3)$; $F_2(0, -3)$

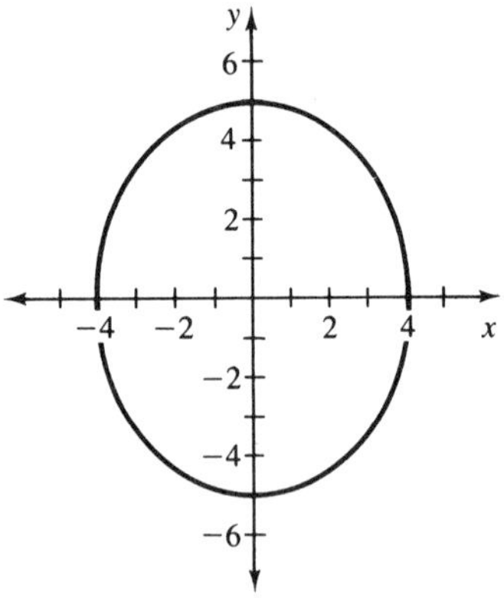

11. $x^2 + 12y^2 = 36$ **13.** $9x^2 + 49y^2 = 441$ **15.** $9x^2 + 25y^2 = 225$

17.

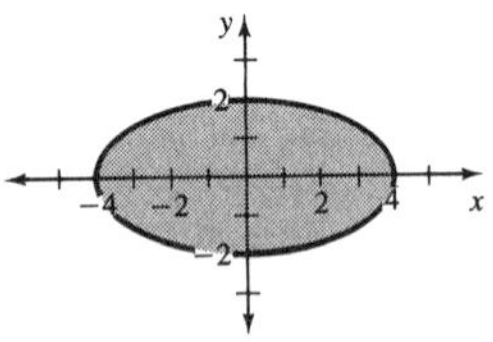

23.

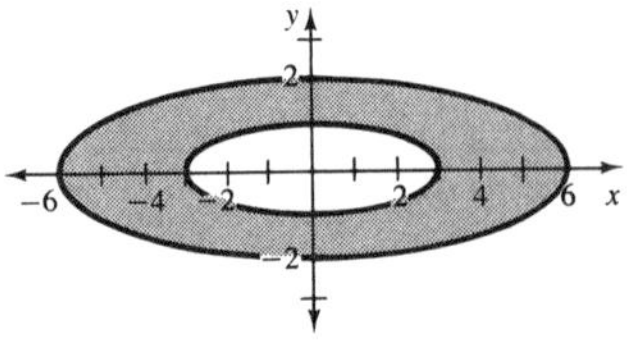

27. About 1.3 ft above and below the center of the board and 3 ft in from each edge; $16x^2 + 9y^2 = 36$

Exercise Set 6–4

3.

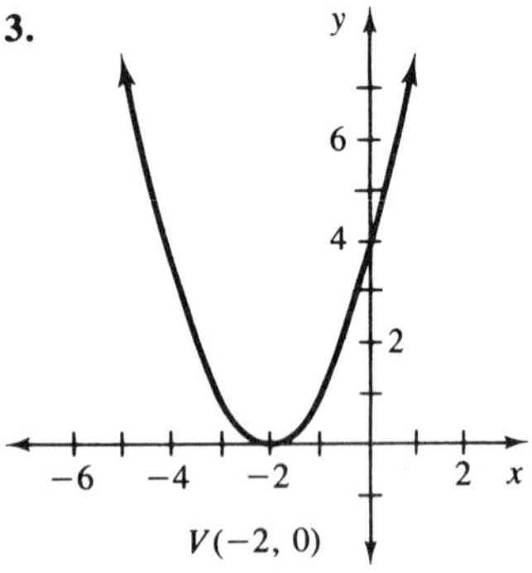

11.

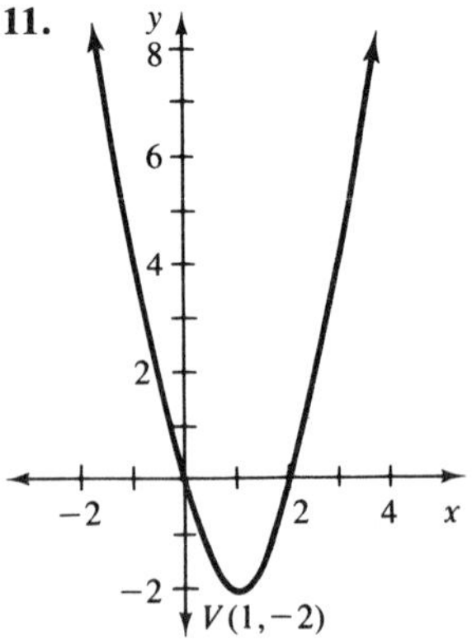

15.

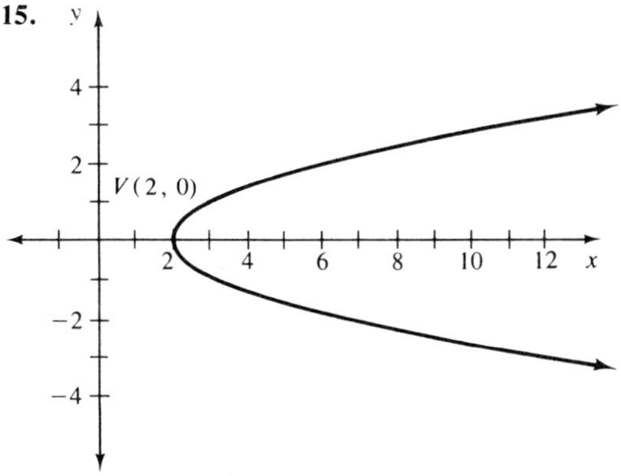

17.

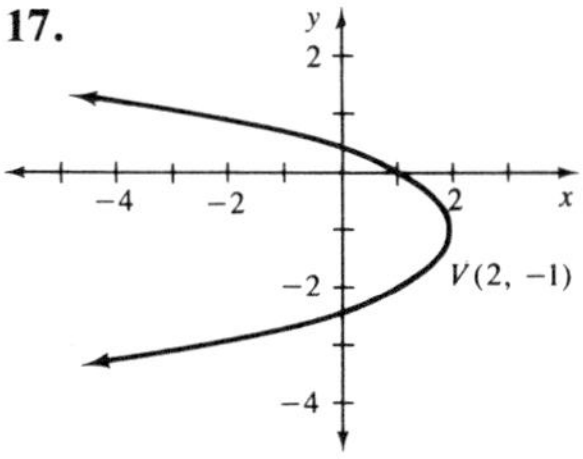

21.

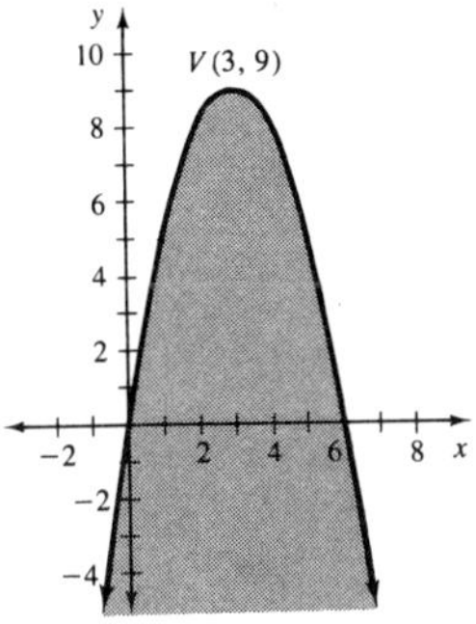

23.

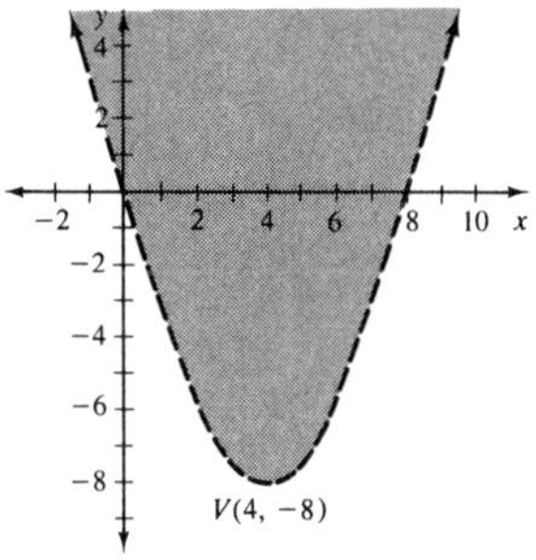

25. $y = 2x^2$ **27.** $y = 3 - 4x^2$ **29.** $8y = x^2$ **31.** $8y = x^2 - 2x + 1$

Exercise Set 6–5

3.

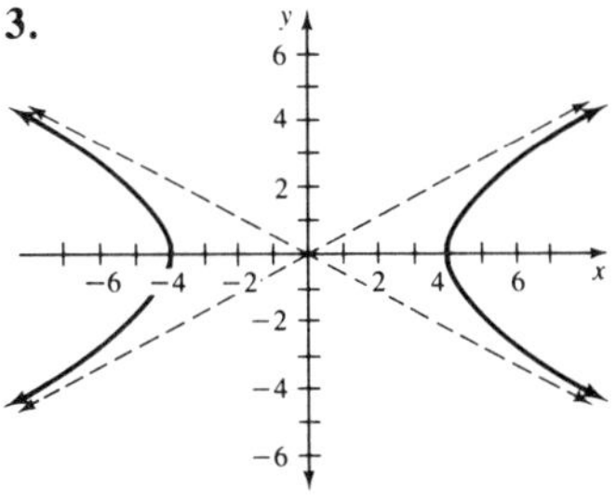

5.

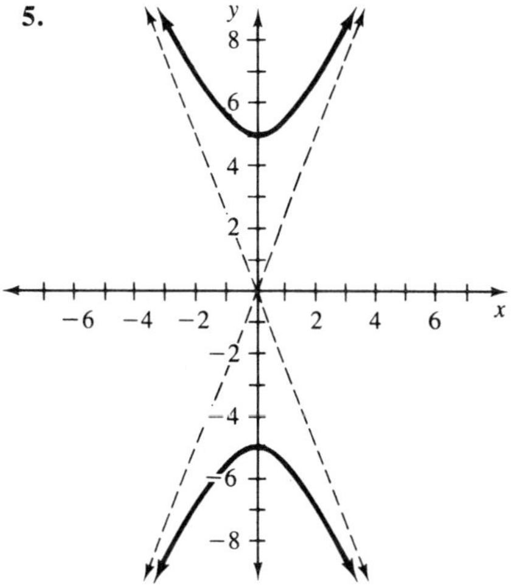

9. $5x^2 - 4y^2 = 20$ **11.** $7y^2 - 9x^2 = 63$ **13.** $16x^2 - 25y^2 = 400$ **15.** $9y^2 - 4x^2 = 36$
17. $9x^2 - 16y^2 = 144$

25.

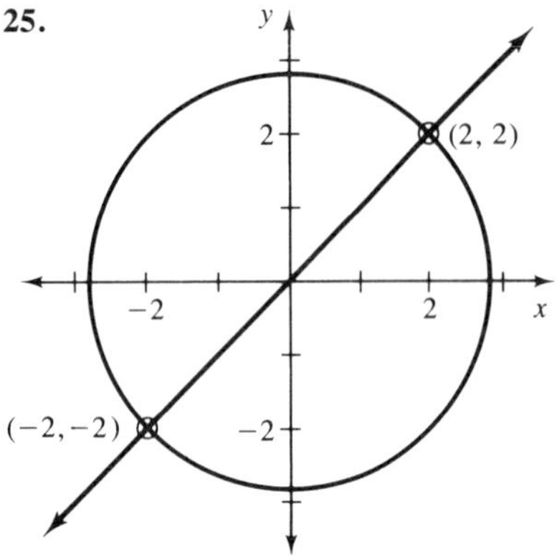

29.

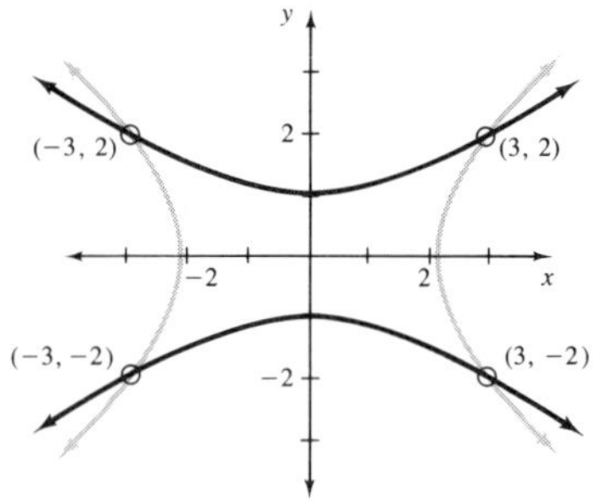

Exercise Set 6–6

1. (b) $\{x \mid -3 \le x \le 5\}$; $\{y \mid -6 \le y \le 2\}$ **3.** (b) $\{x \mid x \in \text{reals}\}$; $\{y \mid y \le 1\}$

5. (a)

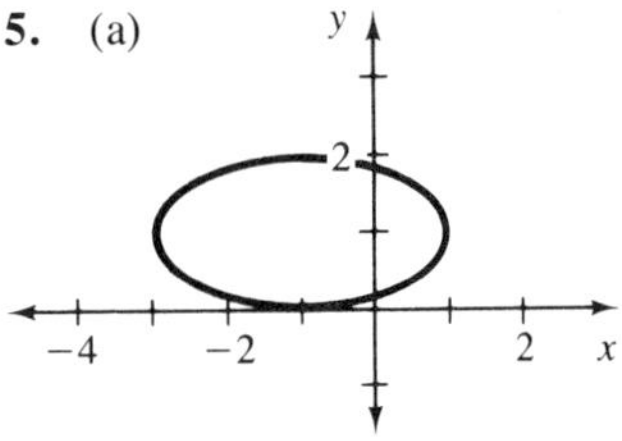

(b) $\{x \mid -3 \le x \le 1\}$; $\{y \mid 0 \le y \le 2\}$

7. (a)

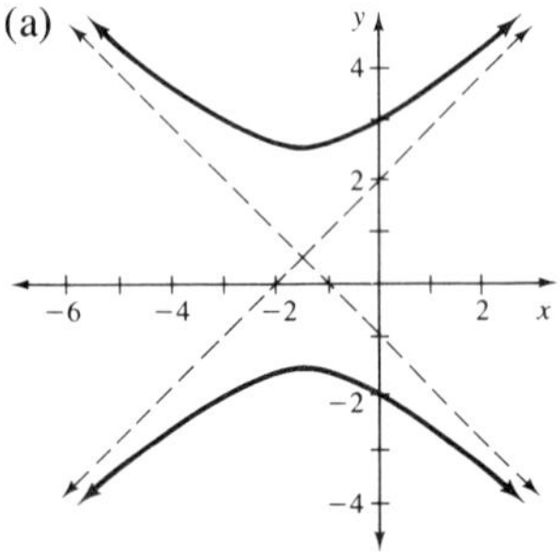

(b) $\{x \mid x \in \text{reals}\}$; $\{y \mid y \le -1.5 \text{ or } y \ge 2.5\}$

9. (a)

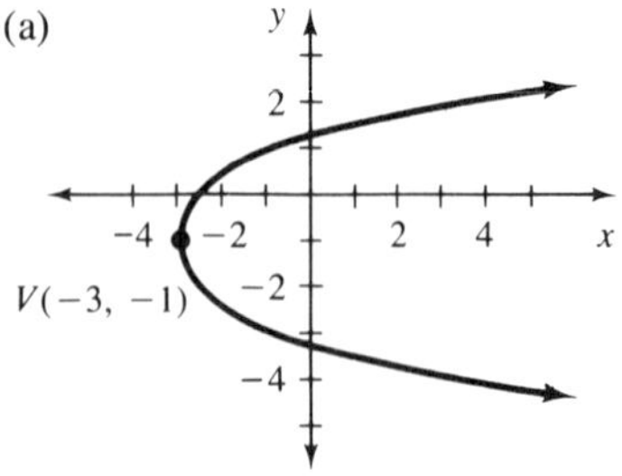

(b) $\{x \mid x \ge -3\}$; $\{y \mid y \in \text{reals}\}$

11. (b) $\{x \mid -1 \le x \le 5\}$; $\{y \mid -8 \le y \le 2\}$
13. (b) $\{x \mid x \in \text{reals}\}$; $\{y \mid y \ge 0 \text{ or } y \le -2\}$
15. (b) $\{x \mid 0.5 \le x \le 3.5\}$; $\{y \mid -0.5 \le y \le 2.5\}$

17. (a)

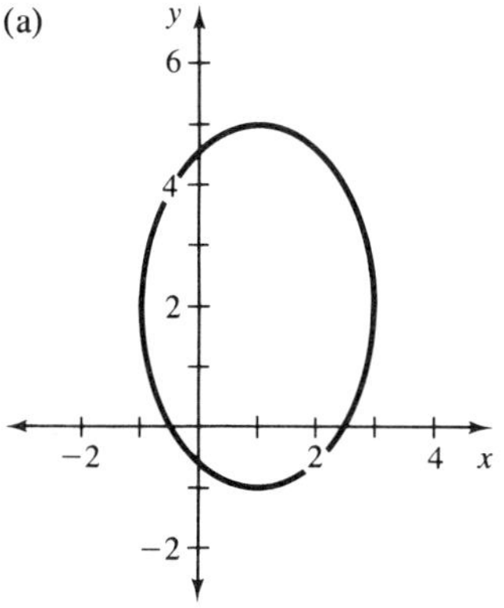

(b) $\{x \mid -1 \leq x \leq 3\}$; $\{y \mid -1 \leq y \leq 5\}$

19. (b) $\{x \mid x \geq -5\}$; $\{y \mid y \in \text{reals}\}$

Exercise Set 6–7

1. (a) Linear-quadratic (d) $\{(-1, -4), (3, 4)\}$ **3.** (a) Linear-quadratic (d) $\{-1/5, -18/5), (2, 3)\}$ **5.** (a) Linear-quadratic (d) $\{(3/4, 4/3), (1, 1)\}$ **7.** (a) Quadratic-quadratic (d) $\{(2, \sqrt{5}), (2, -\sqrt{5}), (-2, \sqrt{5}), (-2, -\sqrt{5})\}$ **9.** (a) Quadratic-quadratic (d) $\{(6, 0), (-6, 0)\}$ **11.** Quadratic-quadratic (d) $\varnothing$ **13.** (a) Quadratic-quadratic (d) $\{(2, 1), (-2, 1)\}$ **15.** (a) Quadratic-quadratic (d) $\{(2, \sqrt{3}), (2, -\sqrt{3}), (-2, \sqrt{3}), (-2, -\sqrt{3})\}$ **17.** (a) Linear-quadratic (d) $\{(1, 3), (13/5, 9/5)\}$ **19.** (a) Linear-quadratic (d) $\{(1, 0), (-3, 4)\}$ **21.** (a) Quadratic-quadratic (d) $\{(-2, 0), (2, 0)\}$

CHAPTER 7

Exercise Set 7–1

1. $\pi/3$ **3.** $3\pi/4$ **5.** $4\pi/3$ **7.** $7\pi/4$ **9.** 1/6 **11.** 1/2 **13.** 5/8 **15.** 11/12 **17.** 0.16 **19.** 0.48 **21.** 0.64 **23.** 0.80 **25.** $\cos s = -3/5$; $\sin s = 4/5$; $\tan s = -4/3$; $\cot s = -3/4$; $\sec s = -5/3$; $\csc s = 5/4$ **27.** $\cos s = 7/25$; $\sin s = 24/25$; $\tan s = 24/7$; $\cot s = 7/24$; $\sec s = 25/7$; $\csc s = 25/24$ **29.** $\cos s = 1/2$; $\sin s = -\sqrt{3}/2$; $\tan s = -\sqrt{3}$; $\cot s = -\sqrt{3}/3$; $\sec s = 2$; $\csc s = -2\sqrt{3}/3$ **31.** $\cos s = -8/17$; $\sin s = 15/17$; $\tan s = -15/8$; $\cot s = -8/15$; $\sec s = -17/8$; $\csc s = 17/15$ **33.** $\cos s = 12/13$; $\tan s = 5/12$; $\cot s = 12/5$; $\sec s = 13/12$; $\csc s = 13/5$ **35.** $\cos s = -\sqrt{2}/2$; $\sin s = \sqrt{2}/2$; $\cot s = -1$; $\sec s = -\sqrt{2}$; $\csc s = \sqrt{2}$ **37.** $\cos s = -\sqrt{3}/2$; $\sin s = -1/2$; $\tan s = \sqrt{3}/3$; $\sec s = -2\sqrt{3}/3$; $\csc s = -2$ **39.** $\sin s = -\sqrt{3}/2$; $\tan s = -\sqrt{3}$; $\cot s = -\sqrt{3}/3$; $\sec s = 2$; $\csc s = -2\sqrt{3}/3$ **41.** $9\pi/4$; $-7\pi/4$ **43.** $5\pi/2$; $-3\pi/2$ **45.** $13\pi/4$; $-3\pi/4$ **47.** 2π; -2π **49.** $\pi/6$ **51.** $2\pi/3$ **53.** $5\pi/4$ **55.** $\pi/4$

Exercise Set 7–2

1. 1/4 **3.** 3/8 **5.** 5/6 **7.** 1/12 **9.** 5/12 **11.** 3/4 **13.** 90°; $\pi/2$ **15.** 120°; $2\pi/3$ **17.** 80°; $4\pi/9$ **19.** 150°; $5\pi/6$ **21.** 252°; $7\pi/5$ **23.** 1560°; $26\pi/3$ **25.** 112.5° **27.** −72° **29.** 114.65° **31.** 57.32° **33.** 1.13 **35.** 3.35 **37.** 9.14 **39.** −5.20 **41.** 0; 30°; 45°; $\pi/3$; $\pi/2$; $2\pi/3$; 135°; $5\pi/6$; 180°; 210°; 225°; $4\pi/3$; $3\pi/2$; 300°; 315°; $11\pi/6$; 360°; $13\pi/6$ **43.** $\cos\theta = -4/5$; $\sin\theta = -3/5$; $\tan\theta = 3/4$; $\cot\theta = 4/3$; $\sec\theta = -5/4$; $\csc\theta = -5/3$ **45.** $\cos\theta = 7/25$; $\sin\theta = 24/25$; $\tan\theta = 24/7$; $\cot\theta = 7/24$; $\sec\theta = 25/7$; $\csc\theta = 25/24$ **47.** $\cos\theta = 1/2$; $\sin\theta = \sqrt{3}/2$; $\tan\theta = \sqrt{3}$; $\cot\theta = \sqrt{3}/3$; $\sec\theta = 2$; $\csc\theta = 2\sqrt{3}/3$ **49.** $\cos\theta = -5/13$; $\sin\theta = 12/13$; $\tan\theta = -12/5$; $\cot\theta = -5/12$; $\sec\theta = -13/5$; $\csc\theta = 13/12$ **51.** $\cos\theta = 4/5$; $\tan\theta = 3/4$; $\cot\theta = 4/3$; $\sec\theta = 5/4$; $\csc\theta = 5/3$ **53.** $\cos\theta = -\sqrt{2}/2$; $\sin\theta = -\sqrt{2}/2$; $\cot\theta = 1$; $\sec\theta = -\sqrt{2}$; $\csc\theta = -\sqrt{2}$

55. $\sin\theta = -8/17$; $\tan\theta = 8/15$; $\cot\theta = 15/8$; $\sec\theta = -17/15$; $\csc\theta = -17/8$ **57.** (a) 433° (b) −287° **59.** (a) 235° (b) −485° **61.** (a) $13\pi/5$ (b) $-7\pi/5$ **63.** (a) $8\pi/7$ (b) $-20\pi/7$ **65.** (a) 10.05 (b) −2.51 **67.** 0° = 1, 0, 0, und, 1, und; 90° = 0, 1, und, 0, und, 1; 180° = −1, 0, 0, und, −1, und; 270° = 0, −1, und, 0, und, −1; 360° = 1, 0, 0, und, 1, und **69.** 12.5 **71.** 175/9

Exercise Set 7–3

1. 0.2419 **3.** 1.157 **5.** 0.4147 **7.** 0.5398 **9.** 18°20′ **11.** 11°00′ **13.** 0.8436 **15.** 0.3171 **17.** cos 70° **19.** csc 8° **21.** sec 1.3439 **23.** $\csc\frac{\pi}{18}$ **25.** IV; 80° **27.** II; 20° **29.** III; $\pi/4$ **31.** II; $5\pi/12$ **33.** II; 1.1416 **35.** 1.122 **37.** −0.2588 **39.** 1.428 **41.** 0.8732 **43.** 0.9757 **45.** −0.5774 **47.** $-\cos\frac{\pi}{4}$ **49.** csc 80° **51.** −tan 30° **53.** IV **55.** I

Exercise Set 7–4

1.

θ in Degrees	θ in Radians	$\sin\theta$	$\cos\theta$	$\tan\theta$	$\cot\theta$	$\sec\theta$	$\csc\theta$
0°	0	0	1	0	und	1	und
30°	$\pi/6$	1/2	$\sqrt{3}/2$	$\sqrt{3}/3$	$\sqrt{3}$	$2\sqrt{3}/3$	2
45°	$\pi/4$	$\sqrt{2}/2$	$\sqrt{2}/2$	1	1	$\sqrt{2}$	$\sqrt{2}$
60°	$\pi/3$	$\sqrt{3}/2$	1/2	$\sqrt{3}$	$\sqrt{3}/3$	2	$2\sqrt{3}/3$
90°	$\pi/2$	1	0	und	0	und	1
120°	$2\pi/3$	$\sqrt{3}/2$	−1/2	$-\sqrt{3}$	$-\sqrt{3}/3$	−2	$2\sqrt{3}/3$
135°	$3\pi/4$	$\sqrt{2}/2$	$-\sqrt{2}/2$	−1	−1	$-\sqrt{2}$	$\sqrt{2}$
150°	$5\pi/6$	1/2	$-\sqrt{3}/2$	$-\sqrt{3}/3$	$-\sqrt{3}$	$-2\sqrt{3}/3$	2
180°	π	0	−1	0	und	−1	und
210°	$7\pi/6$	−1/2	$-\sqrt{3}/2$	$\sqrt{3}/3$	$\sqrt{3}$	$-2\sqrt{3}/3$	−2
225°	$5\pi/4$	$-\sqrt{2}/2$	$-\sqrt{2}/2$	1	1	$-\sqrt{2}$	$-\sqrt{2}$
240°	$4\pi/3$	$-\sqrt{3}/2$	−1/2	$\sqrt{3}$	$\sqrt{3}/3$	−2	$-2\sqrt{3}/3$
270°	$3\pi/2$	−1	0	und	0	und	−1
300°	$5\pi/3$	$-\sqrt{3}/2$	1/2	$-\sqrt{3}$	$-\sqrt{3}/3$	2	$-2\sqrt{3}/3$
315°	$7\pi/4$	$-\sqrt{2}/2$	$\sqrt{2}/2$	−1	−1	$\sqrt{2}$	$-\sqrt{2}$
330°	$11\pi/6$	−1/2	$\sqrt{3}/2$	$-\sqrt{3}/3$	$-\sqrt{3}$	$2\sqrt{3}/3$	−2
360°	2π	0	1	0	und	1	und

3. $\dfrac{-\sqrt{2}-\sqrt{6}}{4}$ **5.** 1 **7.** 1/2 **9.** $-\sqrt{3}/2$ **11.** 0 **13.** 1 **15.** $-2\sqrt{3}/3$ **17.** $\sqrt{2}/2$ **19.** $2\sqrt{3}/3$ **21.** 0 **23.** 0° **25.** 30° **27.** 45° **29.** (a) 270° (b) 270° + 360°n, $n \in$ {integers} **31.** (a) 135°; 225° (b) 135° + 360°n; 225° + 360°n, $n \in$ {integers} **33.** (a) 225°; 315° (b) 225° + 360°n; 315° + 360°n, $n \in$ {integers} **35.** (a) 45°; 135°, 225°, 315° (b) 45° + 90°n, $n \in$ {integers}

Exercise Set 7–5

1. (a) 1 (b) $2\pi/3$ **3.** (a) 1/2 (b) 2π **5.** (a) 4 (b) π **7.** (a) 1/9 (b) $2\pi/5$ **9.** (a) 3 (b) 2 **11.** (a) π (b) π

15.

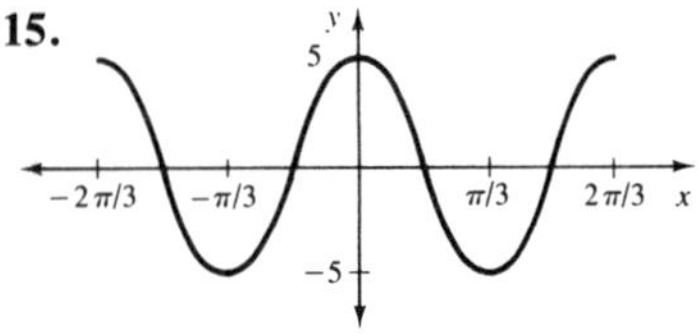

17.

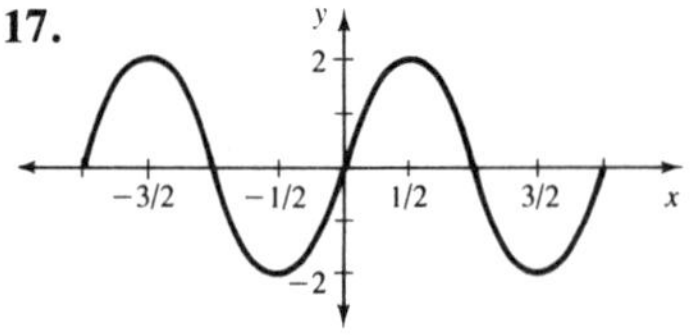

21.

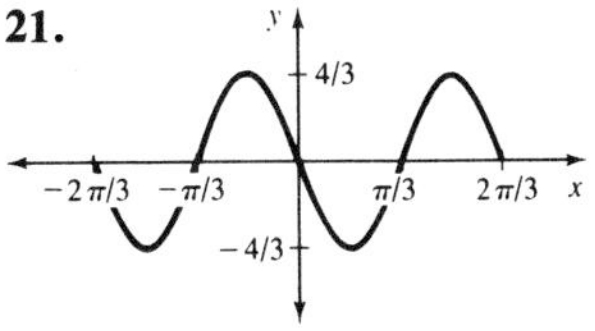

23. Points of intersection: -2π, $-\pi$, 0, π, 2π

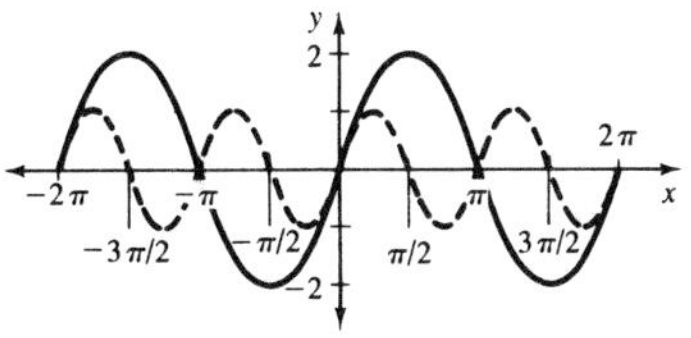

25. $y = 2 \sin x$ **27.** $y = \dfrac{3}{2} \cos \dfrac{2x}{3}$

Exercise Set 7–6

1. (a) 1 (b) 2π (c) $\pi/2$ right **3.** (a) 1 (b) 2π (c) $\pi/2$ left **5.** (a) 3 (b) π (c) $\pi/2$ right **7.** (a) 1 (b) $12\pi/5$ (c) $3\pi/5$ left **9.** (a) 2 (b) π (c) $\pi/4$ left **11.** (a) 3/2 (b) 2 (c) 1 left **13.** (a) 3 (b) $\pi/2$ (c) $\pi/2$ left

15.

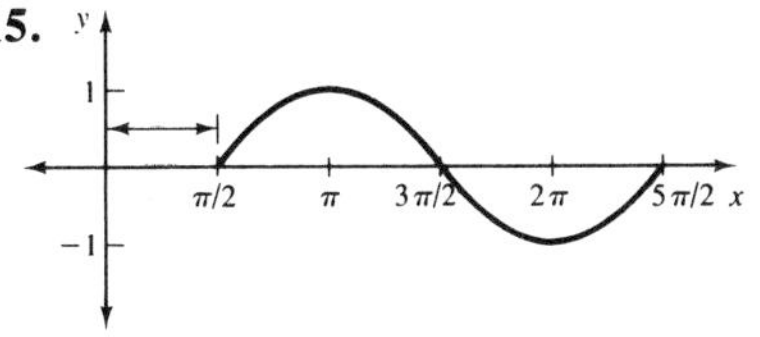

17.

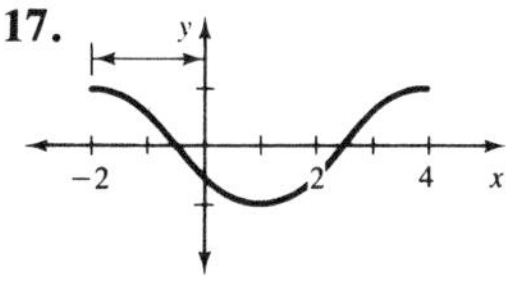

23.

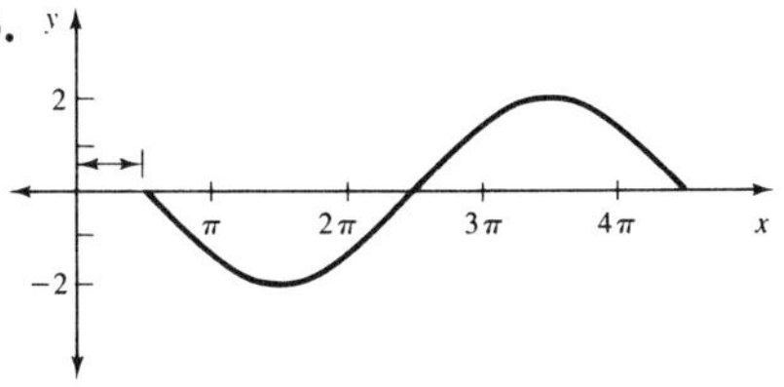

27. $y = 3 \sin(2x - \pi)$ **29.** $y = 4 \sin\left(\dfrac{6x}{5} - \dfrac{\pi}{5}\right)$ **31.** $y = \dfrac{3}{2} \cos\left(\dfrac{4x}{3} - \dfrac{\pi}{3}\right)$

33. $y = 3 \cos\left(\dfrac{4\pi x}{3} + \dfrac{4\pi^2}{15}\right)$

Exercise Set 7–7

1. (a) 3π (b) 0 **3.** (a) $\pi/2$ (b) $\pi/2$ right **5.** (a) $\pi/4$ (b) 0 **7.** (a) $\pi/2$ (b) $\pi/6$ left **9.** (a) 0, $\pi/2$, π, $3\pi/2$, 2π (b) $\pi/4$, $3\pi/4$, $5\pi/4$, $7\pi/4$ **11.** (a) $3\pi/8$, $7\pi/8$, $11\pi/8$, $15\pi/8$ (b) $\pi/8$, $5\pi/8$, $13\pi/8$ **13.** (a) $5\pi/6$ (b) $11\pi/6$ **15.** (a) Disjoint branches separated by vertical asymptotes (b) Domain = $\{x \mid x \in \text{reals}, x \neq n\pi/b, n \text{ an integer}\}$ (c) Range = $\{y \mid |y| \geq |a|\}$ (d) Period = $\dfrac{2\pi}{|b|}$ (e) Asymptotes at . . . , $\dfrac{-2\pi}{|b|}$, $\dfrac{-\pi}{|b|}$, 0, $\dfrac{\pi}{|b|}$, $\dfrac{2\pi}{|b|}$, . . . **17.** (a) $2\pi/3$ (b) . . . , $-2\pi/3$, $-\pi/3$, 0, $\pi/3$, $2\pi/3$, . . . **19.** (a) 5π (b) . . . , $-15\pi/4$, $-5\pi/4$, $5\pi/4$, $15\pi/4$, . . .

23.

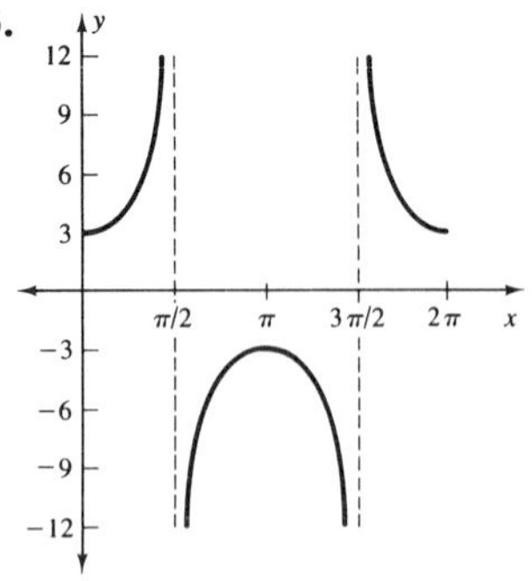

25.

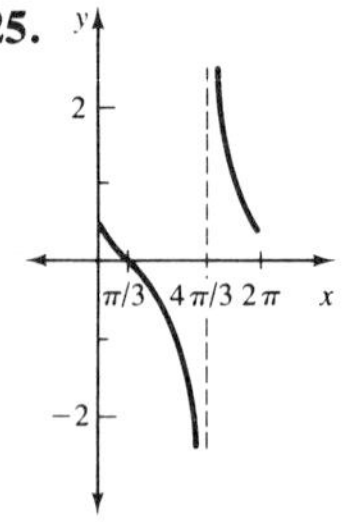

27.

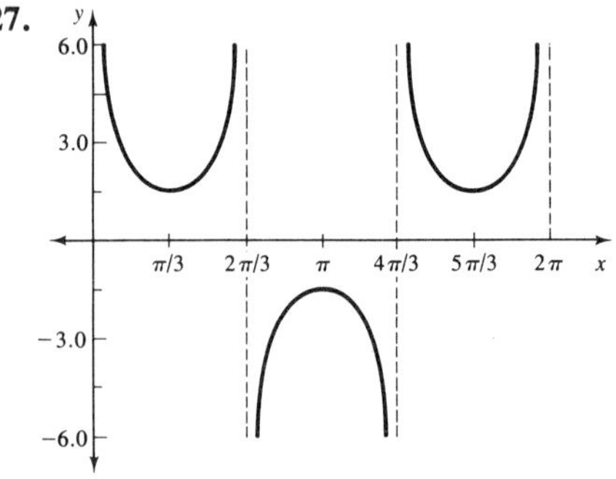

29.

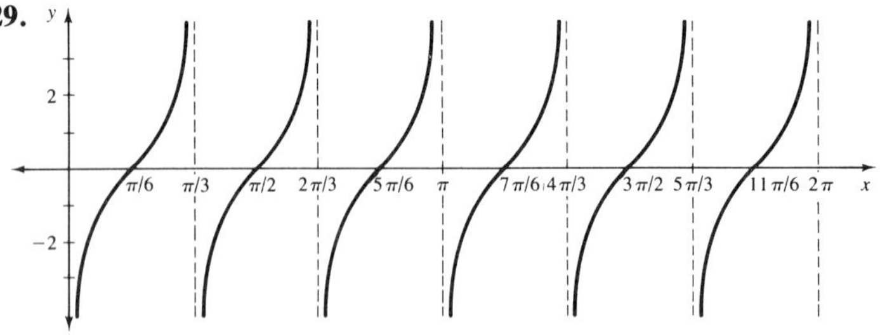

33. $\pi/4 \le x < \pi/2$ or $5\pi/4 \le x < 3\pi/2$ **35.** $\pi/3 \le x < \pi/2$ or $3\pi/2 < x \le 5\pi/3$ **37.** $\pi/2 \le x \le \pi$ or $3\pi/2 \le x \le 2\pi$ **39.** $0 \le x \le \pi/4$ or $\pi \le x \le 5\pi/4$ **41.** $0 \le x < \pi/2$ or $\pi \le x < 3\pi/2$

CHAPTER 8

Exercise Set 8–1

1. Answers can vary: $\sin^2 x = 1 - \cos^2 x$; $\cos^2 x = 1 - \sin^2 x$; $\cos^2 x - 1 = -\sin^2 x$ **3.** Answers can vary: $\cot^2 x = \csc^2 - 1$; $1 = \csc^2 x - \cot^2 x$; $\cot^2 x - \csc^2 x = -1$ **5.** 1 **7.** $\sin x$
9. 1 **11.** 1 **13.** 1 **15.** 1

17. $\dfrac{\pm\sin x}{\sqrt{1 - \sin^2 x}}$ **19.** $\dfrac{\pm 1}{\sqrt{1 - \sin^2 x}}$ **21.** $1/\tan x$ **23.** $\dfrac{\pm 1}{\sqrt{1 + \tan^2 x}}$ **25.** $\dfrac{\pm\tan x}{\sqrt{1 + \tan^2 x}}$
27. $\csc x$ **29.** $\sec x$ **31.** $\cot x$ **33.** $-2 \sin x\sqrt{1 - \sin^2 x}$ or $-2 \cos x\sqrt{1 - \cos^2 x}$ **35.** $\sin x$
37. $-\sin x$ **39.** $y = 2 \sec \theta$ **41–59.** Answers can vary.

Exercise Set 8–2

1. $\sin x$ **3.** $-\tan x$ **5.** $-\cos x$ **7.** $\dfrac{(\cos x - \sqrt{3}\sin x)}{2}$ **9.** $\dfrac{\tan x - 1}{\tan x + 1}$

11. $\cos x$ **13.** $\dfrac{\sqrt{6}+\sqrt{2}}{4}$ **15.** $\sqrt{3} - 2$ **17.** $\dfrac{-\sqrt{6}-\sqrt{2}}{4}$ **19.** $\dfrac{-\sqrt{6}-\sqrt{2}}{4}$

21–29. Answers can vary.

Exercise Set 8–3

1–9. Answers can vary. **11.** $\sqrt{3}/2$ **13.** $\sqrt{2}/2$ **15.** $\dfrac{\sqrt{2+\sqrt{2}}}{2}$ **17.** $\dfrac{\sqrt{2-\sqrt{3}}}{2}$

19. $-\sqrt{3}/2$ **21.** $-7/25$ **23.** $4/5$ **25.** -2 **27.** $-4/5$ **29.** $-336/625$

31–37. Answers can vary. **39.** $\frac{1}{2}(\sin 74° + \sin 10°)$ **41.** $\frac{1}{2}(\cos 40° + \cos 72°)$

43. $\frac{1}{2}(\cos x - \cos 5x)$ **45.** $2 \sin 11° \cos 2°$ **47.** $-2 \sin 77° \sin 42°$ **49.** $2 \cos 3x \cos x$

51. $\sqrt{6}/2$ **53.** $\dfrac{1+\sqrt{3}}{4}$

Exercise Set 8–4

1. $\{\pi/2, 3\pi/2\}$ **3.** $\{\pi/4, 5\pi/4\}$ **5.** $\{2\pi/3, 5\pi/3\}$ **7.** $\{n\pi\}$, n an integer
9. $\{3\pi/4 + n\pi\}$, n an integer **11.** $\{\pi/2 + 2n\pi\}$, n an integer **13.** $2\pi/3$
15. $\{\pi/9, 2\pi/9, 7\pi/9, 8\pi/9, 13\pi/9, 14\pi/9\}$ **17.** $\{\pi/8, 5\pi/8, 9\pi/8, 13\pi/8\}$
19. $\{3\pi/8, 7\pi/8, 11\pi/8, 15\pi/8\}$ **21.** $\{3\pi + 4n\pi\}$, n an integer
23. $\{5\pi/9 + 4n\pi/3, 7\pi/9 + 4n\pi/3\}$, n an integer **25.** $\{\pi/2, 3\pi/2\}$ **27.** $\{\pi 4, \pi/2, 5\pi/4, 3\pi/2\}$
29. $\{\pi/2, 7\pi/6, 11\pi/6\}$ **31.** $\{\pi/4, 1.89, 5\pi/4, 5.03\}$ **33.** $\{0°, 45°, 135°, 180°, 225°, 315°\}$
35. $\{90°, 270°\}$ **37.** $\{0°, 30°, 150°, 180°, 210°, 330°\}$ **39.** $\{0°, 45°, 180°, 225°\}$
41. $\{0.00, 6.28\}$ **43.** $\{1.57, 4{,}71\}$

Exercise Set 8–5

1. $\{90°, 270°\}$ **3.** $\{0°, 45°, 135°, 180°, 225°, 315°\}$ **5.** $\{270°\}$ **7.** $\{45°, 90°, 225°, 270°\}$
9. $\{30°, 150°, 210°, 330°\}$ **11.** $\{0, \pi\}$ **13.** $\{\pi/2, 3.78\}$ **15.** $\{3\pi/2\}$ **17.** $\{\pi/2, \pi\}$
19. $\{\pi/6, \pi/3, 5\pi/6, 5\pi/3\}$ **21.** $\{\pi/4 + n\pi/2\}$, n an integer **23.** $\{\pi/4 + n\pi/2\}$, n an integer
25. $\{0.25 + 2n\pi, \pi/2 + 2n\pi, 2.89 + 2n\pi\}$, n an integer
27. $\{0.64 + 2n\pi, 2\pi/3 + 2n\pi, 4\pi/3 + 2n\pi, 5.64 + 2n\pi\}$, n an integer **29.** $\{4.07 + 2n\pi\}$, n an integer

Exercise Set 8–6

1. $\{\pi/4, 3\pi/4, 5\pi/4, 7\pi/4\}$ **3.** $\{0, \pi/3, 2\pi/3, \pi, 4\pi/3, 5\pi/3\}$ **5.** $\{0, \pi/2, 3\pi/2\}$
7. $\{0, \pi/3, \pi/2, 2\pi/3, \pi, 4\pi/3, 3\pi/2, 5\pi/3\}$ **9.** $\{0, \pi/4, \pi/2, 3\pi/4, \pi, 5\pi/4, 3\pi/2, 7\pi/4\}$
11. $\{0, \pi/18, 5\pi/18, 13\pi/18, 17\pi/18, \pi, 25\pi/18, 29\pi/18\}$ **13.** $\{\pi\}$

CHAPTER 9

Exercise Set 9–1

1. $\{\pi/2\}$ **3.** $\{-\pi/6\}$ **5.** $\{\pi/6\}$ **7.** $11°00'$ **9.** $-1°00'$ **11.** $29°40'$ **13.** 1.25
15. -0.84 **17.** -0.49 **19.** $\pi/5$ **21.** $3/4$ **23.** $1/2$ **25.** -1 **27.** $\{\text{Sin}^{-1}(-0.25)\}$
29. $\{\text{Sin}^{-1}(2/3)\}$ **31.** $\{\pi/6, \text{Sin}^{-1}(1/3)\}$ **33.** $\{-\pi/2, \pi/2, \text{Sin}^{-1}(-0.2)\}$ **35.** $\{\text{Csc}^{-1}(-1.5), \text{Csc}^{-1}4\}$

37. $\left\{-\pi/6, \frac{1}{3}\ \text{Sin}^{-1}0.25\right\}$ **39.** $\left\{-\pi/24, \frac{1}{4}\ \text{Csc}^{-1}3\right\}$

Exercise Set 9–2

1. $\{\pi/3\}$ **3.** $\{2\pi/3\}$ **5.** $\{5\pi/6\}$ **7.** 53°30′ **9.** 122°00′ **11.** 78°10′ **13.** 0.07 **15.** 0.48 **17.** 1.91 **19.** 3/2 **21.** $-7/8$ **23.** $\sqrt{3}/3$ **25.** -3 **27.** $\{\text{Cos}^{-1}(-1/3)\}$ **29.** $\{\text{Sec}^{-1}(76/9)\}$ **31.** $\{\pi/3, \text{Cos}^{-1}(-2/3)\}$ **33.** $\{\text{Sec}^{-1}2.5, \text{Sec}^{-1}(-3)\}$ **35.** $\left\{\pi/9, \frac{1}{3}\ \text{Cos}^{-1}0.4\right\}$ **37.** $\left\{\pi/12, \frac{1}{4}\ \text{Sec}^{-1}(-2.5)\right\}$

Exercise Set 9–3

1. 0.45 **3.** 0.58 **5.** -0.84 **7.** 0.02 **9.** -0.75 **11.** $\pi/4$ **13.** $-\pi/3$ **15.** $3\pi/4$ **17.** 1/3 **19.** $-2/5$ **21.** 3/2 **23.** 1 **25.** $\sqrt{3}/3$ **27.** $\{\text{Tan}^{-1}1.5\}$ **29.** $\{-\pi/4, \pi/4\}$ **31.** $\{\pi/2, \text{Cot}^{-1}(-4/3)\}$ **33.** $\left\{0, \frac{1}{2}\ \text{Tan}^{-1}5\right\}$ **35.** $\left\{3\pi/16, \frac{1}{4}\ \text{Cot}^{-1}0.4\right\}$

Exercise Set 9–4

1. $\cos\theta = 4/5$; $\sin\theta = 3/5$; $\tan\theta = 3/4$; $\cot\theta = 4/3$; $\sec\theta = 5/4$; $\csc\theta = 5/3$ **3.** $\cos\theta = 7/25$; $\sin\theta = -24/25$; $\tan\theta = -24/7$; $\cot\theta = -7/24$; $\sec\theta = 25/7$; $\csc\theta = -25/24$ **5.** $\cos\theta = 5/13$; $\sin\theta = 12/13$; $\tan\theta = 12/5$; $\cot\theta = 5/12$; $\sec\theta = 13/5$; $\csc\theta = 13/12$ **7.** $\cos\theta = -1/\sqrt{10}$; $\sin\theta = 3/\sqrt{10}$; $\tan\theta = -3$; $\cot\theta = -1/3$; $\sec\theta = -\sqrt{10}$; $\csc\theta = \sqrt{10}/3$ **9.** $\cos\theta = -4/\sqrt{17}$; $\sin\theta = -1/\sqrt{17}$; $\tan\theta = 1/4$; $\cot\theta = 4$; $\sec\theta = -\sqrt{17}/4$; $\csc\theta = -\sqrt{17}$ **11.** $\sin x = 5/13$; $\tan x = -5/12$; $\cot x = -12/5$; $\sec x = -13/12$; $\csc x = 13/5$ **13.** $\cos x = 8/17$; $\sin x = -15/17$; $\cot x = -8/15$; $\sec x = 17/8$; $\csc x = -17/15$
15. $\sin x = \sqrt{91}/10$; $\tan x = \sqrt{91}/3$; $\cot x = 3/\sqrt{91}$; $\sec x = 10/3$; $\csc x = 10/\sqrt{91}$
17. $\cos x = 3\sqrt{11}/10$; $\tan x = -1/3\sqrt{11}$; $\cot x = -3\sqrt{11}$; $\sec x = 10/3\sqrt{11}$; $\csc x = -10$
19. $\cos x = 3/\sqrt{13}$; $\sin x = 2/\sqrt{13}$; $\tan x = 2/3$; $\sec x = \sqrt{13}/3$; $\csc x = \sqrt{13}/2$ **21.** $-\pi/4$
23. $\pi/6$ **25.** 15/8 **27.** 3/4 **29.** π **31.** $-2/5$ **33.** $-244/255$ **35.** -0.88
37. $(-28/13, -29/13)$ **39.** $\sqrt{1-z^2}$ **41.** $\sqrt{1+z^2}$ **43.** z

Exercise Set 9–5

1. 240/289 **3.** $\sqrt{3}$ **5.** $-7/9$ **7.** $3/\sqrt{10}$ **9.** $1/\sqrt{7}$ **11.** $\sqrt{\dfrac{1-x}{2}}$ **13.** 33/65

15. $\sqrt{3}/3$

(*Note:* Answers can vary.)

17. If

$$\text{Sin}^{-1}\left(\frac{1}{2}\right) + \text{Cos}^{-1}\left(\frac{1}{2}\right) = \frac{\pi}{2}$$

then

$$\sin\left(\text{Sin}^{-1}\left(\frac{1}{2}\right) + \text{Cos}^{-1}\left(\frac{1}{2}\right)\right) = \sin\frac{\pi}{2}$$

$$\left(\frac{1}{2}\right)\left(\frac{1}{2}\right) + \left(\frac{\sqrt{3}}{2}\right)\left(\frac{\sqrt{3}}{2}\right) = 1$$

$$\frac{1}{4} + \frac{3}{4} = 1 \quad \text{(True)}$$

19. If $\text{Tan}^{-1}1 + \text{Cot}^{-1}(-1) = \pi$, then

$$\tan(\text{Tan}^{-1}1 + \text{Cot}^{-1}(-1)) = \tan \pi$$

$$\frac{1 + (-1)}{1 - (1)(-1)} = 0$$

$$\frac{0}{2} = 0 \qquad \text{(True)}$$

21. If

$$\text{Sin}^{-1}\left(\frac{3}{\sqrt{13}}\right) + \text{Tan}^{-1}\left(\frac{3}{2}\right) = \text{Cos}^{-1}\left(\frac{-5}{13}\right)$$

then

$$\cos\left(\text{Sin}^{-1}\left(\frac{3}{\sqrt{13}}\right) + \text{Tan}^{-1}\left(\frac{3}{2}\right)\right) = \cos\left(\text{Cos}^{-1}\left(\frac{-5}{13}\right)\right)$$

$$\left(\frac{2}{\sqrt{13}}\right)\left(\frac{2}{\sqrt{13}}\right) - \left(\frac{3}{\sqrt{13}}\right)\left(\frac{3}{\sqrt{13}}\right) = \frac{-5}{\sqrt{13}}$$

$$\frac{4}{13} - \frac{9}{13} = \frac{-5}{13} \qquad \text{(True)}$$

23. If $2\,\text{Tan}^{-1}2 - 3\pi/4 = \text{Tan}^{-1}(-1/7)$, then

$$2\,\text{Tan}^{-1}2 = \text{Tan}^{-1}(-1/7) + 3\pi/4$$

$$\tan(2\,\text{Tan}^{-1}2) = \tan(\text{Tan}^{-1}(-1/7) + 3\pi/4)$$

$$\frac{2(2)}{1 - 4} = \frac{-1/7 + (-1)}{1 - (-1/7)(-1)}$$

$$-4/3 = -4/3 \qquad \text{(True)}$$

25. $\{\sqrt{2}/2\}$ **27.** $\{x \mid |x| \geq 1\}$ **29.** $\{0\}$

CHAPTER 10

Exercise Set 10–1

1. $A = 51°$; $a = 56$; $B = 39°$; $b = 45$; $C = 90°$; $c = 72$ **3.** $A = 37°50'$; $a = 70$; $B = 52°10'$; $b = 90$; $C = 90°00'$; $c = 114$ **5.** $A = 51°00'$; $a = 41.9$; $B = 39°00'$; $b = 33.9$; $C = 90°00'$; $c = 53.9$ **7.** $A = 40°20'$; $a = 13.1$; $B = 49°40'$; $b = 15.4$; $C = 90°00'$; $c = 20.2$ **9.** 87 feet **11.** 774.9 feet **13.** 1420 meters **15.** 9.0 meters long **17.** No triangles **19.** 2 triangles **21.** $A = 20°$; $a = 2.2$; $B = 80°$; $b = 6.3$; $C = 80°$; $c = 6.3$ **23.** $A = 98°00'$; $a = 422$; $B = 24°30'$; $b = 177$, $C = 57°30'$; $c = 359$ **25.** $A = 28°$; $a = 2.4$; $A^* = 28°$; $a^* = 2.4$; $B = 45°$; $b = 3.6$; $B^* = 135°$; $b^* = 3.6$; $C = 107°$; $c = 4.9$; $C^* = 17°$; $c^* = 1.5$ **27.** $A = 27°30'$; $a = 109$; $A^* = 27°30'$; $a^* = 109$; $B = 53°20'$; $b = 189$; $B^* = 71°40'$; $b^* = 224$; $C = 99°10'$; $c = 233$; $C^* = 80°50'$; $c^* = 233$ **29.** 46 inches and 48 inches **31.** 41°

Exercise Set 10–2

1. $A = 52°$; $a = 28$; $B = 78°$; $b = 35$; $C = 49°$; $c = 27$ **3.** $A = 41°20'$; $a = 4.0$; $B = 55°50'$; $b = 5.0$; $C = 82°50'$; $c = 6.0$ **5.** $A = 30°00'$; $a = 5.0$; $B = 60°00'$; $b = 8.7$; $C = 90°00'$; $c = 10.0$ **7.** 15.7 **9.** 101°00′ **11.** 123 **13.** 18.8 **15.** 88°50′ **17.** 63.7 feet **19.** 9 meters **21.** 18.7 meters

Exercise Set 10–3

1, 3, 5.

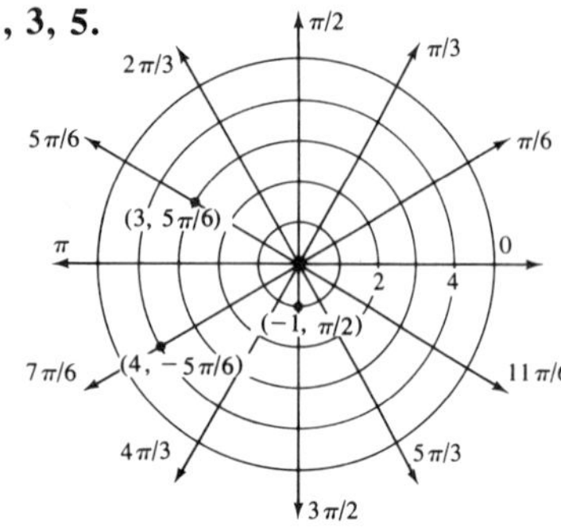

7. $(-\sqrt{2}, -\sqrt{2})$ **9.** $(0, 3)$ **11.** $(-\sqrt{3}, 0)$

13. $(2\sqrt{2}, \pi/4)$ **15.** $(2, 0)$ **17.** $(1, \pi/3)$

19. Answers can vary: $(\sqrt{2}, \pi/4)$; $(-\sqrt{2}, -3\pi/4)$; $(\sqrt{2}, -7\pi/4)$; $(-\sqrt{2}, 5\pi/4)$ **21.** $y = 2$ **23.** $y = x$

25. $4y = (x^2 + y^2)^{3/2}$ **27.** $2xy = (x^2 + y^2)^{3/2}$

29. $r^2 = 16$ **31.** $r \cos\theta = 4$ **33.** $r \sin^2\theta = 2 \cos\theta$

Exercise Set 10–4

1. (a) No (b) No (c) Yes **3.** (a) Yes (b) No (c) No

15. $r = 2 + \cos\theta$

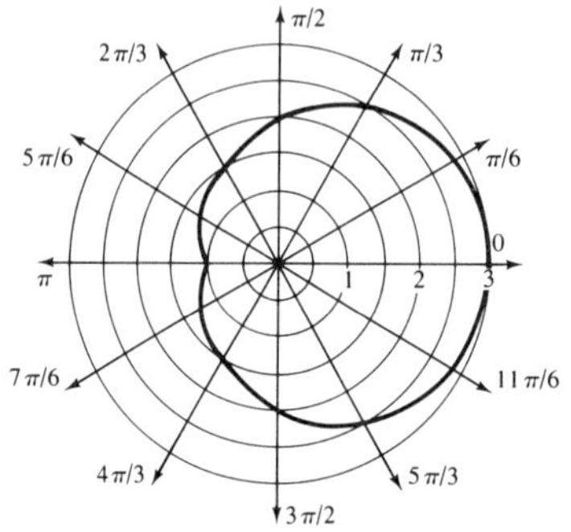

17. $r^2 = 16 \cos 2\theta$

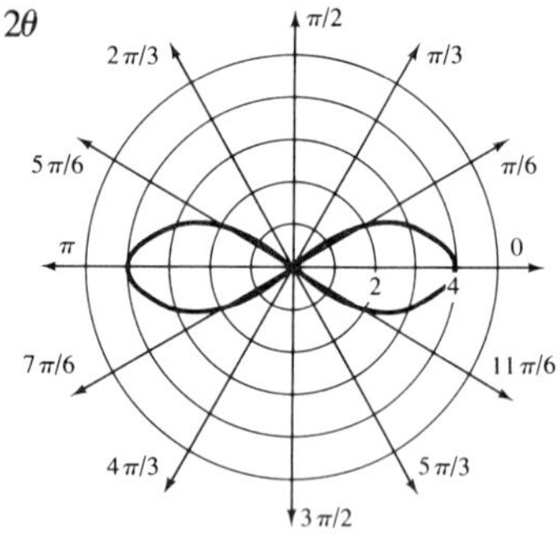

19. $r = 2\theta$

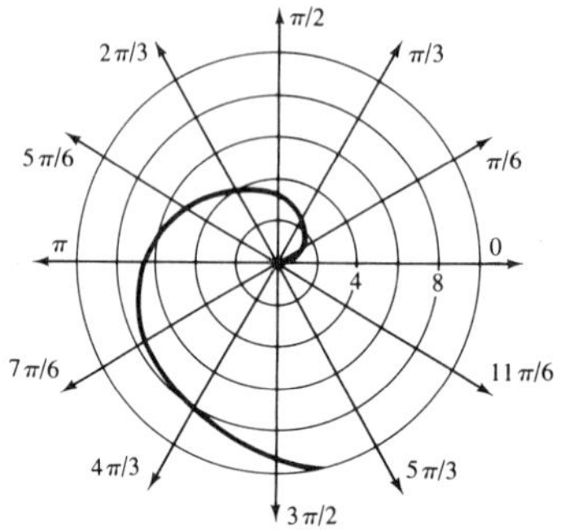

21. $r = \dfrac{1}{1 - \cos\theta}$

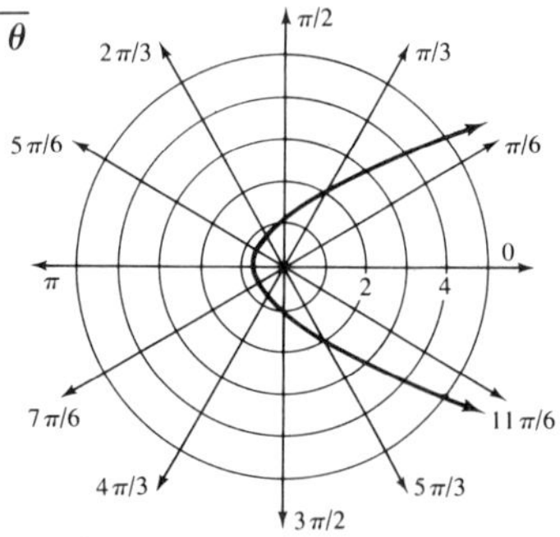

Exercise Set 10–5

1. $x = 6t + 2$; $y = 2t$

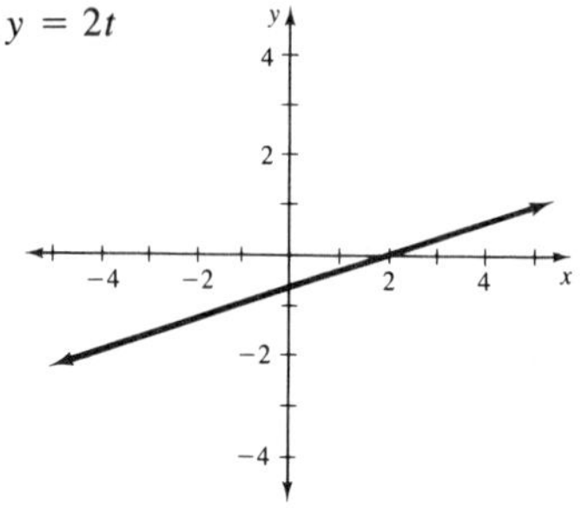

3. $x = 2t + 4;\ y = t^2$

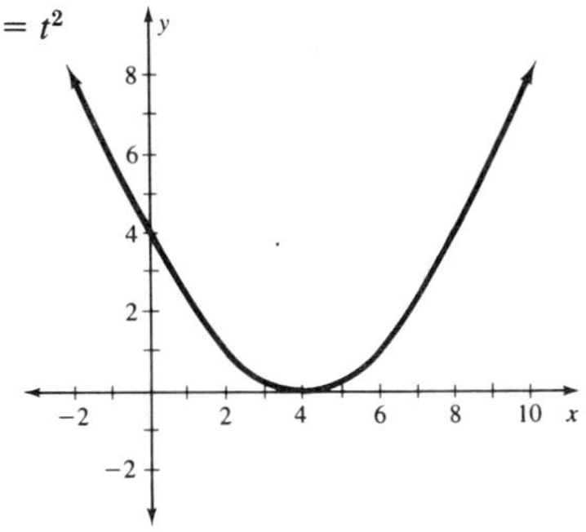

5. $x = 2t;\ y = t^3$

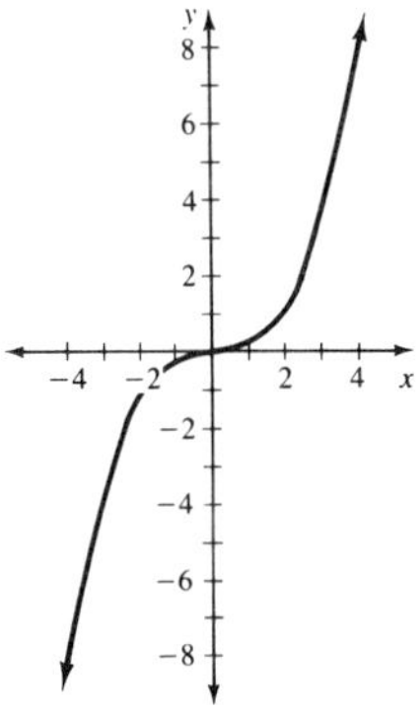

11. $x = \tan\theta;\ y = \sec\theta$

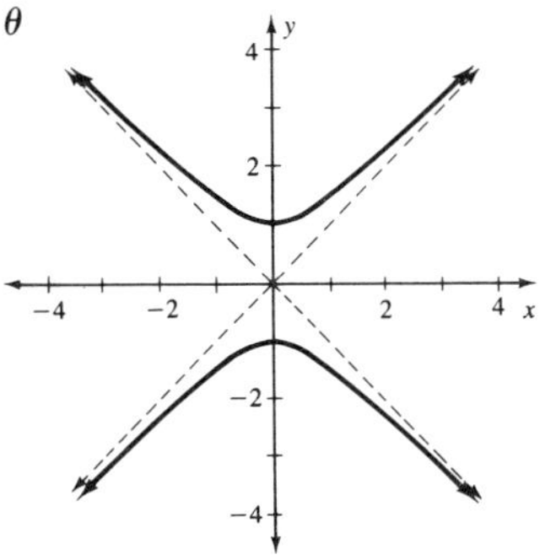

15. $x = 4\cos\theta;\ y = 3\sin\theta$

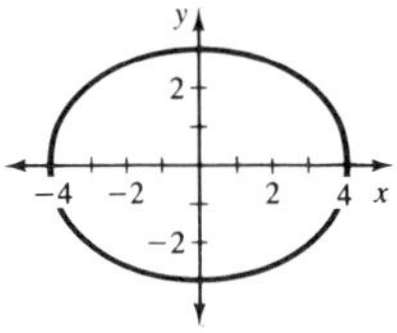

17. $x = (y + 2)^2$ **19.** $x = 1 - y^2$

21. $9x^2 - 4y^2 = 36$ **23.** $(x - y)^3 = 4(x + y)$ **25.** $y + 4 = (x - 1)^{3/2}$ **27.** $100x^2 - 36y^2 = 225$
29. $9x^2 + 25y^2 = 225$

Exercise Set 10–6

1. (a) 13.0 (b) 67°20′ **3.** (a) 25 (b) −73°40′ **5.** (a) 10.4 (b) −16°40′ **7.** (8.7, 5)
9. (−3.5, 3.5) **11.** (−4.0, 6.9) **13.** (a) (5, 7) (b) 8.6 (c) 54°30′ **15.** (a) (0, 2) (b) 2.0
(c) 90°00′ **17.** (a) (7, 3) (b) 7.6 (c) 23°10′ **19.** (a) (4.2, 1.4) (b) 4.4 (c) 18°30′
21. (a) (−3.5, −2.6) (b) 4.4 (c) −143°20′ **23.** 213 pounds **25.** $\text{Sin}^{-1}0.6000 \approx 36°50'$
27. 164°; 89 mph **29.** $F_4 = (-40, -60)$

Exercise Set 10–7

1. $(2\sqrt{2}, 4\sqrt{2})$; (2.8, 5.7) **3.** $\left(\dfrac{-1 - 3\sqrt{3}}{2}, \dfrac{-\sqrt{3} + 3}{2}\right)$; (−3.1, 0.6) **5.** $(5\sqrt{3} - 3, 5 + 3\sqrt{3})$;
(5.7, 10.2) **7.** $\left(\dfrac{3\sqrt{3} - 5}{2}, \dfrac{5\sqrt{3} + 3}{2}\right)$; (0.1, 5.8) **9.** (a) $(x^*)^2 + 2(y^*)^2 = 8$ (b) ellipse
11. (a) $5(x^*)^2 + (y^*)^2 = 4$ (b) ellipse (c) $2x^2 + \sqrt{3}xy + y^2 = 2$

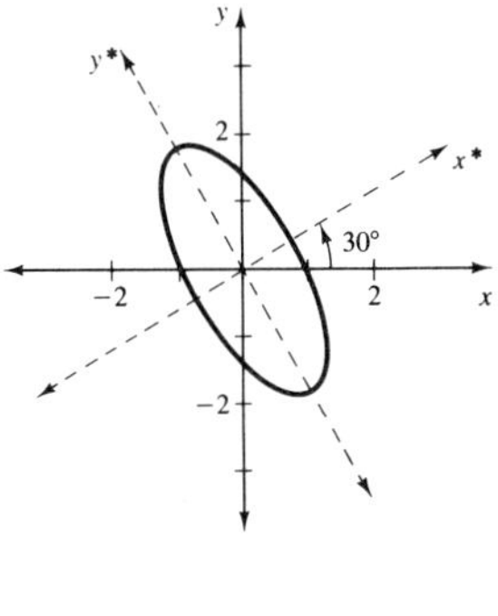

13. (a) $4(y^*)^2 - (x^*)^2 = 4$ (b) hyperbola (c) $11x^2 - 10\sqrt{3}xy + y^2 = 16$

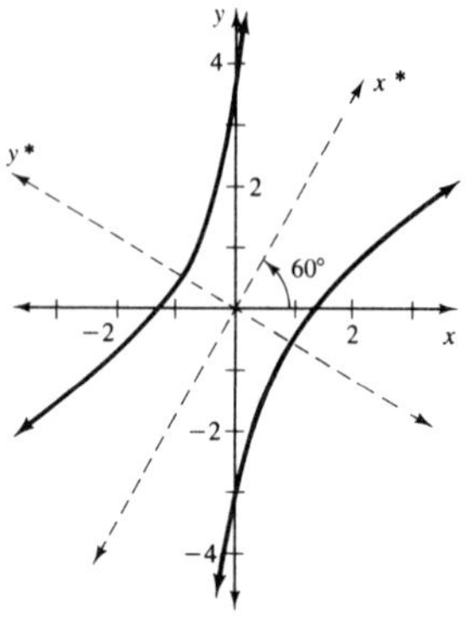

15. (a) $8x^* = -(y^*)^2$ (b) parabola (c) $x^2 - 2xy + y^2 + 8\sqrt{2}x + 8\sqrt{2}y = 0$

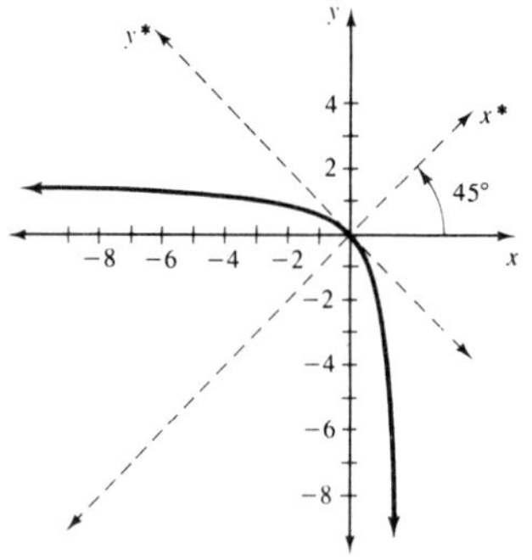

17. Ellipse **19.** Hyperbola **21.** Parabola

CHAPTER 11

Exercise Set 11–1

1. $-i$ **3.** -1 **5.** 1 **7.** $4i\sqrt{2}$ **9.** $13i$ **11.** $-5\sqrt{6}$ **13.** $-147i$ **15.** $24i\sqrt{2}$
17. $6 + 11i$ **19.** $-8\sqrt{2} + 4i$ **21.** $103 + 11i$ **23.** $-21 + 20i$ **25.** $-2 + 2i\sqrt{3}$
27. $\frac{3}{5} - \frac{4}{5}i$ **29.** $\frac{-2}{65} + \frac{16}{65}i$ **31.** 5 **33.** $3\sqrt{2} - 2i\sqrt{3}$ **35.** $\sqrt{3} + i\sqrt{2}$
37. $x = 3; y = -2$ **39.** $x = 3; y = 1$

Exercise Set 11–2

1. 3. 5. 7. 9.

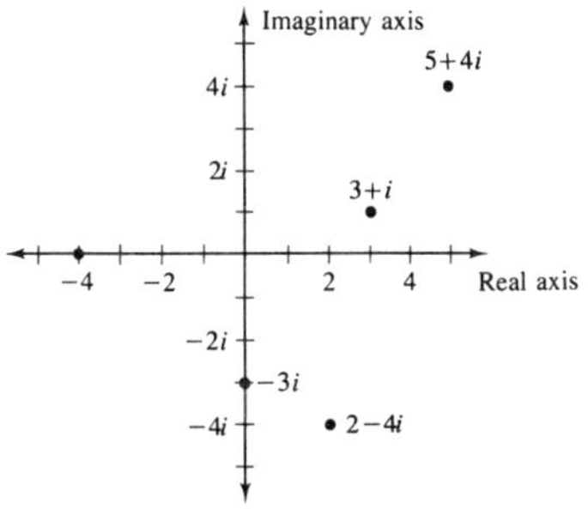

11. 2 cis 60° **13.** 2 cis 180° **15.** $\sqrt{41}$ cis 38° 40′ **17.** $2\sqrt{7}$ cis 240° **19.** 5 cis 233°10′

21. $\frac{3\sqrt{3}}{2} + \frac{3}{2}i$ **23.** $-4 + 4i\sqrt{3}$ **25.** 3 **27.** $-6.4 + 7.7i$ **29.** $-5i$ **31.** $-3\sqrt{2} + 3i\sqrt{2}$

35. -4 **37.** $1.3 + 2.1i$ **39.** $-115.2 + 20.3i$ **41.** cis $\frac{\pi}{2}$ **43.** $6\sqrt{2}$ cis $\frac{3\pi}{2}$ **45.** cis $\frac{19\pi}{12}$

Exercise Set 11–3

1. 8 cis 180° **3.** 9 cis 80° **5.** 125 cis 4.84 **7.** 27 cis $\frac{\pi}{5}$

9. $-2 + 2i$ **11.** $16\sqrt{3} + 16i$ **13.** $6 - 6i\sqrt{3}$ **15.** -64 **17.** 3 cis 15°; 3 cis 195°

19. 2 cis 50°; 2 cis 140°; 2 cis 230°; 2 cis 320°

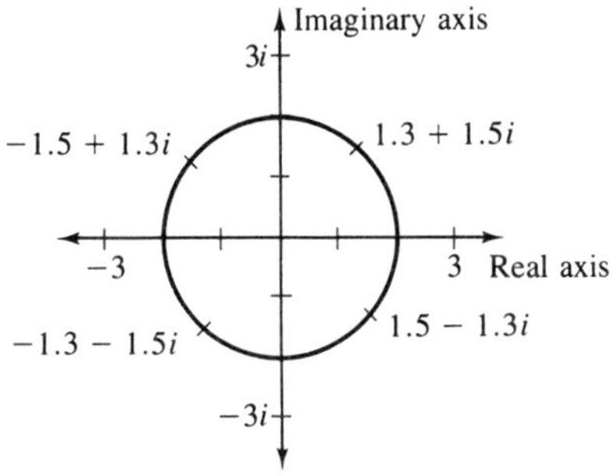

21. 4 cis 70°; 4 cis 190°; 4 cis 310°

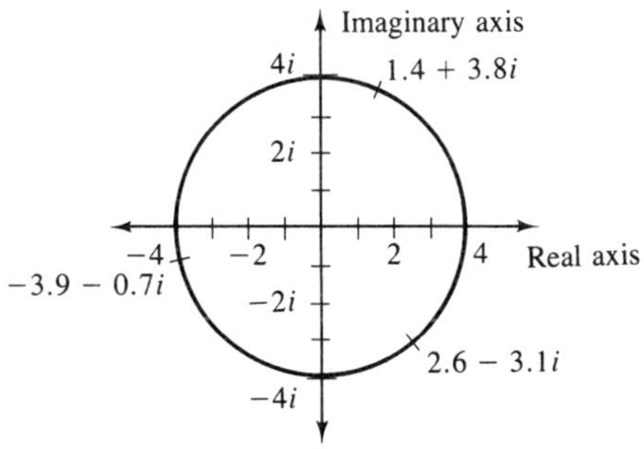

23. 2 cis 0°; 2 cis 60°; 2 cis 120°; 2 cis 180°; 2 cis 240°; 2 cis 300° **25.** $\{1, i, -1, -i\}$

27. {2 cis 36°, 2 cis 108°, 2 cis 180°, 2 cis 252°, 2 cis 324°}

29. {cis 30°, cis 90°, cis 150°, cis 210°, cis 270°, cis 330°}

CHAPTER 12

Exercise Set 12–1

1. 5, 7, 9, 11, 13 **3.** 3, 7, 11, 15, 19 **5.** −2, 4, −8, 16, −32 **7.** 3, 5/4, 1, 9/10, 11/13
9. 0, 1/5, −2/7, 1/3, −4/11 **11.** $-1/2, -1, -5/4, \ldots, \frac{1-2n}{1+n}, \ldots$
13. $3, 8, 15, \ldots, n^2 + 2n, \ldots$ **15.** $-1, 1/4, -1/9, \ldots, \frac{(-1)^n}{n^2}, \ldots$
17. $\sqrt{3}, 2, \sqrt{5}, \ldots, \sqrt{n+2}, \ldots$ **19.** $4n + 2$ **21.** 2^{n-1} **23.** $\frac{(-1)^n}{2^n}$ **25.** $\frac{n+1}{n}$
27. $3(1 - 2n)$

Exercise Set 12–2

1. −1 + 2 + 7 + 14 + 23 = 45 **3.** −1 + 8 − 27 + 64 − 125 + 216 = 135
5. log 1 + log 2 + log 3 + log 4 + log 5 = log 120 **7.** −2 + 11 − 24 + 85 = 70
9. 0 + 2 + 6 + 12 + 20 + 30 + 42 + 56 + 72 + 90 = 330 **11.** $-1 + 4 + 21 + \cdots + (n^3 - 2n)$
13. $1/4 + 4/9 + 9/14 + \cdots + \frac{n^2}{5n-1}$ **15.** $2x + x^2 + \frac{2x^3}{3} + \cdots + \frac{2x^n}{n} + \cdots$
17. $1 + 5 + 19 + \cdots + (3^n - 2^n) + \cdots$ **19.** $1/2 + 2/3 + 3/4 + \cdots + \frac{n}{n+1} + \cdots$
21. $\Sigma_{i=1}^{6} 2^i$ **23.** $\Sigma_{j=1}^{6} (3j - 2)$ **25.** $\Sigma_{k=1}^{6} (-3)^k$ **27.** $\Sigma_{i=1}^{5} (2x)^{i-1}$

Exercise Set 12–3

1. 5, 8, 11, 14, 17 **3.** 20, 13, 6, −1, −8 **5.** 6, 12, 18, 24, 30
7. 10, 12.5, 15, 17.5, 20 **9.** 76 **11.** 33/4 **13.** 29.1 **15.** 21 **17.** 20 **19.** 34
21. 790 **23.** −147 **25.** 820x **27.** −440 **29.** 165 **31.** 35.4 **33.** 583.8

Exercise Set 12–4

1. 5 **3.** 1/4 **5.** 0.3 **7.** 5/2 **9.** $a^{3/2}$ **11.** −16 **13.** 0.243 **15.** −5/16
17. 10 **19.** 80/567 **21.** 16/5 **23.** ±3 **25.** 0.1 **27.** $\pm\frac{b^{1/2}}{3}$ **29.** 1365
31. 1640/27 **33.** 999.999 **35.** 2059/48 **37.** 31/8 **39.** 18,915/64 **41.** 1270
43. (a) 88.23 meters to two decimal places (b) 131.27 meters to two decimal places

Exercise Set 12–5

1. 10/3 **3.** −2 **5.** ∞ **7.** −32 **9.** 40/3 **11.** 1.5 **13.** −6 **15.** −10/3
17. ∞ **19.** −30/17 **21.** 4/9 **23.** 2/33 **25.** 58/11 **27.** 452/111 **29.** 1

CHAPTER 13

Exercise Set 13–1

1. 10^9 or 1,000,000,000 **3.** 10^5 or 100,000 **5.** 216,000,000 **7.** 1,209,600 **9.** 4320
11. 720 **13.** 720 **15.** 11,880 **17.** 2,598,960 **19.** 20 **21.** 1 **23.** 5040
25. 120 **27.** 15,120 **29.** 720 **31.** 120 **33.** 720

Exercise Set 13–2

1. 1 **3.** 780 **5.** 210 **7.** 1 **9.** 56 **11.** 924 **13.** (a) 1 (b) 5 (c) 10 (d) 10 (e) 5 (f) 1 (g) 32 **15.** 220 **17.** (a) 2,598,960 (b) 1 (c) 1 **19.** (a) 560 (b) 120 (c) 270 (d) 150 (e) 20 **21.** (a) 300 (b) 120 (c) 144 (d) 36 **23.** (a) 45 (b) 210 (c) 252 (d) 1

Exercise Set 13–3

1. (a)

r_1r_2	r_2r_1	r_3r_1	b_1r_1	b_2r_1
r_1r_3	r_2r_3	r_3r_2	b_1r_2	b_2r_2
r_1b_1	r_2b_1	r_3b_1	b_1r_3	b_2r_3
r_1b_2	r_2b_2	r_3b_2	b_1b_2	b_2b_1

(b) $P(E_1) = 3/10$; $P(E_2) = 3/5$; $P(E_3) = 1/10$

3. (a)

g g g g
g g g b
g g b g
g g b b
g b g g
g b g b
g b b g
g b b b
b g g g
b g g b
b g b g
b g b b
b b g g
b b g b
b b b g
b b b b

(b) $P(E_1) = 1/16$; $P(E_2) = 4/16$; $P(E_3 = 6/16$; $P(E_4) = 4/16$; $P(E_5) = 1/16$

5. (a)

d_1d_2	d_2d_1	d_3d_1	n_1d_1	n_2d_1	n_3d_1
d_1d_3	d_2d_3	d_3d_2	n_1d_2	n_2d_2	n_3d_2
d_1n_1	d_2n_1	d_3n_1	n_1d_3	n_2d_3	n_3d_3
d_1n_2	d_2n_2	d_3n_2	n_1n_2	n_2n_1	n_3n_1
d_1n_3	d_2n_3	d_3n_3	n_1n_3	n_2n_3	n_3n_2

(b) $P(E_1) = 1/5$; $P(E_2) = 3/5$; $P(E_3) = 0$; $P(E_4) = 1$

7. (a)

i_1i_2	i_2i_1	i_3i_1	d_1i_1	d_2i_1	ri_1
i_1i_3	i_2i_3	i_3i_2	d_1i_2	d_2i_2	ri_2
i_1d_1	i_2d_1	i_3d_1	d_1i_3	d_2i_3	ri_3
i_1d_2	i_2d_2	i_3d_2	d_1d_2	d_2d_1	rd_1
i_1r	i_2r	i_3r	d_1r	d_2r	rd_2

(b) $P(E_1) = 1/5$; $P(E_2) = 1/15$; $P(E_3) = 0$

9. (a) 1/13 (b) 1/4 (c) 1/2 (d) 5/13 (e) 3/13
11. (a) 19/40 (b) 1/8 (c) 21/40 (d) 7/8

Exercise Set 13–4

1. Exclusive **3.** Inclusive **5.** Inclusive **7.** Exclusive **9.** (a) 2/9 (b) 1 (c) 1/3 (d) 5/9 (e) 2/3 **11.** (a) 7/15 (b) 2/3 (c) 13/15 (d) 13/15 **13.** (a) 0.5 (b) 0.4 (c) 0.6 (d) 0.7 (e) 0.7 (f) 0.7 (g) 0.5 (h) 0.6 (i) 0.8 (j) 0.9 (k) 0.8 **15.** (a) 1/5 (b) 2/5 (c) 7/10 (d) 3/5 **17.** (a) 2/3 (b) 11/15 (c) 1 (d) 23/30

Exercise Set 13–5

1. Independent **3.** Dependent **5.** Dependent **7.** Independent **9.** (a) 1/2 (b) 1/2 (c) 1/3 (d) no **11.** (a) 1/17 (b) 11/221 (c) 1/221 (d) 1/442 (e) 0 **13.** (a) 228/1309 (b) 570/1309 (c) 420/1309 (d) 91/1309 (e) 1 **15.** (a) 0.4 (b) 0.5 (c) 0.4 (d) 0.7 (e) 0.6 (f) 0.7 (g) 0.2 (h) 0.2 (i) 0.2 (j) E_1 and E_2; E_2 and E_3 **17.** (a) 3/190 (b) 21/190 (c) 45/190 (d) 21/190 (e) 30/190 (f) 70/190 (g) 1 **19.** (a) 385/1015 (b) 462/1015 (c) 154/1015 (d) 14/1015 (e) 1

Exercise Set 13–6

1. $32x^5 + 80x^4 + 80x^3 + 40x^2 + 10x + 1$ **3.** $81y^4 - 108y^3 + 54y^2 - 12y + 1$
5. $x^6 + 12x^5y + 60x^4y^2 + 160x^3y^3 + 240x^2y^4 + 192xy^5 + 64y^6$
7. $m^5 + m^4n + \frac{2m^3n^2}{5} + \frac{2m^2n^3}{25} + \frac{mn^4}{125} + \frac{m^5}{3125}$
9. $m^{14} + 14m^{12}n^3 + 84m^{10}n^6 + 280m^8n^9 + 560m^6n^{12} + 672m^4n^{15} + 448m^2n^{18} + 128n^{21}$ **11.** $11{,}520r^8s^2$
13. $\frac{1001x^{10}y^4}{16}$ **15.** $8008x^{20}y^{18}$ **17.** $\frac{4455k^6}{256}$
19.

1 11 55 165 330 462 462 330 165 55 11 1
1 12 66 220 495 792 924 792 495 220 66 12 1
1 13 78 286 715 1287 1716 1716 1287 715 286 78 13 1

21. $x^{12} - 12x^{11} + 66x^{10} - 220x^9 + 495x^8 - 792x^7 + 924x^6 - 792x^5 + 495x^4 - 220x^3 + 66x^2 - 12x + 1$

Index

BASIC IDENTITIES

$$\csc\theta = \frac{1}{\sin\theta}$$

$$\sec\theta = \frac{1}{\cos\theta}$$

$$\cot\theta = \frac{1}{\tan\theta}$$

$$\tan\theta = \frac{\sin\theta}{\cos\theta}$$

$$\cot\theta = \frac{\cos\theta}{\sin\theta}$$

$$\sin^2\theta + \cos^2\theta = 1$$

$$\tan^2\theta + 1 = \sec^2\theta$$

$$1 + \cot^2\theta = \csc^2\theta$$

NEGATIVE ANGLE IDENTITIES

$$\sin(-\theta) = -\sin\theta \qquad \csc(-\theta) = -\csc\theta$$

$$\cos(-\theta) = \cos\theta \qquad \sec(-\theta) = \sec\theta$$

$$\tan(-\theta) = -\tan\theta \qquad \cot(-\theta) = -\cot\theta$$

COFUNCTION IDENTITIES

$$\sin\left(\frac{\pi}{2} - \theta\right) = \cos\theta \qquad \csc\left(\frac{\pi}{2} - \theta\right) = \sec\theta$$

$$\cos\left(\frac{\pi}{2} - \theta\right) = \sin\theta \qquad \sec\left(\frac{\pi}{2} - \theta\right) = \csc\theta$$

$$\tan\left(\frac{\pi}{2} - \theta\right) = \cot\theta \qquad \cot\left(\frac{\pi}{2} - \theta\right) = \tan\theta$$

ADDITION IDENTITIES

$$\cos(A + B) = \cos A \cos B - \sin A \sin B$$

$$\cos(A - B) = \cos A \cos B + \sin A \sin B$$

$$\sin(A + B) = \sin A \cos B + \cos A \sin B$$

$$\sin(A - B) = \sin A \cos B - \cos A \sin B$$

$$\tan(A - B) = \frac{\tan A - \tan B}{1 + \tan A \tan B} \qquad \tan(A + B) = \frac{\tan A + \tan B}{1 - \tan A \tan B}$$

DOUBLE-ANGLE IDENTITIES

$$\cos 2\theta = \cos^2\theta - \sin^2\theta$$

$$\cos 2\theta = 2\cos^2\theta - 1$$

$$\cos 2\theta = 1 - 2\sin^2\theta$$

$$\sin 2\theta = 2\sin\theta\cos\theta$$

$$\tan 2\theta = \frac{2\tan\theta}{1 - \tan^2\theta}$$

HALF-ANGLE IDENTITIES

$$\sin\frac{\theta}{2} = \pm\sqrt{\frac{1 - \cos\theta}{2}}$$

$$\cos\frac{\theta}{2} = \pm\sqrt{\frac{1 + \cos\theta}{2}}$$

$$\tan\frac{\theta}{2} = \frac{1 - \cos\theta}{\sin\theta}$$